Laila Prager, Michael Prager,
Guido Sprenger (Eds.)

Parts and Wholes

PARTS AND WHOLES

Essays on Social Morphology, Cosmology, and Exchange

in Honour of

J. D. M. Platenkamp

edited by

Laila Prager, Michael Prager,
Guido Sprenger

LIT

Cover photo:
Pu Nyoe and Nya Nyoe, guardian spirits of Luang Prabang, Laos
Copyright and courtesy: Hans Georg Berger, Berlin
www.hansgeorgberger.de

Bibliographic information published by the Deutsche Nationalbibliothek
The Deutsche Nationalbibliothek lists this publication in the Deutsche Nationalbibliografie; detailed bibliographic data are available on the Internet at http://dnb.d-nb.de.

ISBN 978-3-643-90789-9

A catalogue record for this book is available from the British Library

Klosbachstr. 107
CH-8032 Zürich
Tel. +41 (0) 44-251 75 05 Fax +41 (0) 44-251 75 06
E-Mail: zuerich@lit-verlag.ch http://www.lit-verlag.ch

Distribution:
In the UK: Global Book Marketing, e-mail: mo@centralbooks.com
In North America: International Specialized Book Services, e-mail: orders@isbs.com
In Germany: LIT Verlag Fresnostr. 2, D-48159 Münster
Tel. +49 (0) 2 51-620 32 22, Fax +49 (0) 2 51-922 60 99, e-mail: vertrieb@lit-verlag.de
In Austria: Medienlogistik Pichler-ÖBZ, e-mail: mlo@medien-logistik.at
e-books are available at www.litwebshop.de

Contents

EXCHANGE

COSMOLOGY

List of Figures

List of Tables

Preface

This collection is at once a felicitation volume for Josephus D.M. Platenkamp and a collection connecting his various fields of interest – which are indeed all connected: social morphology, exchange, cosmology, history and the practical application of social anthropology.

On June 23rd, 2016, Josephus D.M. Platenkamp is celebrating his 65th birthday. The next winter term will mark his final teaching semester at the University of Münster, where he has been professor of social anthropology since 1993. We present this volume to him as an expression of gratitude for the intellectual companionship, the great inspiration, the enthusiasm, and the warm-hearted sympathy he provided us with over many years. Quite a number of us can honestly state that they would not be what they are and where they are now without Jos.

For Jos Platenkamp, the most constitutive relationships of society are based on exchange. At the same time, these relationships encompass both human and non-human beings, the social and the cosmological domain. Thus, the study of social morphology, cosmology, exchange and the values informing these relationships merge into a single project. Jos Platenkamp applies these concepts to Southeast Asian societies, both in the insular region – specifically Halmahera – and the mainland – in particular Laos. However, his intellectual curiosity truly exceeds regional limitations, revealing him as a scholar whose scope and vision are ultimately grounded in general anthropology. In the course of his career, this has brought him in touch with numerous distinguished scholars, committed colleagues and enthusiastic students, many of whom have contributed to the present volume.

Our thanks go, first and foremost, to all contributors to the present volume which expressed their sympathy for Jos Platenkamp by delivering outstanding and inspiring work. Special thanks are due to Veit Hopf of LitVerlag, Felix Bregulla who helped to prepare the manuscript, and Melanie Duch for typesetting. We also thank Hans Georg Berger for supplying the cover picture, depicting Pu Nyoe and Nya Nyoe, the guardian spirits of Luang Prabang, Laos.

Laila Prager
Michael Prager
Guido Sprenger

Jos D.M. Platenkamp (2016)

Traditions, Achievements, and Debates

Holism, Value Systems, and Comparison: Key Concepts in the Anthropology of J.D.M. Platenkamp

MICHAEL PRAGER

After Jos Platenkamp had been appointed Professor at the Institute of Ethnology of the University of Münster in 1993, he began to introduce a variety of theoretical and thematic concepts, which up to then had rarely been discussed in German anthropology. His Münster students were suddenly confronted with notions such as "holism", "circulating connubium", "house societies", the "hierarchy of values", and theories about exchange and "the concept of the person" – topics well known in certain traditions of Dutch, British, and French anthropology but largely absent from the study programme of ethnological institutes in Germany during this period.[1] Rather than being only taught about "mainstream" approaches in British and American anthropology, as was common practice at the time, Jos' students were also introduced to the French anthropological tradition, from the pioneering work of the *Année sociologique* to the modern period featuring theoreticians such as Claude Lévi-Strauss, Louis Dumont, and Maurice Godelier. Moreover, the students were familiarised with theoretical concepts deriving from the work of various Dutch anthropologists whose research and theoretical insights had been largely neglected, not only in Germany but in many other countries of Europe and the world at large. Judged by the usual topics covered in German anthropology classes during this period, Jos' teaching was thus rather "exotic". However, it was precisely because of the incorporation of additional theoretical traditions deriving from other countries than Great Britain and the USA, that the Münster students acquired a broad theoretical basis, enabling them to find their way through the dense jungle of anthropological theory.

The theoretical and thematic foci characterising the lectures and courses taught by Jos in Münster can to some extent be attributed to his intellectual "upbringing" and involvement within an academic environment that is commonly called "the Leiden tradition in structural anthropology" (see Vermeulen, this volume). While studying in Leiden, Jos had been made familiar with the pioneering research of scholars such as F.D.E. van Ossenbruggen (1869-1950), W.H. Rassers (1877-1973), J.P.B. de Josselin de Jong (1884-1969) and F.A.E. van Wouden (1908-1987), whose work had focussed on the analysis of Indonesian cosmologies and their underlying symbolic classification systems, (asymmetric) marriage exchange, and (double) descent systems.

1 A notable exception is Georg Pfeffer who taught social anthropology in Heidelberg (1971-1985) and Berlin (1985-2008).

These scholars had later become known as the "Dutch precursors" of Lévi-Straussian structuralism (De Heusch 1968). Jos' teacher and supervisor, P.E. de Josselin de Jong (1922-1999), himself was a famous representative of the Leiden tradition (Vermeulen 1997), who preserved many of the insights his predecessors had achieved but who also had been instrumental in modernising Leiden structuralism and bringing it into dialogue with other structuralist approaches in England and France.

It is not my intention to present a detailed list of all correspondences between the Leiden tradition and Jos' own work. Some basic features, however, are clearly standing out. The Leiden tradition was well known for its focus on intra-regional comparison as a method of studying societies as transformations or variations of each other (P.E. de Josselin de Jong 1980a, 1984). The "Leiden version of the comparative method" (Barnes 1985) is firmly associated with the idea that single societies should be studied within the larger regional framework of so called "fields of anthropological study" (FAS), of which Indonesia was considered as one of many possible manifestations.[2] Jos' adherence to the idea that comparison forms an essential tool of anthropological reasoning, largely absent from contemporary anthropology, is surely one of the major characteristics and strengths of his researches and writings. However, he has not only produced analyses subjecting Indonesian societies to an intra-regional comparison – as the FAS model initially suggested –, he is also engaged in inter-regional comparisons by comparing Indonesian with mainland Southeast Asian societies. The latter type of comparison particularly emerged after Jos had begun to carry out research in Laos in the second half of the 1990s. Ultimately, as Jos maintains during lectures and discussions, any anthropological reflection is based on a comparison between the "self" and the "other", making comparison a universal feature of perception and cognition.

Another concept figuring prominently in the Leiden tradition is that of holism. The early Leiden ethnologists from van Ossenbrugggen onwards were strongly influenced by the work of the French *Année sociologique* (see P.E. de Josselin de Jong 1977; P.E. de Josselin de Jong and Vermeulen 1989). Dutch scholars had inherited a certain bias for sociological determinism from Durkheim (later receding to the background), but, more importantly, from Marcel Mauss they had learned that societies should be studied as wholes, that kinship, exchange, religious beliefs and actions, laws and economic activities are all forming part of an "organic whole" or system which should be acknowledged as such and analysed in its own right. Parts should thus always

2 J.P.B de Josselin de Jong had introduced the concept of FAS in 1935 as "certain areas of the earth's surface with a population whose culture appears to be sufficiently homogenous and unique to form a separate object of ethnological study, and which at the same time apparently reveals sufficient local shades of differences to make internal comparative research worth while" (J.P.B. de Josselin de Jong 1977 [1935]: 168).

be studied in relation to that whole. In Jos' professional understanding, likewise, holism is one of the key concepts guiding his ethnographic analyses up to the present.

Moreover, Leiden anthropologists were among the first to explore (and criticise) the work of Lévi-Strauss and to establish a scientific dialogue between Leiden and Paris (Prager 1996). Thus, Jos obtained parts of his intimate knowledge of the Dutch and French tradition in structural anthropology from his student days in Leiden, where both traditions were discussed and kept alive. As holds for other Leiden anthropologists, Lévi-Strauss remained a strong inspiration for Jos, even if he also has his reservations, particularly regarding the fact that the analysis of singular societies as holistic configurations increasingly receded to the background in Lévi-Strauss' later publications.

Another important feature of Jos' writings, related to Lévi-Strauss' concept of structure, is based on one of the main principles of structural anthropology, the idea that the meaning of elements is defined by their relations within a system. As Jos (1988a: 7)[3] put it:

> "Structural anthropology learned from linguistics that the meaning of elements which are part of a system derives from the relationships perceived between them. In order to elucidate the meaning of social facts the anthropologist analyses the relationships which order these facts into a system."

In Leiden structuralism, the emphasis on the importance of relations became manifest in the course of the reception of the work of Lévi-Strauss, who himself had been inspired by the structural linguistics of de Saussure, Jakobson and Trubetzkoy. When Jos studied in Leiden, structural linguistics was considered as belonging to the canon of anthropological theory. His encounter with Dumontian theory at a later stage would further strengthen his conviction that the unveiling of social and cosmological *relationships* governing the *system* of ideas and values of individual societies (as well as the comparison of the latter) constitutes one of the essential tasks of ethnographic analysis.

In addition, there is Jos' never-ending enthusiasm for the study of exchange systems – whether they become manifest in rituals, myths, or in the field of "economic" action including tax-paying. Together with Mauss Leiden anthropologists were the first to acknowledge the importance of exchange for the constitution and cohesion of societies, particularly in the field of marriage, thereby anticipating insights that would later become solely associated with the work of Lévi-Strauss. The focus on exchange is surely one of the main ideas Jos had assimilated in Leiden, but the manner in which he elaborated his own views of exchange has been moulded by many other influences.

---

3 References to Jos' publications follow the order of his bibliography at the end of this section (see *Publications of J.D.M. Platenkamp*).

Finally, mention should be made of the fact that Jos is an expert in kinship analysis. From the times of J.P.B de Josselin de Jong and van Wouden onwards, Leiden anthropologists were steeped in kinship studies and thus familiar with the most complex forms of framing and translating relationship terminologies and marriage systems into abstract diagrams illustrating the structural dimensions of the systems in question. Having learned kinship studies in Leiden, and having perfected his skills in formal kinship analysis in the course of his career, Jos belongs to a diminishing group of specialists who are able to draw and decipher such diagrams. Thanks to Jos, however, some of his students will carry this expertise on into the future.

However, having been educated within the Leiden tradition, and having adopted some of its basic ideas, does not mean that Jos is merely a recipient and uncritical proponent of this academic tradition. While retaining some of its essential insights he further developed and refined Leiden structuralism by incorporating and critically reflecting on a variety of other theoretical approaches. One of the most formative influences in this respect was his encounter with the theory of the "hierarchy of values" advocated by Louis Dumont and his students. However, before scrutinising Jos' reception of Dumontian theory, some information on his earlier achievements are in place.

From Leiden to Halmahera, to Paris and back to Leiden

The influence exerted by Dumont on Jos' thinking – at least as it emerges from his publications – became manifest for the first time in his PhD dissertation (Platenkamp 1988a). Before this dissertation materialised, however, Jos had already published several articles testifying to his capacity of gathering rich ethnographic data and subjecting these to a rigorous theoretical analysis. As a graduate student, Jos had come forward with two ambitious publications (1979, 1982)[4] in which he was daring enough to criticise Leach's theory on animal categories and tabooed words (Leach 1972) by comparing it with ethnographic data he had obtained during fieldwork among Scottish fishermen in 1975 (see Vermeulen, this volume). As it turned out, the animal classification system used by these fishermen did not correspond to the analytical categories suggested by Leach, thus shedding doubts on the universal validity of the theory in question. These bold conclusions put forward by a young scholar indicated that Jos was not likely to develop into an anthropologist who would be satisfied by producing ethnography only.

In the same year that the first of these articles was published (1979) Jos had begun to undertake ethnographic fieldwork among the Tobelo of North

4 In the framework of this article not every publication of Jos will be addressed in detail. I shall concentrate on those publications which I deem to be representative for the general themes covered in Jos' anthropological analyses.

Halmahera. In 1977 the Indonesian Institute of Sciences (LIPI) had inaugurated the Halmahera Research Project with the purpose of generating basic knowledge on the societies and cultures of Halmahera and adjacent islands, an area about which – as E.K.M. Masinambow, the project coordinator later recalled – "nothing had been written since the end of the second world war" (Masinambow 1994: 213). The project involved anthropologists from Indonesia and the Netherlands as well as American and Japanese researchers (Visser 1994). Together with other Dutch anthropologists (see Vermeulen, this volume) Jos conducted his field research within the framework of the Halmahera Research Project. His research was focussing on the Tobelo, who inhabit the eastern part of the northern peninsula of Halmahera. At the time of his research, the Tobelo numbered about 20.000 people. Jos stayed in the field for a period of 18 months (July 1979 – December 1980) and returned for an additional period of 3 months in 1982, the coastal village of Paca serving as the main research site. Initially, Jos had planned to focus his fieldwork on the study of shamanism (see Platenkamp 1980) but it soon turned out that a more comprehensive study was necessary in order understand the complexities of Tobelo ideology and society.

What made the Tobelo special when compared to other Indonesian societies studied by earlier generations of Leiden anthropologists was that they are speaking a non-Austronesian language. This raised the question whether they should be classified as belonging to the Indonesian FAS, as the majority of the Indonesian societies speak Austronesian languages.[5] Over the years Jos repeatedly addressed this important question in his publications.

Having returned to Leiden, Jos published several articles giving a first glimpse on Tobelo kinship and some of the major rituals (1983, 1984c). Moreover, in 1982 P.E. de Josselin de Jong convened an international conference with the purpose of critically reviewing the concept of the Field of Anthropological Study and comparing it with (then) recent anthropological research undertaken in Indonesia by non-Leiden scholars. Several discussants had been invited from abroad (France, Switzerland, England, Germany, etc.) to comment on the papers. Jos contributed with a paper entitled "The Tobelo in the Context of the Field of Anthropological Study" (Platenkamp 1984a), in which he raised the question whether and to what extent the Tobelo who are speaking a non-Austronesian language should be considered as belonging to the FAS.

As Jos aptly remarked, the early proponents of the FAS approach, such as J.P.B. de Josselin de Jong and F.A.E. van Wouden, never had defined the boundaries of the FAS in terms of linguistic criteria, but with regard to the presence or absence of certain structural core-elements of which "double descent", asymmetric marriage alliance and "socio-cosmic dualism" were considered to be the most important ones (J.P.B. de Josselin de Jong 1977 [1935]).

5 This also applies to other non-Austronesian speaking societies of North Halmahera, such as the Galela, Loloda, and Sahu.

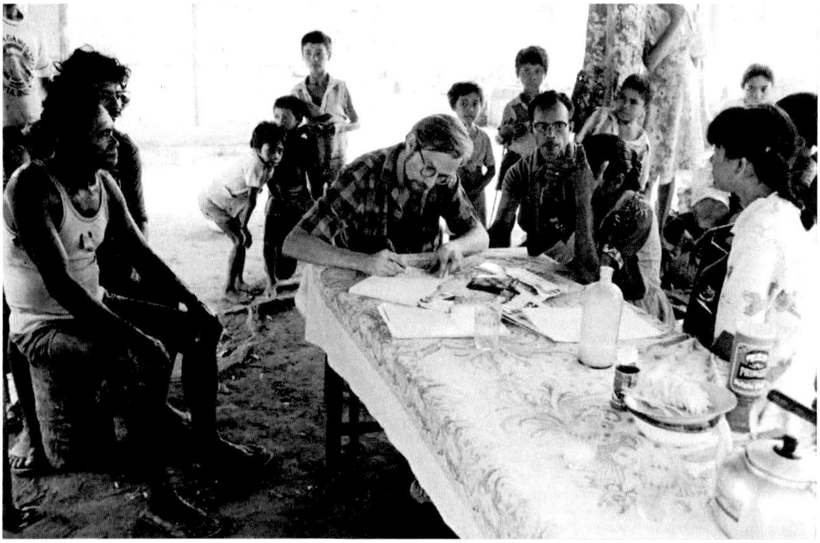

Fig. 1: Jos Platenkamp taking notes in Tobelo, 1982
(Photo: Frans Rijoly/Dirk Nijland)

Fig. 2: "Let's Dance" – Jos Platenkamp participating in a Tobelo dance, 1982
(Photo: Frans Rijoly/Dirk Nijland)

In the case of the Tobelo the answer to the question as to their position within the FAS remained undecisive. On the one hand, as Jos argued, the Tobelo have no uni-lineal (or double-unilineal) descent groups but kin groups who consider themselves to be "people of one house", membership of which is defined in terms of cognatic kin ties. Moreover, there are no repetitive forms of marriage alliance linking wife-giving and wife-taking groups on a permanent basis. Marriage between cognatically related kin was forbidden, so that a marriage could only be contracted among "strangers". Whereas in the Tobelo case asymmetric alliance and double-unilineal descent – as core-elements – are thus clearly absent, wife-giver and wife-taker on the other hand are classified as "female" and "male" respectively and the marriage ritual – as Jos demonstrated – is accompanied by a type of gift exchange strongly reminiscent of the "classic" circulation of "male" and "female" goods as described for the asymmetric marriage systems within the Indonesian FAS. The delineation of the Tobelo's position within the FAS thus depends on the question whether the presence of all core-elements is considered to be essential and whether they all must be interrelated.

Jos' paper was commented upon by Cécile Barraud (Barraud 1984), a former student of Louis Dumont and member of the then newly established research team ERASME (Équipe de Recherche en Anthropologie Sociale – Morphologie, Échanges). Barraud, who had undertaken ethnographic research in the South Moluccan Kei islands (1979) argued that the task of anthropology is not to compare isolated elements, but "ideologies" as understood as systems of "ideas and values" in the sense of Louis Dumont (1979, 1980). As Barraud explained, the so-called core-elements may be absent at one level of the ideology but present at another one, depending on the prevailing "hierarchy of values". Asymmetric alliance, for instance, may occur in a particular society but can be subordinated to the value of "ancestor-making". The search for core-elements thus only makes sense if one determines their place within the overall ideology of a given society (Barraud 1984: 194).

At this time, the analytical vocabulary Barraud put forward in her comment such as "ideology" and "hierarchy of values" was more or less alien to Leiden anthropologists, whose major sources of inspiration from France – as mentioned before – had been Durkheim, Mauss, and Lévi-Strauss, while the work of Louis Dumont was only marginally received. Given the difficulties of translating one scientific paradigm into another, it does not come as a surprise that Jos and Cécile became involved in scholarly discussions exceeding those of the conference proper. As a follow up of these conversations Jos was invited to come to Paris. Thus, in 1984 he spent a year in Paris where he was introduced to the members of ERASME, the research group which had been founded by Louis Dumont and his collaborators in 1983. When Jos came to Paris, ERASME was directed by Daniel de Coppet (1933-2002) who became an important source of inspiration for Jos and a dear friend. By the time of

Jos' arrival the group had published several key texts (Coppet 1981; Itéanu 1983; Barraud et al. 1984) which, together with Dumont's own publications on values (Dumont 1977, 1979, 1980, 1983), formed the theoretical backbone of the group's ethnographic investigations. Apart from Daniel de Coppet and Cécile Barraud the group consisted of André Iteanu, Denis Monnerie, Raymond Jamous, Dominique Casajus, Serge Tcherkézoff and others. Until the group's dispersal after Daniel de Coppet had passed away (2002) Jos remained an active member of ERASME by giving lectures in Paris and by collaborating in publication projects such as the volume on Eastern Indonesian rituals jointly edited by Cécile Barraud and himself (Barraud and Platenkamp 1989a and 1990d).

After Jos had returned to the Netherlands, he began to process the Dumontian concepts of value, hierarchy, and ideology in his own work and introduce them to Leiden anthropology. In the years to follow, other Dutch anthropologists based at Leiden, such as Leontine Visser (1988), Danielle Geirnaert (1987), Jarich Oosten (1993), and Sjoerd Zanen and Bert van den Hoek (1987) began to explore Dumontian theory in their own writings. Scientific relations were further strengthened when the Leiden research group CASA (Cognitive and Structural Anthropology) and ERASME set up a collaboration under the designation CASA-ERASME, which led to the mutual exchange of scholars (Platenkamp 1991d). By introducing Dumontian value theory to Leiden, Jos had opened another chapter in what P.E. de Josselin de Jong (1980b: 256), following Georges Condominas, had designated as the "mouvement structuraliste Paris-Leyde".

Apart from inspiring fellow scholars, Jos in 1988 came forward with a comprehensive monograph on the Tobelo, which he submitted as his PhD dissertation (Platenkamp 1988a). As indicated by its title, the focus was on the "ideas and values" of the Tobelo, thus immediately evoking the theoretical vocabulary used by Dumont and his followers. Already in the introduction Jos made clear that while the FAS approach from Leiden remained important as a heuristic device for the analysis of North Halmaheran societies, another major source of inspiration had been the studies written by Dumont and ERASME (Platenkamp 1988a: 1-9). The following quotation from Jos' thesis aptly summarises the underlying theoretical programme:

> "Values, as configurations of ideas about the existence, are ordered in hierarchical systems. Because each value derives its significance from the position which society grants it within the system as whole, the value system of the Tobelo, like that of any other society, should be studied as a whole" (Platenkamp 1988a: 254).

In his dissertation Jos presented an analysis of the major socio-cosmic concepts governing Tobelo ideology, such as *ma dutu* ("to be owned or related to someone or something") *gurumini*, and *gikiri*, the latter two denoting the main spiritual components of which human beings (and society) are thought to be composed of, namely, "image" and "life". Other chapters are concerned with the house and its cosmological dimensions, with kinship and land use, and with the role played by ancestors and other spiritual beings. A major part of the thesis is devoted to the description and analysis of the exchanges structuring Tobelo ritual life, from the birth- and marriage-ritual to the first and secondary mortuary rituals. Jos concluded his analysis by arguing that the major socio-cosmic concepts of *gurumini* and *gikiri* also constitute two contrasting social values, the former being associated with the value of "reputation", the latter with that of "fertility". Both values, however, are ordered in a hierarchical manner:

> "Globally, Tobelo society presents itself at two levels of value: one at which fertility contrasts with reputation, and another at which reputation encompasses this contrast. These are the two perspectives in which society conceives of itself [...]" (Platenkamp 1988a: 255).

In the years to come Jos further elaborated on the analysis of value systems, not only with regard to the Tobelo material but also in the context of his more recent ethnographic researches in Laos. Before directing our attention to these undertakings mention should be made that the dissertation and the previously discussed articles were not the only outcomes of the Halmahera research. Another result was the film "Tobelo marriage" which had been recorded by the Dutch ethnocinematographer Dirk Nijland[6] in 1982 in collaboration with Jos (see Nijland, this volume). The film was released in 1985 and awarded the "Grand Prix Nanook" at the Ethnographic Film Festival in Paris (1987).

Moreover, in 1990 Jos was involved in a museum exhibition on Tobelo ceremonial gifts which was presented at the Museon of The Hague (December 1990 – March 1991). The exhibition was bringing together basketry, weapons and textiles Jos had collected during his fieldwork. For this occasion, Jos compiled a highly informative museum catalogue (Platenkamp 1990b) in which he contextualised the displayed objects within the framework of Tobelo ritual exchange, focussing above all on the ceremonial objects circulating during the marriage ritual. By explaining how these objects are embedded in the society and cosmology at large, Jos was elaborating on a scientific tradition initiated by scholars such as Rassers and Gerbrands (1917-1997) in Leiden who were exploring the "language of things".[7] Quite recently, the theme of ethnographic exhibitions has become an issue in Jos' thinking again when he and his colleague Andreas Hartmann made a plea for the foundation of a Westphalian Museum of Ethnography (Platenkamp and Hartmann 2015a).

---

6 Nijland himself has produced a series of seminal publications on the Tobelo, particularly in connection with ethnocinematography, see Nijland (1989, 2006).

7 In an article from 1995 written for the symposium "Der Mensch als homo pictor?" Jos again addressed the question as to the way in which Tobelo objects express socio-cosmic values (Platenkamp 1995b).

After Jos had successfully defended his thesis, he was exploring the possibility to undertake additional fieldwork in another part of the Moluccas. In 1989 he made a trip to the island of Seram (Central Moluccas) where he undertook a pilot study in several coastal and inland villages of the Taluti Bay. The preliminary data collected by Jos[8] testified to the existence of moiety-like village structures with "houses" classified either as "male" or "female" and a complex system of social and political segmentation stretching over various inter-connected villages. A thorough investigation of these socio-political configurations would have been more than worthwhile, particularly if undertaken by Jos, but due to several circumstances further research on Seram did not materialise.[9]

In the same year, Jos presented a lecture to the *Oostersch Genootschap* in Leiden (22 May) in which he again addressed the important question as to the position of the non-Austronesian societies within the wider comparative field of the Austronesian speaking societies of Indonesia.

> "Firstly, should these North Halmaheran societies be considered as part of the 'Austronesian' field of anthropological study if comparison reveals fundamental similarities with other (eastern) Indonesian societies even though the latter are not related linguistically? Secondly, if this should be the case, what does that tell us about the criterion of linguistic relatedness for the inclusion – for comparative purposes – of societies in such a field of study?" (Platenkamp 1990e: 5-6).

Jos critically reviewed contemporary attempts at defining "fields of anthropological study" such as Indonesia or Austronesia in terms of linguistic boundaries. He argued that the ideologies of the non-Austronesian speaking societies of Indonesia such as the Sahu and the Tobelo of North Halmahera, and the Bunaq of Timor, in many respects share similar metaphors, models, and sociocomic ideas with their Austronesian counterparts, such as the idea of the "House", the "differential origin of society", and botanical metaphors (tree, stem, etc.) expressing basic social relations. However, if comparable models and metaphors are fundamental to both Austronesian and non-Austroneasian speaking societies of Indonesia, the prevalence of such common ideas and values across language borders can hardly be attributed to a common origin defined by linguistic criteria. Ultimately, as Jos rightly pointed out, "one has to account for the shared socio-cultural representations and metaphors in terms of an 'area-independent" general theory' (Platenkamp 1990e: 27).

Whereas in the lecture presented to the "Oostersch Genootschap" the question as to the intricacies of such an "area-independent theory" was more or less left open, its shape became more clearly expressed in a programmatic

8 See "Results Seram Survey, jan/feb 1989" (unpublished), and "Hierarchie in Zuidoost Seram: het concept oorsprong" (unpublished).

9 Jos briefly referred to some of these data in his article on "Solidarität und Austausch in Indonesien" ["Solidarity and Exchange in Indonesia"] (Platenkamp 2000a: 124-126).

paper that Jos (in collaboration with Cécile Barraud) published on the comparison of Eastern Indonesian rituals in 1990. Beforehand, Jos had been asked by the editors of the journal *Bijdragen tot de Taal-, Land- en Volkenkunde* to edit a special issue on a theme deemed relevant for the modern anthropological study of Indonesia. Jos and Cécile Barraud, who was acting as co-editor, had selected the topic of ritual. The special issue, titled "Rituals and Socio-Cosmic Order in Eastern Indonesian Societies" was published in two volumes bringing together the majority of the anthropologists working on Eastern Indonesia in this period. In his own contribution to the special issue Jos focussed on a ritual called "The Severance of the Origin" (Platenkamp 1990c) performed by the Tobelo in those rare cases when people belonging to the same "House" wish to marry. In order to avoid the disastrous consequences associated with "incest" the ritual effectuates the severance of the ancestral origin shared by the relatives from the bride's and groom's side. As a result, the bride receiving side has been ritually transformed into "foreigners", who now may encounter the groom's family – with which they are no longer related – as "ordinary" wife takers.

In addition to this article Jos and Cécile Barraud wrote a programmatic piece on "Ritual and the Comparison of Societies". Both authors argued that the study of rituals offers a privileged insight into the overall ideas and values of a society and that is the latter that should be subjected to a comparative analysis: "Because societies differ from one another in their systems of ideas and values, the comparison of societies requires a comparison of ideologies and not of elements a priori selected" (Barraud and Platenkamp 1990: 104). Rituals – as Jos and Cécile argued – express the very relationships on which societies are founded, relations between the living and the ancestors/deities, between the internal morphological order of a society and the domains situated "beyond". Rather than articulating demarcation lines, rituals bring the "inside" and the "outside", society and cosmos, into communication by effectuating the "circulation of beings and things" along the relations that constitute the socio-cosmic universe as a whole (Barraud and Platenkamp 1990: 117).

Having thus sketched a tentative programme for the comparison of eastern Indonesian societies – at least as far as the study of rituals was concerned – during the following years did Jos not further engage in comparative analyses but focussed his attention on the analysis of his Tobelo material. Here, the topic of ritual again figured prominently in an article on "Transforming Tobelo Ritual" (Platenkamp 1992). As Jos observed, many ceremonial actions performed by the Tobelo are nowadays shrouded in Christian terms such as "village church auctions" and "New Year parades" in honour of the dead, but when comparing these performances with the pre-Christian ritual cycle, particularly with the first and secondary mortuary rituals and the marriage ritual, it turns out that these modern forms of "ritual" have retained much of their initial logic by involving ideas about the transfer and circulation of money and rice along the socio-cosmic relations deemed necessary for the constitution of Tobelo society as a whole. Rather than disappearing, Tobelo rituals – as Jos' analysis suggests – have been transformed in the process of conversion and adapted to changing political and socio-religious circumstances. A similar way of analysing ritual change can be discerned in Jos' more recent analysis of the Pi Mai ritual of Luang Prabang (Platenkamp 2010a).

Meanwhile, however, Jos had been appointed to the professorship in Münster, opening a new chapter in his academic career.

From Münster to Laos and back to the North Moluccas

In 1993 Jos took over the chair of Ethnology in Münster from his predecessor Rüdiger Schott (1927-2012), who had been a specialist in African Studies. Annemarie Fiedermutz, who was likewise specialising on Africa, had been appointed to the other professorship at the institute in 1992.[10] Indonesia, and Southeast Asia more generally, as a region for ethnographic research was thus a novelty at Münster University.

Jos quickly adopted the German language, while also giving many of his lectures and courses in English. He always emphasised that students of ethnology should become familiar with scientific English as soon as possible because most of the academic literature is written in this language, and, moreover, because they should be able to present their research at international conferences. As far as teaching was concerned one of the most urgent tasks after Jos' appointment had been the creation of a standard curriculum in Ethnology in order to guarantee that each student would follow the same introductory lectures and courses. As part of this curriculum Jos henceforth gave introductory lectures on kinship and economic anthropology at the BA level and courses on scientific models at the MA level. In addition, Jos regularly offered courses on a variety of regional topics, ranging from "Indonesian house societies" to the comparative analysis of Southeast Asian rituals.

The latter topic, ritual as a mode of studying "the hierarchical order of relationships" (Platenkamp 1992: 74), remains a dominant theme in Jos' lectures and writings to the present day. However, in the years following his appointment at Münster, the analysis of myth also became a prominent field of research. Coming from a structuralist background, Jos' style of myth analysis is informed by the Leiden tradition as well as by theoretical insights achieved by Lévi-Strauss and the Dumont school.

In his contribution to the *Festschrift* for his teacher P.E. de Josselin de Jong (1988b), Jos had already put forward an analysis of several North Halmaheran myths in which he demonstrated that these narratives share the same idea/value of *ma datu* (see above). His analysis revealed that the myths can be

10 In 2006 Fiedermutz was succeeded by Helene Basu who specialises on South Asia and the Indian Ocean.

understood as transformations of each other – transformations which are leading to inversions among the hierarchical order of the values expressed in these narratives.

In another fascinating analysis (Platenkamp 1993), Jos demonstrated how historical events leading to the virtual extinction of one of North Halmahera's indigenous societies during the 16th and 17th century are reflected in Tobelo mythology. As he argued, the respective myths only selectively draw on these historical events and rearrange them in accordance with pre-existing socio-cosmic ideas about spirit-domains and the "external origin of political rule". "History" as understood in the Western sense is thus appropriated and reinterpreted within the framework of Tobelo cosmology and oral history.

The theme of the external origin of political rule also comes to the fore in Jos' article "In Search of a Name" (Platenkamp 1995a), in which he examined a myth from the Tobaru, a neighbouring society of the Tobelo. The myth narrates how the Tobaru leave their original dwelling place (on Sangir Island) and migrate to North Halmahera. The story thus corresponds to a category of narratives widely spread among Island Southeast Asian and Pacific societies stipulating that the local society initially has its origin (or one of its origins) "abroad", i.e. on other islands or in cosmological domains situated beyond the world inhabited by ordinary human beings. In a painstaking analysis, Jos identified the various geographical (north/south axis, seaward/landward, upward/downward) and numerical codes underlying the myth, scrutinising their transformations emerging from the Tobaru's migration from the "Sea" to the "Land".

Apart from the analysis of myths,[11] the anthropology of illness and healing also became a topic in Jos' writings. In two articles Jos deals with the contextualisation of Tobelo "medicines" within the overall system of ideas and values. In Tobelo thought, healing involves the idea that a patient must be reconnected to an encompassing cosmological order, particularly to its life-giving principle, from which he or she has been disconnected. This is done by constructing part-whole relations by which the patient is ultimately related to the "origin of life". Or in other words, whereas health is considered as a social condition, illness is conceptualised as desocialisation (Platenkamp 1998b: 61). Furthermore, Jos showed how in Tobelo ideology healing is subordinated to

11 In a later article (Platenkamp 2004b, see also 2005b) Jos presented a comparative analysis of several Indonesian myths dealing with the image of the "half-person" or "unilateral figure". These myths relate that the protagonist – the first human being or another person – is born with the deficiency of consisting only of a half-body, i.e. the lateral half of a symmetrical body parted longitudinally. As Jos argues, these myths attribute the protagonists' deficiency to the absence of certain primordial socio-cosmological relationships. In the course of the narration the myths explain how these relations consecutively come into being one after another, and only after this process has been accomplished, the protagonist will have become complete. Though varying considerably from one society to another, these relationships "constitute a kind of blueprint of the social order in itself" (Platenkamp 2014: 2).

the idea of exchange. In order to be effective, medicines (or the knowledge about them) must have been acquired through gift exchange preferably with "strangers", and the patient in turn is expected to reciprocate the practitioner with the same type of gift (Platenkamp 1996b: 328).

Other publications from this period deal with Tobelo kinship terminology (Platenkamp 1998a), the scientific paradigms used in the early ethnographic description of Indonesian societies (Platenkamp and Prager 1994), and the contrasting ways of conceptualising "nature" in modern and non-modern societies (Platenkamp 1999).

In the second half of the 1990s Jos and his colleagues undertook efforts to establish a comparative and interdisciplinary research project on Southeast Asia involving various disciplines from the University of Münster (Ethnology, History, Political Science). Beginning in 1999, the project was titled "Cultural Diversity and the Construction of the Polity in Southeast Asia: Continuity, Discontinuity, and Transformation" (FOR 362, 1999-2002) and financed by the German Research Foundation. In the course of the preparations for this project, Jos had decided to extend his ethnographic expertise to mainland Southeast Asia and was planning to undertake new field research in Laos, a country which up to then had been rather inaccessible to foreign researchers. During preparatory talks in Vientiane, Jos managed to obtain the consent of the Laotian government for the setting up of such an anthropological project involving not only his own field research but also that of two of his Ph.D students.

Jos' research focussed on the ritual structure of Luang Prabang, the old royal centre of Laos. From 1996 on, he regularly returned to Luang Prabang for longer and shorter field trips to collect data on various rituals and the socio-cosmic ideas and values of which the latter form part. Up to the present, only a part of the collected data has been published.

One of the outcomes of the Laos research is Jos' analysis of the Pi Mai (New Year) ritual. The New Year ritual, performed at the end of the dry season in April before the coming of the monsoon rains, is based on the idea that water from the nearby Mekong and Nam Khan rivers – where powerful *phi* spirits reside, who are said to be the "owners" of the water – is ceremonially transferred to the city where it is distributed among the inhabitants and along different levels of the religious hierarchy in order to protect the domain against the intrusion of evil spirits. It is above all a ritual of regeneration and purification by means of which the city and its inhabitants are cleansed from the misconduct, misfortunes, and impurities amassed during the last year. Jos gives a detailed description of the ritual topography of the domain, and of the major sequences characterising the Pi Mai ritual.

In a careful analysis, he also explores to what extent the ritual has been subject to changes from 1975 onwards due to the abolition of kingship and the rise of the communist government. Whereas such changes are clearly visible, particularly with regard to the articulation of hierarchical relationships, the ritual, as Jos' analysis demonstrates, still shows a "remarkable conceptual continuity" (Platenkamp 2010a :194).

Fig. 3: Jos taking fieldnotes in Luang Prabang, Laos, 1996

While Jos' explorations into the New Year and the canoe racing rituals (Platenkamp 2008a) are concerned with ceremonial acts forming part of the annual ritual cycle and being of importance for the society and the domain as a whole, his analysis of the Lao birth ritual (Platenkamp 2010b) addresses a context situated at the level of individual families and houses, although in the course of the ritual a more encompassing set of socio-cosmic relations has to be mobilised as well. Jos carefully scrutinises the ceremonial cycle surrounding pregnancy and birth and shows how with each consecutive ritual act a newborn child is incorporated within a progressively expanding range of social and cosmological relations, thereby providing the emerging Lao person with his or her essential corporeal and spiritual constituents. Rather than being conceptualised as a discrete and self-contained entity, the Lao person as being consecutively completed in the course of the birth ritual is above all a relational construct.

Fieldwork in Laos also stimulated Jos to engage in the comparison between mainland and island Southeast Asian societies, by focussing on spirit representations (Platenkamp 2007a) or by comparing how the category of the "strange(r)" is valued in Tobelo and Lao ideology (Platenkamp 2004a). By undertaking such comparisons Jos was elaborating on an old idea of P.E. de

Josselin de Jong (1965) that mainland and island Southeast Asia could be considered as a larger "Southeast Asian Field of Anthropological Study" (Platenkamp 2007a: 99). Up to the present, continental and island Southeast Asian societies only rarely have been subjected to a systematic comparative analysis. Jos thus belongs to a small group of pioneers who have ventured into this field of research.

The theme of the "stranger" remains an important topic in Jos' work. In several of his articles, Jos has dealt with the way in which "strangers" are classified and become part of local societies by incorporating them into the prevailing system of kinship and ritual exchange (Platenkamp 2000a) and/or by attributing to them – as representatives of an external non-autochthonous origin – specific political offices and cosmological roles. Jos argues that Southeast Asian societies as the Tobelo und Lao valorise the stranger as a necessary part of their own society, complementing the socio-cosmic order as a whole. In doing so they differ from Western societies like Germany where the stranger is not valued as a constituent part of the society. A case in point are the Turkish "guest labourers", who came to Germany in the wake of the *Wirtschaftswunder* during the 1960s and who – together with their German-born children and grand-children – are officially designated by the state as "people with a migration background". As Jos observes, the migrant workers were valued because they contributed to the national economy, not because they represented different cultural identities. On the contrary, the latter – up to the present – are rather considered as an impediment to "integration" (Platenkamp 2004a, 2009: 8-12, 2014).

In one of his most recent contributions to the topic of the stranger Jos concludes with a remarkable passage, which contains a solemn warning against the discourse of "exteriorisation":

> "In view of the mere presence of millions of citizens of foreign origin in west European states interacting daily with other citizens in various social contexts, the exteriorising definitions of the 'stranger' create fundamental problems. They deprive citizens of different origins of the means to conceive of their own cultural identity as a valuable contribution in itself to the society of which they have become part. The real and potential results are political indifference and social alienation. It therefore is an urgent question whether such exteriorising discourses are an inevitable corollary of any construction of socio-political identity and a necessary condition for the effective functioning of multicultural nation-states. The examples from Southeast Asia show that alternative modes of identity construction are in evidence in other, non-European societies. It may be socially and politically advantageous to take this fact properly into account" (Platenkamp 2014: 9).

Given the influx of refugees in Germany and other European countries, these observations have gained additional relevance, and German politicians would be well advised to ponder about the potential consequences resulting from such exteriorising discourses.

During the last few years, Jos has focussed his attention on the North Moluccas again,[12] particularly on the way in which the local societies (and former sultanates) have been shaped and transformed in confrontation with "world" history. In this context Jos undertook research on the question of how, and to what extent, North Moluccan societies in their historical encounter with Iberian Catholicism and Dutch Protestant Christianity were able to appropriate and accommodate these foreign sources of knowledge and belief within their indigenous socio-cosmic systems, particularly with regard to local conceptualisations of "cosmological sovereignty" and "diarchy" (Platenkamp 2015b, Platenkamp f.c.).[13] From a similar perspective Jos has demonstrated that North Moluccan societies in the course of history adopted various foreign currencies while reinterpreting their respective "values" in accordance with indigenous notions about the circulation of "living" and "dead money" (Platenkamp 2013, 2016). Clearly emerging from these analyses is the cultural "resilience" of these societies, their capacity to creatively adapt to the imponderabilities of globalisation and socio-political change. This research is still going on and related publications are expected to appear in the near future.

Among his colleagues Jos is known as a scholar whose work has a strong theoretical orientation, particularly in the field of structural analysis. Besides applying such theoretical models to his own ethnographic material he has also written introductory texts on structuralism (Platenkamp 1984b, 2003) and co-edited a volume on theoretical anthropology (Platenkamp 1996c). Being considered as a "theoretician" rather than a practical anthropologist, it is often overlooked that Jos over the years frequently has addressed questions of applied anthropology, either by offering courses about the use of ethnological knowledge in the post-Tsunami crisis management of Indonesia, or by writing about the relevance of anthropology in social practice (Platenkamp 2004a). In another contribution, (Platenkamp 2007b) Jos examines how a self-reflexive anthropological perspective may contribute to the successful implementation of peacekeeping missions. The text is a vibrant plea against ethnocentrism and the idea that Western culture by definition is morally superior to other societies. It is only by respecting their values and codes of conduct that one wins

12 In an earlier article (Platenkamp 2006a) Jos had already begun to re-examine his Tobelo material by focussing on the concept of "visibility" as a means of articulating basic ideas about sociality.

13 Parts of this research were undertaken in the context of the research project "Local Traditions and World Religions: The Appropriation of 'Religion' in Southeast Asia and Beyond" (funded by the German Research Foundation and the French National Research Agency) of which Jos became a member in 2011.

the "hearts and minds of the people" (Platenkamp 2007b: 104), the latter being a precondition for the establishment of a successful communication.

Having reviewed some of Jos' publications it has become evident that his analyses are addressing a wide array of subjects. Despite this thematic range (healing, myths, rituals, kinship, exchange, applied anthropology, history of anthropology, material culture, etc.), there are some basic ideas running through his work, key concepts which were already mentioned at the beginning of this article:

- the idea that the phenomena under scrutiny can only be properly analysed if set in relation to society as a whole
- such wholes are predicated on systems of ideas and values, which may be subject to a hierarchical order
- the similarities and differences pertaining to these wholes only come to the fore through comparison

Holism, value systems, and comparison are thus pivotal concepts in Jos' research which he never loses out of sight. At a time when the concept of the social is increasingly on the retreat (Kapferer 2005), Jos is one of few scholars who adhere to the idea that society constitutes a focal value and that comparison is the "raison d'être" of social anthropology (Platenkamp 2009: 8). Such a perspective, however, as Jos argues by drawing on Lévi-Strauss (1985), requires a "view from afar", for "only when observed through the lenses of other societies may one perceive the cultural idiosyncrasy of our own ideas and values more sharply" (Platenkamp 2014: 8).

These are basic insights which he also conveys to his students, leading us to say a few words about Jos as a teacher.

Teaching and Supervising

Jos is well remembered by his students for his vivid style of teaching, constantly moving back and forth between the desk and the blackboard, covering the latter with myriads of diagrams and technical terms. Even when giving the same courses again for a couple of times – as part of a repetitive curriculum – each year anew he presents the various topics with an enthusiasm as if it were for the first time. This is probably one of the most remarkable characteristics that makes him stand out as a teacher, his enthusiasm for the discipline of anthropology and the ability to convey this enthusiasm and dedication to the students. After all these years of having been in the teaching "business", when he is unfolding a specific theoretical problem in front of the students and eventually presenting its solution, he still manages to evoke the impression among the listeners that they are witnessing the discovery of an important scientific fact. Not many teachers can compete with him in this respect, particularly after having taught for more than three decades.

Jos also enjoyed teaching in a special programme on Southeast Asian Studies. At the beginning of the 2000s Jos and several colleagues from other disciplines (History of Southeast Asia, Political Science, and Sociology) made plans to establish an interdisciplinary Master programme in Southeast Asian Studies at the University of Münster. The Master programme started in 2002 and drew international students from Asia, Europe, and the United States. The programme was dear to Jos, as he greatly enjoyed teaching students from various cultural backgrounds, including Southeast Asia. With its interdisciplinary character and with English being the main language of instruction, the Master programme of Southeast Asian Studies was unique in Germany at this time. Due to budget cuts, however, the Master programme was cancelled in 2006.

Apart from teaching in Münster and Leiden, Jos while holding visiting professorships has also taught students in Aarhus, Helsinki, Bergen, Strasbourg, Ljubljana, and other places.

Not only as a teacher but also as a supervisor Jos has been very successful throughout the years. From his appointment at the University of Münster on, he has guided a large number of BA and MA students through their studies and theses. In Münster he has supervised 16 PhD theses written by German students as well as students from Indonesia, Slovenia, Sierra Leone, and Jordan (a list is provided at the end of this section). These theses show a high degree of regional variability. Apart from Southeast Asia they involve researches undertaken in South Asia, the Middle East, Africa, and Europe. This testifies to the fact that while Jos' own ethnographic work is focussed on Southeast Asia, his curiosity and intellectual depth truly exceed regional limitations, revealing him as a scholar whose scope and vision are ultimately grounded in general anthropology. Being a scholar of international renown Jos was repeatedly requested to act as an external examiner or co-supervisor for PhD candidates in the Netherlands, France, Australia, England, and other countries.

Meanwhile, many of Jos' former MA and PhD students have become scholars and teachers themselves. Two of his PhD students have been appointed Professor of Ethnology/Anthropology at German universities, two others as Professors of Ergotherapy and Religion respectively. Others are teaching as lecturers at European, Asian and Middle Eastern Universities. Still others are employed as researchers in various national and international research projects. Moreover, he has supervised three Habilitation theses focussing on Africa, India, and Indonesia. Finally mention should be made that two of the PhD theses supervised by Jos were awarded the prestigious Frobenius Research Prize. To sum up, as a teacher, supervisor, and mentor, Jos has been extremely successful and instrumental in engendering and inspiring a new generation of scholars following in his footsteps and spreading the news that anthropology is one of the most fascinating disciplines in the world.

Apart from having been deeply immersed in social anthropology for more than four decades, Jos is pursuing many other interests. Keeping track with major developments in the humanities is one; an interest in ornithology another. Moreover, he is a passionate reader of classical and contemporary literature. He is well versed in recent developments of modern fiction, always keeping track of literary trends by consulting the *Times Literary Supplement* and other relevant publications. Mention should also be made of his love for music. He is a connoisseur of classical music, Johann Sebastian Bach and Johannes Brahms being among his favourite composers. In addition, he is steeped in different styles of rock and folk music, singer-songwriter James Taylor probably being his most cherished musician in this respect. Having learned to play guitar in his youth (later in a band!) he not only enjoys listening to music, he also has a refined ear and is able to perceive the finest compositional details and nuances. Listening music together with Jos is a joy, both in an intellectual and an aesthetical sense.

When talking to him about the mysteries and beauty of music and when listening to his smooth guitar playing, one could easily get the impression that in addition to anthropology he would also have been able to pursue a career as a professional musician. It is a blessing for anthropology that he chose otherwise.

• • •

While this article is published in a *Festschrift* in his honour, Jos Platenkamp is engaged in a variety of ongoing researches, in the supervision of BA, MA and PhD theses, and in many publication projects. This overview of some of the key concepts characterising his way of "doing anthropology" and the brief outline of major landmarks highlighting his professional career are only a temporary "snapshot" of what Jos as a scholar has achieved at the age of 65. In 2017 he is going to retire from his professorship, but he will continue to leave his mark on anthropology both in its theoretical and regional-ethnographic dimensions.

Jos often refers to Salman Rushdie's (1999) words that "the only one who sees the whole picture [...] is the one who steps out of the frame". It is to be hoped that Jos Platenkamp in years to come keeps staying inside anthropology but remains outside the frame. Anthropologists like Jos encourage us to strive for the whole and not be content with the parts only. Anthropology is thus still strongly in need of him.

References

BARNES, R.H. (1985) The Leiden Version of the Comparative Method in Southeast Asia. *Journal of the Anthropological Society of Oxford* 16(2): 87-110.

BARRAUD, C. (1979) *Tanebar-Evav. Une Société des Maisons Tournée vers le Large.* Paris/Cambridge: École des Hautes Études en Sciences Sociales/Cambridge University Press.

BARRAUD, C. (1984) Comment on J.D.M. Platenkamp's Paper. In JOSSELIN de JONG, P.E. de (ed.), *Unity in Diversity: Indonesia as a Field of Anthropological Study*, pp. 190-197. Dordrecht: Foris.

BARRAUD, C., COPPET, D. de, ITEANU, A. and JAMOUS, R. (1984) Des relations et des morts. Quatre sociétés vues sous l'angle des échanges. In GALEY, J.C. (ed.), *Différences, valeurs, hiérarchie*, pp. 421-520. Paris: Éditions EHESS [English translation: *Of Relations and the Dead: Four Societies Viewed from the Angle of Their Exchanges*. Oxford/Providence: Berg, 1994].

COPPET, D. de (1981) The Life-Giving Death. In HUMPHREYS, S. and KINGS, H. (eds.), *Mortalitiy and Immortality: The Anthropology and Archaeology of Death*, pp. 175-204. New York: Academic Press.

DUMONT, L. (1977) *Homo aequalis I: Genése et épanouissement de l'idéologie économique*. Paris: Gallimard.

DUMONT, L. (1979) *Homo hierarchicus. Le système des castes et ses implications*. Paris: Gallimard.

DUMONT, L. (1980: *Homo Hierarchicus: The Caste System and its Implications.* Chicago: University of Chicago Press.

DUMONT, L. (1983) *Essais sur l'individualisme. Une perspective anthropologique sur l'idéologie moderne.* Paris: Éditions du Seuil.

GEIRNAERT, D. (1987) Hunt Wild Pig and Grow Rice: On Food Exchanges and Values in Laboya, West Sumba (Eastern Indonesia). In RIDDER, R. de and KARREMANS, J.A.J. (eds.), *The Leiden Tradition in Structural Anthropology: Essays in Honour of P.E. de Josselin de Jong*, pp. 106-122. Leiden: Brill.

HEUSCH, L. de (1968) Les points de vue structuralistes en anthropologie et leur principaux champs d'application. In *The Social Sciences: Problems and Orientations*, pp. 33-46. The Hague/Paris: Mouton/Unesco.

ITEANU, A. (1983) *La ronde des éxhanges. De la circulation aux valuers chez les Orokaiva*. Paris/Cambridge: Éditions de la Maison des Sciences de l'Homme/Cambridge University Press.

JOSSELIN de JONG, J.P.B. de (1977[1935]) The Malay Archipelago as a Field of Ethnological Study. In JOSSELIN de JONG, P.E. de (ed.), *Structural Anthropology in the Netherlands: A Reader*, pp. 164-182. The Hague: Nijhoff.

JOSSELIN de JONG, P.E. de (1977) Introduction. Structural Anthropology in the Netherlands: Creature of Circumstance. In JOSSELIN de JONG, P.E. de (ed.), *Structural Anthropology in the Netherlands: A Reader*, pp. 1-29. The Hague: Nijhoff.

JOSSELIN de JONG, P.E. de (1980a) The Concept of the Field of Ethnological Study. In FOX, J.J. (ed.), *The Flow of Life: Essays on Eastern Indonesia*, pp. 317-326. Harvard: Harvard University Press.

JOSSELIN de JONG, P.E. de (1980b) The Netherlands: Structuralism before Lévi-Strauss. In DIAMOND, S. (ed.), *Anthropology: Ancestors and Heirs*, pp. 243-257. The Hague/Paris: Mouton.

JOSSELIN de JONG, P.E. de (1984) A Field of Anthropological Study in Transformation. In JOSSELIN de JONG, P.E. de (ed.), *Unity and Diversity: Indonesia as a Field of Anthropological Study*, pp. 1-10. Dordrecht: Foris.

JOSSELIN de JONG, P.E. de and VERMEULEN, H.F. (1989) Cultural Anthropology at Leiden University: From Encyclopedism to Structuralism. In OTTERSPEER, W. (ed.), *Leiden Oriental Connections, 1850-1940*, pp. 280-316. Leiden: Brill.

KAPFERER, B. (ed.) (2005) *The Retreat of the Social: The Rise and Rise of Reductionism*. New York/Oxford: Berghahn.

LEACH, E.R. (1972) Anthropological Aspects of Language: Animal Categories and Verbal Abuse. In MARANDA, P. (ed.), *Mythology*, pp. 39-67. Harmondsworth: Penguin.

LÉVI-STRAUSS, C. (1985) *The View From Afar*. Chicago: University of Chicago Press.

MASINAMBOW, E.K.M. (1994) The Halmahera Research Project and Regional Development. In VISSER, L.E. (ed.), *Halmahera and Beyond: Social Science Research in the Moluccas*, pp. 213-224. Leiden: KITLV Press.

NIJLAND, D.J. (1989) *Schaduwen en werkelijkheid. Cultureel bepaald kennen, waarnemen, nonverbaal communiceren, en onderzoek daarvan via de terugvertoning van de etnografische film "Tobelo Marriage"*. PhD Thesis, Leiden University.

NIJLAND, D.J. (2006) Film and Non-verbal Expressions of Culture Amongst the Tobelo. In POSTMA, M. and CRAWFORD, P.I. (eds.), *Reflecting Visual Ethnography: Using the Camera in Anthropological Research*, pp. 52-77. Leiden: CNWS Publications/Højbjerg: Intervention Press.

OOSTEN, J.G. (1993) Cosmological Cycles and the Constituents of the Person. In DUFOUR, M. and THÉRIEN, F. (eds.), *Looking for the Future: Papers of the Seventh Inuit Studies Conference*, pp. 185-196 (Inuit Studies Occasional Papers 4).

PLATENKAMP, J.D.M.; see *Publications of J.D.M. Platenkamp* attached to this section.

PRAGER, M. (1996) *Strukturale Anthropologie in Leiden. Ursprung und Entwicklung eines wissenschaftlichen Forschungsprogramms*. PhD Thesis, University of Heidelberg.

RUSHDIE, S. (1999) *The Ground Beneath her Feet*. New York: Henry Holt and Company.

VERMEULEN, H.F. (1987) P.E. de Josselin de Jong and the Leiden Tradition: A Short History. In RIDDER, R. de and KARREMANS, J.A.J. (eds.), *The Leiden Tradition in Structural Anthropology: Essays in Honour of P.E. de Josselin de Jong*, pp. 4-84. Leiden: Brill.

VISSER, L.E. (1988) The Elder and the Younger: Dual Opposition and a Hierarchy of Values. In MOYERS, D.S. and CLAESSEN, H.J.M. (eds.), *Time Past, Time Present, Time Future: Essays in Honour of P.E. de Josselin de Jong*, pp. 136-147. Dordrecht: Foris.

VISSER, L.E. (1994) Introduction. In VISSER, L.E. (ed.), *Halmahera and Beyond: Social Science Research in the Moluccas*, pp. 1-9. Leiden: KITLV Press.

ZANEN, S.M. and HOEK, A.W. van den (1987) Dinka Dualism and the Nilotic Hierarchy of Values. In RIDDER, R. de and KARREMANS, J.A.J. (eds.), *The Leiden Tradition in Structural Anthropology: Essays in Honour of P.E. de Josselin de Jong*, pp. 170-196. Leiden: Brill.

Jos Platenkamp and the Leiden Tradition in Structural Anthropology[1]

HAN F. VERMEULEN

When Jos Platenkamp enrolled at Leiden University as a student of cultural anthropology in September 1970, few signs indicated he would become one of the foremost continuators of the Leiden tradition in structural anthropology. However, upon graduating it was clear that Jos was not just another student. By the time of his doctorate it had become evident that he was developing into an outstanding practitioner of comparative structuralism, putting the Leiden research programme to the test in Southeast Asia and rejuvenating it with ideas borrowed from Paris. Subsequently, after being appointed professor of anthropology at Münster in 1993, Jos Platenkamp continued the Leiden tradition within a new academic environment, transmitting the Leiden-Paris research programme to students and refining it further.

Student Days

Born into a middle-class Roman Catholic family in Voorburg (The Netherlands) on 23 June 1951, Jos attended school in The Hague and in September 1969 moved on to the Catholic University of Nijmegen to study cultural anthropology. His cousin John Franssen had shown him the way by enrolling as a student at Nijmegen, then a centre of Catholic liberalism and a bulwark of radical socialism. Ethnology as an academic subject had been admitted to Nijmegen with B.A.G. Vroklage in 1948. He and his successor R.J. Mohr practiced historical ethnology, a specialism common in Germany until World War II but regarded as out-dated in the western part of the Netherlands after the war. Following the independence of Indonesia in 1949, ethnology in the Netherlands was renamed "cultural anthropology" and the former Indology course for prospective colonial civil servants was restructured under the name "sociology of non-western peoples" or non-western sociology in 1952-53 (Vermeulen 2002: 109-112). The first to hold a chair in non-western sociology at

1 This article was written at the invitation of Michael Prager and is based on documents and publications acquired in Leiden during the 1970s, 1980s and 1990s. Consultation with its main protagonist was undesirable in order to keep this *Festschrift* a surprise. I am grateful to Roy Jordaan, Dirk Nijland, Michael Prager, Peter Richardus, Leontine Visser and Sjoerd Zanen for their comments and suggestions. Leontine Visser also kindly provided the two illustrations herewith. Many thanks to Robert Busschots for digitally "cleaning" the photographs in this section and preparing them for reproduction. Needless to say, any remaining inaccuracies are my responsibility.

Nijmegen was Elisabeth Allard, a former professor of ethnology in Indonesia. She held this chair from 1958 until her retirement in 1969. The lecturers Albert Trouwborst and Geert van den Steenhoven assisted Mohr and Allard, teaching social anthropology and anthropology of law, respectively. When Gerrit Huizer succeeded Allard, he became a professor and the founding director of the Third World Centre. Apart from Trouwborst, two other lecturers, Alex van der Leeden and Jan Pouwer, had studied at and were awarded their PhDs from Leiden University. Anthropology at Leiden had a longer tradition – its first chair being founded in 1877 – and more staff members. After a year of studies at Nijmegen, resulting in a first-year degree (*propedeuse*), Jos switched to cultural anthropology (CA) at the University of Leiden.

At Leiden, Jos found anthropology to be dynamic and vibrant. Including cultural, structural, religious, economic, political, museum and visual anthropology, it was embedded in a wide field of studies covering comparative law, sociology, philosophy, literature studies, general history and many Oriental (later: Non-Western) language and culture studies. In 1955-56, G.W. Locher had founded the Institute of Cultural Anthropology (ICA) in Leiden and assembled a team of active, young scholars around two chairs: Cultural Anthropology, in particular of Southeast Asia and the Pacific, and Sociology of Non-Western Peoples (later: Societies). While the first chair was held by J.P.B. de Josselin de Jong until 1956 and, after his retirement, by his nephew P.E. de Josselin de Jong, the second field was covered by R.A.J. van Lier and G.W. Locher (Vermeulen 1987, 1997, 2002). This new field is now called Development Sociology. The ICA still serves as a platform for both study programmes.

During the 1970s, debates at the ICA revolved around two main oppositions: anthropology and non-western sociology on the one hand; structural and evolutionist anthropology on the other. Whereas Locher tried to bridge the gap between anthropology and non-western sociology (Vermeulen 1999), Rudi van Lier and his successor, Benno Galjart, promoted a large-scale sociology aimed at developing Third-World countries, among others focusing on *dependencia* theories and "access and participation" in South America. The general view among anthropologists was: How could such societies be developed without a thorough knowledge of their cultures? They felt development aid should be based on in-depth studies of local societies, which only anthropology was able to provide.

Within the anthropology section itself, however, there was an opposition between practitioners of structuralism and those favouring evolutionism and materialism. The latter subjects were developed in Leiden by Hans Claessen and Peter Kloos from the University of Amsterdam, and (on a temporary position) Petr Skalník from Prague. Basing their analyses on socio-political and economic data, they promoted a worldwide comparison in order to establish patterns of evolution. A third group developed around the chair of social anthropology of Africa, first held by Hans Holleman, then by John Beattie and his successor, Adam Kuper. Finally, a group of feminist anthropologists, empowered by Els Postel-Coster and Joke Schrijvers, was developing the subject "Women and Development".

Inspired by the teachings of A.A. Gerbrands and P.E. de Josselin de Jong, Jos sided with the structural anthropologists. Faced with a choice between Marx, Weber and Durkheim, he clearly opted for the latter. Jos was particularly interested in Durkheim's work with Marcel Mauss, which combined ethnology and sociology. There was room for debate about whether the social order determined religion, as Durkheim and W.H. Rassers held, or the realm of ideas influenced social organisation, as J.P.B. de Josselin de Jong viewed the relation during the 1920s and 1930s. By and large, however, Rassers and JPB adopted viewpoints developed by Durkheim and Mauss in Paris in order to interpret ethnographic material from Southeast Asia. F.D.E. van Ossenbruggen (1917) was the first to apply Durkheim and Mauss' essay on *Primitive Classification* (1903) to Javanese classification systems. It is generally assumed that the work of Rassers and J.P.B. de Josselin de Jong initiated the Leiden tradition (*de Leidse richting*), both at the Museum of Ethnology and at the University of Leiden.[2] However, from a historical point of view it would be more correct to consider this tradition as resulting from the work of Mauss, Van Ossenbruggen, Rassers and J.P.B. de Josselin de Jong.[3]

Jos' theoretical interests sprang from Durkheim and Mauss' essay (1903), Mauss' essay on exchange (1925), Mauss' essay with Beuchat (1906), the structural linguistics of Ferdinand de Saussure (1916) and Claude Lévi-Strauss' structural analyses in *Les Structures élémentaires de la parenté* (1949), *Anthropologie structurale* I&II (1958, 1973), *Le Totémisme aujourd'hui* (1962a), *La Pensée sauvage* (1962b) and *Mythologiques* (1964-71). He also studied the work of Edmund Leach (1961, 1976), read Bertrand

2 See among others, P.E. de Josselin de Jong (1972, 1977a, 1980b-c, 1984), Moyer (1976, 1978), Locher (1978, 1990), Miyazaki (1980, 1987), Niessen (1982), Vermeulen (1987, 1997, 1999, 2002), De Josselin de Jong and Vermeulen (1989), Platenkamp and Prager (1994), Prager (1996), Oosten (2006), Beaufils (2009, 2015).

3 P.E. de Josselin de Jong has consistently emphasised the importance of Marcel Mauss' work for both the Leiden tradition and Lévi-Strauss. However, it is only recently becoming clear how strong Mauss' impact on Dutch scholarship has been. David Moyer (1978: 457) pointed out that Durkheim sent his nephew Mauss to The Hague in 1898 to arrange an exchange between the *Bijdragen* of the KITLV and the *L'Année sociologique*, mentioning that Mauss would stay in Holland for several months (Visser and Moyer 1999: 177-8). Additional data on Mauss' reviews of J.J.M. de Groot, Van Ossenbruggen and A.C. Kruyt, and on the contacts between Mauss, Van Ossenbruggen and probably De Groot, were published in De Josselin de Jong and Vermeulen (1989: 297-8). See also Platenkamp (1984b, 1991d) and Oosten (2006). Thomas Beaufils (2015) adds to our knowledge of the exchange between Mauss and his Dutch colleagues by analysing correspondence dated between 1898 and 1938.

Russell's lectures on *The History of Western Philosophy* (1945) and was impressed by David Bidney's *Theoretical Anthropology* (1953), finding the latter's analysis of values in systems of culture revealing.

The study programme of cultural anthropology at Leiden, which despite its name came closer to social anthropology (Moyer 1975), consisted of introductions to a variety of thematic and regional subjects all geared to one aim: to conduct ethnographic fieldwork. Students in both cultural anthropology and non-western sociology were trained in anthropological and sociological methods and techniques (M&T), which in Leiden were provided by a section led by Hans Speckmann and Adri Gerbrands. All students were required to do fieldwork. During their BA (*candidaats*) they would carry out a three-week study in the Netherlands; during their MA (*doctoraal*) three months of research elsewhere in Europe or further afield. For his BA fieldwork Jos travelled to the town of Emmen in Drenthe, a province in the northeast of the Netherlands. Together with his fellow-student Bert van den Hoek he studied the social background of young adults, who had begun to work, and their need for schooling (6-26 January 1972). This research resulted in a joint report comprising 56 pages.

Apart from the professors Gerbrands and De Josselin de Jong, Jos was inspired by lecturers such as Jarich Oosten on the anthropology of religion, Hans van den Breemer on economic anthropology and Dirk Nijland on visual anthropology.[4] Whereas Gerbrands presided over the BA, De Josselin de Jong officiated as the main professor during the MA. Teaching ethnography, museum anthropology and visual anthropology, Gerbrands developed "ethno-communication: the study of non-verbal communication systems" (Gerbrands 1990). De Josselin de Jong lectured on structural anthropology and structural linguistics, especially within the Field of Anthropological Study (FAS), a concept introduced by J.P.B. de Josselin de Jong in 1935 and further developed by his successor (P.E. de Josselin de Jong 1977, 1980a-c, 1984). Gerbrands had carried out fieldwork in West New Guinea and West New Brittany and was an expert in ethno-aesthetics, the "language of things" and ethno-cinematography, recording ethnographic films at both fieldwork sites. P.E. de Josselin de Jong had conducted fieldwork in Western Indonesia and Malaysia. His main interest lay in the field of theory, developing the structural anthropology of Southeast Asia by focusing on kinship and political myths. He taught that it is vital to think in totalities (holism), search for relations between cultural categories, observe their transformations, and work within the framework of a comparative study of cultural wholes (FAS) on subjects such as kinship, politics and mythology. Another of P.E. de Josselin de Jong's contributions was his focus on "The Participants' View of Their Culture" (1956, 1977b), adopting an *emic* point of view avant-la-lettre. The debate on models

4 On Gerbrands and his work, see Vermeulen (1985), Locher (1990); on P.E. de Josselin de Jong, see Vermeulen (1987), Visser and Moyer (1999), Suzuki and Vermeulen (2000).

in anthropology (Burling 1964, Schneider 1965, Caws 1974, Fox 1980) was repeatedly dealt with during lectures. One of De Josselin de Jong's statements in this context was: "representations precede institutions" (*de voorstellingen gaan aan de instellingen vooraf*) – clearly indicating his cognitive orientation.[5]

Absorbing these views, Jos studied structural anthropology with a focus on the comparative study of small-scale societies, bound by a shared culture and tradition. After passing his "Atheneum" exams in 1974 and his *candidaatsexamen* in cultural anthropology in 1975, Jos entered the *doctoraal* level (now MA), which included fieldwork abroad. Being offered the choice between rural Spain and rural Scotland, Jos opted for the latter. Along with his fellow-student Henk Molenaar he stayed in Scotland for three months during the spring of 1975. Jos studied the village community of Portsoy on the northeastern coast of Scotland with strong divisions between those living off the land (farmers) and those working at sea (fishermen). He arrived at a structural analysis of a system of animal classification in a paper titled "The Cod and the Corn Don't Mix". It was of such quality that De Josselin de Jong accepted it as an MA thesis and offered to publish it as an article in a volume he was editing together with Erik Schwimmer, entitled *Symbolic Anthropology in the Netherlands* (1982).

Through a course in economic anthropology, provided by Hans van den Breemer, Jos came into contact with Marius Jacobs, a botanist at Leiden, who gave a course in human ecology, which Jos also followed. Jacobs initiated ecological and anthropological studies into the possible consequences of a project in Sudan to divert the White Nile from the southern *Sudd* marshes by constructing the so-called Jonglei Canal. The Dutch Ministry of Foreign Affairs in The Hague funded studies related to the construction of the canal. In 1976-77, Jos analysed the canal's potential impact on "an integrated system in the Southern Sudan" – a "well-balanced but easily disrupted socio-ecologic system". Cooperating with the ecologist C.J. Post, who was to visit the region, he studied the literature and interviewed experts in Great Britain. Jos concluded that the impact would be dramatic and at least "crossing-points" should be created in order for the Nuer and Dinka pastoralists to reach their dry-season pastures and to prevent an overconcentration of people and cattle around the few inland pastures, which would increase political friction between various tribes and tribal sections (Platenkamp 1978: ix, xiii, 96). The Leiden anthropologists Bert van den Hoek and Sjoerd Zanen conducted an ethnographic field study in Sudan in the spring of 1978. After consulting the population of

5 This formula may well have been borrowed from Lévi-Strauss, who made a similar point in his introduction to the collection of Mauss' articles (Lévi-Strauss 1950: xxxxii, cited in Platenkamp 1988a: 7).

15 court-centres, they recommended an "eastern trajectory" of the canal. However, this alignment was only partly accepted and, due to civil war, never realised.

Another thesis for a subject in the anthropology of religion dealt with the concepts of suppression and taboo in the work of Edmund Leach, one of the adherents of structural anthropology in Great Britain. Titled "Out of Mind, Out of Sight", this thesis was published as an article in the journal *Bijdragen tot de Taal-, Land- en Volkenkunde* (1979). In it, Jos pointed to logical inconsistencies in Leach's theory of "animal categories and verbal abuse" (Leach 1972 [1964]). Leach had argued that taboo names form an overlap of categories, which are "suppressed" by un-naming. On the basis of his data from Scotland, however, and the fact that the names of land animals such as pig and rabbit were prohibited among fishermen at sea, Jos argued that these animal taboo names are not "un-named" but "emphasized" and "accentuated" (Platenkamp 1979: 177-79). Thus, the question could be raised as to the level with which Leach was concerned: "ideal order" or "empirical fact" (Kuper 1973: 96), or, as Jos rephrased the former, as part of a "cultural socialization process, and therefore located at an unconscious cognitive level" (Platenkamp 1979: 176).

On 19 June 1978, Jos graduated in cultural anthropology, with ecology and economic anthropology as subsidiary subjects. Combining insights from specialists in Leiden and Paris, as well as in Oxford and Cambridge, he had published a research report on the Sudan, an article on his fieldwork in Scotland, in which he applied structural anthropology to an unexpected region, and a critical analysis of the theories of Leach. With these publications Jos demonstrated he was able to integrate several fields of knowledge in an innovative way.

In earlier days he would immediately have been appointed as a lecturer. But these were the late 1970s. Competition for financial resources was fierce at an institute where competing factions were fighting for dwindling resources, baby-boomers had occupied all positions, and research was increasingly being carried out on the basis of part-time contracts. In response to the introduction of a national system of "conditional funding" specialised research groups were formed in Leiden, one of which attracted Jos' interest: "Cognitive Anthropology and Structural Anthropology" (CASA). Led by De Josselin de Jong from 1976 on, and later by Jarich Oosten and Jos Platenkamp, CASA encouraged students to present and discuss their research results during weekly seminars (Oosten 1979). Gradually, this group took over the position previously fulfilled by the Ethnological Debating Society WDO.

In November 1978, Jos chaired an international conference dealing with the theory and history of Leiden anthropology. It was held on the occasion of WDO's 50th anniversary. Founded in 1928 by the first-generation students of J.P.B. de Josselin de Jong, this society served and still serves as a platform where students meet each other for discussions after lectures. Together with his fellow-students, Franklin Tjon Sie Fat, Gérard Geurten, Danielle Geirnaert and Leontine Visser, Jos invited an impressive number of lecturers, both from first- and second-generation students of J.P.B. de Josselin de Jong and from abroad, to contribute to three days of discussions on "Fifty Years of the Leiden Tradition in Cultural Anthropology". The papers were published in a volume co-edited by P.E. de Josselin de Jong, entitled *The Nature of Structure* (1981). In the course of this conference Leiden anthropology clearly established itself as a holistic, structural and comparative discipline. They are the main features of what P.E. de Josselin de Jong and others called "the Leiden tradition".

Several years after the coup leading to Suharto's seizure of power, relations between Indonesia and the Netherlands improved and a cultural agreement was signed in 1968. Subsequently, new research could be undertaken in Indonesia: by J.D. Speckmann and a team of non-western sociologists in Serpong, West Java (1971-75); by P.E. de Josselin de Jong and a team of anthropologists in Minangkabau, Sumatra (1973-74). After the cultural agreement had come into effect in 1975, Indonesian students again enrolled at Leiden to study. Under the Programme of Indonesian Studies (PRIS), Dutch scholars carried out research in Indonesia, together with their Indonesian counterparts, under the auspices of the Indonesian Institute of Sciences (LIPI), founded in 1967. PRIS comprised an Indonesian-Dutch scientific collaboration between LIPI in Jakarta and KITLV in Leiden.[6] It offered anthropologists and sociologists from Dutch universities the opportunity to conduct doctoral field research in Indonesia. From Leiden, fieldwork was carried out by Chris van Fraassen on Ternate and Tidore (1976-77); Anke Niehof and Roy Jordaan on Madura (1977-79); Jos Platenkamp and Leontine Visser on Halmahera (1979-81). In the same period, Sandra Niessen conducted fieldwork on Sumatra (1979-80), funded by the National Research Council of Canada.

Reacting to these developments, De Josselin de Jong had procured a full-time lectureship in 1975. The American anthropologist Robert H. Barnes, focusing on Eastern Indonesia and serving as a lecturer at the University of Edinburgh, was a candidate. When Barnes opted for Oxford, the post went to David S. Moyer from Canada, who in 1971-75 had written a thesis on "the logic of the laws" from Sumatra, which De Josselin de Jong had supervised. Moyer taught structural anthropology and the anthropology of Southeast Asia during the academic year 1976-77. Well versed in British and French structuralism, he wrote interesting articles on Dutch structuralism (Moyer 1976, 1978). After his return to Canada the position was occupied on a part-time or temporary basis.

6 KITLV: The Koninklijk Instituut voor de Taal-, Land- en Volkenkunde was founded in 1851 in Delft and later moved to The Hague to be newly established at Leiden in 1967. It is the publisher of two journals, including *Bijdragen tot de Taal-, Land- en Volkenkunde*, as well as several book series.

Rejuvenating the Leiden Tradition

In 1978 Jos accepted the position of a PhD researcher in the Halmahera Research Project, one of two research projects in Indonesia conducted under the Programme of Indonesian Studies, the other being the Madura Research Project. Profiting from the increased interest in Indonesian studies, Jos opted for a post entailing fieldwork on Halmahera in the Northern Moluccas, Eastern Indonesia. The Halmahera Research Project included not only the study of Halmahera but also that of the Raja Ampat Islands, located between Halmahera and the Bird's Head Peninsula of Irian Jaya. Dr. E.K.M. Masinambow, director of the National Economics and Social Institute (LEKNAS) in Jakarta, coordinated this project. It was part of an ambitious research effort of LEKNAS/LIPI to encourage the participation of Dutch scholars in research in Indonesia within the framework of the already existing PRIS (Visser 1994: 1).[7]

Preparations for the research were intensive. Jos and his colleague, Leontine Visser, also employed to do research on Halmahera, had to learn the Indonesian national language, study the literature on Southeast Asia and write research proposals for WOTRO in The Hague, the Netherlands Foundation for the Advancement of Tropical Research funding their research. They collaborated with two Indonesian researchers, sent to Leiden to prepare for their own fieldwork on Halmahera: Piet Koagouw (Menado) focussed on shamanism and Tjak Soukotta (Ambon) on agriculture. Lex van der Leeden, who had already begun his research on the Raja Ampat Islands in the context of the same project, mentored them. In the summer of 1979 Jos set off for a period of fieldwork that was to last 18 months. He was driven by the desire to make a contribution to structural anthropology by studying systems of classification among the Tobelo. His research was funded by WOTRO between 1979 and 1982.

Jos' initial focus was on systems of religious classification, in particular shamanism. In a paper entitled "The Structural Study of Shamanism", presented at the first Halmahera conference convened by Dr. Masinambow in Ternate, 10-20 July 1979, Jos presented a theoretical outline, arguing that individual or psychological views of shamanism as "ecstasy" or "mental deviance" were partial. He concluded that a structural analysis of "the society's system of religious, cosmological and moral ideas" was required in order to establish "the cognitive, classificatory order on which [...] social order is based" (Platenkamp 1980: 344-45).

7 The scope was broad: "Between 1977 and 1990 a total of seven Indonesian and twelve foreign scholars carried out studies and published on the North Moluccas (Halmahera, Ternate, Tidore, Bacan and Makian)" (Visser 1994: 1). The Halmahera Research Project resulted in a large number of publications mainly in the fields of anthropology, linguistics and history, including eight PhD theses, two films and a dictionary (Visser 1994: 1-2).

Fig. 1: Jos Platenkamp holding a Tobelorese sword and shield in the museum of the Sultan's palace on Ternate, July 1979
(Photo: Leontine E. Visser, Wageningen University)

With this basic scheme in mind Jos arrived in Tobelo, settling in the village of Paca on the eastern coast of North Halmahera, where he worked from July 1979 until December 1980. Leontine Visser worked in Sahu, on the western coast of North Halmahera, investigating "social and territorial aspects of swidden cultivation" (Visser 1984, 1989). They had to learn the local languages, which are non-Austronesian and thus differ from languages such as *bahasa* Indonesia and *bahasa* Melayu. They had difficulty communicating as Sahu and Tobelo are separated by a mountain chain, forcing them to travel by ship or meet in Ternate. Gerbrands, a visiting professor of museum anthropology at the Universitas Indonesia in Jakarta in 1979-80, travelled to Halmahera to pay Jos and Leontine a visit on their field sites.

During his fieldwork, Jos gradually expanded his research, collecting data not only on religious categories, including Christianity, but also on kinship and marriage, residence and housing patterns, the ideological system, the ancestors, social and spiritual exchanges, the first and second mortuary rituals, indigenous medicine, objects of material culture, etc. All in all, he conducted or at least attempted to conduct research as comprehensively as possible, both in four coastal villages and in four inland villages of Tobelo district. He also acquired data from historical reports, written by Dutch Protestant missionaries between 1870 and 1940, not only before undertaking fieldwork but also after returning to Leiden, in 1981-83. Ultimately, he viewed his research as dedicated to the study of kinship, social organisation and cosmology.

In the summer of 1982, Jos returned to Halmahera to assist Dirk Nijland in recording, on 16 mm film, a marriage ritual in Paca and to carry out additional research. From the end of July until the middle of October 1982, Jos and Dirk, along with their Indonesian counterparts Lis Asinoella, a graduate in cultural anthropology at the Universitas Indonesia in Jakarta, and Frans Rijoly, assistant curator of the Ethnographic Museum "Siwa Lima" on Ambon, collaborated in making the film *Tobelo Marriage* (Nijland 1985). Gerbrands had acquired Hfl. 120.000 from WOTRO for a project of his choice and it was decided to produce ethnographic films on Halmahera by two teams of ethnographic filmmakers and anthropological fieldworkers: Dirk Nijland and Jos Platenkamp on a marriage ritual in Tobelo (Nijland 1985); Saskia Jouwersma and Leontine Visser on a harvest festival in Sahu, in collaboration with their counterparts I Made Seraye and I Gusti Ngurah Arrington (Jouwersma 1985).

Jos had selected the marriage ritual because it expressed core issues of Tobelo society. The film aimed at revealing the complexity of the marriage ritual, involving exchanges between the group of the wife-taker and that of the wife-giver. These exchanges took place beforehand, during the negotiations and preparations, and during the three-day ritual itself, along with ceremonial exchanges of valuables, weapons, basketry and foodstuffs, as well as communal meals and ritual dances. In all exchanges, items from the "male side" were exchanged for those from the "female side". On the first day, the groom was presented to his future in-laws with sagu, fish and palm wine being exchanged for cooked rice and vegetables, consumed in a communal meal. On the second day, in the absence of both groom and bride, money, weapons, sagu and smoked fish were exchanged for basketry, plaited mats and rice seeds, again leading to the exchange of food stuffs in a communal meal and to the performance of dances in the evening. The third day saw the bride being escorted to "wade across the river to the other side" of the village. Here relatives of the groom's group awaited her and the bride was presented in full richness to her family-in-law (the groom being absent). Another communal meal and dances of various types concluded the marriage ritual.

The film *Tobelo Marriage* premiered at the ICA-Festival in Leiden on 8 June 1985 and drew international acclaim. It was selected for the sixth Bilan du Film Ethnographique, held under the auspices of the Centre National du Recherche Scientifique (CNRS) in Paris (1987), and awarded the "Grand Prix Nanook" for

> "the high ethno-cinematographic quality with which it presents a precise and humane picture of the preparations for and execution of a marriage among the Tobelo, Halmahera, Indonesia".

The film was also selected for the Margaret Mead Film Festival in New York (1989) and the annual meeting of the American Anthropological Association in Chicago (1991), where it received an "Honorable Mention". It was distributed for educational purposes in the Netherlands, Indonesia, the USA (Berkeley) and Japan. Dirk Nijland was to return to Tobelo in January 1988, together with his wife and counterpart Frans Rijoly, to view the film together with the people of Paca and to study their responses. This "feedback research" was discussed in Dirk's PhD thesis (Nijland 1989).

This ethno-cinematographic interlude, important as it is for understanding the value of linking filmic images and ethnographic texts (see Nijland, this volume), has diverted us from Jos' and Leontine's resettlement in Leiden in December 1980 and January 1981, respectively.

Upon Jos' and Leontine's return, De Josselin de Jong was keen on incorporating both their and Sandra Niessen's findings in the "Field of Anthropological Study" (FAS) approach developed by his predecessor, which he critically evaluated, expanded and redefined as a heuristic tool. In retrospect, this must be considered one of P.E. de Josselin de Jong's most important contributions to the Leiden tradition (De Josselin de Jong 1977, 1980a-c). The results of the international FAS conference held at Leiden, 22-26 November 1982, were published in a volume of essays entitled *Unity in Diversity: Indonesia as a Field of Anthropological Study* (De Josselin de Jong 1984; see also Barnes 1985 and De Josselin de Jong 1985).

To this volume Jos contributed a paper on the Tobelo, in which he argued that although the Tobelo and other Halmahera societies were non-Austronesian speakers, they were culturally part of the FAS of Southeast Asia. Analysing Tobelo concepts of kinship and marriage, he noted that kinship was cognatic and that no fixed alliance relationships between groups existed, although an exchange of siblings was possible (Platenkamp 1984a). Both conclusions would invalidate the inclusion of Tobelo in the FAS model. According to J.P.B. de Josselin de Jong (1935, 1977) and his PhD student Frans van Wouden (1935, 1968), societies in the Malay Archipelago were distinguished by a "structural core" including double descent, asymmetric connubium, socio-cosmic dualism and a specific form of resilience against foreign influences. However, adopting James Fox's findings that East-Indonesian societies employ "crucial social categories" such as the "house", the "ideal of a systematic alliance" and the MB/ZS relationship (Fox 1980), and endorsing P.E. de Josselin de Jong's heuristic view that cultures in the Indonesian field of anthropological study could "become more comprehensible" when studied as part of that field of study (De Josselin de Jong 1980b), Jos concluded that "the Tobelo are linked to the cultures of the Eastern Indonesian Field of Study" by means of "transformations rather than by imperfect resemblances" (Platenkamp 1984a; the latter citation being a direct quote from P.E. de Josselin de Jong 1980b: 322) (see Prager, this volume).

Another paper was presented during the second Halmahera conference, convened by Masinambow in Jakarta in June 1981, titled "Gia-Dutu: A True Relationship" (Platenkamp 1983). During the third Halmahera conference, held on Ambon in January 1984, Jos presented a paper on "Marriage and Death: Social Change in Tobelo" (Platenkamp 1984c).

Fig. 2: Jos Platenkamp, Leontine Visser and Shouichi Yoshida during the second Halmahera conference in Jakarta, June 1981
(Photo: C.L. Voorhoeve, Wageningen)

In 1983, Jos served as a "doctoral assistant" at the ICA, teaching an introduction to cultural anthropology and assisting in courses on Indonesia. In April 1983, he presented a theoretical exposition of "Structural Anthropology and the Study of Indonesian Cultures" at the Netherlands Institute of Advanced Studies (NIAS) in Wassenaar (Platenkamp 1984b).

In 1984, he obtained funding from the Netherlands Organisation for Scientific Research (ZWO) for a year of advanced studies at the research group ERASME (*Équipe de Recherche d'Anthropologie Sociale: Morphologie, Échanges*) in Paris. Building on the work of Louis Dumont, ERASME was part of CNRS and directed by Daniel de Coppet from 1983 on. It combined the study of social forms (morphology) with that of exchange (rituals). Jos was invited at the initiative of Cécile Barraud, who had been a discussant of his paper during the 1982 FAS conference held in Leiden (see Prager, this volume). In Paris, Jos lectured on his research in Indonesia and studied the ways in which the ideas of Mauss, Lévi-Strauss and Dumont had been further developed. In their fieldwork, the anthropologists taking part in ERASME tested central concepts of Dumont (1979, 1980), such as: values as important components of ideology, hierarchy as the "encompassment of the contrary" (*l'englobement du contraire*), and the exchange of "*idées-valeurs*" (idea values) in rituals. Jos found this period inspiring for analysing his own field data and important for complementing the structuralism developed at Leiden, which, in response to the early impulses from Durkheim and Mauss, was holistic and structurally focused, but now again was reinforced by crucial ideas from Paris.

The contacts between Durkheim and Mauss and scholars in the Netherlands had been seminal for initiating *le mouvement structuraliste Paris-Leyde*.[8] By travelling to Paris after his return from Indonesia, Jos did the reverse of what Mauss had done in 1898. Whereas Mauss had visited the Netherlands to contact colleagues having fresh data from the East Indies, Jos travelled to Paris for theoretical comments on his data from Eastern Indonesia. The central concepts forwarded by Dumont, especially the unity of values and ideas, enhanced Jos' view of the comparative study of societies, particularly in Southeast Asia (Platenkamp 1991d).

Immediately after returning from France, Jos began to process these insights in his own work and introduce them to his colleagues in Leiden. Collaborating with Jarich Oosten, he played an active role in the CASA research group, pointing out that Leiden anthropologists shared their holistic approach and their focus on systems of exhange with their colleagues in Paris (Platenkamp 1991d: 44). In 1985 a CASE-ERASME workgroup was established, in 1986 followed by the cooperative project CASA-ERASME, which envisaged comparative research with a focus on ideas and values and aimed at directly linking theory and fieldwork.

During his time in Leiden, Jos valued the discussions he had with Bert van den Hoek, who in the 1970s and 1980s carried out post-graduate research in South India and Nepal, and in 1977 had studied in Paris under Dumont. Before

8 The term "*mouvement structuraliste Paris-Leyde*" was coined by Georges Condominas (De Josselin de Jong 1995: 37). De Josselin de Jong (1980c) mentioned the term, not the name of the scholar who introduced it.

and after Jos' visit to Paris they discussed the work of Dumont and exchanged experiences. Jos thanked Bert in his dissertation "for sharing his perceptive insights during the many hours discussing the topic of this research". Jos worked in temporal assignments at the University of Leiden, among others at the Bureau of Indonesian Studies (BIS). In July 1987, he accepted a part-time position as lecturer in the cultural anthropology of Indonesia. Less than half a year later he submitted his dissertation, titled *Tobelo: Ideas and Values of a North Moluccan Society* (Platenkamp 1988a).

Leiden University awarded him a doctoral degree in social sciences on 26 May 1988. The commission led by P.E. de Josselin de Jong, and including D. de Coppet and C. Barraud (both Paris), J.J. Fox (Canberra, Australia), A.A. Gerbrands, G.W. Locher, E. Postel-Coster and J.D. Speckmann, evaluated both the dissertation and the defence as excellent, conferring the doctorate with distinction (*cum laude*). Locher characterised the book as "a first class achievement in the field of cognitive anthropology" and was particularly impressed by Jos' findings concerning "the second mortuary ritual" in Tobelo, which reminded him of his own fieldwork on Timor in 1940-41 (Vermeulen 1999) and of similar reports in the literature.

The dissertation of Jos Platenkamp is a major contribution to the ethnography of Indonesia and to anthropological theory. It analyses the system of ideas and values (cosmology) from the perspectives of social organisation and exchange, dealing with "social morphology, cosmology and exchange" (the first and the third concept being borrowed from the ERASME group in Paris). As to the Leiden tradition, it tested the field of anthropological study approach both on the basis of historical material (which was innovative) and of other non-Austronesian speaking groups on Halmahera. However, although the Leiden approach remained important as a heuristic device, another major source of inspiration was the work of Dumont and members of ERASME (Platenkamp 1988a: 7-8). He summarised the outcome of his research as follows: The values of Tobelo society should be considered in relation to each other, "these relations are ordered into models by the society itself", and "these values were expressed through the exchange of gifts" (Platenkamp 1988a: 297).

After his promotion, Jos was active in teaching, lecturing, organising and editing. He also carried out new fieldwork in Eastern Indonesia, albeit briefly. As early as the summer of 1985 he had submitted a proposal to the Bureau of Indonesian Studies in Leiden to conduct research on "social morphology and systems of exchange" in coastal and inland villages on Buru. This research was to be conducted in collaboration with Dr. Mus Huliselan from the Universitas Pattimura on Ambon, who had recently obtained his doctorate under Dumont in Paris. For several reasons, this project never materialised. However, in January and February 1989 Jos was able to carry out exploratory fieldwork on Seram (Central Moluccas). The results were promising (see Prager, this volume), but the research was not continued.

In May 1989, Jos presented an invited lecture to the Leiden-based *Oostersch Genootschap in Nederland*. In it he returned to the position of Halmaheran and other non-Austronesian speaking societies among Austronesian speaking ones, previously addressed in his chapter in *Unity and Diversity* (Platenkamp 1984a). It was selected for publication (Platenkamp 1990e) and reprinted as Platenkamp (1996c).

In 1989-90, Jos Platenkamp and Cécile Barraud co-edited two special issues of the journal *Bijdragen tot de Taal-, Land- en Volkenkund*, published at Leiden by KITLV. This project started when Hans Claessen, one of the editors of the *Bijdragen*, suggested Jos to edit a special anthropological issue with recent research on Indonesia. Jos and Cécile selected the topic of *Rituals and Socio-Cosmic Order in Eastern Indonesian Societies*. They invited so many researchers that the project grew into two issues dealing with two Indonesian regions: Nusa Tenggara Timur (Part I) and Maluku (Part II). Jos added a chapter on "The Severance of the Origin" (Platenkamp 1990c) and, together with Barraud, an afterword on "Rituals and the Comparison of Societies" (Platenkamp 1990d).[9] The latter was a programmatic article, summarising the research of all contributors to the special issues (see Prager, this volume).

Together with Jarich Oosten, Jos organised seminars and conferences under the auspices of CASA. Jos also convened the IXe Dutch Conference of Asian Studies (KOTA), which dealt with "Hierarchy, Equality and Individualism", reflecting the themes developed by Dumont and the ERASME group, novel in the Netherlands.[10] This KOTA conference was held at the Centre of Non-Western Studies (CNWS) in Leiden, 29-30 June 1989.

The CNWS had been founded in September 1988 by the Faculties of Arts and Social Sciences in order to concentrate all non-western studies by staff members and PhD students in a single centre of research. Oosten was its founding secretary and he continued to cooperate with Jos in directing CASA and in convening the first joint CASA-ERASME seminar, which also took place at the CNWS, on 12-14 March 1990. This seminar dealt with "The Study of Values in Ritual" and included members of both research teams, from Paris: Daniel de Coppet, Cécile Barraud, Annie Bruyer, Raymond Jamous, André Iteanu, Simone Pauwels; from Leiden: Jos Platenkamp, Jarich Oosten, Danielle Geirnaert, Franklin Tjon Sie Fat, Trudeke Vuyk, Ed van Hoven and Bert van den Hoek. The results were partly published in a special issue of the journal *Antropologische Verkenningen*, edited by Oosten (1990).

---

9 A review essay of both special issues by Michael Prager was published in *Anthropos* (Prager 1992).

10 KOTA: Kontaktgroep Tropisch Azië. In the period 1989-92, Jos often worked with Bert van den Hoek, working on his dissertation about the ritual cycle of Kathmandu (Van den Hoek 2004, 2014).

In the midst of these developments a commission in Leiden, chaired by Claessen, reached a decision regarding the vacancy resulting from P.E. de Josselin de Jong's retirement on 1 August 1987. Reimar Schefold (University of Basel, Free University of Amsterdam) was appointed professor of the cultural anthropology and sociology of Indonesia on 1 March 1989. He was the 7th occupant of this chair, inaugurated by P.J. Veth in 1877.[11]

Jos continued his post as lecturer in anthropology at Leiden, teaching the anthropology of Indonesia and the anthropology of religion. In 1992-93 he succeeded Oosten as secretary of the Research School CNWS, working closely with its recently appointed director, Dirk Kolff. It was an interesting period as the CNWS was expanding and would soon harbour over a hundred scholars and PhD students from four Faculties, including Law and Theology.

Attending the first conference of the European Association of Social Anthropologists (EASA) in Coimbra, Portugal, August 1990, Jos read a paper "Transforming Tobelo Ritual" at the invitation of De Coppet (Platenkamp 1992). He was a visiting scholar in Canberra (Australia) for two months in 1990 and in Aarhus (Denmark) for one month in 1991.

An opportunity for broadening his academic base presented itself when Jos was invited to teach ethnology at the University of Münster (Germany) during the summer semester of 1992. The idea was to fill the void due to a professorship of ethnology that had become vacant (*Vertretung*). Apparently, his performance impressed the students and faculty alike. As a result, Dr. Josephus D. M. Platenkamp was appointed full professor at the University of Münster (*C-4 Professor der Ethnologie*) as of October 1993. He took over the chair of Rüdiger Schott, a specialist in African studies, and set out to add the structural anthropology of Southeast Asia to the academic landscape of Münster, Westphalia.

Expanding and Refining the Leiden Tradition

In Münster Jos directed the small but budding Institute of Ethnology (*Seminar für Völkerkunde*), teaching structural and cognitive anthropology in the Paris-Leiden-Paris tradition. From 1995 on, he embarked upon new research in Laos, mainland Southeast Asia (see Prager, this volume). In April-October 1997, he was a visiting professor at the National Museum of Ethnology in Osaka, Japan, which led to new contacts. Together with colleagues from other departments he set up an interdisciplinary research project, "Cultural Diversity and the Construction of the Polity in Southeast Asia: Continuity, Discontinuity, and Transformation" (1999-2002), and an MA programme for Southeast Asian Studies, which began in 2002.

11 Schefold retired on 1 March 2003. Patricia Spyer (University of Chicago, University of Amsterdam) succeeded him on the very same day; she resigned on 31 December 2015.

Jos continued to expand his relations with colleagues in France, Australia, Japan and the Netherlands. Jarich Oosten and Sabine Luning invited him to give lectures in Leiden. A "Daniel de Coppet Memorial Symposium" took place at Leiden in December 2002. When Dirk Nijland retired in 1999, Jos was invited to present his views on the relation between text and film during the conference "Evaluating Visual Ethnography: Research, Analysis, Representation, and Culture" in Leiden. The results were published as *Reflecting Visual Ethnography: Using the Camera in Anthropological Research* (Postma & Crawford 2006). Jos' lecture was an analysis of Tobelo ritual but also a tribute to Nijland's project to set up a cooperation between an anthropologist-fieldworker and an anthropologist-filmmaker in order to arrive at an optimal combination of an ethnographic text and an ethnographic film.

In Münster, Jos was involved in new research projects, the most recent of which is the French-German project on "Local Traditions and World Religions: The Appropriation of 'Religion' in Southeast Asia and Beyond", funded by the French National Research Agency (ANR) and the German Research Foundation (DFG). Jos joined this project in 2011 with historical research on "The Emergence of 'Religion' in Halmahera" in order to study the introduction and transformative impact of Christianity in the North Moluccas (1500-1900). To this end, he regularly returns to Leiden to study historical texts in the library of the KITLV.

Would Jos have had the same impact in Leiden as in Münster? There is no doubt that he would have tried to pursue the Leiden tradition and that the circumstances would have been good, given the critical mass of scholars and students in many Oriental and Non-Western studies at Leiden. Of course, there were and are other bearers of the Leiden tradition. At the ICA, we need only to recall the work of Jarich Oosten, Hans van den Breemer, Franklin Tjon Sie Fat and Sabine Luning, the dissertations of Trudeke Vuyk and Ed van Hoven, or the contributions of Dirk Nijland, Metje Postma and Erik de Maaker to visual ethnography.

At the conclusion of my article on "Anthropology and Other Non-Western Studies in Leiden" the following third-generation PhD students from Leiden, who were appointed professor at universities outside Leiden, were mentioned: Peter Suzuki (Omaha, USA), Jacob Vredebregt (Jakarta, Indonesia), David Moyer (Victoria, Canada), Joke Schrijvers (Amsterdam), Anke Niehof (Wageningen), Leontine Visser (Wageningen), Isa Baud (Free University Amsterdam), Danielle Geirnaert (Paris, France), Victoria Baker (Florida, USA), Jos Platenkamp (Münster, Germany), Koji Miyazaki (Tokyo, Japan), Jan Brouwer (Shillong, India), Wasif Shadid (Tilburg) and Ruben Gowricharn (Tilburg) (Vermeulen 2002: 151).

Of the few Leiden students of this generation to be appointed professor in Leiden, Carla Risseeuw, Cees van Dijk, Gerard Persoon and Jan Slikkerveer, only Carla Risseeuw and Gerard Persoon served as a professor at the home department (ICA). Hereto may be added those who continued the Leiden tradition without being appointed professor, for instance, Bert van den Hoek (specialising in Nepal), Roy Jordaan (specialising in Indonesia), Sandra Niessen (specialising in Indonesia, initially in Canada, now in the Netherlands), or Sjoerd Zanen (applying structuralism to development anthropology). Yet few anthropologists trained in Leiden have pursued and innovated the characteristic trend of anthropology developed at Leiden in such a consistent and ingenious manner as has Jos Platenkamp.

It will be interesting to observe, on the basis of this *Festschrift*, as well as the 19 PhD and Habilitation theses supervised by Jos Platenkamp in Münster, the extent to which he has been successful in transmitting the programme of cognitive and structural anthropology, cosmology and social morphology, exchange and ritual to students and colleagues. He developed this research programme by means of a fascinating mixture of French and Dutch structuralism with contributions from Australia, Great Britain, Japan, Indonesia and Laos.

Characteristic of the Leiden tradition was its holistic, structural and comparative approach. During the past 40 years Jos Platenkamp has added, in collaboration with others, the hierarchy of values and the importance of "idea values" as formulated by Dumont; the combination of social morphology and exchange as emphasised by the members of ERASME in Paris; the extension of the FAS method by including historical data and by comparing Halmahera with other parts of Eastern Indonesia and other parts of Southeast Asia; the study of rituals, in which ideas and values are expressed, with the help of both texts and film reinforcing each other; and the insistence on cosmology and its relation to social structure.

Jos Platenkamp has been and is one of most consistent continuators of the Leiden tradition in structural anthropology, i.e., *le mouvement structuraliste Paris-Leyde* initiated by Mauss, Van Ossenbruggen, Rassers and J.P.B. de Josselin de Jong, which P.E. de Josselin de Jong had enhanced and elaborated. Jos Platenkamp has likewise continued the tradition of his predecessors, rejuvenating it by his research in Indonesia and France, and expanding it in Germany. Thanks to Jos Platenkamp, the Paris-Leiden tradition evolved into the Paris-Leiden-Münster tradition. Thus, after a complex exchange of ideas between Paris and Leiden, Leiden and Halmahera, Münster and Laos, and again between Halmahera and Leiden, the question arises: How will the Paris-Leiden-Münster tradition develop and where will it lead to next?

References

BARNES, R.H. (1985) The Leiden Version of the Comparative Method in Southeast Asia. Review of P.E. de Josselin de Jong (ed.), *Unity in Diversity: Indonesia as a Field of Anthropological Study* (1984). *JASO: Journal of the Anthropological Society of Oxford* 16(2): 87-110.

BEAUFILS, T. (2009) "Tache aveugle". L'absence d'un contrepoint visuel a-t-elle été fatale au "structuralisme" hollandaise? In *Histoires de rendez-vous manqués: J.P.B. de Josselin de Jong et l'anthropologie structurale*. Special issue *Deshima: Revue française des mondes néerlandophones* 3: 21-46.

BEAUFILS, T. (2015) Marcel Mauss, la Hollande et les Hollandais. Correspondance de 1898 à 1938. *Deshima: Revue d'histoire globale des pays du nord* 9: 47-86.

BIDNEY, D. (1953) *Theoretical Anthropology*. New York: Columbia University Press. New York: Shocken Books 1967.

BURLING, R. (1964) Cognition and Componential Analysis: God's Truth or Hocus-Pocus? *American Anthropologist* 66(1): 20-28.

CAWS, P. (1974) Operational, Representational, and Explanatory Models. *American Anthropologist* 76(1): 1-10.

DUMONT, L. (1979) Vers une théorie de la hierarchie. In *Homo hierarchicus. Le système des castes et ses implications*. Postface pour l'édition "Tel". Paris: Gallimard.

DUMONT, L. (1980) On Value. Radcliffe-Brown Lecture in Social Anthropology. *Proceedings of the British Academy* 64: 207-241.

DURKHEIM, É. and MAUSS, M. (1903) De quelques formes primitives de classification: Contribution à l'étude des repésentations collectives. *L'Année sociologique* VI (1901-1902): 1-72. English transl. *Primitive Classification*. London: Cohen and West 1963.

FOX, J.J. (1980) Models and Metaphors: Comparative Research in Eastern Indonesia. In FOX, J.J. (ed.), *The Flow of Life: Essays on Eastern Indonesia*, pp. 327-333. Cambridge, MA and London: Harvard University Press.

GEIRNAERT, D.C. (1992) *The Woven Land of Laboya: Socio-Cosmic Ideas and Values in West Sumba, Eastern Indonesia*. PhD thesis Leiden University. Leiden: Centre of Non-Western Studies (CNWS Publications 11).

GERBRANDS, A.A. (1990) Made by Man: Cultural-Anthropological Reflections on the Theme of Ethno-Communication. [Valedictory Lecture, 13 June 1984]. In KEURS, P. ter and SMIDT, D. (eds.), *The Language of Things: Studies in Ethnocommunication. In honour of Professor Adrian A. Gerbrands*, pp. 45-74 Leiden: Rijksmuseum voor Volkenkunde.

HOEK, A.W. van den (2004) *Caturmāsa: Celebrations of Death in Kathmandu, Nepal*. Edited by J.C. Heesterman, Bal Gopal Shrestha, Han F. Vermeulen and Sjoerd M. Zanen. Leiden: Research School CNWS (CNWS Publications 133).

HOEK, A.W. van den (2014) *Caturmāsa: Celebrations of Death in Kathmandu, Nepal*. Edited by Jan C. Heesterman, Bal Gopal Shrestha, Han F. Vermeulen and Sjoerd M. Zanen. 2nd edition with a foreword by David N. Gellner. Kathmandu: Vajra Books.

HOEK, A.W. van den, ZANEN, S. and DENG, P.L. (1978) *Social-Anthropological Aspects of the Jonglei Development Projects in South Sudan (field-work report)*. Leiden: Instituut voor Culturele Antropologie en Sociologie der Niet-Westerse Volken (ICA Publication 67). Reprinted with a new introduction 1985.

HOVEN, E. van (1995) *L'oncle maternel est roi. La formation d'alliances hiérarchiques chez les Mandigues du Wuli (Sénégal)*. PhD thesis Leiden University. Leiden: Research School CNWS (CNWS Publications 36).

JOSSELIN de JONG, J.P.B. de (1935) *De Maleische Archipel als ethnologisch studieveld*. Inaugural lecture. Leiden: Universiteitsboekhandel en Antiquariaat J. Ginsberg.

JOSSELIN de JONG, J.P.B. de (1977) The Malay Archipelago as a Field of Ethnological Study. In JOSSELIN de JONG, P.E. de (ed.), *Structural Anthropology in the Netherlands: A Reader*, pp. 164-182. The Hague: Martinus Nijhoff (2nd ed. Dordrecht/Cinnaminson: Foris Publications 1983).

JOSSELIN de JONG, P.E. de (1956) De visie der participanten op hun cultuur. *Bijdragen tot de Taal-, Land- en Volkenkunde* 112(2): 149-168. [Special issue for J.P.B. de Josselin de Jong on his retirement.] English transl. 1967, 1977; Japanese transl. 1987.
JOSSELIN de JONG, P.E. de (1972) Marcel Mauss et les origines de l'anthropologie structurale hollandaise. *L'Homme* 12(4): 62-84.
JOSSELIN de JONG, P.E. de (ed.), (1977) *Structural Anthropology in the Netherlands: A Reader*. The Hague: Martinus Nijhoff (KITLV Translation Series 17). 2nd ed. Dordrecht: Foris 1983.
JOSSELIN de JONG, P.E. de (1977a) Structural Anthropology in the Netherlands: Creature of Circumstance. In JOSSELIN de JONG, P.E. de (ed.), *Structural Anthropology in the Netherlands: A Reader*, pp. 1-29. The Hague: Martinus Nijhoff. Japanese transl. 1987.
JOSSELIN de JONG, P.E. de (1977b) The Participants' View of Their Culture. In JOSSELIN de JONG, P.E. de (ed.), *Structural Anthropology in the Netherlands: A Reader*, pp. 231-252. The Hague: Martinus Nijhoff. Japanese transl. 1987.
JOSSELIN de JONG, P.E. de (1980a) Supplementary Notes. Chapter XIII of: *Minangkabau and Negri Sembilan: Socio-Political Structure in Indonesia*, pp. 214-231. 3rd ed. The Hague: Martinus Nijhofff.
JOSSELIN de JONG, P.E. de (1980b) The Concept of the Field of Ethnological Study. In FOX, J.J. (ed.), *The Flow of Life: Essays on Eastern Indonesia*, pp. 317-326, 355. Cambridge, MA: Harvard University Press.
JOSSELIN de JONG, P.E. de (1980c) The Netherlands: Structuralism before Lévi-Strauss. In DIAMOND, S. (ed.), *Anthropology: Ancestors and Heirs*, pp. 243-257. The Hague/Paris: Mouton.
JOSSELIN de JONG, P.E. de with MOYER, G.A. and MOYER, D.S. (eds.), (1981) *The Nature of Structure*. Leiden: Institute of Cultural and Social Studies (ICA Publication No. 45). Reprinted 1985.
JOSSELIN de JONG, P.E. de and SCHWIMMER, E. (eds.), (1982) *Symbolic Anthropology in the Netherlands*. The Hague: Martinus Nijhoff (Verhandelingen van het KITLV 95).
JOSSELIN de JONG, P.E. de (ed.), (1984) *Unity in Diversity: Indonesia as a Field of Anthropological Study*. Dordrecht/Cinnaminson: Foris (KITLV Verhandelingen 103).
JOSSELIN de JONG, P.E. de (1985) The Comparative Method in Southeast Asia: Ideal and Practice. *Journal of the Anthropological Society of Oxford* 16(3): 199-208. [Rejoinder to BARNES, R.H. 1985].
JOSSELIN de JONG, P.E. de (1995) Circa vijftig jaar anthropologie beleefd. Een niet autobiografisch, maar egocentrisch opstel. In WENGEN, G.D. van et al. (eds.), *Waar Dromers Ontwaken. Terugblik van oud-leden van het Leids Ethnologisch Dispuut W.D.O. op 45 jaar culturele antropologie*, pp. 33-41. Leiden: P.E. Bijvoet.
JOSSELIN de JONG, P.E. de and VERMEULEN, H.F. (1989) Cultural Anthropology at Leiden University: From Encyclopedism to Structuralism. In OTTERSPEER, W. (ed.), *Leiden Oriental Connections, 1850-1940*, pp. 280-315. Leiden: E.J. Brill.
JOUWERSMA, S. (1985) *Harvest Festival in Sahu*. 16mm, colour, optic sound, 90 min. An ethnographic film by S. Jouwersma, research by L.E. Visser, in collaboration with I Made Seraye and I Gusti Ngurah Arrington. Leiden: Institute of Cultural and Social Studies.
KUPER, A.J. (1973) *Anthropologists and Anthropology: The British School, 1922-1972*. London: Allen Lane/New York: Pica Press. Reprint London: Penguin 1975. 2nd revised and expanded edition London: Routledge 1983, 3rd ed. 1996, 4th ed. 2014.

KUPER, A.J. (1979) Regional Comparison in African Anthropology. *African Affairs* 78(310): 103-113.

LEACH, E.R. (1961) *Rethinking Anthropology*. London: The Athlone Press.

LEACH, E.R. (1972[1964]) Anthropological Aspects of Language: Animal Categories and Verbal Abuse. In MIRANDA, P. (ed.), *Mythology*, pp. 39-67. Harmondsworth: Penguin.

LEACH, E.R. (1976) *Culture and Communication: The Logic by which Symbols are Connected. An Introduction to the Use of Structuralist Analysis in Social Anthropology*. Cambridge: Cambridge University Press.

LÉVI-STRAUSS, C. (1949) *Les Structures élémentaires de la parenté*. Paris: Presses Universitaires de France. 2nd ed. La Haye/Paris: Mouton et Cie., 1967. [English transl. *The Elementary Structures of Kinship*. Boston: Beacon Press, 1969].

LÉVI-STRAUSS, C. (1950) Introduction à l'oeuvre de Marcel Mauss. In MAUSS, M. *Sociologie et anthropologie*. Paris: Presses Universitaires de France. [English transl. *Introduction to the Work of Marcel Mauss*. London: Routledge and Kegan Paul, 1987].

LÉVI-STRAUSS, C. (1958) *Anthropologie structurale*. Paris: Librairie Plon. [English transl. *Structural Anthropology*. New York: Basic Books, 1963].

LÉVI-STRAUSS, C. (1962a) *Le Totémisme aujourd'hui*. Paris: Presses Universitaires de Paris. [English transl. by Rodney Needham, *Totemism*. Boston: Beacon Press, 1963].

LÉVI-STRAUSS, C. (1962b) *La Pensée sauvage*. Paris: Librairie Plon. [English transl. *The Savage Mind*. London: Weidenfeld and Nicolson, 1966].

LÉVI-STRAUSS, C. (1964-71) *Mythologiques*. 4 vols. Paris: Plon. Vol. 1, *Le Cru et le cuit*, 1964; vol. 2, *Du Miel aux cendres*, 1967; vol. 3, *L'Origine des manières de table*, 1968; vol. 4, *L'Homme nu*, 1971. [English transl. *Mythologiques*, New York: Harper and Row, 1969-81].

LÉVI-STRAUSS, C. (1973) *Anthropologie structurale deux*. Paris: Librairie Plon. [English transl. *Structural Anthropology*, vol. 2, New York: Basic Books, 1976].

LOCHER, G.W. (1978) *Transformation and Tradition. And Other Essays*. The Hague: Martinus Nijhoff (Koninklijk Instituut voor Taal-, Land- en Volkenkunde Translation Series 18).

LOCHER, G.W. (1990) From Museum to University: The Scientific Career of Adrian A. Gerbrands. In KEURS, P. ter and SMIDT, D. (eds.), *The Language of Things: Studies in Ethnocommunication. In Honour of Professor Adrian A. Gerbrands*, pp. 13-22. Leiden: Rijksmuseum voor Volkenkunde (Mededelingen van het Rijksmuseum voor Volkenkunde, Leiden No. 25).

LUNING, S. (1997) *Het binnenhalen van de oogst. Ritueel en samenleving in Maane, Burkina Faso*. PhD thesis Leiden University. Leiden: Research School CNWS, School of Asian, Africa and Amerindian Studies.

MAUSS, M. with BEUCHAT, H. (1906) Essai sur les variations saisonnières des sociétés eskimos. Étude de morphologie sociale. *L'Année sociologique* 9 (1904-1905). [Reprinted in MAUSS, M. *Sociologie et anthropologie*, Paris: Presses Universitaires de France, 1950].

MAUSS, M. (1925) Essai sur le don. Forme et raison de l'échange dans les sociétés archaïques. *L'Année sociologique*, n.s. 1 (1923-1924), pp. 30-186. [Reprinted in MAUSS, M. *Sociologie et anthropologie*. Paris 1950: 143-279. English transl. by Ian Cunnison, *The Gift*. New York, 1967].

MAAKER, E. de (2006) Negotiating Life: Garo Death Rituals and the Transformation of Society. PhD thesis Leiden University.

MASINAMBOW, E.K.M. (ed.), (1980) *Halmahera dan Raja Ampat. Konsep dan strategi penelitian*. Jakarta: Lembaga Ekonomi dan Kemasyarakatan Nasional/Lembaga Ilmut Pengetahuan Indonesia (LEKNAS/LIPI).

MASINAMBOW, E.K.M. (ed.), (1983) *Halmahera dan Raja Ampat sebagai Kesatuan Majemuk: Studi-Studi terhadap suatu Daerah Transisi*. Jakarta: Lembaga Ekonomi dan Kemasyarakatan Nasional/Lembaga Ilmut Pengetahuan Indonesia. *Buletin LEKNAS* 2(2), 1983. Printed in 1987.

MIYAZAKI, K. (1979) *Javanese Classification Systems: the Problem of "Maximal Correspondence"*. With an Introduction by P.E. de Josselin de Jong. Leiden: Institute of Cultural and Social Studies (ICA Publications 35).

MIYAZAKI, K. (1980) Oranda kozoshugi to bunruitaikei kenkyo' [Dutch Structuralism and the Study of Classification Systems]. *Minzokugáku kenkyo* [The Japanese Journal of Ethnology] 45(1): 52-59.

MIYAZAKI, K. (1987) with ENDOH, H. and GOH, T. (eds.), *Oranda Kouzou Jinruigaku/Oranda kôzô jinruigaku*. Japanese transl. and edition of P.E. de Josselin de Jong (ed.), *Structural Anthropology in the Netherlands: A Reader* (The Hague 1977, 2nd ed. 1983) by Koyi Miyazaki, Hisahi Endôh and Taro Gôh. Tokyo: Serika Shobô.

MIYAZAKI, K. (1988) The King and the People: The Conceptual Structure of a Javanese Kingdom. PhD thesis Leiden University.

MOYER, D.S. (1975) *The Logic of the Laws: A Structural Analysis of Malay Language Legal Codes from Bengkulu*. PhD thesis Leiden University.'s-Gravenhage: H.L. Smits (Verhandelingen van het KITLV 75).

MOYER, D.S. (1976) *Paris-Leiden and Other Oppositions*. Leiden: Instituut voor Culturele Antropologie en Sociologie der Niet-Westerse Volken (ICA Publication 16). (2nd ed. 1980).

MOYER, D.S. (1978) A Dutch Treat: Structural Anthropology in the Netherlands. Review of P.E. de Josselin de Jong (ed.), *Structural Anthropology in the Netherlands: A Reader* (The Hague 1977). *Reviews in Anthropology* 5(4): 457-463.

NIESSEN, S.A. (1982) *De "Leidse Richting" toen en thans*. Leiden: Instituut voor Culturele Antropologie en Sociologie der Niet-Westerse Volken (ICA Publicatie no. 49).

NIESSEN, S.A. (1985) *Motifs of Life in Toba Batak Texts and Textiles*. PhD thesis Leiden University. Dordrecht: Foris Publications (Verhandelingen van het KITLV 110).

NIJLAND, D.J. (1985<av>) *Tobelo Marriage*. 16mm, colour, optic sound, 108 min. An ethnographic film by Dirk Nijland, research by Jos Platenkamp, in collaboration with Lis Asinoellah and Frans Rijoly. Leiden: Institute of Cultural and Social Studies.

NIJLAND, D.J. (1989) Schaduwen en werkelijkheid. Cultureel bepaald kennen, waarnemen, nonverbaal communiceren, en onderzoek daarvan via de terugvertoning van de etnografische film "Tobelo Marriage". PhD thesis Leiden University.

NIJLAND, D.J. (2006a) Ritual Performance and Visual Representation: The Body, the Brain and Video Supported Observation. In POSTMA, M. and CRAWFORD, P.I. (eds.), *Reflecting Visual Ethnography: Using the Camera in Anthropological Research*, pp. 26-49. Leiden: CNWS Publications/Højbjerg: Intervention Press.

NIJLAND, D.J. (2006b) Film and Non-verbal Expressions of Culture Amongst the Tobelo. In POSTMA, M. and CRAWFORD, P.I. (eds.), *Reflecting Visual Ethnography: Using the Camera in Anthropological Research*, pp. 52-77, ill. Filmfragments. Leiden: CNWS Publications/Højbjerg: Intervention Press. Earlier published in Dutch, 1994.

NIJLAND, D.J. (2006c) The Bilan Adventure: The Shadow-Journey. Openings lecture at the 25th Bilan du Film Ethnographique, Colloque international du cinéma ethnographique à l'anthropologie visuelle, Paris, 25 March 2006. In *Comité du Film Ethnographique, Festival International du Cinema Ethnographique Jean Rouch*, URL http://comitedufilmethnographique.com/dirk-nijland/.

OOSTEN, J.G. (1976) *The Theoretical Structure of the Religion of the Netsilik and Iglulik*. PhD thesis University of Groningen. Meppel: Krips Repro.

OOSTEN, J.G. (1979) Voorwoord. In *CASA – 1979: Bijdragen tot de Cognitieve- en Structurele Antropologie*, pp. 1-2. Leiden: Instituut voor Culturele Antropologie en Sociologie der Niet-Westerse Volken (ICA Publicatie 36).

OOSTEN, J.G. (ed.), (1990) *Antropologische Verkenningen* 9(3). Special issue including the papers of T. Vuyk, A.W. van den Hoek and J.G. Oosten presented during the CASA-ERASME seminar on "The Study of Values in Ritual" at CNWS, March 1990.

OOSTEN, J.G. (1999) *De waarde van het verschil/Verschillen in waarden*. Leiden: Onderzoekschool voor Aziatische, Afrikaanse en Amerindische Studies (CNWS).

OOSTEN, J.G. (2006) "A Privileged Field of Study": Marcel Mauss and Structural Anthropology in Leiden. *Etudes/Inuit/Studies* 30(2): 51-71.

OSSENBRUGGEN, F.D.E. van (1917) De oorsprong van het Javaansche begrip montjåpat, in verband met primitieve classificaties. *Verslagen en Mededeelingen der Koninklijke [Nederlandsche] Akademie van Wetenschappen*, Afdeeling Letterkunde, 5e reeks, derde deel, eerste stuk, pp. 6-44. Amsterdam: Johannes Müller. [Indonesian transl. 1975, English transl. 1977, Japanese transl. 1987].

OSSENBRUGGEN, F.D.E. van (1977) Java's monca-pat: Origins of a Primitive Classification System. In JOSSELIN de JONG, P.E. de (ed.), *Structural Anthropology in the Netherlands: A Reader*, pp. 32-60. The Hague: Martinus Nijhoff (KITLV Translation Series 17). 2nd ed. Dordrecht: Foris 1983.

PLATENKAMP, J.D.M. *see* Publications of J.D.M. Platenkamp attached to this section.

PLATENKAMP, J.D.M. and PRAGER, M. (1994) A Mirror of Paradigms: Nineteenth- and Early Twentieth-century Ethnology reflected in *Bijdragen*. In POEZE, H.A. (ed.), *150 Volumes of Bijdragen: A Backward Glance and a Forward Glimpse*. Special issue *Bijdragen tot de Taal-, Land- en Volkenkunde* 150(4): 703-727.

PRAGER, M. (1992) Structure, Process, and Performance in Eastern Indonesian Rituals: A Review Article. *Anthropos* 87(4-6): 548-555.

PRAGER, M. (1996) Strukturale Anthropologie in Leiden, 1917-1956. Ursprung und Entwicklung eines wissenschaftlichen Forschungsprogramms. PhD thesis University of Heidelberg.

POSTMA, M. and CRAWFORD, P.I. (eds.), *Reflecting Visual Ethnography: Using the Camera in Anthropological Research*. Leiden: CNWS Publications/Højbjerg: Intervention Press.

RASSERS, W.H. (1922) *De Pandji-roman*. PhD thesis Leiden University. Antwerpen: D. de Vos-van Kleef.

RASSERS, W.H. (1959) *Pañji, the Culture Hero: A Structural Study of Religion in Java*. Introduction by J.P.B. de Josselin de Jong. The Hague: Martinus Nijhoff (KITLV Translation Series 3) [2nd ed. with an introduction by P.E. de Josselin de Jong, 1982].

RIDDER, R. de and KARREMANS, J.A.J. (eds.), *The Leiden Tradition in Structural Anthropology: Essays in Honour of P.E. de Josselin de Jong*. Leiden: E.J. Brill.

RUSSELL, B. (1945) *A History of Western Philosophy and Its Connection with Political and Social Circumstances from the Earliest Times to the Present Day*. New York: Simon and Schuster.

SAUSSURE, F. de (1916) *Cours de linguistique générale*, publié par Charles Bally et Albert Sechehaye. Lausanne-Paris: Payot [3rd. ed. 1931].

SCHNEIDER, D.M. (1965) Some Muddles in the Models: Or, How the System really Works. In BANTON, M. (ed.), *The Relevance of Models for Social Anthropology*, pp. 25-85. London: Tavistock.

SUZUKI, P.T. and VERMEULEN, H.F. (2000) Patrick Edward de Josselin de Jong (1922-1999). Obituary. *American Anthropologist* 102(3): 577-581.

VERMEULEN, H.F. (1985) De Leidse lineage van museum-directeuren en antropologie-hoogleraren: Gerbrands en de studie van etnocommunicatie. In BOEREN, A., BRINKGREVE, F. and ROELS, R. (eds.), *Teken van Leven: Studies in Etnocommunicatie. Liber amicorum bij het afscheid van Professor Dr. A.A. Gerbrands*, pp. 33-71. Leiden: Instituut voor Culturele Antropologie en Sociologie der Niet-Westerse Volken (ICA Publicatie No. 71).

VERMEULEN, H.F. (1987) P.E. de Josselin de Jong and the Leiden Tradition: A Short History. In RIDDER, R. de and KARREMANS, J.A.J. (eds.), *The Leiden Tradition in Structural Anthropology: Essays in Honour of P.E. de Josselin de Jong*, pp. 4-84. Leiden: E.J. Brill.

VERMEULEN, H.F. (1997) Van Instituut voor CA en SNWV tot Vakgroep CA/SNWS (1955-1997): professionalisering van de antropo-sociologie in Leiden. In CLAESSEN, H.J.M. and VERMEULEN, H.F. (eds.), *Veertig Jaren Onderweg*, pp. 13-52. Leiden: DSWO Press.

VERMEULEN, H.F. (1999) Structuur en verandering: In memoriam Gottfried Wilhelm Locher (Waspik, 23 maart 1908 – Leiden, 3 december 1997). *Bijdragen tot de Taal-, Land- en Volkenkunde* 155(1): 1-44.

VERMEULEN, H.F. (2002) Contingency and Continuity: Anthropology and Other Non-Western Studies in Leiden, 1922-2002. In VERMEULEN, H. and KOMMERS, J. (eds.), *Tales from Academia: History of Anthropology in the Netherlands*, Part 1, pp. 95-182. Nijmegen/Saarbrücken: Verlag für Entwicklungspolitik (Nijmegen Studies in Development and Cultural Change 39).

VISSER, L.E. (1979) *Verslag oriëntatiereis Halmahera.* Internal report: Leiden.

VISSER, L.E. (1984) Mijn tuin is mijn kind. Een antropologische studie van de droge rijstteelt in Sahu (Indonesië). PhD thesis Leiden University. Leiden: Repro Psychologie.

VISSER, L.E. (1989) *My Rice Field is My Child: Social and Territorial Aspects of Swidden Cultivation in Sahu, Eastern Indonesia.* Dordrecht: Foris (Verhandelingen van het KITLV 136).

VISSER, L.E. (ed.), (1994) *Halmahera and Beyond: Social Science Research in the Moluccas.* Leiden: KITLV Press (Proceedings 1).

VISSER, L.E. (2006) In Memoriam E.K.M. Masinambow. *The Asia Pacific Journal of Anthropology* 7(2): 173-177.

VISSER, L.E. and MOYER, D.S. (1999) In Memoriam P.E. de Josselin de Jong (8 July 1922-1 January 1999). *Bijdragen tot de Taal-, Land- en Volkenkunde* 155(2): 175-186.

VUYK, T. (1991) *Children of One Womb: Descent, Marriage and Gender in Central African Societies.* PhD thesis Leiden University. Leiden: Centre of Non-Western Studies (CNWS Publications 3).

WOUDEN, F.A.E. van (1935) *Sociale structuurtypen in de Groote Oost.* PhD thesis Leiden University. Leiden: J. Ginsberg [English transl. 1968; Indonesian transl. 1985].

WOUDEN, F.A.E. van (1968) *Types of Social Structure in Eastern Indonesia* [English transl. by R. Needham. Preface by G.W. Locher]. The Hague: Martinus Nijhoff (KITLV Translation Series 11).

Fieldwork, Visual Ethnography, and Written Texts: A Letter to Jos Platenkamp about the Making and Significance of the Film "Tobelo Marriage"

DIRK J. NIJLAND

Dear Jos, It has been a while since we last saw each other. However, alerted by Michael Prager, you and I will soon meet again on Friday 24 June 2016. This is the day immediately after your 65th birthday on the occasion of which we plan to surprise you with this *Festschrift*. In it, I very much like to be represented, too! Not by way of an article but in the form of this open letter containing a special salute to the views you have developed on deduced "idea-images" as carriers of complex ideas and values, which present an intriguing contrast with the significance of "empirical images" as recorded by means of photographs or film and their interpretation by the outsider public.

In this context it is only natural for me to recall – and I think you will do so too when reading this letter – the "Tobelo Marriage" film project executed in conjunction with your dissertation entitled "Tobelo: Ideas and Values of a North Moluccan Society" (Platenkamp 1988). The film "Tobelo Marriage" (Nijland 1985) met with some acclaim and received several awards.[1] It was screened, for instance, during the Margaret Mead Film Festival in New York. And it was distributed for educational purposes in the Netherlands, Indonesia, the USA (Berkeley) and Japan.

You had just completed your fieldwork in Paca, Tobelo, on the eastern Indonesian island of Halmahera, where you worked from June 1979 until December 1980. Professor Adrian A. Gerbrands, who occupied the chair of cultural anthropology, in particular ethnography, at Leiden University from 1966 on, was well disposed toward us. He had visited you in Tobelo in 1980 and had appointed me to teach and conduct research about the opportunities audiovisual media offer to cultural anthropology in 1971. Gerbrands had been able to acquire the sum of Hfl. 60,000 from the Netherlands Foundation for the Advancement of Tropical Research (WOTRO) in 1981, in order to realise the film project "Tobelo Marriage".

1 Note from the editors: The film "Tobelo Marriage" was selected for the Sixième Bilan du Film Ethnographique (Paris, 1987) where it was awarded the "Grand Prix Nanook" and for the Margaret Mead Film Festival (New York, 1989). During the annual meeting of the American Anthropological Association (Chicago, December 1991) it received an "Honorable Mention" from the Society for Visual Anthropology.

We embarked upon an intensive period of preparation. For you, Jos, marriage as a *rite de passage* was an appropriate subject because it highlighted core issues of Tobelo society. And you instructed me in a serious manner by discussing texts, studying photographs, maps and diagrams. These we then linked to the chronology of the Tobelorese marriage ceremonies, including negotiations and preparations taking place on the side of the bride givers and on that of the bride takers. Of course, I also read your papers and attended your lectures on Tobelo. Needless to say, they were more than adequate. The way you answered questions impressed me and made me recall the Hungarian fencing master Adam Bela's instructions, whom I had experienced as a student in Utrecht: "En garde [...] right leg forward, etc." During the discussions you acted in a similar way, with concentration and sharpness. It was only later, after watching the Tobelorese *hol*a*, a solo war dance aimed at impressing the enemy, that I understood that this dance had perhaps also inspired the manner in which you express your points of view during discussions.

Arriving at Tobelo in July 1982, we were accompanied by our Indonesian counterparts. As a woman, Lis Asinoella, a graduate student from the Universitas Indonesia in Jakarta, was important for contacting the Tobelorese ladies. Frans Rijoly, assistant-curator of the Ethnographic Museum "Siwa Lima" in Ambon, mostly recorded the sound, but on occasion you did so, too, for instance when the "Bata brothers" were singing the *Lelehe*. When we landed at Paca, Halmahera, it was striking to see how happy the people were to meet you, Tuan Jos, once again. Adat elders realised the usefulness of a descriptive film about Tobelorese marriage because several of its elements were already changing. A film could inform young people how a marriage should take place and which negotiations, preparations and exchanges are involved. Thus, the elders were prepared to collaborate and the people of Tobelo were very helpful. For me it was an entirely new experience. However, when I felt overwhelmed you took matters in hand and inspired me to carry on. Finally, by carefully distributing the film rolls over the subjects and scenes of the marriage ritual – the editing of the material was already on our minds while recording – and with much discipline we managed to produce a 108-minute film, based on twice as much footage (the tight budget did not allow for more; compare this with the digital cameras now available with inexpensive video tapes, smart cards, etc).

Back in the Netherlands, the issue concerning the commentary surfaced during editing, as was to be expected. Your initial reaction was to present far too much information. This is quite a common reaction. However, you soon understood the problem: a film is in the first place something to look at. Comments must be limited to no more than 30% of the length of a film. If longer, it will be impossible to follow the flow of information.

Fig. 1: The bride in her ritual dress. The bride Henny Rope is solemnly escorted from her parental house to that of the groom: "She wades across the river to the other side" (Photo: Asinoella/Rijoly)

Once again it became clear to me that an ethnographic film cannot stand on its own. Films can capture expressions, but the underlying ideas and values, which these expressions communicate, can be very complex. Only a written text can describe or interpret these ideas and values in a satisfactory manner. Therefore, a film, photographs and a written text must always go side by side and need to be studied together. Combined with your dissertation "Tobelo: Ideas and Values of a North Moluccan Society" (1988), the film "Tobelo Marriage" (1985) forms an informative, documenting whole. In it, one can submerge oneself in order to gradually be informed of the expressions and the underlying ideas and values by means of alternatingly consulting the written and the filmed material (see also Nijland 2006b).

Later on, you developed this approach in a number of articles. Here "Visibilty and Objectification in Tobelo Ritual" (2006) and especially "True to the Original" (1995) must be mentioned. Every visual ethnographer, ethnographic photographer or ethnographic filmmaker should read the latter article. With Hans Jonas and his theory of *homo pictor* (Jonas 1961), you underlined the universal cultural phenomenon that human beings can mentally develop perceptible forms into meaningful carriers of ideas and value complexes, which are so much richer than the alien, non-participating observer can observe and presume on the basis of the perceptible forms alone (see also Nijland 2006a and Platenkamp 2006).

When screening the film "Tobelo Marriage" to the participants in Paca in January 1988 (see Nijland 1989), I was able to witness the very intense reality and "symbolic value" of the bride's ritual dress and necklace (to the participants), which to me (as an outside observer) first of all had an "aesthetic" value. It was only because you had explained to me that the bride on becoming a new member of the House of the bride takers was wearing textiles and the necklace, adorned with lozenge, masculine, motifs (in which she herself represented fertility) that I grasped some of the fuller significance of the ritual dress in a rational manner.

Jos, like many of your students, I am grateful to you for learning me more about anthropology and Tobelo society, both in the field and at the Institute, and for your friendship and inspiration. Ankie and I wish you all the best for the coming new stages!

References

NIJLAND, D.J. (1985) *Tobelo Marriage*. 16mm. colour, 108 min. Leiden: Institute of Cultural and Social Studies. A film by Dirk Nijland, research by Jos Platenkamp, in co-operation with Lis Asinoellah and Frans Rijoly.

NIJLAND, D.J. (1989) *Schaduwen en werkelijkheid. Cultureel bepaald kennen, waarnemen, nonverbaal communiceren, en onderzoek daarvan via de terugvertoning van de etnografische film "Tobelo Marriage"*. PhD thesis. Leiden University.

NIJLAND, D.J. (2006a) Film and Non-verbal Expressions of Culture among the Tobelo. In POSTMA, M. and CRAWFORD, P.I. (eds.), *Reflecting Visual Ethnography: Using the Camera in Anthropological Research*, pp. 52-77. Leiden: CNWS Publications/Højbjerg: Intervention Press.

NIJLAND, D.J. (2006b) The Bilan Adventure: The Shadow-Journey. Openings lecture at the 25th Bilan du Film Ethnographique 25, Paris, Colloque international du cinéma ethnographique à l'anthropologie visuelle, 25 March 2006. In *Comité du Film Ethnographique, Festival International du Cinema Ethnographique Jean Rouch*, [Online] http://comitedufilmethnographique.com/dirk-nijland [Accessed: 9 April 2016].

JONAS, H. (1961) Die Freiheit des Bildens – *Homo pictor* und die *differentia* des Menschen. *Zeitschrift für Philosophische Forschung* 15(2): 161-176. [Reprinted in Hans Jonas, *Zwischen Nichts und Ewigkeit – Zur Lehre vom Menschen*, Göttingen: Vandenhoeck & Ruprecht, 1987, pp. 26-43. English translation "Homo Pictor and the Differentia of Man." *Social Research* 29 (1962)].

PLATENKAMP, J.D.M. (1988) *Tobelo: Ideas and Values of a North Moluccan Society*. PhD thesis. Leiden University. Leiden: Repro Psychologie.

PLATENKAMP, J.D.M. (1995) True to the Original. In KÄMPF, H. and SCHOTT, R. (eds.), *Der Mensch als Homo Pictor?* Special issue: Die Kunst traditioneller Kulturen aus der Sicht von Philosophie und Ethnologie. *Zeitschrift für Ästhetik und Allgemeine Kunstwissenschaft* 1: 20-35.

PLATENKAMP, J.D.M. (2006) Visibility and Objectification in Tobelo Ritual. In POSTMA, M. and CRAWFORD, P.I. (eds.), *Reflecting Visual Ethnography: Using the Camera in Anthropological Research*, pp. 78-102. Leiden: CNWS Publications/Højbjerg: Intervention Press.

Precedence and Hierarchy: Meditations on an Unanswered Ethnological Question

E. DOUGLAS LEWIS

CHARLES IVES, 1984 (1930-1935)

The 1980s and 1990s were decades of what can be described as a Canberra-Leiden (or Leiden-Canberra) axis in Indonesian ethnology. At the Australian end were ethnologists, prehistorians, linguists, James J. Fox, and a succession of his students. At the Leiden end were Professors P. E. de Josselin de Jong and, later, Reimar Schefold and a number of their colleagues and students, including Jos Platenkamp. In a valuable memoir, Moyer (1976) described an opposition between Paris and Leiden that arose from a difference of opinion about, in Moyer's words, "the relative importance of exchange as opposed to double descent" (1976: 1). The difference of opinion about this problem, which dated to the publication of Claude Lévi-Strauss's thesis, *Les Structures Élémentaires de la Parenté* (1949), was that between Lévi-Strauss in Paris and J. P. B. and P. E. de Josselin de Jong in Leiden. While the contention had to do with how to explain differences between south Sumatran and Minangkabau social organisation, the issue had much broader implications for the methods of structural analysis and Indonesian, hence Austronesian, ethnology. Moyer rightly pointed out that "the significance of this difference in viewpoint has never been resolved or even discussed in detail in the scientific literature" (Moyer 1976: 1), a judgment that holds in 2016.

As a student in Leiden in the 1980s, Jos Platenkamp would have been well acquainted with the Sumatra problem and with the streams of structural analysis as they had developed in Paris, Leiden, Oxford, and Cambridge in the 1960s and 1970s.

In those years, the work of another French anthropologist, Louis Dumont, at Oxford in the 1950s and later director of the École des Hautes Études en Sciences Sociales, was attracting attention in Leiden and, in the 1980s, in Canberra. Of special interest was *Homo Hierarchicus: Essai sur le système des castes* (1966). As an incompetent Francophone during my early student years, I would not read Dumont's book until after I began graduate studies, by which time it had appeared in English translation. Preoccupied as I was with Indonesia, Dumont did not become a matter of importance to me or, I would hazard, to Platenkamp until the publication of the revised editions of 1979 and 1980 with their peculiar seven page "Postface: Toward a Theory of Hierarchy". By 1990, both he and I were intrigued by Dumont's idea about the logic of hierarchy.

In 1982, within three years of the publication of the "Postface" (Dumont 1979), I completed an account of alliance and exchange among the people of the domain of Wai Brama in the Tana 'Ai region of the regency of Sikka on Flores.[1] Professor Patrick de Josselin de Jong of the University of Leiden read that work and, in his capacity as a member of the editorial Board of the Koninklijk Instituut voor Taal-, Land-, en Volkenkunde, was instrumental in its publication.[2] In these works I identified precedence, the ordering of social groups by reference to the sequence of their founding, as a concept in the thought of the Ata Tana 'Ai and a principle fundamental for understanding the social order of the domain. The proposition was supported by an analysis of field data demonstrating that "the social and ritual precedence of clans in the domain [of Wai Brama] and houses in clans [is] established by the sequence of their founding" (Lewis 1988: 354). I provided neither a general definition of precedence nor considered explicitly the larger implications of this modest idea for the comparison of Austronesian social systems. What I had done was to translate the word *oda* in the Tana 'Ai dialect of Sara Sikka (SS)[3] as "precedence" and to set out how precedence, as a principle of organisation, orders the clans, houses, and ceremonial relations of the people of Wai Brama.[4] During the years that followed, talking, thinking and writing about precedence in Austronesia grew into something of a cottage industry in the anthropological hamlet of the Coombs Building at the Australian National University and it appeared as a theme in the theses of a number of A.N.U. Ph.D. scholars working in the eastern Lesser Sunda Islands in subsequent years (see Fox 2009: 3, 11 n. 4). What began as a translation into English of a pesky, three phoneme word in the Tana 'Ai language came to be recognised, at least among a small group of anthropologists in Canberra, as a principle pertinent in various ways in the Austronesian societies investigated by Fox and his students.

---

1 Lewis (1982).

2 I learned later that Professor de Josselin de Jong may also have been responsible for the title under which it appeared: *People of the Source* (Lewis 1988).

3 Henceforth, Sara Sikka = SS and Bahasa Indonesia = BI when translating words.

4 See Lewis (1997) for an account of unraveling the meaning of the word *oda* during a second season of field work in Wai Brama in 1980.

With *People* in press in Leiden, Dr (later Professor) James J. Fox of the A.N.U. delivered a "Public Lecture in connection with De Wisselleerstoel Indonesische Studien given on the 17 of March 1988" as the Tweede Bezetter van de Wisselleerstoel Indonesische Studien.[5] The lecture was entitled "Origin, Descent and Precedence in the Study of Austronesian Societies" (Fox 1988). In it, Fox spoke of descent and why it is important, in the comparative study of Austronesian societies, to attend to what he called "origin structures."[6] While origin structures provide fundamental points of reference in the varied status systems of the Austronesians (Fox 1988: 15), he did not explore the concept of precedence, commenting only that it results from origin structures which, in these societies, determine who is to be first, foremost, elder, superior, greater, or to occupy the center.

1988 was also the year that J. D. M. (Jos) Platenkamp completed his Ph.D. in the Social Sciences under Professor Dr. P. E. de Josselin de Jong, whose contributions to Leiden University Fox acknowledged fulsomely in the closing paragraph of his 1988 lecture. Fox had become interested in the generality of precedence as an idea common to or, at least, widespread among the peoples of Austronesian societies. He had known Platenkamp and his work at Leiden and had attended Platenkamp's Ph.D. examination. Thus, in 1990, when Fox convened a conference on *Hierarchy, Ancestry and Alliance*, Platenkamp was invited to attend and to contribute to discussions about the concepts of hierarchy and precedence. In January, the conferees gathered at Kioloa, the Australian National University's rustic bush campus on the coast of New South Wales. It was then that I met Jos Platenkamp.

Fox presented a paper on "Hierarchy and Precedence" in which he argued that precedence is a form of hierarchy quite distinct from that of the caste system of India, on the analysis of which Dumont had constructed, in the "Postface" to *Homo Hierarchicus*, what we came to call his general theory of hierarchy as the "englobement du contraire" (Dumont 1979: 397), the "encompassment of the contrary."[7]

---

5 Professor Fox was in the Department of Cultural *Anthropology* at the University of Leiden for a few months while the university sought an appointment to the chair from which Professor de Josselin de Jong retired. Professor Dr Reimar Schefold held the chair in from 1988 to 2003.

6 Fox illustrated his theme with ethnographic material from the Rotenese, the Atoni, and Mambai and the Weyewa of Sumba. I still find it odd that the people of Wai Brama were not mentioned since, in 1988, the relations of descent, alliance, and precedence among them had been set out more clearly than in the literature on any other society of the eastern Lesser Sunda Islands.

7 I will not here recount in detail Fox's argument since it is available in slightly revised form as Fox (1994). In 1990, the paper as delivered at the Kioloa conference was distributed as a Working Paper of the Comparative Austronesian Project, an initiative of the Anthropology Department of the Research School of Pacific Studies in the A.N.U., and is the version to which I will refer here.

Platenkamp's Rejoinder to Fox[8]

Platenkamp disagreed with Fox. The next day, presumably having worked through most of the night, Platenkamp presented a dissenting rejoinder to Fox in a brief and thoughtfully constructed paper in which he argued that the logic of precedence in the ideologies and ordering of societies of the eastern Lesser Sunda Islands is that of Dumont's hierarchy in its general form. Thus, there is not hierarchy and precedence as different concepts, nor as two kinds of hierarchy (India's and that of the eastern Lesser Sunda Islanders), but only one kind of hierarchy that manifests itself in the institutions and practices of societies in various ways.

In his first paragraph, Platenkamp challenged Fox's main point:

> "In a recent working paper [...] Fox (1990) examines the applicability of [...] Dumont's theory of hierarchy to the study of Indonesian societies. He asserts that 'the Dumontian notions of hierarchy appear singularly inappropriate for our understanding of Indonesian societies' (Fox 1990:7). This assertion seems to run counter to the research findings of various anthropologists studying eastern Indonesian societies (e.g. Barraud 1979; Pauwels 1990; Geirnaert-Martin [forthcoming]; Platenkamp 1988). They all claim that the application of Dumont's paradigm enhances our understanding of the systems of representations of the societies under study. Therefore, Fox's argument deserves critical attention and serious discussion.
>
> One of the basic tenets of Fox's criticism of Dumont's concept of hierarchy is that this concept [...] has emerged from, and is therefore only applicable to, Indian society. [...] However, this does not imply that the concept is therefore applicable to Indian society only. To clarify this issue it is convenient to distinguish between what may be labeled a 'specific theory of hierarchy' and a 'general theory of hierarchy'" (Platenkamp 1990).

The specific theory deals with the hierarchy of relationships between castes in Indian society "and the way in which purity as value informs these relationships."

8 An historiographic note is required here. In 1994, I was putting together a volume of my essays on kinship and gender in Wai Brama. I decided to address the hierarchy and precedence question, and wrote to Jos to ask if he would allow me to reproduce his Kioloa paper in my book. He agreed, but I was then unable to find my copy of his paper in my files. I wrote to him again: Douglas, he replied, I have no record of the paper, nor can I find my copy. It had apparently vanished somewhere between the termini of the Canberra–Leiden axis. Then, in about 2000, one of Platenkamp's students, Ms (now Dr.) Claudia Damhuis, arrived in Melbourne to begin her Ph.D. with me. As part of her preparation for fieldwork, I pointed her to the files of my research material on Flores and suggested that she go through them. She did, whereupon she found my copy of Platenkamp's paper. I scanned the paper and sent a copy to Platenkamp. The book on gender has not been published. Given the secretive nature of *festschriften*, I have not asked for his permission to quote the paper again here, but I assume that his original permission is still in effect.

> "The second, 'general' theory of hierarchy [...] is a generalization of what Dumont considers to be the theoretical properties of hierarchical systems of values (Dumont 1979, 1980). *These properties are primarily logical properties* [my italics], which can be described independently of the particular identity of the values of a given society. [...] Thus, the question whether 'purity' [...] represents the encompassing value in Indian society [...] has no immediate bearing on the applicability of the general theory to, for instance, Indonesian societies.
> The confusion about the 'general theory' seems to stem primarily from Dumont's equivocal use of the term 'hierarchy'. As part of the 'specific theory' – the one about Indian society – 'hierarchy' refers to the relationship between social phenomena and an encompassing whole valorized in terms of 'purity'. Therefore, the term 'hierarchy' refers *both* to the logical structure of the configuration of relationships (a structure which was later generalized into a general theory) and to the values that are related in this structure" (Platenkamp 1990).

There is thus an ambiguity in the way that Dumont uses the term *hierarchy*. Section 3. of Platenkamp's paper is on the question, 'Hierarchy or precedence?' "[Fox's] argument," Platenkamp wrote,

> "may be summarized as follows. In eastern Indonesian societies, and particularly in forms of ritual speech (see Fox 1989), there is a whole range of culturally given linguistic categories that are paired in oppositions. Some of these categories are part of an ancient Austronesian heritage, others are of an almost universal nature" (Platenkamp 1990).

Platenkamp reviewed Fox's case of the elder/younger distinction among the people of Termanu on Roté and the way it illustrates precedence in Rotenese society. He then asks:

> "Thus, rather than speaking of 'hierarchy', Fox proposes to speak of 'precedence'. But are the phenomena designated by Dumont's concept of 'hierarchy' and Fox's concept of 'precedence' of the same order" (Platenkamp 1990)?[9]

Platenkamp's last paragraph:

> "The question, then, whether Dumont's theory of hierarchy applies to Indonesian (or Austronesian) societies may be rephrased as follows: Do the value configurations, that emerge from careful analyses of the relationships between contexts, exhibit the particular logical properties that according to Dumont are characteristic of value systems in general" (Platenkamp 1990)?

Platenkamp was inclined to join Dumont in linking the logic of hierarchy to value systems, but, in his Kioloa paper, he leaves the last question unanswered.

9 The phrasing here is slightly ambiguous: I infer that Platenkamp asks, are Dumont's concept of "hierarchy" and Fox's concept of "precedence" a single concept?

The ensuing discussion of precedence continued into the 1990s. In 1996, a conference on precedence and hierarchisation in Leiden brought together anthropologists from several countries.[10] While those attending the conference reached no consensus on the question of precedence and Dumont's hierarchy, papers by nine of the participants and a solicited paper by D. J. Butterworth appeared in *Precedence: Social Differentiation in the Austronesian World* (Vischer 2009). Thereafter, the question of whether or not precedence is Dumont's general concept of hierarchy in disguise languished. By 1993 Platenkamp was in Münster, Fox had intensified his work in Java, and my work had shifted away from Wai Brama and toward the peoples of central Sikka. Most of the interlocutors, having taken up other questions, set the precedence and hierarchy problem aside without a resolution.

Nevertheless, questions about hierarchy persisted subliminally in my research on Sikka's communities perhaps, I suspect, as a result of conversations I had with Platenkamp before he left Leiden for Münster and in 1994. I propose to set out here some of my still inconclusive meditations on the problem as an acknowledgment of the influence on my work of conversations with Jos Platenkamp during the too brief decade during which I worked for a year in The Netherlands and otherwise made several visits to the University of Leiden.[11]

Dumont's Theory of Hierarchy as "Encompassing the Contrary"

I find two reasons for scepticism for thinking of precedence as a reflex of Dumont's conception of hierarchy. The first arises from the logic of his theory of hierarchy as the "englobement du contraire" (Dumont 1979: 397).[12] The second arises from the question of whether or not Dumont's ideas help us to understand the society and culture of Tana Wai Brama and, indeed, whether or not wherever they may occur there can be any general theory that explains all hierarchies.

In *Homo Hierarchicus*, Dumont shifts his subject from the hierarchy of the values *pure* and *impure* and its relation to the hierarchy of the varnas (the four groups or categories of people in society that make up an Indian community) to a general theory of hierarchy. The shift presents problems. To begin with,

10 The conference was sponsored by the International Institute of Asian Studies in Leiden and was organised by Dr Michael P. Vischer.

11 One particular conversation with Platenkamp remains strong in memory (see Lewis 2009: 161).

12 Rodney Needham was also critical of Dumont's ideas and subjected them to an exacting exegesis (Needham 1987: 102-145). Needham was especially critical of the illogic of the proposition that hierarchy is founded in the encompassment of one value or category by its contrary or opposite: "The harder one scrutinises it [encompassment of the contrary]," he writes, "the more incomprehensible every aspect of it becomes" (Needham 1987: 135).

values are not social groups. Needham has shown how one of a pair of opposite values ("good") cannot encompassment its opposite ("evil") because the subsummation requires a degree of identity in their attributes not shared by good and evil (1987: 130–135). It is also less than clear how a social group encompasses another social group, its contrary, especially since social groups must complement one another if they are the constituent elements of a society. Is Dumont's theory a theory of hierarchies of values or of hierarchies of social groups? Does Dumont intend that the general theory explains both value hierarchies and social hierarchies? On this question, Dumont is unclear. The last appears to be the case, but why should we think that hierarchies, regardless of what they are of, manifest a single logical structure?

Be that as it may, I find it difficult to conceive of how one of two social groups can encompass the other. Fifty-thousand berserk football spectators physically encompass a few dozen players on a field. Indeed, in some cases, rather expensive architecture defines the two groups, spectators and players, by separating them and reinforces that social structure during a game. But encompassing → encompassed is not the social relationship of the two groups, which is one of the complementary and reciprocal exchange of money and noise from spectators to players and two hours of entertainment and a license to behave with mindless silliness from players to spectators. There are complexities. Both the group of spectators and the group of players are divided into two *teams*. The teams (let us call them the Hyenas and the Warthogs), while they are complements making up the group of *players*, are nevertheless in a relation of competition while the spectators are divided into categories whose relationship is also one of competition but may erupt into one of hostility – to employ Apthorpe's (1984) typology of relations.[13] Except when a game results in a tied score, the end of the game creates another pair of groups and categories in the remergence of the spectators and players of each team: winners and losers. While it may be argued (by the winners and their supporters) that victory is a value superior to defeat, it is difficult to find encompassing and encompassment in the relation of the two. Indeed, the victorious and defeated losers, separately and intact as groups, abandon the field to the spectators who, if the game were one of British football, frequently launch a riot.[14]

The example is perhaps not as trivial as it may first appear. It illustrates the mutability of social groups, the mutability of individual identity (which is often achieved by *rites du passage*) in a system of social classification, and, in

13 I should perhaps point out a third, usually minor, category of spectators that consists of a few people who are not players and are not quite spectators, but who suffer a game in a state of mindless boredom. I would belong to the last category.

14 I am reminded of Claude Lévi-Strauss's remark that sport events are not rituals. Rituals, he noted, resolve differences by bringing people who are initially different together. Sport events, in contrast, divide people who are initially of one category into two categories: winners and losers.

general, the difficulty of mapping contrasts of values onto the relations of individuals and social structure. In this case, groups who complement one another to the extent of comprising the single group of *players* become competitors who leave the field as either victors or losers or as just tired, begrudged people, in the case of a tied game.

Apart from Dumont's analysis of India's varna and caste hierarchies, one finds in the literature on the Indian caste system a study that avoids the entanglements implicit in Dumont's theory. I refer to Stephen A. Tyler's *India: An Anthropological Perspective* (1973). Before commenting on the logic of Dumont's theory, I should briefly recapitulate Tyler's approach to the analysis of Indian society.

Tyler's Explication of the Caste System

Stephen A. Tyler's *India: An Anthropological Perspective* (1973) provides us with another account of that "enduringly indigestible kernel of human custom [that] has cast such a persistent spell over the mind of Western man that we can only conclude that caste must strike some deeply discordant note in the Western mind," the Indian caste system (Tyler 1973: 147). Tyler casts his analysis of caste not in terms of purity and impurity, but in terms of purity and pollution. Acknowledging different interpretations of caste, his analysis begins with the observation that "*caste is first and foremost a system of social classification* [italics as in the original] – a means of placing people into named categories" (Tyler 1973: 147), a feature, he says, that "is accurately reflected in the native term *jati*" (*ibid.*)[15] The *jati* system classifies the members of Indian society, and

> "Jati primarily denotes species or kind and consequently refers explicitly to a system of classification. To be a member of a particular jati is thus equivalent to being a special kind or species of human" (Tyler 1973: 147).

More than a case of the human mind arbitrarily classifying things of the world (*ibid.*), including human beings, "because the system of jati is consistent with other features of Indian culture, it must reflect some more fundamental attribute of Indian thought" (Tyler 1973: 147-148). That more fundamental attribute of thought "is [an] expression of that universal Indian mode of thought [that] insists that differences among humans are attributable to inherent degrees of purity and pollution" (Tyler 1973: 148). Tyler's analysis answers the questions, "what [...] are the indices of purity, and what is the logic of an hierarchical structure?" (*ibid.*) quite differently from Dumont.[16]

15 Tyler, an American, thus begins with an idea that derives from *L'année sociologique* and Durkheim and Mauss (1969), who figure in Dumont's sociological ancestry.

16 Indeed, his answer may have been the prompt for Dumont's rather gratuitous remark that hierarchy is not a taxonomic tree (Dumont 1980: 239). I find it rather odd that Tyler twice

However that may be, I think Tyler's work is the better ethnographically, which has led me to think about exactly what it is I find unconvincing about hierarchy as encompassing the contrary and its logic. The difficulty can be resolved by leaving aside the ideology of purity and impurity and considering the logic of social classification.

Dumont's Logical Scandal

In the 'Postface' added to the Complete Revised English Edition of *Homo Hierarchicus* of 1980, Dumont proposes to "turn away from India, or rather we shall take what India has taught us about the opposition between the encompassing and the encompassed as a point of departure" toward a general theory of hierarchy. Nevertheless, Dumont's theory of hierarchy derives from his analysis of Indian *jati*s and a value distinction between *purity* and *impurity* that pervades Indian thought and ideology. In the Indian system, *impurity* is the marked element of the two,[17] hence, in Dumont's theory of hierarchy, as "the opposition between the encompassing and the encompassed":

> "Hierarchy is not, essentially, a chain of superimposed commands, nor yet a taxonomic tree, but a relation that can succinctly be called 'the encompassing of the contrary'" (Dumont 1980: 239).

Dumont notes the "logical scandal" that arises when we find a class encompassing its contrary, which produces identity and contradiction (Dumont 1980: 242). There is nothing mysterious here. In systematics, two species are identically of a single taxon. Wombats and koalas are identically *Marsupialia*, but are species of different orders. To be of different species is not a contradiction but a distinction. The logical scandal arises from a paradox or an antinomy lurking in Dumont's argument.

Earlier in *Homo Hierarchicus*, before he reveals his general theory of hierarchy, Dumont first defines hierarchy as "the *principle by which the elements of a whole are ranked in relation to the whole*" (Dumont 1980: 66). This is a very general proposition, one not without logical problems. Wholes are composed by parts – or 'elements' – that constitute the whole of which

cites *Homo Hierarchicus* briefly but does not discuss Dumont's main thesis while Dumont ignores entirely Tyler's work in the 1980 edition of *Homo Hierarchicus*.

17 Cf. Fox (1994: 90). In linguistics, *markedness* is the contrast between two paradigmatically linked linguistic units. Of "units in paradigmatic relationship," "of two units in contrast [...] one will be positive, or *marked*, the other being neutral, or *unmarked*" (Lyons 1968: 79). Lyons was writing of syntactical distinctions such as those between the singular English word *stone* and the plural *stones*. "The plural is positively marked by the final *s*," he remarks, "whereas the singular is unmarked" (Lyons 1968: 79). Further, "in a given context the presence of a particular unit is in contrast with its absence" (Lyons 1968: 79). This can be seen to be the case in the analyses of the contrast *pure/impure* (a lack or reduction of purity) in relation to the classificatory discrimination of persons of different *jati* in the work of Dumont and Tyler.

they are a part by their relationships to one another. To think of parts "in relation to the whole" is a tautology. Parts are not in a relation to the whole, they are *constituents* of the whole. Contrary to Dumont's formulation, elements constitute a whole by their relations to one another. To say "in relation to the whole" is a superfluity, a pleonasm, indeed, a tautology.

If elements contribute to a whole but they constitute a whole by their relations to one another, then an aggregation of human beings is just a mob unless they are related to one another in some way. It is the relations one to another that make them a society. Society arises from the relations and is not simply from the collection and juxtaposition of people. Here, of course, I shift the question from one of values to one about society. Nevertheless, values arise from ideologies. An ideology is not a random heap of ideas, but consists of concepts that stand in some logical relationships to one another. To say that one idea of an ideology is 'ranked' in relation to the whole thus puzzling. Ideas, things or persons may be ranked by some one criterion or set of criteria and thus form a systemic unity – a 'whole'. But to say that they are individually ranked in relation to the whole is simply to say they are ranked to each other and that features of the whole include 'rankedness.'

Dumont makes it clear that the elements of a whole with which he is concerned are those that constitute a value, which is purity. But whether or not purity is a value, it is a quality of things or persons that are more or less pure *in relation to each other*. They are not more or less pure in relation to the whole, except that, in Dumont's argument, purity must be a quality of things and persons and a thing *sui generis*. Dumont's scandal is not one of logic; the solecistic logic of the proposition is the scandal.

What if we shift the subject from the values *purity* and *impurity* as termini of a continuum of *purity* ←→ *impurity* to social groups constituted by a system of social classification?

Fun with Boxes; or, Where To Put the Boxes and What Can Be Put In Them

If the caste system is first a system of social classification, then its logic should accord with the way classification works. I confess that since my first encounter with it, I have suspected something fishy about the logic of a class that encompasses its contrary, or opposite. My suspicion began with the two diagrams with which Dumont (1980: 242) began his argument for the logic of hierarchy:

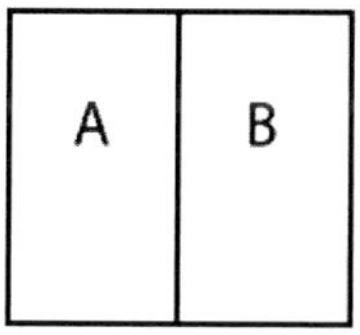

Fig. 1

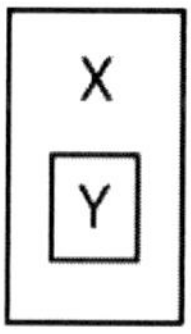

Fig. 2

Fig. 1 and Fig 2: Dumont's representation of complementarity (A+B) and hierarchy (X encompasses Y)

In Figure 1, A and B divide a universe of discourse into two juxtaposed parts which may "be complementary with respect to this universe, or again, contradictory, in the sense that each excludes the other" (Dumont 1980: 241). As complements, A and B depict "the universe of discourse in its composition (structural perspective)" (*ibid.*). As contradictories, the diagram depicts the logical relation of the two "classes", whereby A is not B and B is not A. In this case, the universe of discourse is "only implied in the background of the relation (substantial perspective)" (*ibid.*).

Figure 2 is the case of hierarchy, in which "class X is coextensive with the universe of discourse, and the other is set within the first as a rectangle Y contained in X: Y belongs to X and at the same time differs from it" (Dumont 1980: 242):[18]

> "At the superior level there is unity; at the inferior level there is distinction, there is, as in the first case, complementariness or contradiction. Hierarchy consists in the combination of these two propositions concerning different levels. In hierarchy thus defined, complementariness or contradiction is contained in a unity of superior order. But as soon as we intermingle the two levels, we have a logical scandal, because there is identity and contradiction at the same time" (*ibid.*).

Since, in Dumont's Fig. 2, "Y belongs to X and at the same time differs from it", that is, Y is the contrary of X, we are justified in relabelling Y as –X. Thus:

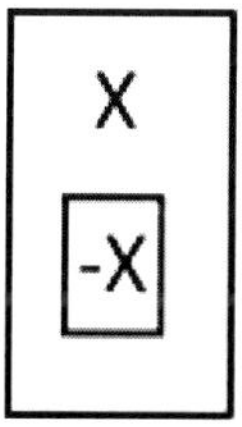

Fig. 3: Dumont's logical scandal: a class that includes elements that are not class members

18 A better way to conceptualise the idea of complex classes such as those encountered by ethnologists is to think in terms of polythetic classes. A polythetic class can contain members that are excluded from the class if it is defined monothetically, that is, by "'successive logical divisions so that the possession of a unique set of features is both sufficient and necessary for membership in the group thus defined'" (Sokal and Sneath (1963) quoted by Needham (1983: 48)), but are subsumable in the class by virtue of Wittgenstein's family resemblances (see Needham 1983: 36-65).

If a category (or class) is defined as a set of things that share a feature or a set of features, then clearly X cannot include as one of its members anything that is not X, that is, –X cannot belong to the class, set or category X. This is Dumont's "logical scandal." It deserves close scrutiny.

The following is a representation of a paradigm, that is, a model in the sense of the word as it is employed, for example, by taxonomists in the classification of plants and animals and in systematics, the study of the principles of scientific classification:

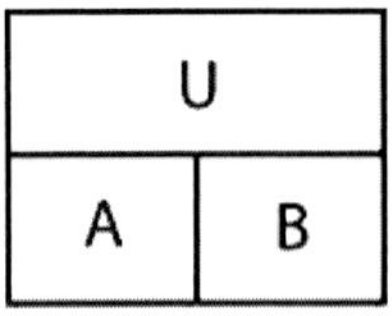

Fig. 4: Classification as a taxonomic paradigm

Here, the universe (of discourse), U, as a genus, consists of two categories or species, A and B. Members of the categories A and B are all members of the genus U, but no members of U are members of both A and B. Thus, members of any category A or B are of U but are of specific kinds of the things that constitute U. This is Dumont's Figure 1, but with the universe of discourse represented in the paradigm. Neither A nor B is "encompassed" by U; the relationship of A to B is one of difference but complementarity whereby, together, they constitute U. Taking a furry example, if the universe of discourse is *cat*, A is *lion*, and B is *tiger*, then it is true that all lions are cats, all tigers are cats, but not all cats are lions and not all cats are tigers. Any number of species (C, D, …) can be added to the taxonomy, so long as their members are cats and are not already included in another species. In Tyler's explication of the *jati* system, U is the caste system, that is, Indian society, while A and B are *jati* whose members are all of Indian society but not all members of Indian society are of *jati* A and not all are of *jati* B.

If the universe of discourse (U) is taken to be the universe as a whole, then matters are formally the same as with the universe of discourse taken to be *cat*, but the introduction of a negative requires explication. If, as a category, A consists of things distinguished monothetically that can be distinguished from everything else, then –A, can be designated in a paradigm as follows:

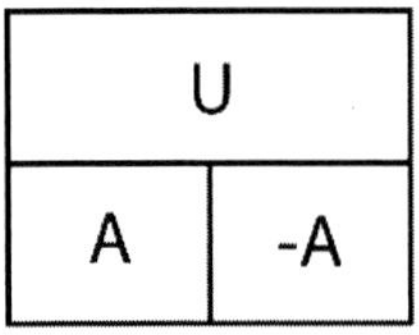

Fig. 5: A universe made up of things that are A and things that are not A

Here, we divide the universe into one set of things (*A*, cats, for example) contrasted with a set of things, –A, that are not cats; in the larger scheme of the universe, a set with a small number of members and a set with a vastly greater number of members.

Now comes the question: is Dumont's Figure 2 a transform of his Figure 1? Dumont has given us two sets of two boxes. Does it matter where they are placed, each with respect to the other? Let us consider that it does not.

In Figure 1, B is not A, which is –B, both of which are subsets of an unrepresented universe of discourse. In Figure 2, Y is not X and is thus –X, but is a subset of X, which is the universe of discourse. If we shift the boxes while maintaining the representation of the universe of discourse, we get the following:

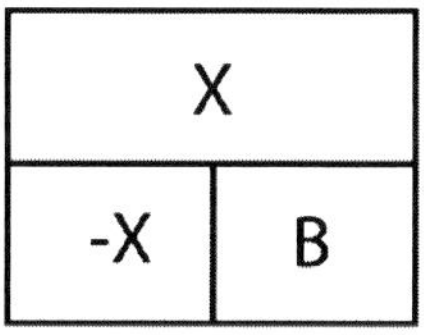

Fig. 6: Genus (X) that includes not-X and B as species

If the logic of classification is such that all species must be components of the genus in which they are subsumed, then we have the awkward case:

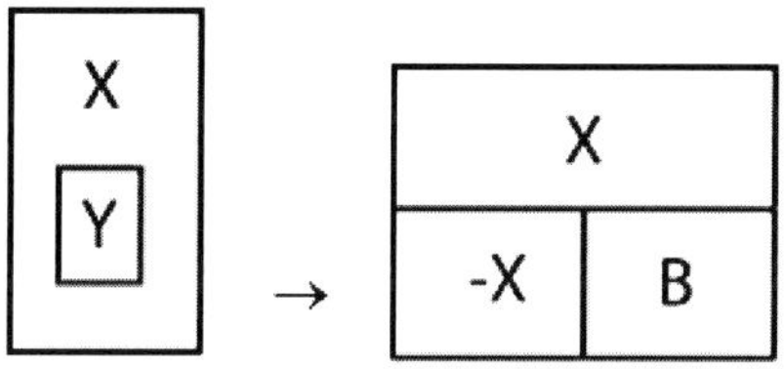

Fig. 7: Dumont's encompassment of the contrary as an antinomy

Dumont, but not Apthorpe in 1984 (see especially pages 290-291), whom Dumont acknowledges as the source of the idea, calls this "the encompassing of the contrary". But, if –X, as a species of X, must also be X, then we have a case in which the genus includes a category that is not of the genus and a species (–X) that is not of its genus. That is, to refer to the concrete example, a genus of cats that includes a category whose members are not cats and a category whose members are not cats but which are cats. Figure 7 asserts that *not-cat* (*–X*) is a species, kind, type, or subset of *cat*. Nothing is a cat and not a cat at the same time. This is that which Dumont calls a "logical scandal". It is indeed a logical scandal, but it is something more specific and more profound: it is either a paradox or an antinomy.[19]

Let us say that we face a paradox, which is a product of language, rather than an antinomy, which is a contradiction arising from the correct application of logic. Referring to the figures, if A is a class or category, it can be opposed to, contrasted with, or complementary with –A (in the last case, A and –A constitute a genus), but –A cannot be among the inclusions of the category A. Nothing can be both a thing and not that thing. Purity, however, is not a thing; it is a quality or condition of things. Hence, a thing can be more or less pure, except that some "cultural" concept or rule posits absolute purity and absolute impurity unadulterated by purity. Is this the Indian case? Dumont calls the encompassment of the contrary a scandal and leaves it at that. I conclude that the scandal is the result of an error of logic.

Dumont credits a 1956 Oxford thesis written by Raymond Apthorpe, a student at Oxford when Dumont was there, as the source of the idea of encompassing/encompassed relations. Apthorpe's thesis remains unpublished except for an excerpt revised and published in Galey's (1984) *festschrift* in honor of Dumont. The excerpt "makes available those parts of that thesis which were, and perhaps still are, pertinent to the issue in question" (Apthorpe 1984: 284). In the opening paragraphs Apthorpe remarks that the thesis

> "was not a work on Indian studies – and no one was more astonished by Louis Dumont's extraordinary reference to it in *Homo Hierarchicus* than I. It was an effort to identify some categorical dimensions of social relations, such as could be found in any society, perhaps indeed in all. To this end it devised a short, and undoubtedly simplistic, sociologic of hostility, hierarchy, competition and cooperation as social relations" (Apthorpe 1984: 284).

Apthorpe's purpose was to describe categories of social relations, that is, features or styles of interaction or 'sociologics'. I read Apthorpe's essay to refer to features or elements of behavior that can be found in any human social interaction. None of the four categories of social relations (hostility, hierarchy, cooperation, competition) can be thought of as governing the whole of a society exclusively, and this was not Apthorpe's argument. Apthorpe was writing, not of the sole sociologic of an entire social structure, society or culture. His interest was the characteristics manifested in individual and fleeting social relations, that is, the interactions of human beings such as, for example, those of a parent who cooperates in getting children out the door to school, who competes in a squash match, who is hostile to the chair of the board's proposal, and who ranks himself hierarchically in relation to children when persuading

19 In 1962, Quine lucidly set out the differences between paradoxes and antinomies (see Quine 1976: 1-21). An antinomy is a class of paradox that "produces a self-contradiction by accepted ways of reasoning" (Quine 1976: 5). A fine treatment of logico-mathematical, definitional, and pragmatic (that is, linguistic and communicative) paradoxes is to be found in Watzlawick at al. (1967), especially Chapter Six: Paradoxical Communication.

them to go to bed – sequentially, but all in the course of a single day. The universes of discourse in which Apthorpe set these sociologics were not those of socio-cultural wholes and cultural macrocosms but of specific, empirically observable interactions such as those of interest to Alfred Schutz and George Herbert Mead.

The basic question put by Platenkamp remains. Given that precedence cannot be a case of Dumont's hierarchy, is it hierarchy of another form? That is, is there a classification of varieties of hierarchy, of which precedence in Wai Brama may be a species different from that of India? Or is precedence a system of ranking – itself reminiscent of hierarchy? Are these questions answerable only by logic and theory? Or can they be answered by reference to ethnology? If so, what sort of theory can be derived from ethnographic facts?

A theory explains a corpus of facts or phenomena.[20] Dumont appeals to logic and values, but human behavior is not always logical (although in analysis it can be given a logic) and humans often act in ways contrary to the 'values' of their community. While behavior can be modeled *a posteriori*, it is non-algorithmic; it evolves but, as is the case of all evolutionary systems, it cannot be predicted with certainty. *Encompassing*, in Dumont's usage, is not a theory but only one of possible characterisations of relationships between categories in a system of classification.[21] Although an explanation, a theory, should be logically coherent, logic itself is not explanatory.

It is one thing for facts to appear inconsistent when set against one another or paradoxical, or for paradoxes or antinomies to arise among the implications of a scientific theory. It is quite another thing for a theory to give rise, in its logical structure, to a paradox or antinomy: how can we learn anything from such a theory?[22] How can such a theory lead to anything other than erroneous explanation? Is it possible that an entire socio-cultural complex, one of the most populous on the planet, could be held enthrall of and governed by a single principle as simplistic, invariable, and inviolable as purity encompassing impurity, its contrary? I am unconvinced. However, untangling an error of logic does not, in itself, explain how the hierarchies of Indian *jati* are different from the precedence of Wai Brama clans and houses. The fundamental differences between the two kinds of social groups and practically every other realm of life in India and life in Tana Wai Brama and the quite different logic of social (that is, group) relations in each convinces me that the answer to this

20 Among the *OED*'s seven definitions of *theory* is: "A scheme or system of ideas or statements held as an explanation or account of a group of facts or phenomena." This is the meaning of the word I shall adopt here.

21 Dumont's "theory of hierarchy" is actually not a theory, but is framed as "hierarchy is [...]" (1980: 66, 239), an expression in the form of a definition, that is, a tautology, rather than that of an *explicans*, such as a theory or an hypothesis.

22 Quine suggests that, in science, logical difficulties arising from the implications of a theory are dealt with by repudiating the theory and replacing it with another, which can amount to "a repudiation of part of our conceptual heritage" (Quine 1976: 9).

question is no. If Dumont's theory of hierarchy as the *englobement du contraire* is specious, then the question, Is precedence Dumontian hierarchy? is answered in the negative; moreover, it does not arise. But to say what something is not is not to say what it is.

Oda and Whiffs of Hierarchy in Tana Wai Brama

In the dialect of Sara Sikka spoken by the people of Wai Brama, the word *oda* means a sequence of events unfolding in time that are or can be construed as causally connected or to which causality can be imputed after the fact. The Wai Brama concept of *oda* accords well with the *Oxford English Dictionary* definition of *precedence*, whereby priority in time conveys priority of rank.[23] In contrast, The *OED* defines *hierarchy* as

> "A body of persons or things ranked in grades, orders, or classes, one above another; spec. in *Natural Science* and *Logic*, a system or series of terms of successive rank (as classes, orders, genera, species, etc.), used in classification."

The difference between the meanings in English of *precedence* and *hierarchy* is that the former is founded in things, events, and persons that succeed one another temporally, whereas the latter is a ranking of things or persons timelessly in "grades, orders, or classes, one above another."

Neither the world and things in it nor human beings and their clans appeared together instantaneously. Rather, in the mythic histories of Wai Brama, events of creation followed seriatim in time. The people of Wai Brama – rather, their ancestors – seized upon the seriality of events as a crux (in both sense of the word) in their thinking about the world and social life. If creations occur sequentially, then everything was once a new thing in the world. If the world was put together of new things, then new things can be expected to appear in the world in the future. As they appear, new things must be fitted into the world in a way that accords with the past, rather than conflicting with it. The Wai Brama world thus preserves *hadat*[24] and is open and adapts to change.[25] We may conclude that *precedence* is founded in diachrony whereas we may suspect that *hierarchy*, once given, arises from a logic of synchrony, persists, and is resistant to change.

23 A second recourse to the *OED*: Precedence, "3. The fact of preceding another or others in time or succession; previous existence or occurrence; priority. (Often with mixture of sense 4.) 4. The fact of preceding another or others in order, rank, importance, estimation, or dignity; higher position, superiority; the foremost place, pre-eminence, supremacy."

24 *Hadat* (SS): a polysemous word encompassing custom, tradition, propriety, etiquette, fitting, true to nature, decorous; term by which the peoples of Sikka refer to their ritual knowledge and customs.

25 Lévi-Strauss may have been tempted to classify Wai Brama as a cold society, but the ethnography implies that some societies are, on close examination, rather warmer than they first appear.

The principle of *oda* is found in the mythic histories of the domain's clans. There is no written or otherwise documentary history of the domain of Wai Brama unless, almost four decades on, one counts my field notes. Myths abound and mythic histories narrated in ritual speech tell of the founding of the domain by the first people to arrive in the mountains as migrants from elsewhere. They and their descendants are of *sukun* (clan) Ipir Wai Brama and the domain they founded they called Tana (land, earth, region of) Wai Brama. Then, in time, four more groups of migrants followed and the people of Wai Brama came to be distributed among five clans. Ipir remained *pu'an*, the "source" clan. In each generation a man of Ipir serves as *tana pu'an*, "source of the domain," whose responsibility is the ritual well-being of the domain, its people, gardens, and animals. The history of clan Ipir recounts the Ipir people's delegation of land and rights and responsibilities in the ceremonial system of the domain to each of the ancestors of the subsequent clans as they arrived.

Clans consist of descent groups of mothers and daughters called *lepo*, or houses. The houses of a clan come into being through the transfer of women from one clan to another following an interclan marriage, whereby one of a man's daughters, who is native to her mother's clan, is transferred to her father's and his sisters' clan. The daughter changes her clan affiliation to that of her father, wherein she founds a new house. Hers and the other houses of a clan are ordered in terms of genealogical seniority. The sequence in which new houses are created in a clan is the *oda* of the clan. While the ordering has little practical effect in daily life, it is important in the conduct of clan rituals and in the rituals of the domain. It is also of force in affinal alliance and exchange.

Children, both girls and boys, belongs to the clans of their mothers. People of a clan can marry, provided they are of a certain genealogical distance, and marriages of people of different clans are frequent. In contrast, the houses of clans are exogamous. Marriage is not marked with ritual although there may be a simple exchange of areca nut between the households of a woman and man who *kawit*, a Tana 'Ai word cognate with the Malay and Bahasa Indonesia *kawin*, to marry, marriage, but which translates best as "to cohabit, form a household, have children." The people of Wai Brama do not exchange bridewealth and its counter prestations and, in this regard, they are distinctly different from the Ata Sikka and Ata Krowé, their cultural and linguistic cousins to the west. Alliance focuses on children and their distribution among clans and the houses of clans. Despite a certain apparent looseness in arrangements, the alliance system is distinctly asymmetrical, as are prestations of *ama 'lo'en*, "father's forelock," the return of the daughter of a man who has married a woman of a clan other than his own, to his sisters. His daughter "replaces his blood of origin" they have lost by his out marriage.[26] On the occasion of a gift of father's forelock, the recipients reciprocate with a gift of two ivory tusks and two bronze gongs to the man's wife's people. As the people of Wai Brama say, here we do not *membeliskan* (BI exchange prestations for) women and men, we *membeliskan* children.

Oda is the order of the alliances of houses in clans and thus articulates the consequences of some (but not) all marriages in the differentiation of groups in a clan through the generation of new houses. *Oda* also defines the relation of women (who are holders of land in Wai Brama) and their brothers, to whom sisters delegate the rites of planting and harvesting. Hence, *oda* is implicated in gender, the relations of brothers and sisters and men and women generally.

Because clan affiliation can be changed by the transfer of a father's daughter from her mother's and his wife's clan to his clan, social classification in Wai Brama is mutable and flexible. Indeed, interclan alliances through marriages, which require the return of a man's daughter to his clan, are fundamental to social cohesion. With daughters who replace their fathers in a clan, access to land shifts as needed with the adult members of the community. Hence, kinship and alliance are not only about distributing people among social groups, but are, in the classical French concept, a total social fact. The asymmetry of alliance and social exchange creates new social groups, allows the incorporation of outsiders into the community, and extends alliance beyond the immediate community when its members marry out. Wai Brama society is an open social universe that is very different from that of India.

While precedence orders clans and their houses, it is inherent in the histories of the domain and in the performance of ritual. In Wai Brama, the sequence of sacrifices by clans in a ritual is explained, by Wai Brama ritualists themselves, by *oda*. *Oda* is the order by which the ancestors of the clans of the ritualists came to the domain. Thereby, it leads, in any particular social gathering, to an order of precedence among an event's participants, an order that is expressed in who speaks what and who does what and when. *Oda*, as precedence, also explains the delegation of tasks in a rite, such as the decapitation of a pig, by ritualists of older clans (those, in the Wai Brama idiom, who are "closer to the source") to ritualists of younger clans or simply from older men to younger men and boys.

Narrations of the precedence of the clans in the chanting of clan histories in ritual language are an integral part of the performance of all rituals. When a ritualist *tota oda*, "seeks the *oda*," he reconstructs the sequence of events that led to some state of affairs: the aetiology of an illness, the reason for a failure of crops, how a man came to fall from a coconut palm, how five clans came to be in the domain of Wai Brama. He constructs a narrative that orders

26 I have documented the complexities that result from this apparently simple system in Lewis (1988); see also Lewis (2003).

the cosmos, society, and individual human lives, a history that wrests meaning from an inchoate and chaotic world. He seeks, in short, the Greek *αἰτία* (*aitia*): cause, origin, reason, motive, responsibility.

Are the Ata Tana Wai Brama Homo Hierarchicus or Homo Aequalis?

Aside from the division of clans, the people of the domain are not divided into social classes in the sense that sociologists employ the term,[27] ranked or otherwise, nor are there marked differences in the wealth of clans, houses, and domestic groups. They are burdened by no bureaucratic institutions and have no secular, political offices. Leadership is fluid and momentary, emerging if at all only when the completion of a communal task requires organisation of people beyond those who live within a cluster of conjointly fenced fields. Ultimate authority in community matters, by *hadat*, rests with the senior women of the senior houses of the domain's clans who delegate the conduct of inter-clan relations to their brothers. This large number of men comprises cohorts from whom ritualists are drawn for the conduct of the rites of the domain's ceremonial system. Positions of ritual authority are not hereditary and are achieved rather than ascribed. A man's knowledge of invocatory ritual language, of the history of the domain, his clan's histories, and rituals is acquired both from ritualists among his classificatory mothers' brothers' and his classificatory fathers. In each generation, a few men manifest a talent for ritual or express sufficient interest and motivation to acquire the knowledge of ritual speech and the liturgy of rites vouchsafed among elder ritualists. Leading ritualists acquire their status voluntarily through achievement and not ascription.

Referring to India, Dumont isolates hierarchy "as purely a matter of religious values" (1980: 66), an absolute somewhat attenuated in the Postface wherein social differentiation, especially in cases of dualism such as the division of "tribal" societies into moieties, "means differentiation of values" (244). On the same page, Dumont adds that it is "understood that in the majority of societies it is religion which provides the view of the whole, and that the ranking will thus be religious in nature" (Dumont 1980: 66). It is difficult to identify 'religion' in Tana Wai Brama except as corpora of oral invocations and narrated myths of origins and a repertoire of sacrificial rites and other ritual practices. Among these, an encompassing value of the kind that Dumont reports for India is not to be found. If a 'whole' can be identified, it is *hadat*, which includes ritual practices and the myths of origins of the domain and its clans. The first to arrive granted land and delegated rights and obligations in the ceremonial system to newcomers. Subsequently arriving ancestors

27 I mean the discrimination of class as a group distinguished from other groups by its economic or social position in a system of ranked levels.

founded clans with equal rights to land in the domain. The ancestral people of Ipir and their living descendants remained 'source' and retained authority in the ceremonial system.

The Sara Sikka word *oda* represents a concept that carries no connotation of hierarchy; indeed, I know of no Sara Sikka word that translates the English *hierarchy*.[28] To derive a hierarchy of values, social groups, offices, or social persons requires a great deal more: a concept of ranking, which requires rankable social statuses, identities, or institutions. In Wai Brama, the five clans of the domain are located in the domain's history according to the sequence of their founding and the houses of clans are related as elder or younger relative to one another by the same principle. But persons are not ranked in this way, except for the categorical distinction in the language between elder (*wué*) and younger (*wari*) same-sex siblings. Members of clan Ipir enjoy no perquisites available to men and women of other clans. Ipir people hold no offices of institutionalised authority over anyone else.

One man of clan Ipir, the *tana pu'an* ("source of the domain"), personifies the origin and *hadat* of the domain. Elsewhere I have characterised the source of the domain as one responsible for the fertility of the land and the fecundity of its plants, animals, and humans. This characterisation is not entirely true. While he represents the source, as origin, of the domain, the *tana pu'an* himself is not that source. Indeed, his attributes include no special links to the deity or spirits of the ancestors – and so he cannot be said to be a priest – and no special authority in the secular life of the community. He himself is not required to conduct rituals attached to his designation as *tana pu'an* and he conducts no rites other than those that a ritualist of any clan can conduct.

Rights, obligation, and delegation prefigure much social action in Wai Brama, especially in the realm of ritual performance. The source of the domain can delegate responsibility for the conduct of rites, as the need for them arises, to other men of his own or other clans. Indeed, the community expects him to delegate. Thus, ritual order emerges, founded on the principle of precedence and acts of delegation.

Even this description is not entirely complete. To delegate is the verb *ma'a*, SS 'to divide'. In Wai Brama, the verb is used equivalently with the verb *himo*, SS 'to receive', and the verb *sélung*, 'to replace'. In contrast to the distinctions of English, to replace, to change, and to divide are, in Wai Brama thought, concinnous facets of a single idea and the words are synonyms and are interchangeable in the syntax of speech. The facets of the idea are also fused in the ritual language phrase *paseng waeng gélur narang*, "to transfer (or delegate) the face, to change the name," which expresses an act of delegation in ritual language.

28 *Ta'ur-ta'ur* is the succession of steps of a house ladder and would serve as a handy trope should the people of Wai Brama ever require a word for hierarchy.

In Sara Sikka "to have a responsibility, obligation" and "to have a right" to something are not distinguished lexically. In other words, in Sara Sikka, one does not distinguish rights and responsibilities. *Tara sanggo* and *hu'u sanggo* mean "to carry the burden" and to "shoulder the burden." In ritual language, *hu'u tutung 'wara galang*, is "to respect (honor, obey) obligation, to shoulder the burden (responsibility).[29] These phrases are denoted more succinctly and are synonymous with the word *hak* (SS and BI "right," in the sense of having a justifiable claim to something). The language of Sikka thus conflates the concepts of responsibility and right. Delegation derives from responsibility which, as a right in Wai Brama, must be distributed and shared. The effect is markedly anti-hierarchical because through delegation authority and power are distributed broadly in the community. Even so, while responsibility is shared by and is the burden of the community, the *tana pu'an*, as center and source, is he who delegates particular elements or tokens of responsibility for the performance of a particular ritual, in which all groups and persons have rights and for which they all have responsibility.

Thus in the *gren mahé*, the culminal rites of the years-long ceremonial cycle in Wai Brama, all clans have rights to obligations, tasks, and responsibilities. This is an odd sentence in English, since Anglophones do not normally speak of people as having rights to responsibilities (although they may agree that responsibilities follow from rights), but it is true of Wai Brama thought, wherein "to honor obligation, to shoulder responsibility," is a right (*hak*). A right and a responsibility are thus equivalent concepts. In delegating tasks in ritual, the *tana pu'an* insures that the equality of the rights-responsibilities-obligations of the clans is respected. Whereas the *tana pu'an* may be mistaken as evidence of hierarchy in Wai Brama social organisation, his job is anti-hierarchical.

The point of this brief summary of a complex social and ceremonial order is that the society of Wai Brama presents very little hierarchy in any of the usual connotations of the word. The hierarchy one finds is circumstantial, that is, it arises in specific situations such as occasions of large-scale ritual performance. The *gren mahé*, the rarely performed culminal rites of the ceremonial system of the domain, is one of a small number of occasions of this kind.

For the *gren mahé*, the community's ritualists must make decisions about sequences of ritual acts to be spoken and to be carried out, and who will do them. The decisions constitute an arena of politics and require the leadership of a ritualist or a small group of ritualists who step forward from among the domain's corps of ritualists and who are accepted by the *tana pu'an* of the

29 *Hu'u* is "to carry on the head; to respect, honor, obey"; *'wara* is "to carry, to bear" especially with a shoulder pole. Only women carry burdens on their heads and men, more frequently than women, use shoulder poles for carrying two burdens.

domain.[30] The result is an apparent hierarchy that is contingent on circumstances and personnel and can be viewed as one manifestation of precedence. At that, it is evanescent and evaporates once the tasks for which it was raised are completed. It may be that this is a kind of hierarchy but, as it hints at no encompassing value and its contrary, it is far from the hierarchy Dumont proposed in his general theory.

In Wai Brama we find a weak structure of 'hierarchy' at the level of social interaction, but not at the level of religious thought, which notably lacks precepts of values as normative rules or strictures. There is, instead, *hadat*, tradition, custom, practice, propriety; the ways of the ancestors. This being the case, it is difficult to depict Tana Wai Brama as a hierarchical society, in either the Indian or the modern western sense, or its people as members of the species *Homo hierarchicus*. They are, rather, of the species *Homo aequalis*.[31]

It is perhaps more useful to avoid thinking that there is one, general, irreducible form of hierarchy and to view hierarchy, in any of its endless forms, not as a fundamental feature of human societies, but as an element more or less forcibly present in every interaction of human beings. These considerations may not answer the question, is precedence hierarchy? They do cast doubt on many of our conceptions of hierarchy and they imply an argument for the protean creativity of our species of social primates and the consequent proliferation of ways in which we can put our talent for sociality to work in the organisation of our communities.

If the community of Tana Wai Brama presents no hierarchy, then we need not worry about whether or not Dumont's – or any other – theory accounts for it. That said, the ethnography of Wai Brama indicates weak ritual, ceremonial, social, and conceptual, that is, classificatory hierarchies, and something of hierarchy in the precedence of clans and houses within clans. The weak social and conceptual hierarchies manifest in the conception of sequences in time. Events, as distinct from mere occurrences and happenings (see Lewis 1989), are nodes in a web of happenings that are marked socially and conceptually; that is, they receive names or other designators in language and speech. They are moments of growth, differentiation, and the creation of new configurations of persons and, indeed, of social groups. They are, in other words, temporal orderings of earlier and later things, precedents and subsequents, antecedents and consequents. The Ata Tana Wai Brama identify them in the unfolding of individual lives and in social life generally. Precedence in Wai Brama is akin to that which Anglophones intend by the word *history*. The people of Wai Brama call it *oda*. Although the forms *oda* produces may be those of the ancestors, in which case established culture is reproduced, the workings of *oda*

30 The *tana pu'an* of the domain is a man of the central house of the central clan of the domain. He does not act or speak in ritual, but from him emanates the authority of the domain's ritual experts. See Lewis (1988) and Lewis, Asch and Asch (1994).

31 I note Dumont's use of the phrase *Homo aequalis* in 1977.

are generative rather than conservative. All the peoples of Flores place a high value on *adat* (to use the Malay and Bahasa Indonesia gloss for the Sara Sikka word *hadat*). Nevertheless, thinking in terms of precedence allows the acceptance of new ideas, new things, new practices, and provides means for their incorporation into the life of the community. The *hadat* of Wai Brama is thus malleable, interpretable, negotiable as occasion requires, adaptive, and constructive. Elements of it can be eliminated from the community's repertoire of cognition and culture without a sense of loss or a nativistic reaction as a result. In these respects, precedence is greatly different from the hierarchy of the varna and jati of India as they are represented in western texts.

The unanswered question is transmogrified from "Is precedence hierarchy?" and becomes "Can a single definition of hierarchy explain both the Indian and Florenese cases as well as the indubitably manifold forms of hierarchy among human societies?" It is to the general study of hierarchy – that is, social organisation – that the study of eastern Indonesian societies contributes.

It was Platenkamp's and Fox's exchange of papers at Kioloa that stirred me to begin thinking about hierarchy in Wai Brama and in Sikka. After many years during which Platenkamp's ideas chased around my nonconscious mind I am now inclined toward a view that the societies of India and those of the Sikkanese peoples are so different that no one theory of hierarchy can explain both. My remarks here are not a criticism of Platenkamp's ideas in the 1990s. Far from it. In the early 1990s, I, too, was intrigued by the idea that one of a pair of contraries might encompass the other. The present meditations on the problem may not be its resolution, but they serve to illustrate the way in which one scholar's thought can influence that of another. In my case, I can say that Platenkamp's cogent and lively advocacy of Dumont's thesis motivated me in my work on Flores, which benefited thereby.

Thus, the following is an encomium for the work of a talented and dedicated scholar and a respected friend. In revisiting my memory of conversations I had with Professor Platenkamp in the first half of the 1990s in Leiden and Münster I am mindful that they were not many, for our paths diverged and did cross after then. But they were intense and they have influenced my thinking about my work as an ethnographer in the years since. Why revisit conversations from a quarter of a century ago? Because I have continued to think about the critique of the concept of precedence that Platenkamp provided and about the nature of hierarchy, if not generally (because I think that, generally, there is no such single thing), then in relation to the societies of the peoples of Wai Brama and Sikka. While I am certain that Jos Platenkamp's has been an abiding and salutary influence on my work as an ethnographer, the question remains unanswered.

References

APTHORPE, R. (1956) *Social Change: An Empirical and Theoretical Study.* D. Phil. Dissertation: Oxford University.

APTHORPE, R. (1984) Hierarchy and Other Social Relations: Some Categorical Logic. In GALEY, J.-C. (ed.), *Différences, Valeurs, Hiérarchie: Textes Offerts a Louis Dumont*, pp. 283-298. Paris: Editions de L'Ecole des Hautes Etudes en Sciences Sociales.

BARRAUD, C. (1979) *Tanebar Evav. Une Société Tournée vers le Large.* Paris and Cambridge: Éditions de Maison des Sciences de l'Homme/Cambridge University Press.

DUMONT, L. (1977) *Homo Aequalis. Genèse et Épanouissement de l'Idéologie Économique*. Paris: Editions Gallimard.

DUMONT, L. (1979) *Homo Hierarchicus. Le Système des Castes et ses Implications.* Paris: Galliard. [rev. ed. of 1966 with the Postface, "Vers une Théorie de la Hiérarchie"].

DUMONT, L. (1980) *Homo Hierarchicus: The Caste System and Its Implications*. Revised English Edition. Chicago: University of Chicago Press.

DURKHEIM, E. and MAUSS, M. (1969) *Primitive Classification.* London: Cohen and West.

FOX, J. J. (1988) *Origin, Descent and Precedence in the Study of Austronesian Societies.* Public Lecture in connection with De Wisselleerstoel Indonesische Studien given on the 17th of March 1988. Leiden University.

FOX, J.J. (1990) *Hierarchy and Precedence. Working Paper no. 3.* Canberra: ComparativeAustronesian Project, Department of Anthropology, Research School of Pacific Studies. The Australian National University.

FOX, J. J. (1994) Reflections on 'Hierarchy' and 'Precedence'. In JOLLY, M. and MOSKO, M. (eds.), Transformations of Hierarchy: Structure, History and Horizon in the Austronesian World, Special Issue of *History and Anthropology* 7(1-4): 87-108.

FOX, J.J. (2009) Precedence in Perspective. In VISCHER, M.P. (ed.), *Precedence: Social Differentiation in the Austronesian World*, pp. 1-11. Canberra: ANU E Press.

GALEY, J.-C. (ed.) (1984) *Différences, Valeurs, Hiérarchie: Textes Offerts à Louis Dumont.* Paris: Editions de l'École des Hautes Etudes en Sciences Sociales.

GEIRNAERT-MARTIN, D.C. (1992) *The Woven Land of Laboya: Socio-Cosmic Ideas and Values in West Sumba, Eastern Indonesia*. Leiden: CNWS Publications.

IVES, C. (1984) *The Unanswered Question*. New York: Peer International.

LÉVI-STRAUSS, C. (1949) *Les Structures Élémentaires de la Parenté.* Paris: Presses Universitaires de France.

LEWIS, E.D. (1982) *Tana Wai Brama. A Study of the Social Organization of an Eastern Florenese Domain*. Ph.D. Thesis (Anthropology). Canberra: Institute of Advanced Studies, The Australian National University.

LEWIS, E.D. (1988) People of the Source. The Social and Ceremonial Order of Tana Wai Brama on Flores. Dordrecht: Foris Publications.

LEWIS, E.D. (1989) Why Did Sina Dance? Stochasm, Choice and Intentionality in the Ritual Life of the Ata Tana 'Ai of Eastern Flores. In ALEXANDER, P. (ed.), *Creating Indonesian Culture*, pp. 175-198. Sydney: Oceania Publications.

LEWIS, E.D. (1996) Origin Structures and Precedence in the Social Orders of Tana 'Ai and Sikka. In FOX, J.J and SATHER, C. (eds.), *Origins, Ancestry and Alliance: Explorations in Austronesian Ethnography,* pp. 157-178. Canberra: Department of Anthropology, Research School of Pacific and Asian Studies, The Australian National University [http://press.anu.edu.au/titles/comparative-austronesian-series/origins-ancestry-and-alliance/].

LEWIS, E.D. (1997) Conversations with Mo'an Koa: Collaboration in Ethnographic Discovery in Tana Wai Brama. *Canberra Anthropology* 20 (1-2): 55-81 [now *The Asia Pacific Journal of Anthropology*]. Special Issue in Memory of Professor Roger M. Keesing, Christine Jourdan (Editor).

LEWIS, E.D. (2003) Ritual, Metaphor, and the Problem of Direct Exchange in a Tana Wai Brama Child Transfer. In NAS, P., PERSOON, G. and JAFFE, R. (eds.), *Framing Indonesian Realities: Essays in Symbolic Anthropology in Honour of Reimar Schefold,* pp. 27-51. Leiden: KITLV Press.

LEWIS, E. D. (2009) Precedence in the Formation of the Domain of Wai Brama and the Rajadom of Sikka. In VISCHER, M. P. (ed.), *Precedence: Social Differentiation in the Austronesian World,* pp. 133-165. Canberra: ANU E Press.

LEWIS, E.D., ASCH, T. and ASCH, P. (1994) *A Celebration of Origins*. 16 mm Film. Watertown, Massachusetts: Documentary Educational Resources.

LYONS, J. (1968) *Introduction to Theoretical Linguistics.* Cambridge: Cambridge University Press.

MOYER, D.S. (1976) *Paris – Leiden and Other Oppositions.* ICA Publication No. 16. Leiden: Institute of Cultural and Social Studies, Leiden University.

NEEDHAM, R. (1983) *Against the Tranquility of Axioms.* Berkeley: University of California Press.

NEEDHAM, R. (1987) *Counterpoints.* Berkeley: University of California Press.

PAUWELS, S. (1990) From Hursu Ribun's 'Three Hearth Stones' to Metanleru's 'Sailing Boat'. A Ritual after the Harvest. In BARRAUD, C. and PLATENKAMP, J.D.M. (eds.), Ritual and the Socio-Cosmic Order in Eastern Indonesian societies, Part II: Maluku, pp. 21-34. Special Issue *Bijdragen tot de Taal-, Land- en Volkenkunde* 146(1).

PLATENKAMP, J.D.M. (1988) *Tobelo. Ideas and Values of a North Moluccan Society.* University of Leiden Ph.D. Thesis.

PLATENKAMP, J.D.M. (1990) Some Notes on Hierarchy in Eastern Indonesia: A Comment on J. J. Fox's 'Hierarchy and Precedence'. Unpublished discussion paper for the Conference on Hierarchy, Ancestry and Alliance, The Research School of Pacific Studies, Australian National University, 25-30 January 1990.

QUINE, W.V. (1976) *The Ways of Paradox and Other Essays.* Cambridge: Harvard University Press.

SOKAL, R.R. and SNEATH, P.H.A. (1963) *Numerical Taxonomy: The Principles and Practice of Numerical Classification.* San Francisco: W. H. Freeman and Company.

TYLER, S.A. (1973) *India: An Anthropological Perspective*. Pacific Palisades, California: Goodyear Publishing Company.

VISCHER, M.P. (ed.) (2009) *Precedence: Social Differentiation in the Austronesian World.* Canberra: ANU E Press.

WATZLAWICK, P., HELMICK BEAVIN, J. and JACKSON, D.D. (1967) *Pragmatics of Human Communication: A Study of Interactional Patterns, Pathologies, and Paradoxes.* New York: W. W. Norton and Co.

Publications of J.D.M. Platenkamp

EDITED BY H.F. VERMEULEN

1978 *The Jonglei Canal: Its Impact on an Integrated System in the Southern Sudan*. Leiden: Institute of Social and Cultural Studies (ICA Publication No. 26). xix+126 pp.

1979a (with L.E. Visser) Review of J.P. Garlick and R.W. Keay (eds.) *Human Ecology in the Tropics. Bijdragen tot de Taal-, Land- en Volkenkunde* 135(1): 192-194.

1979b Out of Mind, Out of Sight: Suppression and Taboo in the "Leach-Theory". *Bijdragen tot de Taal-, Land- en Volkenkunde* 135(1): 172-180.

1980 The Structural Study of Shamanism: Some Reflections on a Model. In: E.K.M. Masinambow (ed.), *Halmahera dan Raja Ampat. Konsep dan strategi penelitian*. Jakarta: LEKNAS/LIPI, pp. 337-346.

1982 "The Cod and the Corn Don't Mix". In: P.E. de Josselin de Jong and Eric Schwimmer (eds.), *Symbolic Anthropology in the Netherlands*. The Hague: Martinus Nijhoff (Verhandelingen van het KITLV 95), pp. 143-155.

1983 Gia-dutu: a True Relationship. Aspects of Kinship and Social Organization of the Tobelo of North Halmahera. In: E.K.M. Masinambow (ed.), *Halmahera dan Raja Ampat sebagai kesatuan majemuk*. Jakarta: LEKNAS/LIPI, pp. 1-31.

1984a The Tobelo of North Halmahera in the Context of the Field of Anthropological Study. In: P.E. de Josselin de Jong (ed.), *Unity in Diversity: Indonesia as a Field of Anthropological Study*. Dordrecht: Foris Publications (Verhandelingen van het KITLV 103), pp. 167-189.

1984b Structurele antropologie en de studie van Indonesische culturen. In: J.D. Speckmann (ed.), *Indonesische Studiën in Nederland*. Wassenaar: Netherlands Institute for Advanced Studies in the Humanities and Social Sciences, pp. 57-66.

1984c Marriage and Death: Social change in Tobelo. In: E.K.M. Masinambow (ed.), *Maluku and Irian Jaya*. Special Issue *Bulletin LEKNAS* 3(1): 105-118.

1988a *Tobelo: Ideas and Values of a North Moluccan Society*. Ph.D. thesis University of Leiden. Leiden: Repro Psychologie. vii + 298 pp.

1988b Myths of Life and Image in Northern Halmahera. In: David S. Moyer and Henry J.M. Claessen (eds.) Time Past, Time Present, Time Future: Perspectives on Indonesian Culture. Essays in Honour of Professor P.E. de Josselin de Jong. Dordrecht: Foris Publications (Verhandelingen van het KITLV 131), pp. 148-167.

1988c *In en rond het ICA. Een overzicht van (interne) publikatiereeksen*. Leiden: Instituut voor Culturele Antropologie en Sociologie der Niet-Westerse Volken, Rijksuniversiteit Leiden.

1989a C. Barraud and J.D.M. Platenkamp (eds.) Rituals and Socio-Cosmic Order in Eastern Indonesian Societies. Part I: Nusa Tenggara Timur. Special Anthropological Issue Bijdragen tot de Taal-, Land- en Volkenkunde 145(4). Leiden: KITLV, pp. 430-566.

1989b Review of Brigitte Renard-Clamagirand, *Marobo: Une société ema de Timor* (Paris: SELAF, 1983). *Bijdragen tot de Taal-, Land- en Volkenkunde* 145(4): 594-598.

1990a *De waarde der dingen. Ceremoniële geschenken van de Tobelo*. Den Haag: Museon, Afdeling Volkenkunde. 48 pp. ill. Catalogue of an exposition in the Museon, 20 December 1990 - 31 March 1991. [Vol. 2 of a series of six expositions on Indonesian textiles from the collections of the Museon, editor: Anne-Marie Boer, Museon]

1990b C. Barraud and J.D.M. Platenkamp (eds.) *Rituals and Socio-Cosmic Order in Eastern Indonesian Societies*. Part II: *Maluku*. Special Anthropological Issue *Bijdragen tot de Taal-, Land- en Volkenkunde* 146(1). Leiden: KITLV, pp. 1-126.

1990c The Severance of the Origin: A Ritual of the Tobelo of North Halmahera. In: Cécile Barraud and J.D.M. Platenkamp (eds.), *Rituals and Socio-Cosmic Order in Eastern Indonesian Societies*. Part II: *Maluku*, Special Anthropological Issue *Bijdragen tot de Taal-, Land- en Volkenkunde* 146(1): 74-92.

1990d (with Cécile Barraud) Rituals and the Comparison of Societies. In: C. Barraud and J.D.M. Platenkamp (eds.), *Rituals and Socio-Cosmic Order in Eastern Indonesian Societies*. Part II: *Maluku*, Special Anthropological Issue *Bijdragen tot de Taal-, Land- en Volkenkunde* 146(1): 103-123.

1990e *North Halmahera: Non-Austronesian Languages, Austronesian Cultures?* Lecture presented to the Oosters Genootschap in Nederland at Leiden on 23 May 1989. Leiden: Oosters Genootschap in Nederland, No. 18. 33 pp. [Reprinted as 1996c].

1991a W.F.L. Buschkens and J.D.M. Platenkamp (eds.) *Studies on China's Contemporary Social Structure by Zhai Shengde*. Leiden: Faculty of Social Sciences.

1991b (with W.F.L. Buschkens) Foreword by the Editors. In: W.F.L. Buschkens and J.D.M. Platenkamp (eds.), *Studies on China's Contemporary Social Structure by Zhai Shengde*. Leiden: Faculty of Social Sciences, pp. i-ii.

1991c Review of Jack Goody, *The Oriental, the Ancient and the Primitive: Systems of Marriage and the Family in the Pre-Industrial Societies of Eurasia*. (New York: Cambridge University Press 1990). *American Anthropologist* 93(4): 1016-1017.

1991d Het samenwerkingsverband CASA-ERASME en de relatie Parijs-Leiden. In: H.F. Vermeulen (ed.), *Recente ontwikkelingen in de Leidse antropologie. Zes lezingen gehouden bij het 12e lustrum van het Interfacultair Ethnologisch Dispuut W.D.O.* Leiden: Instituut voor Culturele Antropologie en Sociologie der Niet-Westerse Volken, Rijksuniversiteit Leiden (ICA Publicatie No. 91), pp. 41-48.

1992 Transforming Tobelo Ritual. In: Daniel de Coppet (ed.) Understanding Rituals. London: Routledge, pp. 74-96.

1993 Tobelo, Moro, Ternate: The Cosmological Valorization of Historical Events. In: Leonard Y. Andaya (ed.), *Centres and Peripheries*. Special Issue *Cakalele* 4: 61-89.

1994 (with Michael Prager) A Mirror of Paradigms: Nineteenth- and Early Twentieth-century Ethnology reflected in *Bijdragen*. In: Harry A. Poeze (ed.), *150 Volumes of Bijdragen: A Backward Glance and a Forward Glimpse*. Special Issue *Bijdragen tot de Taal-, Land- en Volkenkunde* 150(4): 703-727.

1995a In Search of a Name: Reflections on the Sejarah Tobaru. In: Connie Baak et al. (eds.), *Tales from a Concave World: Liber Amicorum Bert Voorhoeve*. Leiden: Leiden University, pp. 440-456.

1995b True to the Original. In: Heike Kämpf and Rüdiger Schott (eds.), *Der Mensch als homo pictor? Die Kunst traditioneller Kulturen aus der Sicht von Philosophie und Ethnologie*. Special Issue *Zeitschrift für Ästhetik und Allgemeine Kunstwissenschaft* 1: 20-35.

1996a Jan G. van Bremen, Vesna V. Godina and Jos D.M. Platenkamp (eds.) *Horizons of Understanding: An Anthology of Theoretical Anthropology in Europe*. Leiden: Research School CNWS (CNWS Publications 23).

1996b (with Jan van Bremen and Vesna Godina) Introduction. In: Jan van Bremen, Vesna Godina and Jos Platenkamp (eds.), *Horizons of Understanding. An Anthology of Theoretical Anthropology in Europe*. Leiden: Research School CNWS (CNWS Publications 23), pp. 1-2.

1996c North Halmahera: Non-Austronesian Languages, Austronesian Cultures? In: Jan van Bremen, Vesna Godina and Jos Platenkamp (eds.), *Horizons of Understanding. An Anthology of Theoretical Anthropology in Europe*. Leiden: Research School CNWS (CNWS Publications 23). [Earlier published as 1990e].

1996d The Healing Gift. In: Signe Howell (ed.), *For the Sake of Our Future: Sacrificing in Eastern Indonesia*. Leiden: Research School CNWS (CNWS Publications 42), pp. 318-336.

1997a Half Persons, Complete Societies. *Minpaku Anthropology Newsletter* 5: 7-8.

1997b Tobelo. In: Paul Oliver (ed.), *Encyclopedia of Vernacular Architecture of the World, Vol. II.* Oxford: Blackwell Publishers, pp: 1098-1099.

1998a Temporality and Male-Female Distinctions in the Tobelo Vocabulary of Relationships. *Social Anthropology* 6(3): 291-307.[Reprinted as 2000b].

1998b Health as a Social Condition. *Folk* 40: 57-70.

1999 Die Natur als Gegenbild der Gesellschaft. Ethnologische Betrachtungen zu einer paradoxen Idee. In: Ruth-Elisabeth Mohrmann (ed.), *Argument Natur – Was ist natürlich?* Münster/Hamburg: LIT Verlag, pp. 5-16.

2000a Solidarität und Austausch in Ostindonesien. In: Günter Best and Reinhart Kößler (eds.), *Subjekte und Systeme. Soziologische und anthropologische Annäherungen. Festschrift für Christian Sigrist zum 65. Geburtstag.* Münster/Hamburg: LIT Verlag, pp. 116-127.

2000b Temporality and Male-Female Distinctions in the Tobelo Vocabulary of Relationships. In: Catherine Alès and Cécile Barraud (eds.), *Sexe relatif ou sexe absolu? De la distinction de sexe dans les sociétés*. Paris: Éditions de la Maison des sciences de l'homme, pp. 241-266. [Reprint of 1998a].

2003 Strukturalismus in der Ethnologie. In: Hans Fischer and Bettina Beer (eds.), *Ethnologie. Einführung und Überblick. Neufassung*. 5th edition. Berlin: Reimer Verlag, pp. 295-308.

2004a Über die gesellschaftliche Relevanz der Ethnologie. In: Ursula Bertels et al. (eds.), *Aus der Ferne in die Nähe. Neue Wege der Ethnologie in die Öffentlichkeit*. Münster/Berlin/New York: Waxmann Verlag, pp. 21-32.

2004b From Partial Persons to Completed Societies. *Zeitschrift für Ethnologie* 129(1): 1-28.

2005a Preface. In: Wassef Al-Sekhaneh, *The Bedouin of Northern Jordan: Kinship, Cosmology and Ritual Exchange*. Berlin: Wissenschaftlicher Verlag, pp. vii-viii.

2005b Des personnes incomplètes aux sociétés accomplies. In: Françoise Héritier et Stephen Headley (eds.), *Moitiés d'Homme*, Special Issue *L'Homme* 174: 125-160. [Extended version of 2004b].

2005-06 Bahnar, Bajau, Bandanesen, Banjar, Buli, Berawan, Bisava, Brao, Bru, Bunak, Ca Tu, Chaobon, Cham, Chor, Dusun, Ema, Ende, Galela, Gie-Trieng, Giay, Giman, Hmong, Hre, Htin, Iban, Jarai, Karen, Kayan, Kedang, Kei, Kenyah, Khmu, Khmer, Kho Mu, Kohor, Kui, Lahu, Lamet, Lao, Lao Isan, Lawa, Lisu, Loven, Ma, Manggarai, Mentawei, Mien, Minangkabau, Muong, Murut, Mnong, Nage, Nung, Oy, Pako, Pak Thai, Paiwan, Penan, Raglai, Rhade, Sahu, San Chay, Sedang, Semai, Semang, Senoi, Sika, So, Stieng, Tanebar-Evav,Tanimbar,Ta Oi, Tengger, Temiar, Ternate, Tidore, Sedang, Tobaru, Tobelo, Xin Munh, Yao, Thailand, Laos. In: *Brockhaus-Enzyklopädie*, 21st ed. Bd. I-XXX, Leipzig: Brockhaus Verlag.

2006a Visibility and Objectification in Tobelo Ritual. In: Metje Postma and Peter Ian Crawford (eds.), *Reflecting Visual Ethnography: Using the Camera in Anthropological Research*. Leiden: CNWS Publications/Højbjerg: Intervention Press, pp. 78-102.

2006b (with Michael Prager) Südostasiatische Völker und Kulturen. In: *Brockhaus-Enzyklopädie*, 21st ed. Bd IX, Leipzig: Brockhaus Verlag.

2007a Spirit Representations in Southeast Asia: A Comparative View. In: Frédéric Laugrand and Jarich Oosten (eds.), Nature des esprits et esprits de la nature dans les cosmologies autochtones/Nature of Spirits and Spirits of Nature in Aboriginal Cosmologies. Quebec: Les Presses de l'Université Laval, pp. 99-129.

2007b Cultural Values, Human Rights and Peacekeeping Tasks: Some Anthropological Considerations. *Militaire Spectator* 176(3): 96-104.

2008a The Canoe Racing Ritual of Luang Prabang. *Social Analysis* 52(3): 1-32.

2008b (with Susanne Feske and Stephan Engelkamp) (eds.) Transformation and Conflict in Southeast Asia. Special Issue Asien: The German Journal of Contemporary Asia 109: 9-87.

2008c (with Susanne Feske and Stephan Engelkamp) Introduction. Transformation and Conflict in Southeast Asia: In: Jos Platenkamp, Susanne Feske and Stephan Engelkamp (eds.), *Transformation and Conflict in Southeast Asia*. Special Issue *Asien: The German Journal of Contemporary Asia* 109: 9-12.

2009 From high up North to Paradise and, by way of the South, back onto the Ice. In: Jan Jansen, Sabine Luning and Erik de Maaker (eds.), *Traditions on the Move: Essays in Honour of Jarich Oosten*. Amsterdam: Rozenberg Publishers, pp. 11-27.

2010a Political Change and Ritual Tenacity: The New Year's Ritual of Luang Prabang, Laos. In: André Iteanu (ed.), *La cohérence des sociétés. Mélanges en hommage à Daniel de Coppet*. Paris: Éditions de la Maison des sciences de l'homme, pp. 193-234.

2010b Becoming a Lao Person: Rituals of Birth and Socialisation in Luang Prabang, Laos. In: Peter Berger, Roland Hardenberg. Ellen Kattner and Michael Prager (eds.), *The Anthropology of Values: Essays in Honour of Georg Pfeffer*. New Delhi: Pearson, pp. 180-200.

2013 Sovereignty in the North Moluccas: Historical Transformations. *History and Anthropology* 24(2): 206-232.

2014 *Strangers, the State and the Self in Germany: A Comparative View*. Vienna/Wien: Austrian Academy of Sciences/Österreichische Akademie der Wissenschaften (AAS Working Papers in Social Anthropology/ÖAW Arbeitspapiere zur Sozialanthropologie 27), pp. 1-9.

2015a (with Andreas Hartmann) *Für ein Museum für Ethnologie Westfalen. Gesellschaftspolitische Rahmenbedingungen und Projektdiagnose zur Neugestaltung des Völkerkundemuseums Forum der Völker*. Werl: Museum Forum der Völker.

2015b On the Confrontation between Perennial Models in 19th Century Halmahera (Indonesia). In: Andreas Hartmann and Oliwia Murawska (eds.), *Representing the Future: Zur kulturellen Logik der Zukunft*. Bielefeld: Transcript Verlag, pp. 73-110.

2016 Money Alive and Money Dead. In: Colin Haselgrove and Stefan Krmnicek (eds.), *The Archaeology of Money. Proceedings of the Workshop "Archaeology of Money," University of Tübingen, October 2013*. Leicester: School of Archaeology and Ancient History, University of Leicester (Leicester Archaeology Monographs 24), pp. 161-181.

f.c. Encounters with Christianity in the North Moluccas (16th-19th Centuries). In: Cécile Barraud, Annette Hornbacher and Michel Picard (eds.), Encountering "Religion" in Southeast Asia and Beyond.

Ethnographic Film

1985 D.J. Nijland, with J.D.M. Platenkamp, *Tobelo Marriage*. 16mm, colour, optic sound, 108 min. A film by Dirk Nijland, research by Jos Platenkamp, in collaboration with Lis Asinoellah and Frans Rijoly. Leiden: Institute of Social and Cultural Studies.

Ph.D. and Habilitation theses supervised by J.D.M. Platenkamp

Habilitation theses (Habilitationen)

1993 **Steinbrich, Sabine**: *Imagination und Realität in westafrikanischen Erzählungen.*

2005 **Hardenberg, Roland**: *Children of the Earth Goddess: Society, Marriage and Sacrifice in the H ighlands of Orissa (India).*

2012 **Prager, Michael**: *Dana Mbojo. Islam, sozio-politische Ordnung und kulturelle Revitalisierung im Sultanat von Bima (Sumbawa, Indonesien).*

Ph.D. theses (Doktorarbeiten)

1998 **Klocke-Daffa, Sabine**: *"Wenn Du hast, musst Du geben". Soziale Sicherung im Ritus und im Alltag bei den/Hai-/Kauan, einer Nama-Gruppe des südlichen Namibias.*

2003 **Postert, Christian**: *Rituelles Handeln Verstehen. Soziale Morphologie und ritueller Zyklus der Person bei den Hmong in Laos und Thailand.*

2004 **Sprenger, Guido**: *Die Männer, die den Geldbaum fällten. Konzepte von Austausch und Gesellschaft bei den Rmeet von Takheung/Laos.*

2005 **Al-Sekhaneh, Wassef**: *The Bedouin of Northern Jordan: Kinship, Cosmology, and Ritual Exchange.*

2007 **Hezel, Sabine**: *Cultural Identity Represented: Celticness in Ireland.*

2010 **Prager, Laila**: *Die "Gemeinschaft des Hauses". Religion, Heiratsstrategien und transnationale Identität türkischer Alawi/Nusairi-Migranten in Deutschland.*

Combey, John: *The Influence of Modernity and Modern Warfare on the Koh Mende Society of Sierra Leone.*

Manan, Abdul: *The Ritual Calendar of South Aceh, Indonesia.*

Rus, Andrej: *Features of Gift Exchange in Market Economy (co-supervision University of Ljubljana).*

2011 **Aksünger, Handan**: *Jenseits des Schweigegebots. Alevitische Migrantenselbstorganisationen und zivilgesellschaftliche Integration in Deutschland und den Niederlanden.*

2012 **Wallner, Ingo**: *Kultur- und Nationalbewusstsein. Zur Gestaltung einer nationalen Identität Madagaskars.*

2013 **Fuhrmann, Berit**: *Die Menschen des Mutterhauses. Soziale Beziehungen, rituelle Prozesse und lokales Christentum bei den Karow in Meghalaya.*

Sugandi, Yulia: *The Notion of Collective Dignity among Hubula in Palim Valley, Papua.*

2015 **Jäger, Kirsten**: *Das Sultanat Jailolo. Die Revitalisierung von "traditionellen" politischen Gemeinwesen in Indonesien.*

Jöckel, Jana: *Vom "Wunder von Bern" zum "Sommermärchen". Fußballweltmeisterschaften und die deutsche Nation.*

2016 **Wulfinghoff, Malte**: *Tour du Faso – Sport und nationale Identität in der multiethnischen Gesellschaft Burkina Fasos.*

Excluding Ph.D. theses of which J.D.M. Platenkamp was external examiner: EHESS Paris and the Universities of Oxford, Canberra, Amsterdam, Leiden, Heidelberg and Göteborg.

Social Morphology

Bedouin Kinship Relations in Jordan: An Examination of Beni Khalid Kinship Terminology and Family Organisation

WASSEF AL-SEKHANEH

The study of kinship/social morphology is one of the subjects that figures prominently in the writings and teaching of Jos Platenkamp. When writing my PhD thesis on the Beni Khalid Bedouins of Jordan which was based on data collected during fieldwork undertaken between 1998 and 2002 in the Hawsha village of the Mafraq (Kingdom of Jordan), he spent countless hours discussing the data with me. In order to express my gratitude to Jos Platenkamp for his supervision and his inspiring comments, I would like to dedicate the following article to him which is based on a chapter of my book "The Bedouin of Northern Jordan" (Al Sekhaneh 2005) focussing on the kinship terminology and core social relations of the Beni Khalid tribe.

Terminology and Structure

Any society classifies its members on the basis of various principles. Many persons to whom an individual may relate will be categorised, thus determining expected reciprocal behaviours forming discrete groups. The totality of such culturally determined relationships forms a society's social structure. Thus, an understanding of social structure necessitates the examination of the ways in which its members form themselves into social groups, of the relationships between the groups and between individual members, as well as of the types of ideal and actual behaviour pertaining to them (Graburn 1971). Kinship relationships are usually differentiated by the terms used by members of the system to address or to refer to one another. Such kinship terms are essential guides to social behaviour, placing people into categories and assigning them statuses and roles. According to Fortes (1969: 54), a kinship term "is a package of definitions, rules and directions for conduct [...] a store of information but also a tool of action". It forms an intricate part of what Murdock (1949) identified as the reciprocal behaviour characterising every relationship between kinspeople. Although many anthropologists have studied the East Arabian Bedouin (e.g. Ashkenazi 1938; Al Tahir 1955; Al-Jameel 1962; Al-Fawal 1974; Meeker 1979; Dickson 1983; Al Zahar 1986; Al-Afenish 1987; Ginat 1987; Lewis 1987; Abu Rabia 1993; Al-Krenawi and Graham 1996), little attention has been paid so far to Bedouin kinship terminology. Yet this forms the permanent core of the 'traditional' Bedouin kinship system. Also the Bedouin notions of kinship and affinity as such, as these are conceptualised in Bedouin representations, are important sources for understanding such principles of classification in Bedouin society.

Bedouin Concepts of Family and Kinship

Kinship in Hawsha is distinguished by four components: Patrilineal descent, patrilocal residence, patriarchal authority, and patrilineal kin group endogamy. Hawsha village, like any other Bedouin village is composed of distinct patrilineal and patrilocal households. Although several households often occupy one block of buildings, the physical and social boundaries between them are never vague. Village and household are the main social units. Both units are based upon the principle of patrilineal kinship (*qarabah* قرابة) which is applied to various kinds of asserted ties and manifest in residential propinquity. The whole village of Hawsha considers itself as a group descending from one common ancestor. They share the firm belief that this ancestor was Sahabi Khalid ben Al-Waleed (صحابي 'Friend of Prophet Mohammed'). There is a close bond and a feeling of true kinship between this tribe and Khalid ben Al-Waleed, whom tribal members refer to as *jiddana* (جدنا 'our grandfather').

Like all other villages belonging to the Beni Khalid tribe, Hawsha village is divided into a number of quarters or wards (*harah* حارة). These have no clear boundaries, but they tend to unite persons belonging to the same lineage or section of a lineage (see figure 1).

Because close neighbours often intermarry and close agnates (and sometimes other close kin) live near each other, the kinship structure of such quarters is reinforced both by patrilineal kinship and by affinity. Each quarter consists of a number of households. The members of each household are descended patrilineally from a common ancestor, located generally in the third or fourth ascending generation. Households normally form local clusters. The rights and duties of members are not precisely defined, and the degree to which members are committed varies greatly between individuals.

Islamic law defines three different kinds of kinship relations: relationships "by blood" (*damm* نسب), of affinity (*mosahara* مصاهرة), and of milk (*rida'aa* رضاعة). The latter term *rida'aa* denotes the relationship between a child and a woman, who is not the birth mother but who nursed the child. These Islamic concepts are also applied in Hawsha and among the Bedouin in North Jordan. In general, Islamic law thus provides the basic categories of kinship and affinity with which to organise the network of Bedouin relationships. Adherence to Islam therefore defines a person not only in relation to Allah but also as a member of a society of believers. These two relationships branch out from the single principle of faith.

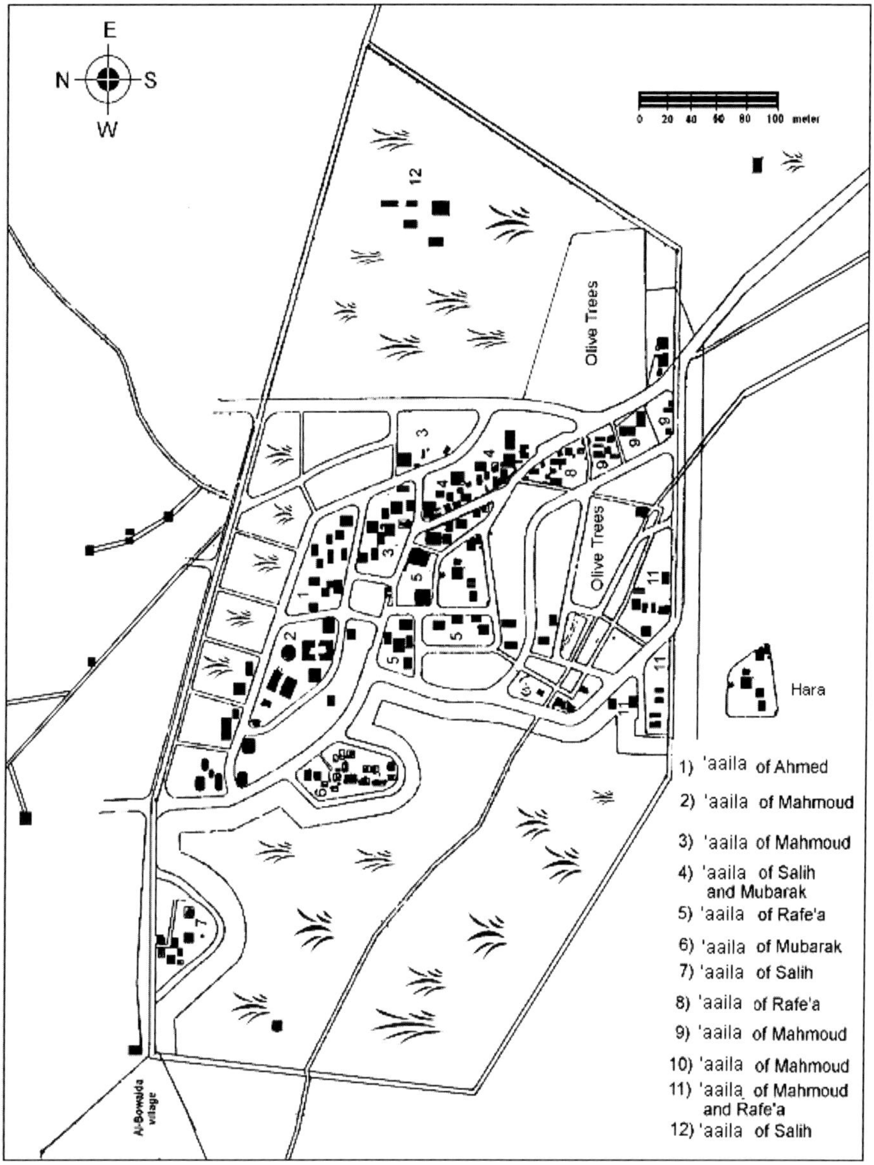

Fig. 1: Distribution of 'aa'ila in hara quarters in Hawsha

Bedouin society recognises several additional types of kinship, based on principles of increasing segmentation and operation of segmentarity. At the highest level people belong to a *qabila*, literally meaning 'acceptance' (from the verb *qabil* (قبل) 'to accept'). It refers to a confederation of lower level patrilineal units. Such units are called *ashira,* literally meaning 'uterus'[1]; it signifies the female abdomen supporting the *qabila* as upper part of the body. The *ashira* ideally consists of a patrilineal unit of maximally five generations in depth (see figure 2).

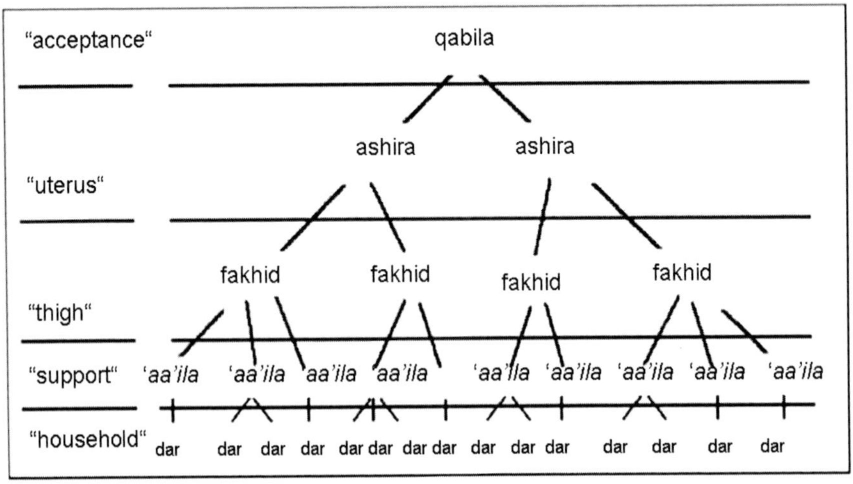

Fig. 2: Level of segmentation of the patrilineal structure

Theoretically, members of an *ashira* are co-residents of a single village but not all members of a same *ashira* live in the same village; they also live in neighbouring ones. In some areas, however, land fragmentation and out-migration have led to the dispersal of many members of the *ashira* to different villages, regions and cities. However, they continue to maintain ties with their natal village and enjoy the legal right of first refusal in any *ashira* land sale.

At a still lower level, the *ashira* divides into *fakhid,* (فخذ) literaly means 'thigh', and a patrilineal unit usually three to four generations in depth. Members of *fakhid* celebrate the major life events together. They are expected to contribute food and to assist guests in the ceremonies accompanying birth, marriage, death, and major religious holidays. The *fakhid* has traditionally served as a combined mutual aid society and welfare agency, arranging loans to members, assisting in finding employment, and contributing to the dowries of poorer families.

1 Denoting not only the human but also the animal uterus, the Bedouin say: the cow is *mo-ashirah* (pregnant معشرة).

Finally, a *fakhid* is composed of various *'aa'ila* (عائلة), literally meaning, 'support [rendered to a dependent person]'. It denotes a group of people who share a common agnatic descent and a common patronym, which includes father, mother, children, sons' wives and their children, unwed father's sisters, and, on occasion, unwed father's brothers. All the male members of the *'aa'ila*, both married and unmarried are expected to support the *'aa'ila* as a whole. The father or grandfather stipulates the contribution from each of them, depending on the size of the individual's income. The *'aa'ila* lives in the same household (*dar* دار). When a son marries, if his father is deceased, he will separate from the household and establish an independent one soon after having children himself. For in addition to the quarrels which take place between the wives of different brothers and between them and their mother-in-law, quarrels also tend to arise between the children of various brothers. Whether or not an extended household endures depends on the preferences of the individuals involved. Quarrels and divisiveness, particularly among a mother- and daughters-in-law, can lead to the premature splitting up of a joint household. Hence an *'aa'ila* may consist of one or more *dar* (see Evans-Pritchard 1940; Smith 1954; Sahlins 1956; Durkheim 1964: 177; Abu-Lughod 1989: 274; Sigrist 1994; Tilly 1998).

Bedouin Kin Terms

Bedouin kin terms employed in Hawsha conform to a Sudanese pattern of kinship classification with minor variations in terminology though not in structure. It is basically the same as that used in the Arab world generally, but the pronunciation of certain terms is unique by reason of the dialect peculiar to this region.

A person addresses his or her father as *yoba* (Arab. *abb* أب), the father reciprocally also addresses both his sons and his daughters as such. The term *ammi* (Arab. *amm* عم) is used to address one's father's brothers, father's father's brothers, and brother's children. *Ammi* can also be used to address any male collateral patrilineal relative in the grandparental and parental generation or any other elderly man in the village. The term *khali* (Arab. *khal* خال) is used to address one's mother's brother. It is a self-reciprocal term that also denotes sister's daughter's child, and daughter's daughter's child. It is also used to address any matrilateral male relative in the paternal generation. The relation with the mother's brothers *fakhid* is designated as a relationship of *mekhwal* (مخول), 'mother's brothership'. The term *ammti* (Arab. *amma* عمة) is used to address father's father's sisters, father's sisters, and all female collateral patrilineal relatives in the grandparental and parental generation.

The term *khalti* (Arab. *khala* خالة) is used to address mother's sister. The term *jeddi* is used to one address all one's male patrilineal relatives of both mother and father in the grandparental and higher ascendant generations. One employs the term *jeddati* (Arab. *Jedda* جدة) to address both father's mother and mother's mother, and the wives of all other men classified as *jeddi* (Arab. *Jedd* جد). The term of address for brother is *khai* (Arab. *akh* اخ), that for sister *khaia* (Arab. *okht* اخت). The terms are not employed for any other lineal or collateral relatives, that is to say, that the terms for brother/sister are singular and not extended to the cousin relationship. Strictly speaking, there are no terms of address for cross cousins and parallel cousins. They may be addressed by the personal names; if that is not known one addresses them as "son"/"daughter" (Arab. ben بن/bent بنت) of the parent's sibling in question, thus:

ben/bent khali : mother's brother's son/daughter
ben/bent khalti : mother's sister's son/daughter
ben/bent ammi : father's brother's son/daughter
ben/bent ammti : father's sister's son/daughter

Mother is addressed as *yomma* (*omm* ام); like the term for father it denotes a singular relationship only (see figure 3).

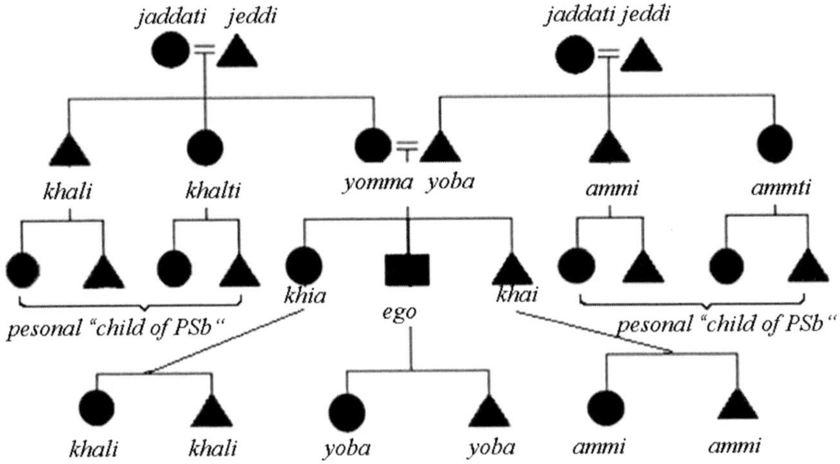

Fig. 3: Consanguinal kin terms

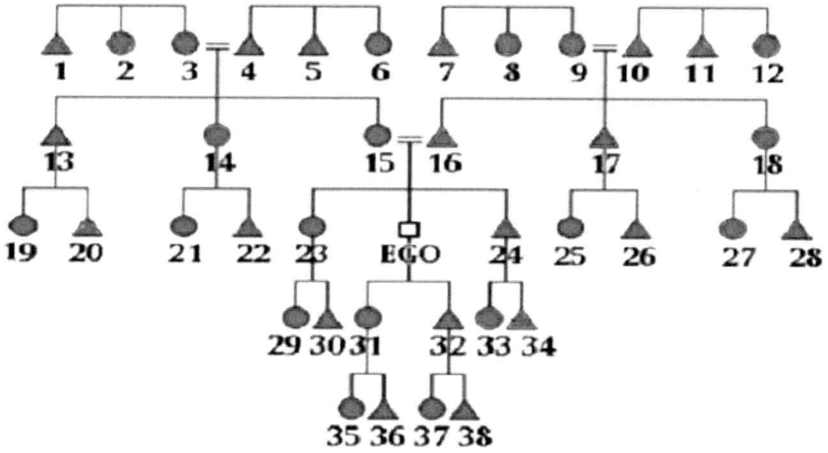

Fig. 4: Reciprocal relations among consanguinal kin in Hawsha village
(see also table 4.1)

Ego	1	↔	*khali*
Ego	2	↔	*khalti*
Ego	3	↔	*jeddati*
Ego	4	↔	*jeddi*
Ego	5	↔	*khali*
Ego	6	↔	*khalti*
Ego	7	↔	*khali*
Ego	8	↔	*khalti*
Ego	9	↔	*jeddati*
Ego	10	↔	*jeddi*
Ego	11	↔	*ammi*
Ego	12	↔	*ammti*
Ego	13	↔	*khali*
Ego	14	↔	*khalti*
Ego	15	↔	*yomma*
Ego	16	↔	*yoba*
Ego	23	↔	*khaia*
Ego	22	↔	*iben khalti*
Ego	21	→	*bent khalti*
Ego	21	←	*ben khalti*
Ego	19	→	*bent khali*

Ego	19	←	*ben ammti*
Ego	20	→	*ben khali*
Ego	20	←	*ben ammti*
Ego	24	↔	*khaii*
Ego	25	→	*bent ammi*
Ego	25	←	*ben ammi*
Ego	26	↔	*ben ammi*
Ego	27	→	*bent ammti*
Ego	27	←	*ben khali*
Ego	28	→	*ben ammti*
Ego	28	←	*ben khali*
Ego	33	↔	*ammi*
Ego	34	↔	*ammi*
Ego	29	↔	*khali*
Ego	30	↔	*khali*
Ego	31	↔	*yoba*
Ego	32	↔	*yoba*
Ego	35	↔	*jeddi*
Ego	36	↔	*jeddi*
Ego	37	↔	*jeddi*
Ego	38	↔	*jeddi*

Table 4.1: Bedouin kinship terms

	Generation +2		
Kin type	Term of address	Term of reference	Arabic
FF	*jeddi*[1]	*jadd*	جدي
MF	*jeddati*	*jadah*	جدتي
FM	*jeddi*	*jadd*	جدي
MM	*jeddati*	*jadah*	جدتي
	Generation +1		
F	*yoba*	*abb*	أبي
M	*yomma*	*omm*	أمي
FB	*ammi*	*amm*	عمي
MB	*khali*	*khal*	خالي
FZ	*ammti*	*ammah*	عمتي
MZ	*khalti*	*khalah*	خالتي
FBW	*mart ammi*	*zawjat amm*	زوجة عمي
FZH	*jouz ammti*	*zawj ammah*	زوج عمتي
MBW	*mart khali*	*zawjat khal*	زوجة خالي
MZH	*jouz khalti*	*zouj khalah*	زوج خالتي
	Generation 0		
B	*khaii*	*akh*	أخي
Z	*khia*	*okht*	أختي
FBS	*ben ammi*	*ben amm*	ابن عمي
FBD	*bent ammi*	*bent amm*	بنت عمي
FZS	*ben ammti*	*ben ammah*	ابن عمتي
FSD	*bent ammti*	*bent ammah*	بنت عمتي
MBS	*ben khali*	*ben khal*	ابن خالي
MBD	*bent khali*	*bent khal*	بنت خالي
MZS	*ben khalti*	*ben khalah*	بن خالتي
MZD	*bent khalti*	*bent khalah*	بنت خالتي
	Generation -1		
S	*yoba*	*ben*	ابني
D	*yoba*	*bent*	ابنتي
BS	*ammi*	*ben akh*	ابن اخي
BD	*ammi*	*bent akh*	بنت اخي
ZS	*khali*	*ben okht*	ابن اختي
ZD	*khali*	*bent okht*	بنت اختي
	Generation -2		
SS	*jeddi*	*eben eben/ hafid*	جدي
DS	*jeddi*	*eben bent*	جدي
DD	*jeddi*	*Hafidah/ bent bent*	جدي
SD	*jeddi*	*bent eben*	جدي

Table 4.2: Bedouin kinship terms

	Kin type	Term of reference	Term of address	Arabic
	DH,DHF,D	*nasib*	*ammi*	نسيبي
	HB, SWF,			
	SWB	*nasibah*	*ammti*	نسيبتي
	DHM,			
	DHZ,			
	SWM, SWZ	*Kana*	*ammti*	
	SW			كنتي
Female Ego representing	H	*zawj*	personal	زوجي
	HF	*amm*	*ammi*	عمي
	HM	*amma*	*ammti*	عمتي
	HB y, HB	*self*	His name	سلفي
	HB e, HB	*amm*	*ammi*	عمي
	HZ	*amma*	*ammti*	عمتي
	HBW	*selfa*	*selfti*	سلفتي
Male Ego representing	W	*zawja*	personal *mrati*	زوجتي
	WF	*amm*	*ammi*	عمي
	WB	*amm*	*ammi*	عمي
	WBS	*ibin nasib*	personal	نسيبي
	WBSS	*ibin nasib*	personal	ابن نسيبي
	WFB	*nasib*	*amm*	ابن نسيبي
	WM	*amma*	*ammti*	عمتي
	WZ	*bent-amah*	personal	عمتي
	WZH	*adil*	personal	عديلي
	ZH	*nasib*	personal	نسيبي
	ZHF			
	ZHB			
	ZHBS			
	BWF			
	BWB			
	BWBS			
	ZHM	*nasiba*	personal	نسيبتي
	ZHZ			
	BWM			
	BWZ			زوجة اخي
	BW	*zawjat akh*	personal	
	MZH	*zawj khala*	personal	زوج خالتي
	MBW	*zawjat khal*	personal	زوجة خالي
	FZH	*zawj amma*	personal	زوج عمتي
	FBW	*zawjat amm*	*mart mmi*	زوجة عمي
	ZSW	*zawjat eben okht*	personal	زوجة ابن اختي
	ZDH	*zawj bent okht*	personal	زوج بنت اخي
	BSW	*zawjat eben akh*	personal	زوجة ابن اخي
	BDH	*zawj bent akh*	personal	زوج بنت اخي

Table 4.3: Affinal kinterms

The system of kin terminology thus differentiates between the following types of reciprocal relationships (see Platenkamp 1998):

abb-eben-ebena (Arab. *obwwah* ابوه)
relationship (1)

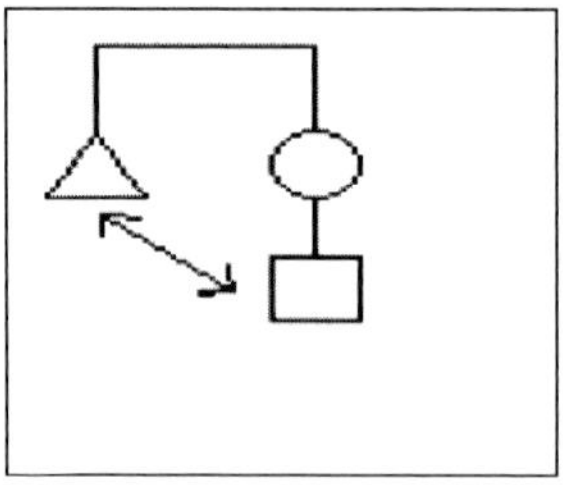

*omm-eben-ebna(*Arab *omwmah* اموهة*)* relationship (2)

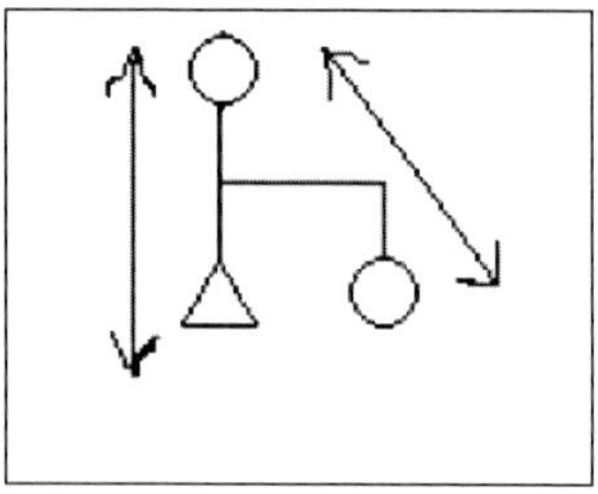

*akh-akh (*Arab. *okhwah* اخوه*)*
relationship (3)

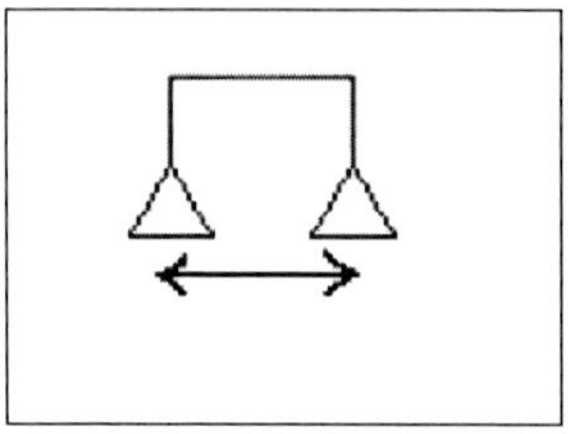

*okt-okht (*Arab.. *okhwah* اخوه*)*
relationship (4)

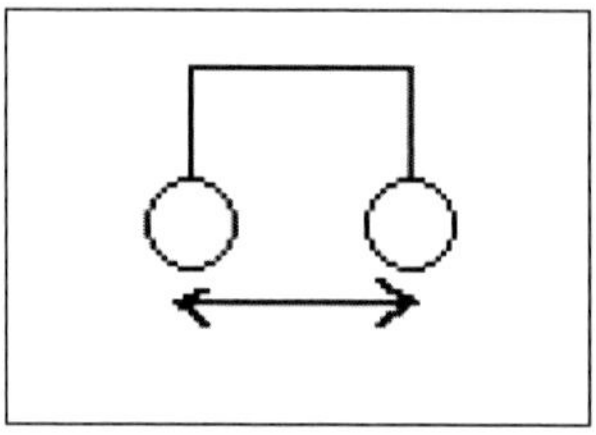

*akh-okht (*Arab. *okhwah* اخوه*)*
relationship (5)

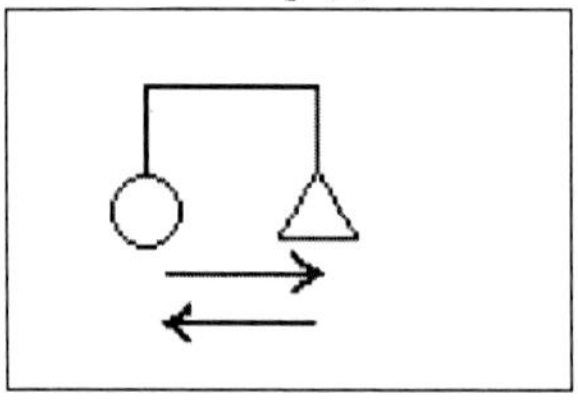

omomah (عمومه*)* relationship (6)

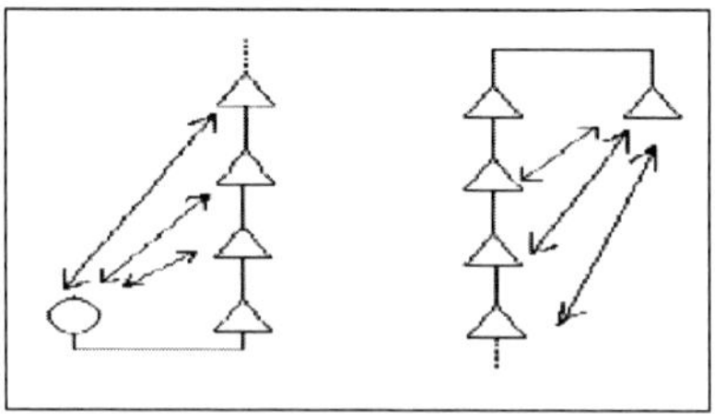

Fig. 5a: Types of reciprocal relationships

abna omoma (ابناء عمومة) relationship

(7)

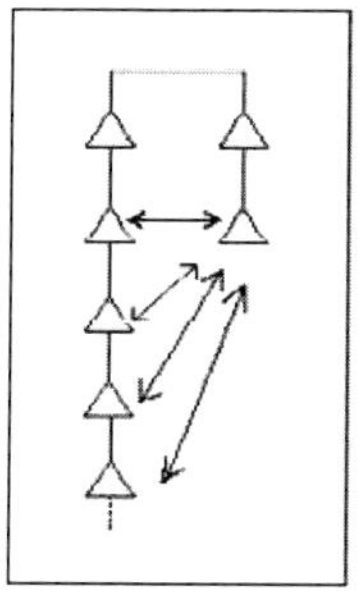

ammti relationship (Arab. *omoma* عمومة)

(8)

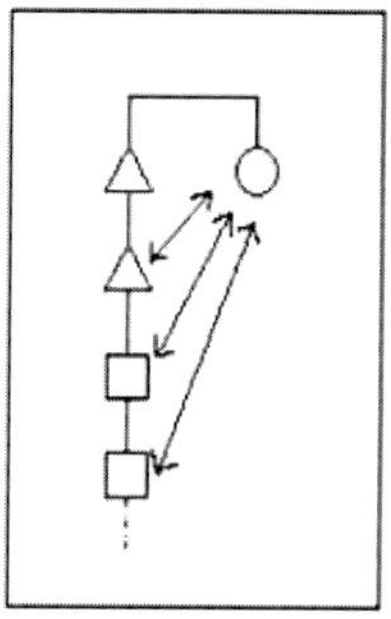

khali relationship (*mekhwal* مخول)

(9)

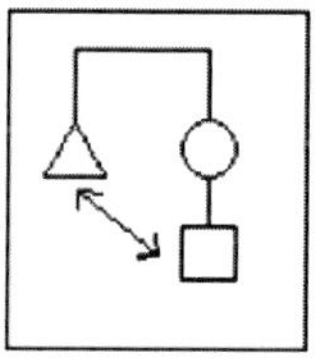

khalti relationship (*mekhwal* مخول)

(10)

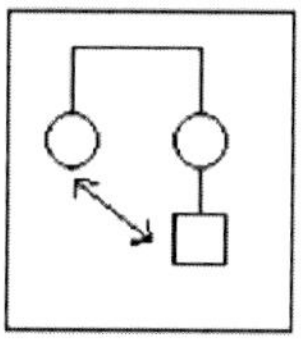

nasab relationship (نسب)

(11)

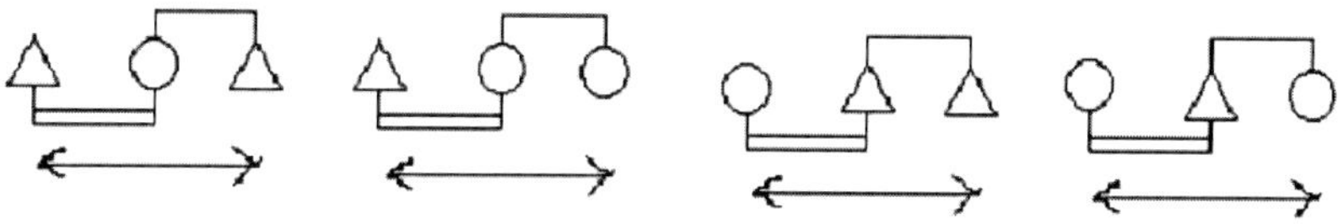

Fig. 5b: Types of reciprocal relationships

Nuclear Family Relationships

In Bedouin society the strongest emotional relationship in the family is not that between husband and wife but between parents and their children. The warmth, trust and affection frequently uniting parents and offspring are reciprocated by the respect shown by the children to both parents. The child depends on the parents, particularly on the mother. Conversely children provide parents with a form of social insurance in illness and old age. The relations, however, between mother and children and between father and children are differently valued. If a son is married, a mother is expected to angrily rebuke her son if he neglects his familial duties and responsibility as a father, and he should never answer her arrogantly. The daughters orient themselves to their mother rather than to their father; from the mother the daughter learns all feminine skills and is believed to take her character from her.

The children are brought up to regard the father as the legal and economic authority. Besides working for him, consulting him in many matters, a child acknowledges his father as the guide in daily affairs. The situation is more pronounced if the child is male. With this obligation in view, parents expect from their children an unquestionable loyalty and respect, which contributes significantly to their dignity.

The first son is the next in status after the father in the family. Parents are called teknonymously as 'father *casu quo* mother of their first born son', for instance, *abu Aajwad* (father of *Ajwad*) or *umm Ajwad* (mother of *Ajwad*). This constitutes a term of address and is used by persons both within and without the own kin group. The eldest son is considered the living descendant of the original first male ancestor of the family, and has a special responsibility. He represents the father, manages his property, helps to control his compound, and inherits the father's status. The first son is the father's representative in his absence and his confidant. He also knows the family's creditors and debtors and the share of the family lands through the laws of Shari'a and of property. After the death of the father, the oldest son inherits the family's property, which he is expected to manage and to share in the best interests of his siblings. As the family guardian, he is expected to assume external responsibilities until his younger siblings stand on their own feet.

One of the characteristic features of Bedouin family life lies in the fact that the relationship between siblings is very strong and that siblings are brought up to live a shared life. While they are young, they sleep together in the same house and often in the same room. They play together, eat together from the same dish, and do everything in common. This intimacy and deep-seated familiarity continues until the age of eight, when boys and girls are tacitly separated from one another and are assigned masculine and feminine duties. The parents start to emphasise the gender and age of the children. The boys spend most of the time with their father in the *shiq* part of the tent/house; the girls stay in the *mahram* part.

The social behaviour among siblings is ordered by a seniority principle. An older sibling has particular responsibilities and is allowed to punish and reprimand younger siblings; the parents treat such actions with deference. As a first wife will help to preserve order and assist the husband, likewise a mother's household duties and responsibilities are alleviated by her eldest children. The sons and daughters herd and feed the domestic animals under the supervision of their elder siblings. They help in building fences and clearing the compound. The girls cook, and do the laundry. This division of duties applies where there are sons and daughters in the family. If there are no daughters, most of their responsibilities fall on the sons. The daughter's duty is said to be trained for motherhood, the bearing and rearing of children and learning to care for a husband. They cook the family food under the guidance of their mother and the elder daughter feeds and looks after her younger siblings. Sisters help to find marriageable girls for their brothers. Conversely the brothers' protection and readiness to defend their sisters forms an important deterrent against a husband's maltreating their wives.

The basic norm of behaviour of sisters towards one another is that of affection and mutual assistance. Sisters help one another in almost every daily task, especially in domestic work. They exchange beads, dresses and trinkets and also advise each other during the menstruation period.

The relationship between sons of the same father is cordial. The senior brother is expected to assist his other brothers in the same way as a father.

Rahim Relationships

Rahim term derived from the Qur'an (Qur'an 4:1), is used by the Bedouin of Hawsha. It is not a kin term proper, so that it cannot be used as a term of address or a term of reference. Literally, meaning "womb" or "vagina", it refers to certain type of relation connecting a male person to women belonging to five different matrilineages. These are (the numbers refer to the lineages in figure 6).

1. Ego's own matrilineage in both ascending and descending generations. The women related are M, MZ, MM, MMM, Z, ZD, ZDD etc.
2. The matrilineage of Ego's wife in descending generations only. The women related are D, DD, DDD, etc.
3. The matrilineage of Ego's father in ascending generations only. The women related are FZ, FM, FMM, etc.
4. The matrilineage of Ego's brother's wife in descending generations only. The women related are BD, BDD, BDDD, etc.
5. The matrilineage of Ego's son's wife in descending generations only. The women related are SD, SDD, SDDD, etc.

Rahim relations thus connect a man to the women of five matrilineages that are all related affinally to his own patrilineage. Once a man is related to a patrilineage by marriage the members of other matrilineage are excluded from marriage.

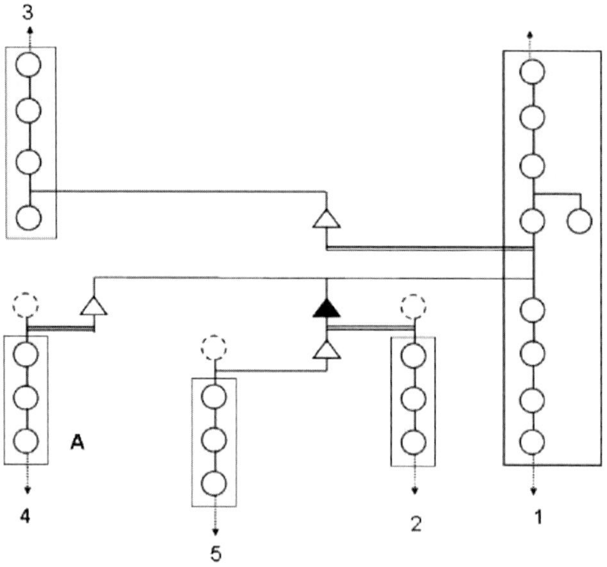

Fig. 6: *Rahim* relationships

Mahram Relationships

Whereas *mahram* relationships connect a man to five different matrilineages, *mahram* relations pertain between a woman and the male members of her own patrilineage, that of her mother and that of her husband; and that of her daughter's husband and that of her sister's husband. The *mahram* relationship that is sanctioned by the Qur'an thus pertains between a woman and:

1. All her male patrilineal relatives in any ascending and descending generation (see figure 7, no. 1).
2. Her mother's patrilineal forefathers in any ascendant generation (7, no. 2).
3. All male ascendant and descendent patrilineal relatives of her husband (7, no. 3).
4. All male patrilineal relatives of her sister's husband descending generation (7, no. 4).
5. All male patrilineal relatives of her daughter's husband in any descending generation (7, no. 5).

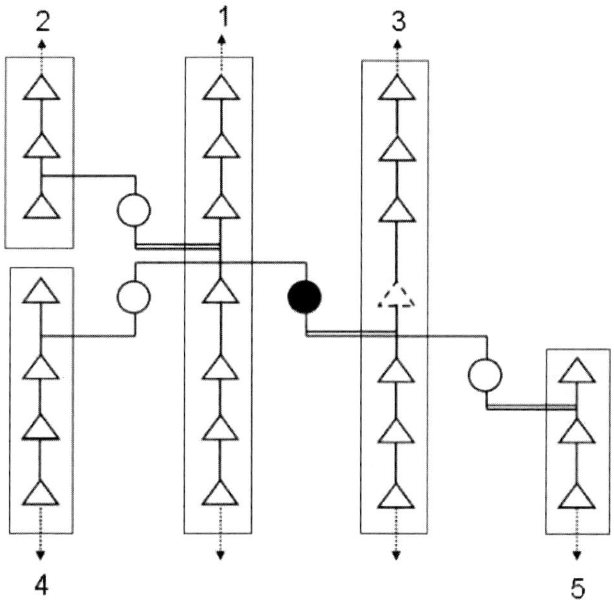

Fig. 7: *Mahram* relationships

A woman is thus related through *mahram* relationships with the male members of her own patrilineage and of those patrilineages that are related by marriage to her own patrilineage in the first ascending own generation. As indicated by the word *mahram* ('forbidden') the men thus related are entitled to enter the *mahram* part of the tent/house. A woman is allowed to show her adornment (*zinat* زينة) to all men thus related. She is, however, not entitled to marry any man to whom she relates as *mahram*. For a woman, therefore, all *mahram* relationships taken together constitute an exogamous category. Once a woman is related to her own patrilineage by marriage the members of other patrilineages are excluded from marriage.

The Relations of Milk Kinship

We have observed that Islam circumscribes three types of kinship relationship: by 'blood', 'marriage' and 'milk'. The last type is referred to as *al rida'aa* (الرضاعة). Children who have been suckled on more than three different occasions by the same woman have a 'milk relationship' with that woman. The Qur'an stipulates "Do not marry women whom your fathers married, except what is past: It was shameful and odious, an abominable custom indeed" (Qur'an, 4 23-24).

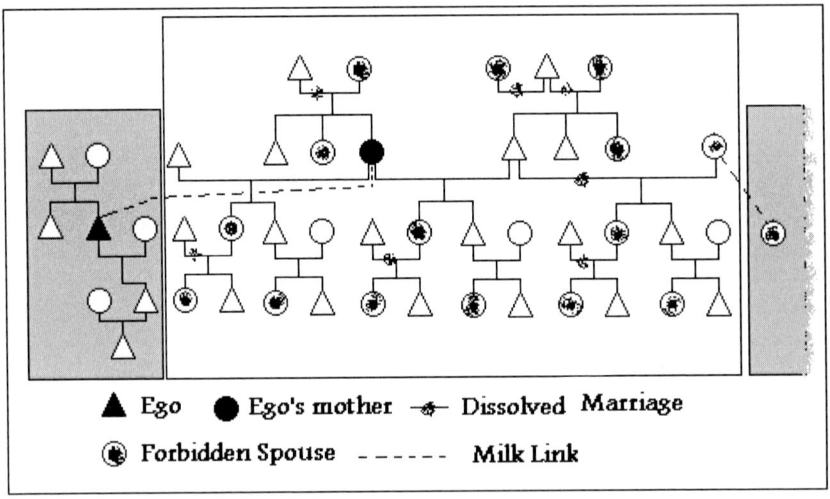

Fig. 8: The Relations of Milk Kinship
(based on Altorki 1980)

Such children are prohibited from marrying any of the following persons:

- The 'milk mother' (*omm bel rida'aa* أم بالرضاعة), including the grandmothers on both sides.
- The husband of the 'milk mother', the 'milk father', who is 'owner of the milk' (*abb al rida'aa* أب الرضاعة).
- The 'milk mother's' father (*abb al omm bel rida'aa* اب الام بالرضاعة) and mother.
- The children born to the 'milk mother' before and after suckling of the child (*akh al rida'aa* أخ الرضاعة).
- The 'milk mother's' descendants whether they are born to her children or her descendants have suckled the children (i.e. they are the descendants' 'milk siblings'(*aqarib al rida'aa* اقارب الرضاعة).
- The 'milk mother's' brothers and sisters, whether real or 'milk siblings' (*khal al rida'aa* خال الرضاعة) and (*khalat al rida'aa* خالة الرضاعة), the paternal and maternal uncles and aunts of the 'milk mother', whether she is the real, or half, or step sister of the father.
- The *consinguinal* children or 'milk children' of the husband of the 'milk mother', born to another wife (i.e. the 'milk mother's' step-children), and their descendants. It is said that "the milk belongs to the husband" (*al haleeb lel zawj* الحليب للزوج).
- The *consinguinal* and 'milk parents' of the 'milk father' and their ascendants (*jadd bel rida'aa* جد بالرضاعة) or (*jadda bel rida'aa* جدة بالرضاعة).
- The 'milk father's' sisters and brothers, whether *consinguinally* or 'milk' related.

- The paternal and maternal aunts and uncles of the 'milk father' and their ascendants, whether *consinguinally* or 'milk' related (see figure 8).

Concluding Remarks

In this article I have delineated the principles of social organisation among the Beni Khalid tribe of Jordan as emerging from a detailed analysis of the kinship system and its terminology by focussing both on the consanguinal and the affinal relationship terms. Furthermore, I have analysed Bedouin kinship concepts such as *rahim, mahram* and the notion of milk kinship in order to highlight the multilateral dimensions of this system that is conventionally described as exclusively patrilineal, although matrilateral relations are also at the core of Bedouin kinship.

References

ABU-LUGHOD, L. (1989) Zones of Theory in the Arab World. *Annual Review of Anthropology* 18: 267-306.

ABU RABIA, A. (1993) Educational Anthropology in the Bedouin Society. *Practicing Anthropology* 15: 21-23.

AL-AFENISH, A. (1987) Process of Changing and Continuity in Kinship System and Family Ideology in Bedouin Society. *Sociologia Rurales* 27: 323-340.

AL-FAWAL, S. (1974) *Bedouin Sociology.* Cairo: The Arab Nahda House.

AL-JAMEEL, M. (1962) *Bedouin and Bedouinism in Arab countries.* Cairo: Social Development Centre.

AL-KRENAWI, A. and GRAHAM, J. R. (1996) Social Work and Traditional Healing Rituals among the Bedouin of the Negev. *International Social Work* 39: 177-188.

AL SEKHANEH (2005) The Bedouin of Northern Jordan: Kinship, Cosmology and Ritual Exchange. Berlin: Wissenschaftlicher Verlag.

AL TAHIR (1955) *Bedouin and Clans in Arab Countries.* Cairo: The Arab League.

ALTORKI, S. (1980) Milk Kinship in Arab Society: An Unexplored Problem in the Ethnography of Marriage. *Ethnology* 19: 233-244

AL-ZAHER, M. (1986) *Arabian East from Bedouinism to Modern State.* Beirut: Arab Development Institute.

ASHKENAZI, T. (1938) *Tribus semi-nomades de la Palestine du Nord.* Paris: Librairie Orientaliste.

DICKSON, H.R. (1983) *The Arab of the Desert.* London and New York: Routledge

DURKHEIM, E. (1964) *The Division of Labor in Society.* New York: Free Press.

EVANS-PRITCHARD, E.E. (1940) *The Nuer.* Oxford: Oxford University Press.

FORTES, M. (1969) *Kinship and the Social Order.* London: Routledge and Kegan Paul.

GINAT, J. (1987) *Blood Disputes among Bedouin and Rural Arabs in Israel.* Pittsburg: University of Pittsburgh Press.

GRABURN, N. (1971) *Readings in Kinship and Social Structure.* New York: Harper Row.

LEWIS, N.N. (1987) *Nomads and Settlers in Syria and Jordan 1800-1980.* Cambridge: Cambridge University Press.

MEEKER, M. (1979) *Literature and Violence in North Arabia.* Cambridge: Cambridge University Press.

MURDOCK, G.P (1949) *Social Structure.* New York: The Macmillan Co.

PLATENKAMP, J.D.M. (1998) Temporality and Male-Female Distinctions in the Tobelo Vocabulary of Relationships. *Social Anthropology* 63: 291-307.

SAHLINS, M. (1956) The Segmentary Lineage: An Organisation of Predatory Expansion. *American Anthropologist* 63: 332-345.

SIGRIST, C. (1994) *Regulierte Anarchie: Untersuchungen zum Fehlen und zur Entstehung politischer Herrschaft in segmentären Gesellschaften Afrikas*. Hamburg: Europäische Verlagsanstalt.

SMITH, M. (1954) On Segementary Lineage Systems. *Journal of the Royal Anthropological Institute* 86: 39-80.

TILLY. C. (1999) *Durable Inequality.* Berkely: University of California Press.

"Those who spread from one father": The Cultural Elaboration of Segmentation and Inclusion in Kïrgïzstan[1]

ROLAND HARDENBERG

Introduction

For many years, Jos Platenkamp personally inspired me in my search for cultural understandings of social categories and forms of action. His own work on relationship terminologies (e.g. Platenkamp 2000) and rituals (e.g. Platenkamp 1992, 2010) provided me with tools of interpretation and new perspectives on my ethnographic material from South and Central Asia. He strongly encouraged me to continue my studies on concepts of tribe, clan and lineage despite a common trend among anthropologists to neglect these topics. In Central Asian Studies, for example, not anthropologists, but political scientists such as Collins (2002, 2004, 2006) or Schatz (2000, 2004) attribute an important role to "clans" or "lineages" as informal institutions in political life. According to Collins, clans are "informal identity networks based on kin or fictive kin bonds" which serve key functions in Central Asian politics (2004: 224). Schatz, on the other hand, distinguishes between the ethnic group, in his case the Kazakh, and various sub-ethnic identities, which he collectively refers to as "lineage identities" (2000: 490). In his view, the sub-ethnic identities survived in Soviet times as networks for acquiring scarce goods (2000: 490). With the collapse of the Soviet power, these identities again entered, according to Schatz, the public sphere due to the process of nation building.

In contrast to these authors, Cambridge anthropologists such as Gullette (2010) and Sneath (2007) dispute that there are (or were) "tribes", "clans" or "segmentary systems" in Central (or "Inner") Asia[2] and claim that the popular image of kinship society is purely a product of western evolutionary theory (Gullette 2010: 54-55; Sneath 2007: 40-43). Gullette denies that the Russian terms (*plemya* and *rod*) or the Kïrgïz terms (*uruu* and *uruk*) for "tribe" and "clan" refer to corporate groups (2010: 5) or that they are a "resource" for receiving help and assistance in present-day Kïrgïzstan (2010: 117). Instead, he argues on the basis of interviews and various written sources that local

1 This paper is based on research generously sponsored by the German Volkswagen Foundation from 2006-2009.

2 Gullette's work is mostly about Kïrgïzstan, while Sneath more generally applies his arguments to "nomadic Inner Asia". He mainly refers to the Mongols but also provides the reader with a "reexamination" of Kazakh and Kïrgïz society (2007: 71-91).

terms refer to "genealogically imagined communities" (2010: 4), which become visible only during life-cycle rituals, especially funerals (2010: 5, 86). The establishment of relatedness is, according to Gullette, based on personal and collective forms of memory, e.g. the remembrance of the seven forefathers (*jeti ata*) or genealogical narratives (*sanjïra*). He uses Andrew Shryock's term "genealogical imagination" (2010: 81) for this process, which socialises persons, creates notions of the self and establishes identity. In his view, political ideologies permeate such notions of the person and the self and thus genealogical imagination plays a key role in political attempts at state- and nation-building (2010: 83).

Sneath, who deals with "more than two millennia of history" (2007: 37) of Inner Asian steppe societies, also argues that anthropological discourse about these societies often operates with a "tribal model"[3] derived from evolutionary theory according to which the nomadic, egalitarian non-state society is defined in opposition to the stratified and territorialised state society. On the basis of various sources from pre-soviet and soviet times he argues that the nomads of Inner Asia had hierarchical, aristocratic orders in which power was decentralised and distributed. In his view, the nomads had no conical clan system and the social and political divisions and sub-divisions of society were not expressions of a segmentary kinship logic but rather a matter of administration and political measures. In particular, he criticises Krader, Sahlins and Barfield for claiming that clans and kinship segmentation are important principles of organisation in non-state societies in Inner Asia and elsewhere (2007: 46-47, 55-56, 60-61,70-71). Using theories on house societies (Lévi-Strauss 1999 [1983] and Carsten and Hugh-Jones 1995), he develops an alternative vision of Inner Asian societies as divided into "sets of ruling lineages/houses with a large number of subjects" (2007: 62). In his view, there existed aristocratic houses who kept genealogical records and often engaged in pragmatic shifting alliances (2007: 61) and commoner families who often had only shallow descent lines (2007: 112). Drawing on Gramsci's and Foucault's approaches to "power", he identifies "state relations" in the nomadic societies of Inner Asia because they were organised on the basis of "aristocratic power and state like processes of administration" (2007: 1, 9-10).

In this paper I argue against this trend in anthropology – exemplified in the writings of Gullette and Sneath – which declares segmentary systems to be merely a product of evolutionary thinking and lacking any direct relevance for "actual" social organisation of past or present societies. Against this general

3 He mostly uses the singular form ("the kinship model at work", "the tribal model", "the segmentary kinship model", "the notion of tribal society," 2007: 14, 20, 42, 52) despite the fact that the words tribe, clan, segmentary lineage have been defined in various ways and that there have been competing models which initiated decades of discussion among anthropologists (see below). His main critique of "the tribal model" (2007: 40-41) is directed against Lewis Henry Morgan's "Ancient Society" from 1877.

rejection of kinship studies in favour of power and politics, I provide an account of segmentary systems on the basis of local ideas and practices I studied during my fieldwork in a north-western village in Kïrgïzstan from 2007 to 2008.

I first give a brief introduction into the history of the construction – and later deconstruction – of so-called segmentary lineage systems. The lesson I learn from this classical debate is that in each particular case people's concepts of segmentation and inclusion, their social ideas and values as well as their motivations to form actual groups in different contexts need to be elaborated. Keeping this in mind, I give an account of the peculiar form of segmentation in Kïrgïz genealogies. Next, I show how segmentation and its opposite, inclusion, relate to ideas about blessings and curses, sharing the same blood, and mothers as originators of segmentary lines. Finally, I examine the importance of notions of segmentation and inclusion in different aspects of life. In particular, I look at contexts of exchange and of organising communal life. My conclusion is that in Kïrgïzstan genealogical segmentation and inclusion play a major role in the way people think about and assess their social and spatial world as well as in the manner they organise certain communal activities.

From Group Theory to Social Values

The concept of social segmentation goes back to the writings of Herbert Spencer on societies as organs (1897 [1873/76]; see also Rumney 2009 [1934]: 72-83) as well as to Lewis Henry Morgan's ideas about the composition of ancient society (1951 [1877]: 91, 99-102; see also Sigrist 2004: 9). Yet, it was most clearly elaborated by Emil Durkheim in his work on the social division of labor (Durkheim 1992 [1893]). The idea of a segmentary lineage system took inspiration from Robertson Smith's description of Arab tribes (see Eickelman 1981: 100, Dresch 1988: 53) and later turned into an influential theory due to Evans-Pritchard's writings on the Nuer (1940) and the Bedouins of Cyrenaica (1949). The further development of the debate saw Gellner's model of segmentary society, Peters' processual approach, Salzman's (2008) functionalist theory of reserve, Caton's symbolic approach and his focus on honor, Dresch's structuralist account on segmentation as a principle of relations and Maynard's cognitive interpretation of segmentary theory (Tobolka 2003: 107-108). While Gellner and Salzman tried to explain the *reasons* why segmentary systems exist (e.g., as an adaptation to ecological conditions), Caton, Dresch and Maynard were rather interested in the *forms* (e.g., symbols, values, relations) they take. Kraus (1998) and Dresch (1986, 1988) initiated an important turn in the debate by distinguishing the different elements segmentary lineage theory is composed of. Both basically argue for a separation of lineage theory and segmentation: "Lineages are one thing, segmentation another" (Dresch 1988: 63). Writing about the Yemen, Dresch shows that a particular cultural value, honor, underlies the segmentary system. In Yemen, honor has to be defended against others and in this case all kinds of groups may form. Thus, Dresch comes to the conclusion that the "segmentary, tribal structure is of honor, not of cohesive groups" (1986: 315).

As a summary of this debate one can say that models of tribal structure consist of different elements such as specific social ideas and values, social categories, actual groups and the different social contexts in which they form. Theories differ in the way they define the relations between these elements (e.g., consistent or contradictory) and whether they generalise from social action (and compare it with ideology) or from social ideas and values (and compare it with social action). Following David Schneider's (1972, 1984) cultural turn in kinship studies one could say that each of these elements as well as the configuration of these elements is subject to cultural elaboration. In other words, social ideology, meanings of descent categories, inclusion into actual groups and the distinction of contexts must be carefully studied in each particular case. With this in mind, I now turn to Kïrgïzstan.

"Those who spread from one father" (bir atadan tarkagandar)

As in the bible, many Kïrgïz express genealogical relatedness with reference to a segmentary order, which rests on the idea that all people descent from one father. In its most inclusive form this idea implies that all Kïrgïz are relatives to each other. In order to know whether another person is a close or a distant relative one only has to identify the common segmentary line (*uruu*)[4] and the point where it segments (*bölünot*).[5] Svetlana Jacquesson calls this the "web of inclusions and divisions" (Jacquesson o. J.: 10) as people ask each other two questions when discussing genealogical relations: "The first one is 'Whom they are a part of?' (Kaysïga giret) and the second one is 'Do they divide? Do they split?' (Bölünöbü)." (Jacquesson o. J.: 8). When we look at the complexities of the segmentary system it becomes clear that answers to these questions about inclusion and segmentation require more than simply calculating generations.

In genealogical narratives (*sanjïra*), which are nowadays not only orally transmitted but also published in various books and narratives, the apical father of the Kïrgïz is often named Dolon Biy. One of the most popular and widely read publications on this topic written by Tüzgön Keneš Jusupov (1993: 82-83) explains that Dolon Biy was an *aksakal* (lit. "white beard", i.e.

4 I have introduced this concept of *segmentary line* in another paper (Hardenberg 2009). "Segmentary" refers to the division of the social category, "lines" to the inclusion into the social category. It can serve as a substitute for the problematic terms "descent line", "clan" or "tribe", which are too loaded with meaning due to their origin in debates about African and Middle Eastern societies. The Kïrgïz terms are *uruu*, *uruk* or *uruu-uruk*.

5 Another term for segmentation is *özünchö bolup* meaning "to become separate".

respected elder) who lived in the 12th century. He had two sons named Ak Uul and Kuu Uul,[6] who were twins. Dolon Biy had to run from Genghis Khan's hordes and fled to the western Tian Šan mountain range but died on the pass between Narïn and Kočkor, which – according to Jusupov – is named after him "Dolon Ašuu". From Ak Uul are said to stem two sons, Adigene and Tagay, and one daughter named Naalï Eje[7] who gave birth to a son called Moŋol. Kuu Uul is believed to have had seven sons[8] who became the originators of seven large segmentary lines (*čoŋ uruu*). All descendants of Ak Uul are considered to belong to the "right wing", those of Kuu Uul to the "left wing".[9] The segmentary lines deriving from these two sons are jointly called the "30 Sons" (*otuz uul*) in Kïrgïzstan. In addition to these descendants of Dolon Biy a third "branch" (*butak*) is also often regarded as part of "the Kïrgïz tribe". These are the so-called Ičkilikter or Kïpčaktar who divide into a number of different segmentary lines jointly called the "10 Sons" (*on uul*) (Jusupov 1993: 82).[10] According to Jusupov they joined the descendants of Dolon Biy when their own political entity (Khanat) collapsed.[11] These segmentary lines (*uruu*) are often identified with "groups" either termed "tribes" or "tribal confederations" whose members lived on different territories (Geiss 2003: 108).

Taking into account the above mentioned deconstruction of the group model of segmentary systems, it seems safer to assume that these were not "groups" but genealogical categories that allowed the contextual formation of groups. These "tribes" did in my view not own or permanently occupy "defined regions" (Geiss 2003: 108). Rather, people claimed, often in a contesting way, rights of use for certain pastures on the basis of genealogies. It is also doubtful whether we are dealing with fixed, uniform genealogies. Thus, due

---

6 Both names have basically the same meaning: *ak* means "white", *kuu* means "(white) swan" and *uul* means "son".

7 She is one the few women recognised in Kïrgïz "national" genealogies. Another is Bugu Ene (see below).

8 Jusupov mentions Kutču, Čoŋ Bagïš, Kïtay, Saruu, Munduz, Basïz, Noygut (1993: 82). In the historical museums in Bishkek and in Orgočor the following names were displayed in graphical representations of the "Kïrgïz segments" (*kïrgïzdardïn uruu-uruk*): Kušču (or Kutču), Saruu, Muŋuz, Jetigen, Kïtay, Basïz, Töböy and Čoŋ Bagïš.

9 Jusupov explains that the reason for this distinction goes back to the twin's mother, who carried Ak Uul on the right and Kuu Uul on the left hip (1993: 82). Geiss quotes a Chinese chronicle according to which a division into two wings, a northern and southern one, was formed after peace talks among the Kïrgïz in the mid of the 16th century (2003: 108).

10 The status of these segments as Kïrgïz is disputed, even today. Thus one of my informants belonged to these Ičkilikter and only after we got to know each other well, she confided to me that she is not a "real Kïrgïz" as her ancestors belong to the Ičkilikter.

11 A well-known myth derives the origins of the Kïrgïz from the "kïrk kïz" meaning "fourty girls". They were, according to the myth, the maidens of a Khan's daughter, who got pregnant and gave birth to one son each.

to strategic manipulations one finds considerable variations of these genealogies, for example concerning the names or the generational levels.[12]

Nevertheless, Kïrgïz genealogies share certain aspects. One of the most interesting features is that the originators of opposed (in the sense of differentiated) segments can – and often do – stem from different generations. For example, in north-eastern Kïrgïzstan people distinguish the Sayak, the Sarïbagïš and the Bugu.[13] These names are well known throughout Kïrgïzstan as they are used by people to identify themselves, especially when meeting relative strangers (see Hardenberg 2009). These three social categories originate from a man named Tagay Biy.[14] In relation to Tagay Biy, Sayak is a son, Sarïbagïš a grandson, and Bugu is the wife of Alseyit, a great-great-grandson.

In other words, segmentation does not follow a simple generational scheme as implied by anthropological diagrams of so-called segmentary lineage systems (see Sahlins 1968: 14, 25; Bastug, 1998: 100, 104). Thus, segmentary orders such as the "conical clan" are often represented as balanced systems in which segmentation occurs repeatedly from generation to generation and divides the children of siblings (often brothers) in a regular and predictable way. Obviously, such a model can hardly correspond to reality due to variations in demographic development. Yet, the problem with such models lies deeper. They imply that segmentation depends on the opposition of siblings or cousins.

According to my analysis, this kind of balanced agnatic opposition does not characterise Kïrgïz segmentary orders. Thus, in the above mentioned case of Sayak, Sarïbagïš and Bugu the segmentary system opposes the descendants of ancestors belonging to quite different generations. In strict generational terms the Sayak segment would be the most encompassing, as it derives from the son of Tagay Biy while the lines of Sarïbagïš and Bugu segment three to four generations later. This feature also occurs on the local level, i.e. in the context of village genealogies. During my fieldwork in Tosor I collected several examples, all of which follow the same logic: segmentation does not only oppose brothers, but also fathers and sons, grandfathers and grandchildren etc. Generational differences are often downplayed and it corresponds with this observation that Kïrgïz relationship terminologies offer many instances of

12 Naalï Eje, for example, is sometimes represented as the sister, sometimes as the mother of Muŋuš.

13 These named segmentary lines (*uruu*) figure prominently as "tribes" in historical writings because of their competition for influence under different leaders (*baatïr*, *biy* or *manap*) (Geiss 2003: 109-112). Semonov (1998 [1856-57]) for example provides interesting reports about the way members of these named segments fought each other for influence in different regions and established strategic alliances with or without the Russian troops.

14 According to my informants, Tagay Biy served the Khan of Kokand and at first lived in what is now Uzbekistan.

what in technical terms is called skewing, i.e. the bracketing of people of different generations under one relationship term.[15]

The peculiar and important feature of the Kïrgïz order of segmentation and inclusion then consists in the fact that these processes do not "naturally" result from generational sequences or the opposition of siblings/cousins. Instead, certain qualities of a person turn someone into a *tüpkü ata*, i.e. make him the "original father" of a new segmentary line and oppose his own descendants to the offspring of another agnate. How do people explain that some are remembered as *tüpkü ata* and others not?

Blessings and Curses

According to David Sneath (2007: 84), *uruu* and *uruk* were "politicial divisions" ruled by Kïrgïz nobility holding titles such as *biy* or *manap*. These were powerful people who have acquired wealth and fame and according to Sneath these ruling houses recorded their genealogies (2007: 89). The narratives I heard in the Ïssïk Köl region indeed tell a lot about leaders (*biy*)[16] as well as warriors and heroes who bravely fought – and sometimes died – for their people. Men are remembered because they were particularly wealthy and helped their community to develop. Yet, this focus on political power and material wealth captures only part of the picture. In interviews and in narratives one often hears other explanations why people became the founders of a segmentary line. The most widespread explanation is that a person had many offspring, that his/her line is strong (*tïŋ* or *mïktï*). This is often interpreted as the result of some kind of blessing (*bata*) or – in the adverse case – a curse (*kargïš*), which continues to affect members of a segmentary line.

A good example for this is the narrative about Bugu, the mythical ancestress of many people living in the eastern part of the lake Ïssïk Köl. According to local legends, the two sons of the famous ancestor Orozbaktï named Asan Mïrza and Kara Mïrza one day climbed the mountains in order to hunt mountain goats (*kiyik*). In the mountains they suddenly saw a deer and Kara Mïrza shot one of them. Another deer did not run away but smelled at the dead animal lying on the ground. Kara Mïrza was about to shoot it too. At that moment the deer turned into a girl and told the hunter that she is Kayberen, another name of Bugu meaning divine mother. After the hunter had touched her, she could not turn back into a deer and was married to a local hero named Alseyit, a grand-son of Orozbaktï.

15 For example *baike* (FB, FBSe, eB), *eje* (FZ, FZDe, eZ) or *taeke* (MB, MBS, MBSS).

16 An 86 year old woman once explained to me the meaning of the term *biy* by comparing it to a political committee during soviet times: "My father was rich, my grandfather was a *biy*. A *biy* is like *raykom* [*rayoni komitet* or district committee] or *rayspolkom* [*rayoni ispolnitelni komitet* or district executive committee] as it was in the USSR. He was a man who was ruling the people" [explanations in brackets, RH].

During my fieldwork I collected a number of versions of this myth which differ in detail yet have in common that they highlight the sacredness of this women who became the ancestress of a large and well know Kïrgïz segment. She is described as a "miraculous" (*kasiettüü*) and "holy" (*ïyïk*) being whose funeral reminds one of the Christian idea of resurrection: after her death she disappears from the yurt where her dead body was placed and the astonished people who watched her corpse through the night only find her footprints (some say of a deer) in the snow which lead into the mountains.

In most versions, Bugu is described as having the power to bless and curse. Thus, when she became pregnant, she requested the breast of a deer from the two sons of Orozbaktï, Asan Mïrza and Kara Mïrza, who only laughed at her. Only their brother, Arïk Mïrza, listened to her and when he shot a deer with a calf he brought her the prey's breast so that she could drink its milk. Afterwards, she cursed the two other brothers by saying that they may never have many offspring, while she blessed Arïk Mïrza who became the father of a fertile segmentary line.

These motives of blessing and curse and their effects, the decline or prosperity of certain segmentary lines, often appear in local narratives and people's memories about the past. Thus, an informant from Tosor once explained to me that the segmentary lines of Toktomat, Ïsa and Atake do not prosper due to a curse (*kargïš*). About 115 years ago, an old person spoke this curse before men from these three segmentary lines murdered him. According to local people, these three segments stopped prospering or in local terms: *tukumdarï öspöy kaldï* ("their offspring did not grow"). When presently members of these lines suffer from alcoholism or are so poor that they have to beg for money this is also attributed to this curse, which is said to spread and affect more and more people in the concerned segmentary lines. People say that if a segment suffers from a curse it is like pouring water in sand – everything vanishes. In contrast, other *uruu* are considered to prosper due to blessings (*bata*) received by their founding ancestor. Thus, it is said that the descendants of Jeldeŋ in the Bugu line prospered and became strong (*mïktï*) because their ancestor ate sacred bread (*kasietüü tokoč*), which was considered to show the signs of prosperity.

The future of a segmentary line is thus considered to depend on the morality of one's ancestors. When the ancestor was brave, blessed and morally good, his descendants could participate in his well-being or otherwise share his bad fate when he had been cursed due to some bad deed. On the other hand, the fact that somebody has many offspring is often interpreted as a sign of his/her blessing and a common reason for choosing someone as the *tüpkü ata*[17] of a line. Thus, one of the largest segments in Tosor is called Tattïkojo after a son of the legendary village founder Muratčan. Tattïkojo had three sons, the

17 Another term used is *tüpkülük* meaning "origin" or "base".

middle one named Alïke who again had two sons, Tokso and Suban. Of these two, only Tokso became the founder of a new line because people say that his offspring are particularly "strong" (*tïŋ*). All descendants of Suban, who would be a collateral line in strictly genealogical terms, nowadays recognise Tokso as their ancestor and consider themselves members of the Tokso *uruu*. This raises the question how relatedness in the sense of belonging to a segmentary line is locally constructed.

"Of One Blood" (kanï bir)

The Kïrgïz segmentary system is often described as being based on "patrilineal descent" (e.g. Yoshida 2005) or "patrilineal linkeages and patrilineal kin" (e.g., Jacquesson 2013: 98). In itself, such a designation does not mean much as the expression "patrilineal" may imply all sorts of ideas. In order to understand the Kïrgïz segmentary system it therefore seems necessary to elaborate some underlying concepts. Several concepts derive from the idea that – at least according to my informants – blood is transmitted through the male line. A father and his children are considered to be of "one blood", *kanï bir*, because the father gives his blood to his children. Among his children, the sons can continue their father's bloodline, not his daughters. As an informant put it: a mother does not share her blood with her children.

In addition to blood, semen (*uruk*) also plays a role in conceptualising this special relationship between men and their descendants. The offspring of both man and woman are called his or her *tukumdar*. In order to refer specifically to a man's offspring one uses the term *urpaktar*. The descendants of a man say that they belong to the same *uruu*, *uruk* or *uruu-uruk*.[18] The term *uruk* means "semen" and this suggests that from a local perspective a segmentary line consist of blood relatives who perpetuate themselves through semen.

How do collateral relatives fit into this concept of "blood relationships"? Cousins are divided into three categories. The first are patrilateral parallel cousins, who are considered to be of "one blood" (*kanï bir*). They refer to each other as elder/younger brother (*baike/ini*) or elder/younger sister (*eje/siŋdi/karïndaš*)[19] depending on the absolute age of alter. They may under no circumstances marry as they are like siblings to each other. The second category is the matrilateral parallel cousins, i.e. children of two sisters, who are not considered to be blood relatives. They refer to each other reciprocally

18 Which of these words is actually used seems to depend on region and context. Thus, in Tosor people said that in their area they mostly use the term *uruu*, not *uruk*. Yet, I also came across the combination *uruu-uruk*, for example in the exhibition of the neighboring museum in Orgočor. Louise Bechtold, who did intensive ethnographic research with women in the Jalalabad region told me that women feel ashamed using the term *uruk* as it refers to the male semen.

19 A woman refers to her younger sister as *siŋdi*, a man as *karïndaš*.

as *bölö*[20] and were, at least in the past[21], not categorically excluded as marriage partners. The third category is the cross cousins, i.e. the children of brothers in relation to the children of sisters.[22] They are also not regarded as blood relatives either, but many informants considered a marriage between these cousins problematic. Yet, according to my inquiries such marriages did occur in the past and the affinal relationship created by this type of marriage was called *kayčïlaš kuda* ("affines like a scissor").[23]

Thus, of the three categories of cousins (and their offspring) only the patrilateral cousins share the same blood as they are related through brothers who had the same father. They are sometimes collectively referred to as *bir atadïn baldarï* ("children of one father") or *bir atadan tarkaganda* ("spread from one father").[24] This may also include several generations of patrilateral parallel relatives and was according to my informants, previously used to refer to those small groups who jointly lived in yurt camps on the summer pastures.

A number of other Kïrgïz kin categories are based on this idea of a special blood relationship between fathers and their offspring. Thus men speak of their *ata jurtu* meaning the settlement where the father and his close relatives reside. A woman has no *ata jurtu* and instead refers to the place of her father as her *törkün*. Another important and even more encompassing concept related to fathers is that of the "seven fathers" (*jeti ata*). Some people say that every Kïrgïz should be able to mention the names of his seven lineal forefathers. This obligation appears to be related to the ego-centric marriage regulation that one should not marry any person with whom one is related "by blood" through any of these seven ancestors.

Another important concept related to fathers is that people have the same *ata tegi* or "father's root" meaning that people share a more or less distant forefather. Yet, this dominant doctrine of a line perpetuated through men explains only a part of the Kïrgïz segmentary system.

Mothers as Originators

In ideology and practice, the segmentary calculus also operates with reference to women. I have already referred above to those prominent women, Naalï Eje and Bugu Ene, who have been included in the Kïrgïz "national" genealogy. In

20 This self-reciprocal term refers to both sexes independently of age and generation.

21 Nowadays, many Kïrgïz refer to scientific studies, which have shown that children born to close relatives are more prone to have certain diseases. Therefore such marriages have come in the focus of critique.

22 The offspring of a sister are referred to as *jeen* (e.g. FFZCh, FZCh, ZCh) by the progeny of a brother. The reciprocal terms used by the sister's children are either *taeke* (in case of a man, e.g. MMB, MB, MBS) or *taeje* (in case of a woman, e.g. MMZ, MZ, MBD).

23 I could not get any definite explanation for this name.

24 Jacquesson translates this expression as "offspring of a single ancestor" or "extended families" and writes that it designates the "real kinship groups on the local level" (o. J.: 8).

the genealogical narratives of regional and local importance women may also play a significant role, especially as co-wives who become the mothers of sons. For example, in the history of the Bugu line told in Tosor almost every important man had more than one wife. These co-wives had their own male offspring whose descendants nowadays distinguish each other as members of different segmentary lines. The alleged status of these mothers is used to negotiate the hierarchy between these segmentary lines. For example, in Tosor a woman once explained to me:

> "My segmentary line is Boor from Toktobek Ene. I am from Alseyit.[25] My father [i.e. Alseyit] went to the Kalmaks, killed the Kalmak Khan and took his daughter, Čeejimjan. She was captured. They gave here to Bugu Ene as a slave. This slave poured water on her head [i.e, washed her hair, RH], she considered Bugu Ene to be sacred, and suddenly the daughter of the Khan became pregnant. So what? This is shame. It is shame to be pregnant without husband. She served Bugu Ene. Čeejimjan, the daughter of the Khan said: 'Bugu Ene gave me her blessing (*batacin berdi*). That is why I gave birth.' Because it was shame, she gave the name Tuuma Kaška to her son. It was a son born with shame, without marriage. That is why his name is Tuuma Kaška. The people of Tosor are from this Kalmak daughter. That is why they have a difficult character." [Translation from Kïrgïz, RH]

From this and other examples it becomes clear that people consider certain moral qualities – good as well as bad – to remain within a segmentary line. As this quote also shows, such qualities can be perpetuated from mother to son. Furthermore, in local narratives, the co-wives of famous heroes are often explicitly mentioned in order to explain the present distribution of their descendants.

These examples clearly show that women as mothers can become the originators of segmentary lines, that they are able to transmit their own qualities to their offspring and that they can actively shape the destiny of their descendants though blessings or curses. Membership in a segmentary line may also be actively constructed through mothers.

In summary then, inclusion in a segmentary line may be claimed with reference to either "fathers" or "mothers". Segmentation (*bölünöt, özünčö bolup*) may occur by either distinguishing the descendants of male forefathers of the same or different generations or by conceptually separating the offspring of a man's co-wives. The founding ancestors of a line may be both men and women, who are usually remembered as having extraordinary qualities. Men are often portrayed as brave, wealthy and smart, while women are praised for their motherly love, their tolerance and their ability to preserve the family's wealth. To become a famous person and originator of a line both men and women must, however, fulfill one condition: they must have many offspring. This is not simply seen as some kind of physical ability but as the result of

25 Alseyit had three wives: 1) Toktobek Ene, the daughter of a rich and influential Biy, 2) the miraculous Bugu Ene, and 3) Čeejimjan, the daughter of a Kalmak Khan.

blessings given to them by influential and renowned people for their righteous behavior. Similar to an inherited physical trait this fertility is believed to continue in the segmentary line just as the bad effects of a curse may be perpetuated through the generations. People are thus related in a way which Sahlins recently called 'the mutuality of being': "But generally considered, kinsmen are persons who belong to one another, who are members of one another, who are co-present in each other, whose lives are joined and interdependent. Ethnography tells repeatedly of such co-presence of kinsmen and the corollaries thereof in the transpersonal unities of bodies, feelings, and experience." (Sahlins 2011: 11) This raises the question, in which contexts do people stress such "mutuality", "co-presence" and "interdependence" and with whom?

The Importance of Segmentation for Contexts of Exchange: Funerals and Marriages

Recent discussion on "clans" in Central Asia (e.g., Collins 2006; Schatz 2004; Starr 2006 often focus in particular on the political relevance of what I call segmentary lines: do people vote along kinship lines, do parties recruit their members from kin pools, do politicians use kin affiliation to attract voters or to distribute their resources etc.? Most recently, Svetlana Jacquesson (2013) enriched this debate by inquiring into the importance of descent for organising economic activities such as pastoralism (in the past) and farming (at present). In my own research in rural northern Kïrgïzstan I also looked at the relevance of genealogies in various realms of village life. Since the issue is very complex, the following can only hint at some of the more general results of my research, however segmentary lines clearly mattered in context of exchange, especially when organising funerals and selecting spouses.

First of all, the segmentary lines are referred to when people form groups called *bir kïrkï* whose responsibility it is to organise the funeral rituals for its members. If someone dies, the members of his *bir kïrkï* will assemble and plan the proceedings of the complex funeral process (see Hardenberg 2010). On these occasions, services and gifts are exchanged between relatives as well as between hosts and guests who may not be relatives but neighbors, friends, colleagues, classmates etc. The *bir kïrkï* act as hosts in all the subsequent events, i.e., they slaughter the animals, cut the meat, cook the food, and serve the tea etc., while the others join the mourning, consume the tea or food offered on this occasion and give blessings[26]. For meeting the expenses of a funeral, the whole village collects money according to a system that is based on the distinction of segmentary lines. The idea is that close relatives pay more than distant ones. For example, if a person of the Kadïr *uruu* dies, then every

26 At a funeral, the guests are not referred to as *konoktor* such as on happy occasions, but as *batačïlar*, literally "those who bless".

household of this line must pay at least 100 som.[27] Kadïr belongs to the larger segment of Atake, which is divided into two more lines, Chïnïbay and Berikbay[28]. Members of these two latter lines have to pay 50 som per household. All other, more distant lines must contribute only 10 som per household. All these financial dues are recorded in writing. Thus in one book all the contributions from the Atake *uruu* will be recorded and in another the money given by the rest of the village households.

Digging the grave and washing the corpse is also organised on the basis of the segmentary system. For this purpose, the village community is divided into five segmentary lines which descent from the first four sons of the legendary village founder Muratčan.[29] Each segment delegates one man for digging the grave. If a man has died, the selected person will also wash the corpse, while his wife will take over this duty in case of a woman's death. This system is locally referred to as "ten heads" (*on baši̇̈*).

Segmentary lines also play an important role for regulating marriages. Jacquesson reports that in the village of her study, the Baybagï, a segmentary line consisting of about 200 households, do not intermarry (o.J.: 11), i.e. form an exogamous unit. Other exogamous units are smaller, comprising 100 or only 15 households (o.J.: 12) In contrast, people often marry a person from a village in the neighborhood or a certain region, which according to Jacquesson "amounts quite often to finding one from the same genealogical line" (o. J.: 11-12). In other words, the territorially dispersed segments of certain lines are linked through marriage exchange. This remarkable finding is confirmed by my own data.

In Tosor, the village where I conducted fieldwork, I recorded 245 marriages. Out of these, 53 men married women who are originally from Tosor village (21,6 %). 119 men brought their wives from a village also located at the southern shore (mostly Jeti Ögüz Rayon, Toŋ Rayon) of the lake Ïssïk Köl (48,6 %), and only 21 men got married to a woman from the northern shore (mostly Tüp Rayon) of the lake (8,6 %). In interviews people expressed the idea that local people prefer marriages either with people from the Ïssïk Köl region or from Narïn district (oblast), and indeed, according to my survey, 35 householders of Tosor found a wife in the Narïn region (14,3 %). Only a very small minority of my sample (6,9 %) had married a woman who was neither from Ïssïk Köl nor from Narïn.

It would be worth further investigating the relationship between marriage exchange and the segmentary system. For example, I found that among those who had married within the village, more than two third of the men had a wife who like them belonged to a segment of the village founder Muratčan while

27 Very close relatives may give more depending on their wealth.
28 This segment was "joined" (*košolgan*) by the descendants of a person named Malay.
29 Since Muratčan's third son, Boogačï, has many descendants, his line split into two segments.

about one-third had married a daughter of the so-called settlers. Some of these intra-village marriages occurred between people who shared the same forefather about six to seven generations ago. This is thus in conformity with the most stated marriage rule that one should not marry within the "seven fathers". One of my informants combined this proscription with a prescription: "Up to seven fathers we do not marry, after seven fathers we must marry" (*Jeti ataga čejin ülönböybüz, jeti atadan kiyin ülönüš kerek*). In other words, an affinal link should be renewed after a lapse of seven generations. Another way to express this is by saying *söök jangïrtïš kerek*, literally meaning "the bones must be renewed".

Jacquesson notes that the generational distance observed by people at marriage has been extended beyond the seven generation rule. I tested the seven generation rule in the village of my study and wrote down in detail the genealogical links between those men and women who had married within their own locality. I found 35 marriages, which occurred within the segmentary line of the village founder Muratčan. In average, husband and wife were 6,5 generations removed from Muratčan, the minimum was five generations, the maximum nine generations. Taking the generation of Muratčan as reference, almost two-third of these marriages occurred between partners belonging to different generations.[30] This can probably attributed to different age developments within the segmentary lines. Although the sample is rather small, the data at least confirms that the seven generation marriage rule has some practical relevance and it would be worth to investigate this marriage practice further with larger samples.

Segmentation and the Allocation of Land

Another context, in which segmentation and genealogies matter, are claims for rights in land. Jacquesson for example states that on the basis of local tradition one particular line (the Čekir Sayak) claims the land between the shores of the lake Soŋ Köl in the north and the Narïn river in the south as their "ancestral land (*yurt*)" (Jacquesson 2013: 100). Similarly, at the nearby lake Ïssïk Köl, the segmentary system was and is used for dividing up land and for allocating rights of usage or ownership to certain people. Thus, the whole Ïssïk Köl region is conceptually divided into two parts: the land at the north- and southwestern shore is said to "belong" to the Sayak *uruu*, while the north- and southeastern territories are said to be the homeland (*bir jurt*) of the Bugu.[31] In-between one finds villages inhabited by people who are neither Bugu or

30 The distribution was (g. = generation; ≈ = marries): 15 x (6.g. ≈7.g).; 7 x (6.g. ≈6. g.); 6 x (7.g. ≈7.g.); 3 x (8.g. ≈7.g.); 2 x (5.g. ≈7.g.); 1x (6.g. ≈5.g.) and 1x (7.g ≈9.g.).

31 I did not make a statistical survey on this but there is a clear notion among people that villages „belong" to certain segmentary lines in the sense that they have been there "since long" and can claim rights in this land.

Sayak, such as the Mundušdar, Kalmaktar, Saruular or Dungan, who reside in the area near Karakol, or the Russians, whose numbers have, however, decreased in recent decades. Within the area claimed by the Bugu there are again further territorial distinctions related to the various segmentary lines within Bugu. For example, the majority of people inhabiting the villages in the modern administrative district (*rayon*) of Jeti Ögüz are said to be descendants of Jeldeŋ, the great-grandson of Bugu's husband Alseyit. Muratčan is said to be the second son of Jeldeng and the descendants of Muratčan claim that they have rights in the land from the river of Tosor up to the river near Tamga, the neighbouring village. If someone from outside wants to use the pastures of Tosor for his animals, he must first take the consent of the village community.

Nowadays bureaucratic forms of land registration and forms of private land ownership exist side by side with this allocation of land rights based on the segmentary system. For example, people of Tosor say that only descendants of Muratčan can build new farms (*saray*) in the mountains, yet everybody who wants to do so must first officially apply for this at a state department. Similarly, if someone wants to buy land and build a house in Tosor, he will have to follow bureaucratic procedures, yet I was assured that without having relatives in the village or without the support of the community no one will be allowed to purchase land. According to my survey, out of 338 houses only 35 belong to families who do not claim descent from Muratčan but have created affinal links with them through marriage. Among the descendants of Muratčan, one finds a certain correspondence between membership in a segmentary line and place of residence. Thus, although members of the four largest *uruu* live interspersed and not in clearly separated areas, the village consists of clusters of closely related relatives. Often four to five or even more houses, whose residents claim descent from the same son of Muratčan, have been built close to each other. If possible, people prefer to build houses near the residence of one's parents, brothers or close agnates.

With respect to the distribution of farming land one comes across a similar clustering which appears to be the result of decisions taken by the villagers after the collapse of the Soviet Union. I was told that after 1991, a cooperative named "Muratčan Kooperativ" or "Komitet" was founded and allowed to use the machines of the Kolkhoz. This cooperative received a portion of land to the southwest of the village and was supposed to cultivate it collectively. The cooperative was headed by the then village chief (see Hardenberg 2012) and its profits were meant to be distributed to the poor people. The initiative was, however, not successful and the cooperative was dissolved after only two years. The rest of the land was divided according to the system of segmentary lines. Those belonging to the same *uruu* were given a plot of land and created their own farm, a so-called *dïykan čarba* (lit. "peasant enterprise).[32] An old man from the village justified this with reference to the past:

> "First they created farms (*dïykan čarba*). For this, people united according to their *uruu*. For example the three sons of one father, Janek, Ïsa, Musa, lived in Bugu Müyüzdüü [winter pasture, RH]. They lived in winter in Bugu Müyüzdüü and in summer they moved to the summer pastures. Just as it was in the past, we also created our farms according to the *uruu*."

Judging from the present land distribution, the greatest part of the land west to the village was given to the sub-segments of the largest *uruu*, Boogachy. Smaller portions were taken by members of the other three *uruu* such as Šapak, Abayïlda and Tattïkojo. The land to the east of the village, which has less access to water, was distributed mainly to the settlers and to Abayïlda *uruu*.

After some time, in many of these newly created *dïykan čarba* quarrels broke out and people decided to divide the land into smaller portions among the various families.[33] Nowadays, the number of *čarba* based on membership in a segmentary line has decreased considerably, although I still noted eleven plots of land owned by a *čarba* in the western part of Tosor.

In other words, land distribution was consciously organised according to a system of genealogical segmentation and when the larger *uruu* did not prove to work effectively, they divided them into smaller segments down to the level of the household which then became the owners of the land.

Conclusions

David Sneath and David Gullette in their recent writings on former as well as present Inner (or Central) Asian societies reject tribal or kinship models as evolutionary speculations and as lacking any relation to real social organisations. I agree with both authors that such models, which have been developed to describe social organisations in other parts of the world (Africa, Middle East), do often not appropriately capture the specific features of Inner (or Central) Asian societies. However, the main problem I see in their writings is of an epistemological nature: in line with earlier Cambridge anthropologists (like Fortes, Gellner and Goody) these authors derive social forms from political and economical aims and combine this with more recent approaches seeing society as a particular configuration of power relations. By substituting culture with power, Sneath for example puts greater emphasis on processes leading to specific social forms rather than on their culturally specific contents. Genealogy and kinship become simply idioms for politics, i.e. the web of power

32 Jacquesson (2013: 107ff) describes a similar procedure in her area of research.

33 Jacquesson provides a number of excellent case studies for these private enterprises, which often failed for various reasons (2013: 108-112).

structuring the real world, and anthropology is given the task to describe the social forms arising from these political realities. My own approach is different: I try to understand how local concepts of genealogical segmentation and inclusion relate to the wider social ideology, i.e. ideas about morality, blood relations, patrilinearity, descent lines and the distribution of people, and how these concepts are referred to when organising communal action in different social contexts.

My analysis starts with genealogies (*sanjïra*), which have become a special genre comprising of narratives about people belonging to the same segmentary line. There are narratives of national importance, which are published in (semi-)historical books about the "ethnogenesis" of the Kïrgïz people as well as local collective memories containing stories about the heroes of only regional or local importance. An important and widely accepted notion of society in these geneaologies starts from the idea that Kïrgïz people either descent from the same father or belong to allied segmentary lines. Recurring features in these narratives are the brave deeds of certain people who fought against the enemies (e.g., Chinese or Kitay, Kalmak, Russians or other Kïrgïz groups) as well as various stories about men and women who either acted according to local norms or behaved unethical. The consequences of such acts concern not only the respective individuals but affect whole descent lines through the mechanism of blessing or curse.

When segmentation of a genealogical category occurs such acts and the fate they generate for the later generations often play an important role. Segmentation does not occur in any mechanical or "natural" way as assumed in classical anthropological models based on the opposition between brothers or cousins. In Kïrgïzstan, segmentation occurs across the generations and may, for example, oppose fathers and sons or grandfathers and grandsons. The person chosen as the founder of a new segmentary line usually has some extraordinary attributes such as bravery, high morality or fertility. The latter quality seems to be of particular importance because a man without many children can hardly become the founder of a new line.

While a line can only be perpetuated through the blood and semen of a man, women play a particular role for segmentation. Thus, it is often women who curse or bless other people and thus cause a kinship unit to divide. In the past, when an influential man could have more than one wife, his offspring often created new lines by distinguishing their different mothers. At present, people establish links to locally important segments through the relationship created by the marriage of an ancestress.

In the final part of this paper I present my data concerning the importance of the Kïrgïz segmentary system in relation to funerals, marriages and spatiality. In his thesis, Gullette argues that funerals are one of the occasions when people gather and interact on the basis of kinship (2010: 5; 85-8). He rather superficially describes the system of help and monetary assistance on this occasion and it remains unclear according to which principles "family" or "relatives" interact. My own data shows that the segmentary system is highly relevant on such events and orders the relationships between groups who perform different functions and contribute different amounts of money depending on genealogical distance.

In terms of marriages, my data confirm Jacquesson's finding that people often seek spouses from the same or a nearby region. In many cases, marriages occur between members of relatively closely related segments. Unlike in Jacquesson's case, the data on intra-village marriages I have suggest that people continue to observe the seven-generation prohibition but also try to renew marriage alliances when possible.

With regard to spatiality, Gullette discusses genealogical imagination with regard to the creation of monuments arguing that these "visible *sanjïra*" promote public interest in people's ancestors and thus help the political project of state or nation building (2010: 145-146). The findings of Jacquesson and my own suggest that collective memories about ancestors are more intimately connected to land divisions and rights of using land as animal pastures or agricultural farms. In the village studied by me, people of the same segmentary line often form neighborhood clusters. At the time of "privatisation" after the collapse of the Soviet Union the villagers divided the land according to the segmentary system. In subsequent years, a number of farms based on genealogical relatedness proved unsuccessful and further divided the land among the families, while others continued as collective farms. In the mountains, the pastures officially belong to the state and one has to register officially if one wants to build a shed there, yet in practice the villagers continue to conceive of the pastures as their ancestral land which can only be used with their permission.

These examples demonstrate that there exists a Kïrgïz segmentary system, which differs, however, in detail from those systems described in the ethnography from Africa and the Middle East. We can use the data from Kïrgïzstan to reformulate certain classical kinship notions such as lineages or segmentation and therefore I see no need to abandon them as unrealistic anthropological inventions. On the contrary, an analysis of practices such as sharing, feasting, farming, exchanging services or spouses shows that kinship categories are an essential aspect of life in present-day Kïrgïzstan, and certainly elsewhere.

References

BASTUG, S. (1998) The Segmentary Lineage System: A Reappraisal. In GINAT, J. and KHAZANOV, A.M. (eds.), *Changing Nomads in a Changing World*, pp. 94-123. Brighton: Sussex Academic Press.

CARSTEN, J. and HUGH-JONES, S. (eds.) (1995) *About the House. Lévi-Strauss and Beyond.* Cambridge: Cambridge University Press.

COLLINS, K. (2002) Clans, Pacts, and Politics in Central Asia. *Journal of Democracy* 13 (3): 137-152.

COLLINS, K. (2004) The Logic of Clan Politics. Evidence from the Central Asian Trajectories. *World Politics* 56: 224-261.

COLLINS, K. (2006) *Clan Politics and Regime Transition in Central Asia.* Cambridge: Cambridge University Press.

DRESCH, P. (1986) The Significance of the Course Events take in Segmentary Systems. *American Ethnologist* 13(2): 309-324.

DRESCH, P. (1988) Segmentation: Its Roots in Arabia and its Flowering Elsewhere. *Cultural Anthropology* 3(1): 50-67.

DURKHEIM, E. (1992 [1893]) *Über soziale Arbeitsteilung: Studie über die Organisation höherer Gesellschaften: Studie über die Organisation höherer Gesellschaften.* Frankfurt am Main: Suhrkamp.

EICKELMAN, D. (1981) *The Middle East: An Anthropological Approach.* Englewood Cliffs. NJ: Prentice-Hall.

EVANS-PRITCHARD, E. (1940) *The Nuer. A Description of the Modes of Livelihood and Political Institutions of a Nilotic People.* New York and Oxford: Oxford University Press.

EVANS-PRITCHARD, E. (1949) *The Sanusi of Cyrenaica.* London and Oxford: Clarendon Press.

GEISS, P.G. (2003) *Pre-Tsarist and Tsarist Central Asia. Communal Commitment and Political Order in Change.* London and New York: Routledge Curzon.

GULLETTE, D. (2010) *The Genealogical Construction of the Kïrgïz Republic: Kinship, State and 'Tribalism'.* Folkestone: Global Oriental.

HARDENBERG, R. (2009) Reconsidering 'Tribe', 'Clan' and 'Relatedness': A Comparison of Social Categorization in Central and South Asia. *Scrutiny: A Journal of International and Pakistan Studies* 1(1): 37-62.

HARDENBERG, R. (2010) The Efficacy of Funeral Rituals in Kïrgïzstan. *Journal of Ritual Studies* 24(1): 29-43.

HARDENBERG, R. (2012) Collective, Communicative and Cultural Memories: Examples of Local Historiography from Northern Kïrgïzstan. *Central Asian Survey* (special issue: Local History as an Identity Discipline) 31(3): 265-276.

JACQUESSON, S. (o. J.) Whom They are a Part of? Do they Split?: On Kinship and Descent among the Kïrgïz, 1-23. http://www.academia.edu/1709577/On_Kinship_and_Descent_among_the_Kïrgïz [Accessed: 27th January 2016].

JACQUESSON, S. (2013) From Herd Breeding to Land Farming. Social Uses of Decent and Kinship in a Kïrgïz Village. In AKÇALI, P. and ENGIN-DEMIR, C. (eds.), *Politics, Identity and Education in Central Asia: Post-Soviet Kïrgïzstan*, pp. 98-117. London and New York: Routledge.

JUSUPOV, T. K. (1993) *Kïrgïzdar: Sanjïra, Tarïx, Muras, Salt.* Biškek: Kïrgïztan Basmasï.

KRAUS, W. (1998) Contestable Identities: Tribal Structures in the Moroccan High Atlas. *Journal of the Royal Anthropological Institute* 4(1): 1-22.

LÉVI-STRAUSS, C. (1999 [1983; French original 1979]) *The Way of the Masks.* Seattle and London: University of Washington Press.
MORGAN, L. H. (1951 [1877]) *Ancient Society.* Calcutta: Bharati Library.
PLATENKAMP, J.D.M. (1992) Transforming Tobelo Ritual. In COPPET, D. de (ed.), *Understanding Rituals*, pp. 74-96. London: Routledge.
PLATENKAMP, J. D.M. (2000) Temporality and Male-Female Distinctions in the Tobelo Vocabulary of Relationships. In ALÈS, C. and BARRAUD, C. (eds.). *Sexe relatif ou sexe absolue? De la distinction de sexe dans les sociétés*, pp. 241-266. Paris: Éditions de la Maison des Sciences de l'Homme.
PLATENKAMP, J.D.M. (2010) Becoming a Lao Person: Rituals of Birth and Socialisation in Luang Prabang, Laos. In BERGER, P. et al. (eds.), *The Anthropology of Values. Essays in Honour of Georg Pfeffer*, pp. 180-200. Delhi: Pearson.
RUMNEY, J. (2009 [1934]) *Herbert Spencer's Sociology*. New Jersey: Transaction Publishers.
SAHLINS, M. (1968) *Tribesmen.* Englewood Cliffs: Prentice Hall.
SAHLINS, M. (2011) What Kinship is (part one). *Journal of the Royal Anthropological Society* 17: 2-19.
SALZMAN, P.C. (2008) The Middle East's Tribal DNA. *Middle East Quarterly*. [Online] Meforum Database 15(1): 23-33. http://www.meforum.org/1813/the-middle-easts-tribal-dna [Accessed: 27th January 2016].
SCHATZ, E. (2000) The Politics of Multiple Identities: Lineage and Ethnicity in Kazakhstan. *Europe-Asia Studies* 52(3): 489-506.
SCHATZ, E. (2004) *Modern Clan Politics. The Power of Blood in Kazakhstan and Beyond.* Seattle and London: University of Washington Press.
SCHNEIDER, D. (1972) What is Kinship all About. In REINING, P. (ed.), *Kinship Studies in the Morgan Centennial Year*, pp. 32-63. Washington, D.C.: Anthropological Society of Washington D.C.
SCHNEIDER, D. (1984) *A Critique on the Study of Kinship.* Ann Arbor: Michigan University Press.
SIGRIST, C. (2004) Segmentary Societies: The Evolution and Actual Relevance of an Interdisciplinary Conception. In STRECK, B. (ed.), *Segmentation und Komplementarität. Organisatorische, ökonomische und kulturelle Aspekte der Interaktion von Nomaden und Sesshaften*, pp. 3-31. Halle: Orientwissenschaftliche Hefte 14; Mitteilungen des SFB „Differenz und Integration" 6.
SNEATH, D. (2007) *The Headless State. Aristocratic Orders, Kinship Society, and Misrepresentations of Nomadic Inner Asia*. New York: Columbia University Press.
SPENCER, H. (1897 [1873/76]). *Principles of Sociology, Vol I.* New York: D. Appleton and Co.
STARR, F.S. (2006) *Clans, Authoritarian Rulers, and Parliaments in Central Asia*. Uppsala and Washington: CACI and SRSP Silk Road Paper.
TOBOLKA, R. (2003) Gellner and Geertz in Morocco: A Segmentary Debate. *Social Evolution & History* 2(2): 88-117.
YOSHIDA, S. (2005) Ethnographic Study of Privatisation in a Kïrgïz Village: Patrilineal Kin and Independent Farmers. *Inner Asia* 7(2): 215-247.

Omaha and all that:
Basic Issues and New Horizons in the Study of Relationship Terminologies

GEORG PFEFFER

Introduction

German socio-cultural anthropology must be exceptionally grateful to Josephus Platenkamp for representing our discipline in Münster as an empirical *Geisteswissenschaft* over the past decades. In retaining the finest traditions of the *Leiden School*, he distinguished himself as a scholar rather than a salesman of academic fashions. Among his many engagements, he also took an interest in relationship terminologies, otherwise an almost uncultivated field in this country. My article in his honour will deal with this subject and initially sketch some basic issues to proceed to new considerations.

Several approaches of our discipline, e.g. that of *cognitive anthropology*, regularly discuss 'kinship terminologies', as if they belonged to our familiar domain. The German civil code[1] neatly defines such 'kinship' by stating, that those who procreate are lineal kin of their procreations, whereas those who have been procreated by the same third person are collateral kin by degrees according to the number of empirical births connecting them. Moreover, in-laws of one spouse are kin of the other (*note the singular*). In short, my kin are substantial individuals connected to me by birth, while my in-laws are my wife's kin and thereby of secondary – or logically inferior – value. Accordingly, a 'kinship terminology' contains the vocabulary, which a person (*ego*) applies to her or his kin and in-laws in a culture like our own one.

Long ago Alfred Kroeber chose a more refined approach. When indirectly criticising Morgan's epochal discovery, he explained that all terminologies classify (1909: 77), though he did not say, what they classify. Do they, like our own one, order substantial individuals connected with *ego* by processes of physical procreation, or do they order a social whole containing formal oppositions of social categories? When dealing with the latter technique, we must note that to this day millions of people identify their community by naming *all* of its members as their respective consanguines or affines in a given context, when knowing perfectly well that most of them are not relatives 'by birth'. Many more refer to such social categories of relationship, rather than

1 §§ 1589 and 1590 of *Das Bürgerliche Gesetzbuch der Bundesrepublik Deutschland* define our notions of kinship (*Verwandtschaft*) and in-laws (*Schwägerschaft*) in juridical language.

personal names, when talking about any regular social contact. In specific cases a speaker may, for example, apply the same term to his genetic mother as well as to certain immature little girls. Obviously, such an appellation is not a metaphorical extension of our familiar 'mother', but rather points to an unusual order of social relationships. Since they are not concerned with genetic kinship, systems classifying such social categories are called 'relationship terminologies' in the discipline.

All of such terminologies, like those of kinship, are *ego*-centric. Relationship terminologies, however, are also *socio*-centric, since *ego* refers to given interrelated sets of mutually opposed social categories, thereby conveying values of a formally assigned structure that is a general feature of the speaker's society. In the following, I propose that the particular complications and variations of those relationship terminologies, which are based upon the opposition of consanguines and affines, arise out of this dual constraint. Relationships are articulated by an individual speaker (*ego*), whose given status must thus be taken into account, and yet all social categories are at the same time interrelated by the formal logic of affinal oppositions.

These structures articulate values implicitly. Speakers convey them without reflecting or reasoning. Accordingly, the values cannot form a part of utilitarian arguments about advantages of the actors involved. Nor are relationship terminologies the outcome of specific behaviour, though they do change in the course of history. Being undoubtedly products of *anthropos*, they are not the result of decisions taken by specific women or men. And yet a wide variety of such terminological systems can be found, even if no intentional human efforts have caused such systematic variation. Accordingly, voluntaristic explanations of the differences are not convincing in these cases. The values are those of a social whole and measurable as such. This very nakedness of 'culture at work' may be the reason, why some anthropologists conceive terminological studies as a paradigmatic discipline, while others reject or belittle such studies in the way clergymen react to heresies.

The implicit values of a relationship terminology result from the combination of eight 'principles', as introduced by Kroeber (1909 78-79). For reasons of space, I will not discuss that of relative *age* and that of *life/death* of the connecting relative. The principles called *sex of the speaker* and *sex of the person through whom the relationship exists* describe the basis of any affinal relationship terminology. Kroeber's principle of *lineality/collaterality* is the basis of another type of system, especially if the entire society is conceived as a conical clan[2]. Regarding Chinese[3] and Turko-Mongolian terminologies[4] this principle does indeed encourage complex efforts in future research, far too

2 Paul Kirchhoff's famous concept (1968) may have been ignored in his native country, because the author refused to live in Nazi-Germany.

3 *See* Feng (1948).

4 A good example is provided by Pfluger-Schindlbeck (2005: 37).

complex, to be applied in this small contribution. I suggest that the opposition of *lineality* and *collaterality* is the basic principle of a terminology, in case that of *consanguinity* and *affinity* is marginal or absent, and vice versa, but this hypothesis is a preliminary one. In the following I will consider Kroeber's remaining three principles[5]:

1. *Generation*: i.e. odd versus even or else exterior and interior levels;
2. *Gender*: i.e. male versus female categories;
3. *Marriageability*, i.e. consanguinal versus affinal relationships.

The most important – though regularly overlooked – message of Kroeber's timeless masterpiece is the observation that any one of his eight principles may not be applied at all in a given terminology, or may be applied only to some of its terms and not to others. Accordingly, values may receive more or less emphasis within a given terminology. Being based upon the contrast of consanguinity and affinity, this opposition as such will, however, always reflect major values.

Contradiction

The mode of diagramming relationship terminologies is a pragmatic rather than a dogmatic issue, though any effort will have to cope with the following basic contradiction: Since terminologies must be studied as wholes, or as systems connecting *all* affinal oppositions, they must show, how *ego* is related to her or his 'father' (F) and 'mother' (M), or if these are affinal or consanguinal categories. At least one 'parent' will not belong to an affinal line, so the fact, that the two of them must be married, will be a cause of concern. If both are consanguines (i.e. those of the non-marriageable line), their marriage is contradicting such a classification. If, however, this status is simply ignored, F and M still require affines within any such system. Who could these be?

In case we recognise M's consanguines as *ego's* consanguines, we would have to explain the status of F's consanguines, who are likely to be related to M as affines. Would F be *ego's* affine as well? I will cut short an extensive debate to postulate that this contradiction *cannot* be resolved! Instead, some 'adjustment' must be made to classify the categories on level +1 convincingly. Since the oppositional pairs of the entire system are logically interrelated, this also implies a feasible 'adjustment' for the reciprocals of M and F on level –1. My point is, that such 'manipulations' of a system result from its dual constraint, or the fact that the *ego*-centric perspective is combined with the *socio*-centric one in any relationship terminology. Based upon this initial observation, my 'grand' hypothesis holds the dual constraint responsible for variation

5 Kroeber does not notice their binary oppositions and writes about 'sex' rather than 'gender'. Instead of 'marriageability' he mentions the difference of 'blood' and affinal relationships.

and change in relationship terminologies and – eventually – for their dissolution or the termination of such a holistic mode of classification based on affinity. To support my hypothesis, I will interpret confirmed empirical data.

Piaroa

Joanna Overing (1975: 131) has recorded the following relationship terms of the Piaroa of Venezuela. On the right side of *figure 1* I have furthermore added their closest correspondents in English, because some readers might feel totally lost when confronted with such a simple holistic classification. The English terms are, of course, never the correct translations, because Piaroa categories do not refer to substantial individuals but to categories that, for example, include many different individuals as a 'mother'.

On the exterior levels +2/-2 this system contains only one basic term with a suffix to mark the gender difference. Thus the question of marriageability is altogether avoided. All 'grandparents' and 'grandchildren' (in a very general sense) are conceived in two unified categories. However, on the odd levels +1/-1 a single category (genderised by a suffix only) in the consanguinal line opposes another single one in the affinal line. In other words: The language as such defines F and M as consanguines in opposition to 'mother's brother' (MB) and 'father's sister' (FZ) as affines. Their reciprocals confirm this pattern: Only one basic term for 'Son' (S) and 'daughter' (D) opposes one for 'brother's son' (BS w.s.) and 'brother's daughter (BD w.s.) of a woman speaker. However, the same kind of assimilation is not exactly repeated on level 0, since two basic terms, ***chirekwo*** (male) and ***chóbiya*** (female) appear in the affinal line thus stressing their special status within the system. This difference adds a unique emphasis to affinity.

Within this simple system all Piaroa are classified by opposed terminological lines of consanguines and affines on the three internal levels. For the people this binary terminology is of great ideological importance and yet in practical behaviour, they frequently marry those defined as their consanguines, though they do not conceive such a breach. In the course of the analytical procedure, formal classification must never be confounded with practical arrangements. The structure of the Piaroa terminology does prescribe[6] symmetric affinal exchange (Sym. E.) between the two lines[7], while empirical marriages may very well avoid any exchange to join two individuals as spouses, who have grown up as 'sister' (Z) and 'brother' (B) in the same round-house.

6 On 'prescription' *see* Needham (1973).

7 To facilitate the reading of the diagrams, I have added black arrows indicating symmetry or asymmetry of exchange, just as I have pointed out the status of consanguinal or of affinal categories in grey coloured letters.

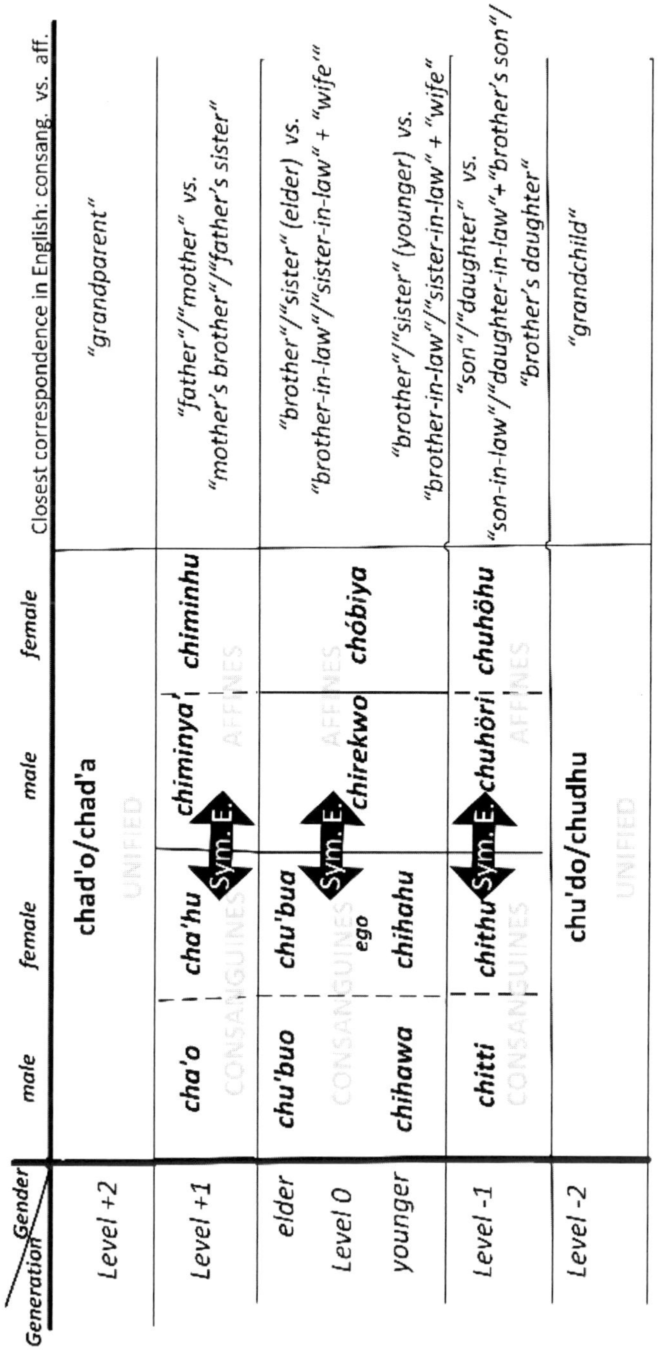

Fig. 1: Piaroa two-line-terminology (*ego* female)
(with data Joanna Overing 1975: 131).

Garo

In the Garo[8] terminology (*see figure 2*) the external levels of +2/-2 are similarly undifferentiated, but compared to the Piaroa structure, the three internal levels (except the affinal line on level -1) contain not one but two basic terms for male and for female categories on each level making up the opposition of the two lines. Thus gender implies a major difference. However, it is far from clear, who is assigned to the affinal, who to the consanguinal line. The opposition as such is prescribed, but by formal standards the status of categories M and FZ, or MB and F, remains unclear.

It is only I, the external observer and not the terminological structure as such, who refers to M and MB as consanguines and FZ and F as their affinal opposites, because I happen to know that Garo society is ordered by matrilaterally assigned clans and moieties. For a male *ego*, a 'sister's daughter' (ZD) and a 'sister's son' (ZS) would – under these circumstances – be consanguines opposing the affinal 'daughter' (D) and 'son' (S). My intervention is arbitrary, as would be one suggesting the only alternative decision, because the 'manipulative' aspect of the structure lies in this latitude. While the overall opposition of consanguines and affines is obvious and the even levels are non-problematic with *jik* as prescribed 'spouse' of a male *ego*, no logical constraint determines, who is to be an affine or a consanguine on the odd levels. The question is left open. As far as I can see, Dumont (1953 [1983]), working on a similar structure in South Indian languages, does not offer a better solution.

8 They live in the NE-Indian province of Meghalaya.

Gender / Generation	female	male		female	male
Level +2	***ambi***	***achu***	UNIFIED	***ambi***	***achu***
Level +1	***ama*** CONSANGUINES (M, MZ, FBW)	***mama*** CONSANGUINES (MB, FZH, WF)	Symm.E.	***mani*** AFFINES (FZ, MBW, WM)	***apa*** AFFINES (F, FB, MZH)
Level 0	***abi*** (eZ) CONSANGUINES **no** (yZ)	***ada*** (eB) CONSANGUINES *ego* **jong** (yB)	Symm.E.	***jik*** AFFINES (FZD, MBD, W)	***boning*** AFFINES (FZS, MBS)
Level -1	***namchik*** CONSANGUINES (ZD, SW)	***gri*** CONSANGUINES (ZS, DH)	Symm.E.	***de*** AFFINES (D)	***de*** AFFINES (S)
Level -2		***angsu***	UNIFIED	***angsu***	

Fig. 2: Garo two-line terminology (*ego* male)
(with data Robbins Burling 1963: 70-85).

Gidjingali

Lester Hiatt's data of the North Australian Gidjingali, diagrammed in the usual manner, indicate the symmetric exchange between the categories prescribed by the structure of the whole. For space reasons, I will only mention its two central aspects: (1.) Cross-gender 'siblings' are sometimes equated (e.g. MMBS/MMBD) and the alternating equations (e.g. FMB/FM, ZH/BW and ZSS/ZSD) on even levels determine, that 'sibling' pairs are always the units opposed to one another within the prescribed exchange. (2.) These alternating equations also determine the consanguinal status of F and FZ. Thus – compared to my first example – different 'adjustments' become evident: Whereas the Piaroa terminology determines that the categories of 'spouses', i.e. F and M in opposition to MB and FZ, represent consanguines and affines on level +1, the Garo and the Gidjingali terminologies classify the 'siblings' F and FZ as the opposites of the 'siblings' M and MB.

Four terminological lines, rather than only two of the Piaroa and Garo schemes, make up the construction of Gidjingali affinal prescription. Of any 'sibling pair' within one of the four lines, the 'brother's children' (i.e. those of the male 'sibling') follow their respective 'father' on the relatively lower level. Moreover, the formal mode of symmetric affinal exchange opposes different terminological lines on even when compared to odd levels, thereby demonstrating, that the affinal relations may vary on different levels within the same terminological whole. In the Gidjingali case, two terminological lines of consanguines (CONSANGUINES 1 and CONSANGUINES 2) oppose the two lines of affines (AFFINES 1 and AFFINES 2). For any single terminological line the respective opposition is shifting to the alternate affinal line on each of the given five levels. In view of this observation it must be legitimate – and indeed desirable – to study prescribed affinal exchanges of the *specific* oppositions on each level of the terminology.

		Symmetric Exchange				Symmetric Exchange		
Gender / Generation	male	female	male	female	male	female	male	female
Level +2	*wala marmunga* (FF, etc) CONSANG. 1	*djala* (FFZ, etc)	*jeramba* (FMB, etc) AFFINES 1	*mangga maga* (FM, etc)	*djaminja* (MF, etc) AFFINES 2	*aragudja* (MFZ, etc)	*djeda* (MMB, etc) CONSANG. 2	*mununa* (MM, etc)
Level +1	*anja* (F, etc) CONSANG. 1	*baba* (FZ, etc)	*djadjadja* (MB, etc) AFFINES 2	*ama* (M, etc)	*ngalinga* (WF/BWF, etc) AFFINES1	*ngalinga* (WFZ/BWFZ, etc)	*djabur* (MMBS, etc) CONSANG. 2	*djabur* (MMBD, etc)
elder Level 0 *younger*	*wala* (eB, etc) *ego* *gurora* (yB, etc) CONSANG. 1	*djala* (Z, etc)	*jeramba* (ZH, etc) AFFINES 1	*mangga* (BW, etc)	*djaminja* (MBS, etc) AFFINES 2	*aragudja* (MBD, etc)	*djeda* (MMBSS, etc) *gangguda* (MMBSS, etc) CONSANG. 2	*mununa* (MMBSD, etc) *gangguda* (MMBSD, etc)
Level -1	*nanja* (S, etc) CONSANG. 1	*bababa* (D, etc)	*djadjadja* (MBSS, etc) AFFINES 2	*ama* (MBSD, etc)	*ngalinga* (ZS, etc) AFFINES1	*ngalinga* (ZD, etc)	*djabur* (MMBSSS, etc) CONSANG. 2	*djabur* (MMBSSD, etc)
Level -2	*gurora* (SS, etc) CONSANG. 1	*djala* (SD, etc)	*jeramba* (ZSS, etc) AFFINES 1	*mangga* (ZSD, etc)	*gangguda* (MMBSSSS, etc) AFFINES 2	*gangguda* (MMBSSSD, etc)	*djaminja* (MBSSS, etc) CONSANG. 2	*aragudja* (MBSSD, etc)
		Symmetric Exchange				Symmetric Exchange		

Fig. 3: Gidjingali four-line terminology (*ego* male)
(with data Lester Hiatt 1965: 48).

Melpa

Having been inspired by an article of Almut Schneider (2010), I have drawn *figure 4* with data supplied by Andrew Strathern and Pamela Stewart (2011). The Melpa[9] terminology contains three important 'messages':

(1.) As in the Piaroa case, the terminology combines F and M (the 'spouses') on level +1 as the categories opposing MB and FZ in symmetric affinal exchange, rather than 'siblings' in Garo and Gidjingali style. Equally, the opposing pairs on the other central levels contain 'siblings', such as HB and HZ/BW as affines of B and Z.

(2.) As in the Gidjingali case, a terminological line of consanguines is opposed to different lines of affines according to the variation of levels. Thus MB and FZ are affinal opposites of F and M, but their 'children' are *not* the opposites of B and Z, just as S and D do *not* appear in affinal opposition to the same terminological line that supplies the opposites of B and Z. The 'message' is clear. Though the terminologically articulated exchanges are symmetric, *they must not be repeated* immediately, as they must in the Piaroa and Garo cases, or on alternating levels, as prescribed by the Gidjingali system.

(3.) Piaroa, Garo and Gidjingali offer prescriptive terminologies, because – by their linguistic logic – they *positively* imply symmetric affinal exchange to be repeated between either two or four terminological lines respectively. By the same linguistic logic the Melpa structure *negatively* implies, that any repetition of affinal exchange between the given four lines is impossible. It does not contain a positive prescription. Given the Melpa case and many similar ones, I suggest that such linguistically defined negative prescription is studied with the same diligence as has been afforded to the study of positive prescription in the past decades.

9 The speakers of this language live on Mount Hagen in highland New Guinea.

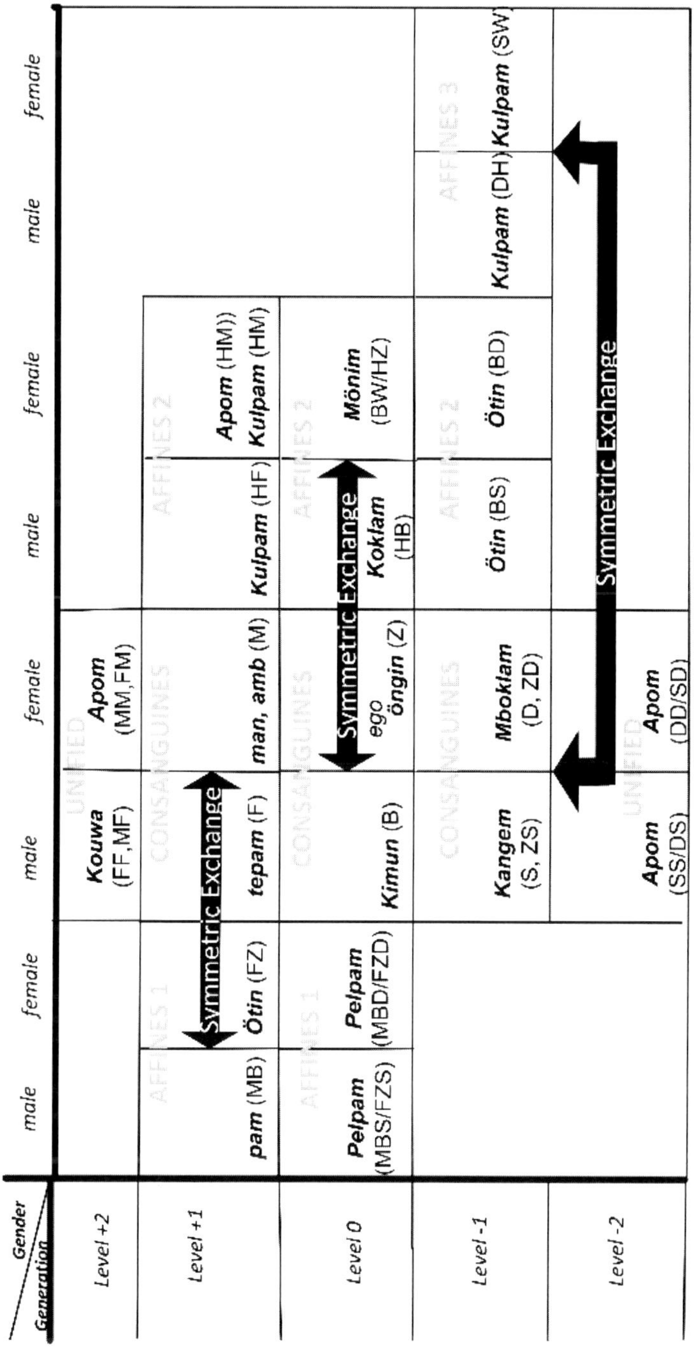

Fig. 4: Melpa four-line terminology (*ego* female)
(with data Andrew Strathern and Pamela J. Stewart 2011: 184-186)

Dongria Kond

The discussion has so far been restricted to terminologies prescribing *symmetric* exchange between terminological lines throughout a system. Categories of 'givers' were the same as those of 'takers'. However, terminologies prescribing *asymmetric* exchange on every level have also been widely discussed in anthropological literature. In their kind of system, the terminological line of 'givers' cannot be that of 'takers' within the same affinal opposition. Terminological 'givers' must differ from 'takers'. In some cases, such as the Jinghpaw terminology as described by Edmund Leach (1961: 41), the 'givers of the givers' are also distinguished terminologically, as are the 'takers of the takers' and *ego's* own terminological line of consanguines, so that altogether five lines, among them four affinal ones, make up the entire prescriptive system. This issue of asymmetry is raised in my next example, which is concerned with Roland Hardenberg's Dongria Kond[10] data (Fig. 5).

As in the Melpa case, one line of consanguines is opposed to several lines of affines (AFFINES 1, 2, 3, and 4) on the three interior levels. The affines of *ego's* 'parents' do not belong to the line of *ego's* affines and affines of *ego's* 'children' represent yet another line. In view of the Melpa data, the significant variation of the Dongria Kond terminology prescribes *asymmetric* exchange on level 0, where *ego's* 'givers' are not the same as *ego's* 'takers', while *symmetric* exchange is prescribed on the odd levels +1/-1. Accordingly, this 'message' of the terminological whole not only excludes any kind of repetition of affinal exchange by way of negative prescription, as does the Melpa whole, but also excludes the idea of symmetric reciprocity on *ego's* level, even if past and future relations of affinity (on levels +1/-1) are viewed as a mutual 'give and take'.

10 They live in the south of the Indian province of Odisha.

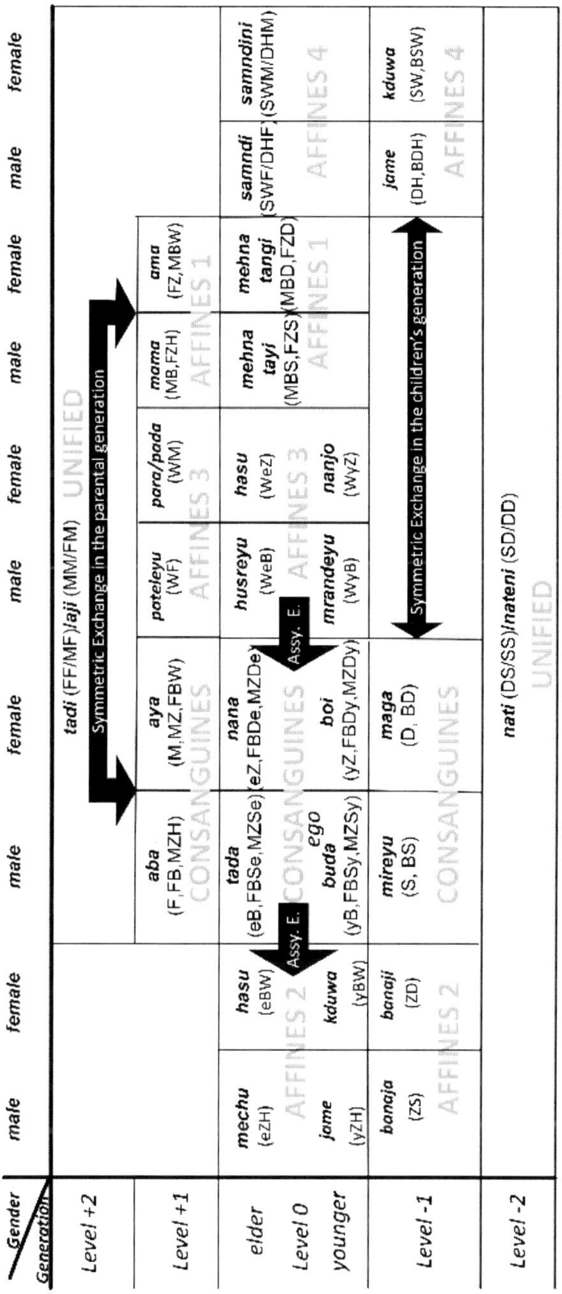

Fig. 5: Dongria Kond five-line terminology (*ego* male)
(with data Roland Hardenberg 2006:170-171)

Omaha

Anthropological literature has devoted much attention to the terminological 'type' known as 'Crow/Omaha' ever since Leslie Spier (1925) had suggested such 'types'. German anthropologists, though usually desisting from terminological studies, all the same like to refer to 'Crow/Omaha' in textbooks, even though decades ago Rodney Needham (1971: 15-16) had clearly demonstrated the fallacy of such 'types'. Rather than taking into account *all* relationship terms of a language, the various 'types' only deal with an extract. Such a selection is then arranged according to our familiar cultural conception of kinship. Implicitly but forcefully, the 'types' also seem to convey the illusion of an ethnographic base.

In fact, Lowie's fine work[11] on the Crow of Montana does *not* claim to contain adequate terminological data and numerous terminologies, subsumed under the 'Omaha type', differ significantly from the system of the historic Omaha of the Great Plains. For these Amerindians the excellent ethnographies of Owen Dorsey (1884: 252-255) and Alice Fletcher and Francis La Flesche (1970 [1911]: 313-317) have provided good terminological data. Robert Barnes (1984: 132-134) must be praised for examining and ordering these relationship terms according to the standards of modern scholarship. He has also numbered them. My *figure 6* offers his numbers, so that anyone interested in the multitude of terminological equations[12] of each designation may gain quick access to them in Barnes' masterpiece.

In the earlier example, I had suggested that the Dongria Kond prescribe symmetric exchange between different terminological lines on odd levels but asymmetric exchange on *ego's* level 0, thereby formally excluding direct reciprocity and any immediate or delayed repetition of such exchange. By contrast, the Omaha prescribe asymmetric exchange on level +1 and symmetric exchange on both levels 0 and level -1. Whereas the structure of the Dongria Kond terminology opposes one terminological line of consanguines to four different terminological lines of affines, the Omaha structure opposes two consanguinal lines (CONSANGUINES 1, 2) to three different affinal ones (AFFINES 1, 2, 3). However, the effect is similar. Any immediate or delayed repetition of an exchange relationship between the same terminological lines is excluded by the linguistic logic of the whole. This proposition will have to be explained at some length.

11 See Lowie 1912, 1917, 1918, 1919, 1935, 1963.

12 For reasons of space I cannot reproduce them here.

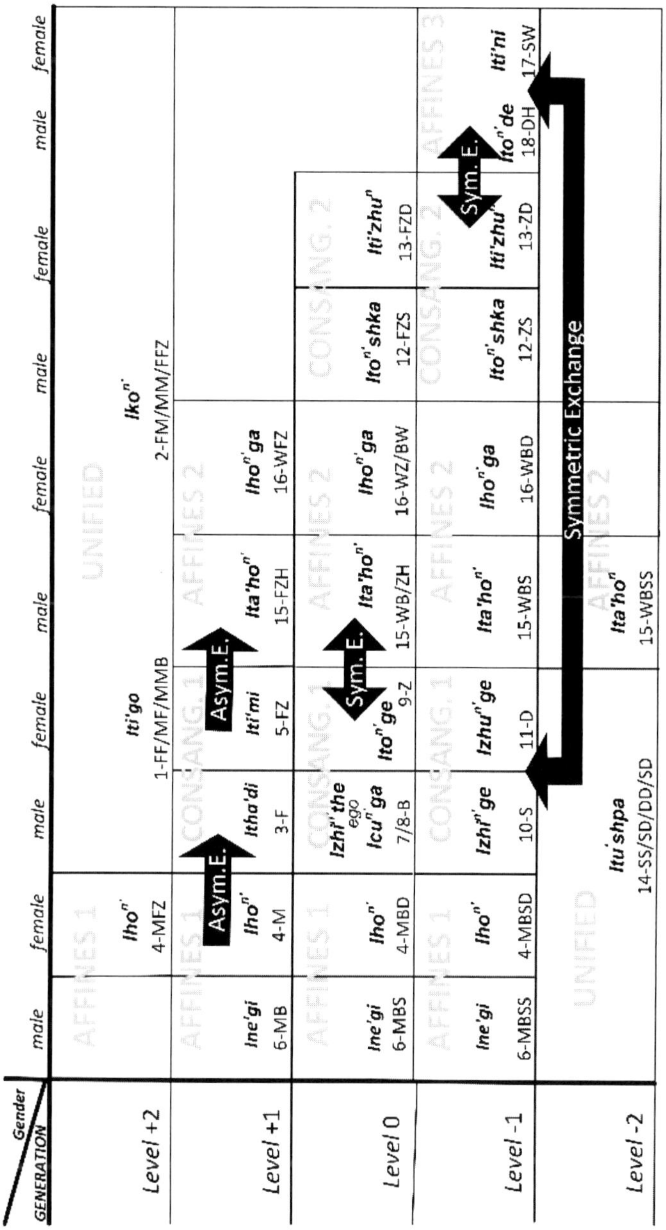

Fig. 6: Omaha five-line terminology (*ego* male)
with data Robert Barnes (1984: 132-134)

Proposition

The Dongria Kond terminology conceives F and M as consanguines (CONSANGUINES), being opposed and identified as such by MB and FZ as affines (AFFINES 1).

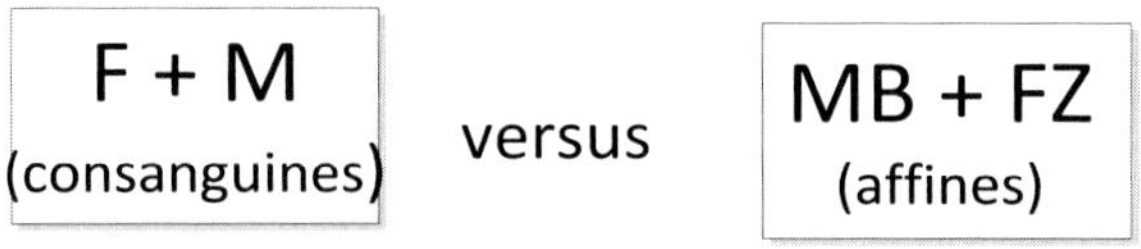

Fig. 7: Dongria Kond: Two pairs of ‚spouses' are opposed to each other

Accordingly, the consanguinal line of the Dongria containing categories F and M on level +1 includes their joint 'children', namely *ego*, B and Z on level 0, as well as *ego's* 'children', S and D, on level -1. In the same style all terminological lines of the Dongria Kond are linguistic constructions containing 'parents' and their respective 'children'. However, this kind of 'adjustment' is not adopted by the Omaha system.

By contrast, the Omaha terminology classifies F and FZ as consanguines (CONSANGUINES 1) who are opposed and identified as such by MB and M as affines (AFFINES 1).

Fig. 8: Omaha: Two pairs of ‚siblings' are opposed to each other

Accordingly, the Omaha terminological line of CONSANGUINES 1 cannot contain 'children' of the 'sibling' pair F and FZ. It contains the 'children' of F only, i.e. *ego*, Z and B, as well as (on the subsequent level) the 'children' of *ego* and *ego's* 'brother', if *ego* is a male. 'Incest', or non-exchange, runs against the basic logic of a system based on affinal exchange and must thus be ruled out. As a consanguine, FZ must have 'children' who are not *ego's* 'siblings'. Accordingly, she must found a second consanguinal line (CONSANGUINES 2). By the same logic the status of ZS and ZD must be that of consanguines, though they are neither 'children' nor 'grandchildren' of FZ. Should they be conceived as a third consanguinal line?

The Omaha terminology offers a different kind of classification. The two categories (No. 12 and No 13 of Barnes's list) which equate FZS with ZS and

FZD with ZD respectively, thus combining two different levels in each, convey the specific classificatory technique. FZS and ZS, or FZD and ZD, do not represent 'parents' and 'children', but by their respective equation the Omaha language *creates* the second consanguinal line (CONSANGUINES 2). A second structural feature does indeed confirm this finding: S (No.10) and D (No.11) are engaged in symmetric affinal exchange with DH (No.18) and SW (No.17), but the latter two affines are *also* engaged in symmetric exchange with the consanguines ZS (No.12) and ZD (No.13).

Thus the terminology differentiates S and ZS as well as D and ZD, but equates their affinal opposites as DH=ZDH and SW=ZSW (Barnes 1984: 134). In short, the linguistic equation (FZS=ZS; FZD=ZD) is posed as a consanguinal line, because ZS and ZD (as CONSANGUINES 2) and S and D (as CONSANGUINES 1) *share* the same affinal opposition vis-à-vis DH/ZDH and SW/ZSW.

The most prominent features of the Omaha terminology are several equations ignoring the difference of levels. MB for example, on level +1 the unilateral 'giver' (AFFINES 1) in the affinal relation to F, is equated with MBS and MBSS on the lower levels, even though the category is terminologically excluded as 'giver' or 'taker' in relation to consanguines of any level other than +1 of MB and F. At the same time, the unilateral 'takers' of FZ (i.e. FZH/No. 15 of AFFINES 2) are *not* excluded as affinal opposites on the lower level. There they are *both* 'takers' and 'givers'. The equations of this category (including WB=ZH/No.15 of AFFINES 2) on *ego's* level prove, that they are no longer just asymmetric 'takers', but have been promoted to the status of 'givers' as well, now engaged in symmetric affinal exchange with B/No.8 and Z/No.9 of CONSANGUINES 1. Why, one may ask, should WB/ZH and WZ/BW form a common terminological line with FZH and WFZ? The crystal clear answer is offered by the Omaha language: Because as males – and separately as females – they are terminologically equated! At the same time, these inter-level linguistic equations do not rule out, but strongly confirm, the different relationships of a single category, depending upon the level of the opposite line, as is expressed by *the latter's* change of category.

Studies of the historic Omaha or of their so-called 'type' usually proclaim that the tribe's system of patrilateral clans is 'reflected' in the remarkable terminology that is equating kin-types of different levels. This, however, cannot be the case. We have seen that, because of their equation, FZH and ZH stand for a single terminological line, just as FZS and ZS share one line, because they are equated, but in both cases substantial individual husbands of *ego's* individual FZ or Z could never share the same exogamic clan. Clearly the reflectionist logic must be fallacious! FZH and ZH, or FZS and ZS, form a common terminological line because they are terminologically equated and because in each one of the two cases, they share the same opposites in an affinal relationship. However, some substantial individuals included by these categories, i.e. those of different generations, cannot share the same clan, since Omaha social norms exclude the repetition of intermarriage.

Just as other terminologies at times ignore the gender difference, the Omaha system partially ignores the difference of levels. This technique is one way to classify a terminological line, thereby 'adjusting' to the basic contradiction of all affinal terminologies and prescribing the primary value, i.e. the exclusion of repeated affinal exchange. Phrased differently, the value implies a plurality of affinal relations from one generation to the next.

Conclusion

Relationship terminologies may be based upon the opposition of lineals and collaterals, or upon that of consanguines and affines. The latter are subdivided into those prescribing symmetric and those prescribing asymmetric exchange. Within each one of these two modes the prescription may be positive or negative. I have exemplified positive symmetric prescription of two-line terminologies by the Piaroa and the Garo cases and four-line prescription by the Gidjingali one. Negative prescription, i.e. the one excluding any repetition of affinal exchange between two terminological lines, was exemplified by the symmetric Melpa scheme. The terminologies of the Dongria Kond and the Omaha combine – with the change of levels – symmetric and asymmetric affinal oppositions in different ways. They too rule out any repetition of exchange. Thus various kinds of prescriptive terminologies (*see figure 9*) represent different 'adjustments' to 'cover up' the basic contradiction. Their linguistic logic implies or denies diachronic affinity. However, the awkwardness of the various 'adjustments' is likely to have promoted an eventual dissolution of such affinal relationship terminologies in numerous cases.

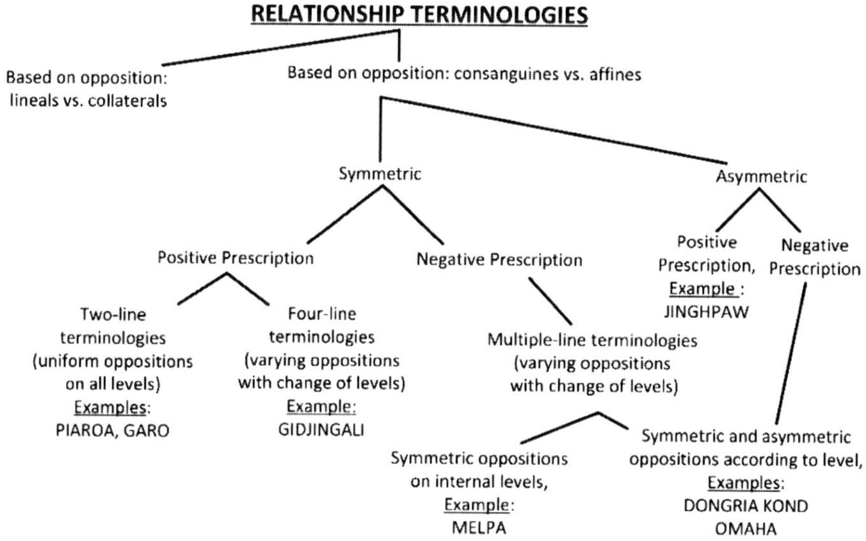

Fig. 9: Relationship terminologies

References

BARNES, R.H. (1984) Two Crows Denies It. A History of Controversy in Omaha Sociology. Lincoln: University of Nebraska Press.

BÜRGERLICHES GESETZBUCH; Bekanntmachung vom 2. Januar 2002.

BURLING, R. (1963) Garo Kinship Terms and the Analysis of Meaning. *Ethnology* 2(1): 70-85.

DORSEY, J.O. (1884) *Omaha Sociology.* Third Annual Report of the Bureau of Ethnology. Washington: Smithsonian Institution.

DUMONT, L. (1983 [1953]) The Dravidian Terminology as an expression of marriage. In DUMONT, L. *Affinity as a Value. Marriage Alliance in South India, with Comparative Essays on Australia*, pp. 3-17. Chicago: University of Chicago Press.

FENG, HAN-CHI [Han-yi FENG] (1948) *The Chinese Kinship System.* Cambridge, Massachusetts: Harvard University Press.

FLETCHER, A.C. and LA FLESCHE, F. (1970 [1911]) *The Omaha Tribe.* 27th Annual Report of the Bureau of American Ethnology. Washington, Smithsonian Institution.

HARDENBERG, R. (2006) *Children of the Earth Goddess: Society, Marriage and Sacrifice in the Highlands of Orissa.* Unpubl. Thesis of Habilitation; Faculty of History and Philosophy, University of Münster.

HIATT, L.H. (1965) Kinship and Conflict. A Study of an Aboriginal Community in Northern Arnhem Land. Canberra: Australian National University.

KIRCHHOFF, P. (1968) The Principles of Clanship in Human Society. In FRIED, M.H. (ed.), *Readings in Anthropology; Volume II Cultural Anthropology*, pp. 370-381. New York: Thomas Y. Crowell Co.

KROEBER, A.L. (1909) Classificatory Systems of Relationship. *Journal of the Royal Anthropological Institute* 39: 77-81.

LEACH, E. R. (1961 [1945]) Jinghpaw Kinship Terminology. An Experiment in Ethnographic Algebra. In LEACH, E.R. (ed.), *Rethinking Anthropology*, 28-53. London Tavistock.

LOWIE, R.H. (1912) Social Life of the Crow Indians. *Anthropological Papers of the American Museum of Natural History. 9(2): 181-248.*

LOWIE, R.H. (1917) Notes on the Social Organization and Customs of the Mandan, Hidatsa, and Crow Indians. *Anthropological Papers of the American Museum of Natural History. 11(1): 1-100.*

LOWIE, R.H. (1918) Myths and Traditions of the Crow Indians. *Anthropological Papers of the American Museum of Natural History. 15(1): 1-308.*

LOWIE, R.H. (1919) The Tobacco Society of the Crow Indians. *Anthropological Papers of the American Museum of Natural History 11(2): 103-200.*

LOWIE, R.H. (1935) *The Crow Indians.* New York: Farrar and Rinehart.

LOWIE, R.H. (1963 [1954]) *Indians of the Plains.* New York: The American Museum of Natural History.

NEEDHAM, R. (1971) Remarks on the Analysis of Kinship and Marriage. In NEEDHAM, R. (ed.), *Rethinking Kinship and Marriage*, pp. 1-34. London: Tavistock.

NEEDHAM, R. (1973) Prescription. *Oceania 43*: 166-181.

OVERING KAPLAN, J. (1975) *The Piaroa A People of the Orinoco Basin. A Study in Kinship and Marriage*. Oxford: Clarendon.

PFLUGER-SCHINDLBECK, I. (2005) *Verwandtschaft, Religion und Geschlecht in Aserbaidschan.* Wiesbaden: Reichert.

SCHNEIDER, A. (2010) The Distinction of Sex and Generation in Gawigl Relationship Terminology: An Exercise. In BERGER, P. et al (eds.), *The Anthropology of Values. Essays in Honour of Georg Pfeffer*, pp. 59-76. Delhi: Pearson.

SPIER, L. (1925) *The Distribution of Kinship Systems in North America.* University of Washington Publications in Anthropology 1(2). Seattle, University of Washington Press.

STRATHERN, A. and STEWART, P.J. (2011) *Kinship in Action. Self and Group.* Boston: Prentice Hall.

From Linearity to Cognation: The Transformation of Alawi Kinship Terminologies (Southeast Turkey)

LAILA PRAGER

Jos Platenkamp is known as an expert in kinship studies, an expertise which he has demonstrated in his publications (Platenkamp 1983, 1988, 2000), lectures, and seminars. In his analyses of kinship the examination of relationship terminologies occupies a prominent place. While still being a student I was intrigued by the great variety of kinship vocabularies to which he introduced us during classes by drawing complex diagrams for which the blackboard usually did not provide sufficient space. Later on, while working on my Master and PhD thesis we regularly had discussions about the comparison of relationship terminologies. Recalling these discussions, I shall express my gratitude to Jos by presenting an analysis of the relationship terminologies of the Alawi of Southeastern Turkey, a society where I conducted extensive fieldwork (2004-2005, 2006-2008, 2009). What makes the Alawi case stand out is that they do not only employ one but four different types of relationship terminology. Moreover, the situation is even more complicated by the fact that these kinship vocabularies are phrased in two different languages: Arabic and Turkish respectively. This raises the question why a society uses four relationship terminologies at the same time. In this context one should notice that these terminologies are not all employed simultaneously by one and the same person, but by different Alawi groups depending on age and/or on the region where they have settled. Moreover, these terminologies are subject to a diachronic order, the Arabic type being the oldest form of kin classification, whereas the Turkish variants are of a more recent origin.

Generally, one can notice that the Alawi relationship terminologies show remarkable differences in the way in which consanguinal and affinal relatives are classified. Furthermore, it will become evident that the transition from the Arabic to the Turkish types of kin classification goes along with a loss of complexity and a decrease in the number of kin terms.

While Lewis Henry Morgan (1871) and other authors had maintained that relationship terminologies are constituting relatively resilient forms of classification, anthropologists such as Eggan (1937), Spoehr (1947), and Friedrich (1966) have drawn attention to the phenomenon that kinship vocabularies can be subject to historical change, the latter being brought about by pressures to adapt to new social environments and/or by changes in the social structure itself. Eventually, such transformations may entail substitutions and/or semantic alterations in the field of the lexemes.

Comparable changes can be observed in the case of the Alawi relationship terminologies, though in this context we are not talking about historical changes within the same terminology as in the examples described by the authors mentioned above, but transformations marking the transition from one terminology to other ones.

In the following I shall first present an analysis of the "Arabic" form of kin classification, and then contrasting the latter with the different variants of the Alawi "Turkish" terminologies. Finally, some concluding remarks are due in order to explain how and why these different forms of kin classification came into being.

The Alawi in Turkey

The Alawi should not be confused with the better-known Turkish Alevi. Although the term Alevi also derives from Ali, the cousin and son-in-law of Muhammad, the beliefs and ritual practices of the Alevi differ greatly from those of the Alawi (cf. Prager 2010b: 3). Since the tenth century, the Alawi – or Nusairi, as they are also known – have inhabited a mountain range (Ǧabel Ansayri) that nowadays forms part of Southeastern Turkey (Çukurova, Mersin and Hatay region), Syria, and North Lebanon. In Syria and Turkey, the Alawi constitute an Islamic "heterodox" religious minority whose most prominent and controversial member is the president of Syria, Bashar al-Assad. According to official statistics, they account for 10-13% of the total population of Syria (Engin and Franz 2000: 157), whereas Lebanon hardly has more than 9000 Alawi (Halm 1982: 383-604). In Turkey, they number between 700,000 and one million (Brawer 1988: 96). Since the 1960s, the Alawi have also settled in Western European countries such as Germany, France, and the Netherlands, as well as in the Arab Gulf States (Prager 2010b: 38-40).

The origin of the Alawi can be traced back to religious movements of 9th century Iraq, where Ibn Nusair, the founder of the Alawi religion and companion of the 11th Imam, heralded his doctrines (Azzi 2002). While there is on-going debate in Islamic studies about whether the Alawi should be classified as belonging either to the Seveners, Eleveners or Twelver Shi'ites (cf. Halm 1982: 284ff; Nimier 1987; Khuri 1991: 49-61), contemporary Alawi clearly associate themselves with the Twelver group (Merwin 2007: 363; Prager 2010b: 41). A major difference between the Shia and the Alawi, however, is that the latter worship Ali as a manifestation of the divine essence and believe in the reincarnation and transmigration of souls (see Prager 2013; Prager 2016). Generally, Alawi religion can be considered as a syncretistic system of ideas, involving Islamic, Christian and Jewish influences along with ideas from Persian, Indian and Greek philosophies (Khuri 1991: 50-51; Prager

2010b: 43-45). Moreover, Alawi religious doctrines are only transmitted in a secretive initiation ritual to young men, while women – at least in theory – are excluded from religious knowledge.

The ethnonym Alawi is nowadays used within the group as an auto-designation, whereas the term Nusairi is only rarely used by Alawi themselves. The term Alawi came into use when the short-lived Alawi state was established in 1925 under French mandate. The core region of the Alawi state was later incorporated into Syria, while a smaller part in the North, along with the Sanjak of Alexandretta, was annexed by Turkey in 1939. Differing greatly from normative Sunni Muslims in religious beliefs and ritual practices, the Alawi have been regarded with suspicion by governments from the age of the Ottoman Empire to the modern Turkish nation state, leading to a variety of religious, social and economic pressures.

The older generations still speak Arabic as the mother language, whereas younger people mostly use Turkish, while partly still speaking and understanding the local Arabic dialect. Bilingualism is important as the Alawi of Turkey – as stated above – not only use one but four different relationship terminologies, depending on the age of the speaker and the region of origin.

Among the Arab speaking Alawi of Hatay there is a clear tendency to conceptualise descent in terms of patrilineality which ranges over not more than three generations. A patrilineal group, *'a'ile*, consists of F, (FeB), B, Z, S, D, the M being part of her father's *'a'ile* (cf. Prager 2010a, 2010b). At the same time the FB is considered to have engendered his own patrilineal group, so that the FBCh are not forming part of one's own *'a'ile*. The *'a'ile* constitutes the basic social group which is obliged to perform the major religious rituals, among them the annual ritual sacrifice which is inherited by a son from his father. Until the 1990s among the Alawi of Hatay there was a preference for FBD-FBS-marriage. In line with this, the postmarital residence rule was (and still is) patri-/virilocality. Among the Alawi from the Mersin and Çukurova-regions (including the city of Adana) who are mainly communicating in the Turkish language the *'a'ile* has lost its patrilineal dimension and is considered basically as a nuclear family. Also FBD-FBS-marriage has receded to the background. Finally, mention should be made that neolocality has become the usual mode of postmarital residence.

The Alawi-Arabic Terminology

The Alawi-Arabic relationship terminology (henceforth AAT) is a classificatory system mainly used by the Alawi people of the Hatay region. Although being a variant of the Arabic terminologies[1] – the latter corresponding to the so-called "descriptive" (Morgan 1871) or "Sudanese Type" (Murdock 1949)

1 For generic forms of Arabic terminologies see Cuisinier and Miquel (1965) and Eickelman (2002).

– it displays certain features testifying to the influence of a genuine Alawi logic of classification.

The most basic differentiation within this system of classification is that between *qarīb* and *yabanği,* between the "own" and the "other", "close" and "distant", and/or "relatives" and "non-relatives", applying to different categories of persons depending on the context. In order to further differentiate the category of *qarīb* (sg.)/*qarībīn* (pl.), the AAT draws on *primary* as well as on *composite kin terms.* A primary kin term can be at the root of a composite term, the latter thus being a combination of different primary kin terms. Contrary to Cuisenier and Miquel (1965: 30) who consider only those terms as "primary kin terms" which explicitly refer to a distinctive radical word and excluding those terms which are genderised by a suffix – in the following I shall include such genderised forms among the primary kin terms, particularly if the latter are understood by the Alawi informants as constituting a distinct category. A case in point is the Alawi distinction between *akh/khaye* (B) and *ikht* (Z), both terms ultimately deriving from Classical Arabic, the root letters Alif (أ) and Ḫ (ح) forming the term *akh* (B). The term *ukht* (Z) [ikht] is the feminine version of *akh* indicated by the suffix T (ت), the latter being a specific phonetic variant of the *ta marbuta* (ة). According to Cuisenier and Miquel's (1965: 30) interpretation of the Classical Arabic terminology, the term *ukht* is simply deriving from *akh* and is thus not a distinctive primary kin term. However, the majority of the Alawi informants in Hatay who are using the AAT on a daily basis are unaware that *akh/khaye* (B) and *ikht* (Z)[2] derive from the same root. Moreover, by simply focusing on the linguistic closeness between gender-differentiated terms (*akh/ikht*; '*amm/'amma*; *khal/khala* see column below), one could easily ignore that societies as the Alawi by using these genderised terms are projecting different uni-lineal descent segments (*'a'ile*), both at the paternal and the maternal side.

In the following I shall describe the consanguinal and affinal primary kin terms used in the AAT (terms of reference):

F	ab/baye	M	imm
B	akh/khaye	Z	ikht
S	ibn	D	bitt
FB; FFB; MFB	'amm	FZ; FFZ; MFZ	'amma
MB; FMB; MMB	khal	MZ; FMZ; MFZ	khala
FF; MF	ğidd	MM; FM	sitt

Fig. 1: Consanguinal primary kin terms (Alawi Arabic Terminology)

2 *Ikht* is the pronunciation of the word *ukht* in the Syrian dialect used by the Alawi.

Six of these primary kin terms are referring only to one genealogical position, whereas six others are classificatory terms merging several kin types. One immediately notices that those kin terms which only apply to one genealogical position (*imm*, *ab*/*baye*, *ikht*, *akh*/*khaye*, *bitt*, and *ibn*) are referring to the direct ascendants of Ego in G 1 (F, M), his/her siblings in G 0 (Z, B), and his/her direct descendants in G -1 (S, D). The other six primary kin terms designate collateral relatives on the paternal and maternal side. While *sitt* and *ğidd* are denoting the grandparents (PP) on both sides – only differentiating among their sex – the terms *khal* and *khala* and *ʹamm* and *ʹamma* indicate – beyond the sex differentiation – the idea of lineality. *ʹAmm* und *ʹamma* occur in G 1 as well as in G 2. In G 1 these terms are referring to the siblings of Ego's father, in G 2 to the siblings of the FF and the MF. In a similar vein, *khal* and *khala* refer to the siblings of Ego's mother in G 1 while denoting the siblings of FM and MM in G 2. In G 2 one thus encounters on the maternal side kin terms which are actually characteristic for the paternal side, *ʹamm* and *ʹamma* which are expressing a FB- or FZ-relation respectively. Likewise the terms *khal* and *khala* – although having a clear matrilateral stress – can be found in G2 on the paternal side. This suggests that from Ego's perspective a fundamental distinction is made between relatives belonging to the mother's and to the father's side in consecutive generations. In this context one encounters one of the major differences between the AAT and many other Arab relations terminologies where the siblings of Ego's grandparents are merged with the grandparents themselves and thus being denoted as *ğidd* and *sitt*. In the Alawi case, however, whether a relative is deriving from the mother's or father's side in consecutive generations can easily be determined within the terminology by adding the suffixes *al-ab*/*baye* or *al-imm* to a given kin term, meaning "from the father" and "from the mother", for example *ʹammi la- immi* means "my uncle (MFB) from the mother's side".

Affines are also denoted by using primary and composite kin terms.

H	ğauze	W	mart
HB, WZH	silfe	WZ	(baldiz)
HZH, DH, WB, BDH, ZDH	sihre	HBW	silfte
HM, WM	hmate	SW	(kante)
HF, WF	ʹamm		
FZH, MZH, FZDH, FFZH, FMZH, MFZH, MMZH	aziz		

Fig. 2: Primary Kin Terms for Affinal Relations (Alawi Arabic Terminology)

When looking at the primary kin terms for affinal relations three categories can be distinguished. First, there are the reciprocal terms used by husband and wife, *ğauze* and *mart*; then, there are terms which are used irrespective of

Ego's sex in order to indicate an affinal relative, for instance *hmate* (HM, WM); finally, one encounters reciprocal terms used by relatives from the side of a husband and a wife respectively, such as *silfe* (WZH) and *sihre* (HZH).

Mention should be made that in G 1, G 2, and G 3 female affines are usually denoted by composite terms, such as *mart ′amm* (FBW, HM, WM, literally meaning "wife of the paternal uncle") or *mart khal* (MBW), the only exception to this rule being the term *hmate* (HM, WM), meaning "mother-in-law". The male counterpart term of *hmate* is *′amm*, the latter also denoting her husband. In the AAT the term *′amm* is the only primary kin term expressing both consanguinal and affinal relations simultaneously. Accordingly, the term for the spouse of an *′amm* (HF, WF) is either *mart ′amm* (HM, WM) or *hmate*, the latter term being more commonly used in everyday life. It should be noted that until recently a *hmate* was often identical with a real *mart ′amm* (FBW), particularly applying in those cases when a patrilateral parallel cousin marriage (FBS = FBD) had been performed. Until recently, FBD-FBS-marriage was considered as the ideal marriage arrangement among the Alawi of Hatay (Prager 2010b, 2015).

Azize (FZH, MZH) is an affinal term simultaneously existing in G 1, G 2, and G 3 at the husband's (*ğauze*) and wife's (*mart*) side to denote the spouses of the parent's female siblings – for instance *khala* and *′ammta* – in ascending generations.

The final four primary kin terms for affines are referring to relations in G 0 and G -1 where they are covering nine *denotata.* For a male Ego the term *silfe* refers to a WZH, whereas for a female Ego it denotes HB. A woman can also call her *silfe* with the composite terms *ibn ′amm* (literally meaning "son of the paternal uncle") or *ibn hmate*, while a man can use the Turkish term *bacanak* (WZH).[3] The phenomenon that WZH and HB are designated by the same term, *silfe*, implies that in the AAT both kin types are conceptualised as being identical. Moreover, if from the perspective of a female Ego a *silfe* is identical with her *ibn ′amm* (patrilateral parallel cousin), the *silfe*-relation seems to correspond to the idea that two brothers have married two sisters – a marital practice which up to the 1990s was still part of the social reality of the Alawi in Hatay. Ideally, this type of marital relation should have taken the form of a double FBD-FBS-marriage.

The ideal of FBD-FBS-marriage is also reflected by the phenomenon that the WZ is referred to as *bit ′amm* (literally: "daughter of the paternal uncle"). However, the HBW is designated as *silfte,* the female form of the term *silfe.*

The primary kin term for the affinal relations of ZH, WB, HZH, DH, ZDH, BDH is *sihre*. However, in social practice this term is rarely used to denote affines in G 0; it is usually substituted by the composite term *ibn ′amm* (FBS)

3 About the influence of Turkish kin terminologies on neighbouring societies, particularly with regard to the appropriation of the term *bacanak,* see Fliche (2008).

or by the primary kin term *akh*/*khaye* (B), thereby transforming affinal into consanguinal relations at the level of G 0. When Alawi explain in-law-relations to "foreigners" the term *sihre* may be used in order to specify ZH, WB, HZH. If the latter, however, are present in the same room it would be considered as a gross violation of the etiquette to designate them by this term. This does not apply, however, for a DH, ZDH and BDH as these are always denoted by the term *sihre*. Other terms of reference which should not be used in the presence of the respective persons are *hmate* (HM, WM) and *kante* (SW), the latter being substituted by the consanguinal terms *mart 'amm* (FBW) and *bit khaye* (BD).

In the AAT the following primary kin terms are used to generate composite terms: *ibn* (S), *bit* (D), *mart* (W), *hmate* (HM, WM), *sihre* (DH, HZH, ZDH) and *kante* (SW). While the combination with *ibn* and *bit* allows for the construction of male and female consanguinal- and affinal composite terms (for instance *ibn 'amm*, FBS, HB, WB, etc.), the terms *sihre, mart, hmate* and *kante* only connote affinal relations. All the children of Ego's lineal and collateral relatives in G 0, G 1, G 2, G 3, G -1, G -2, and G -3 can be indicated by combing a term with *ibn* or *bit*, such as *khal* + *ibn* = *ibn khal* (MBS) or *'amma* + *bit* = *bit 'amma* (FZD).

Ibn and *bit* also occur in combination with the term *hmate* (*ibn hmate*: HPS; WPS or *bit hmate*: HPD, WPD). Moreover, combinations with *sihre* and *kante* are possible (*sihre khal:* MBDH, FMBDH, or *kante 'amma:* FZSW, MFZSW, etc.). Generally, one observes that composite kin terms occur mainly in G 2, G 3, G -2, and G -3, whereas in Ego's closer genealogical reach (G 0, G 1, and G -1) nearly all consanguinal and affinal relations are identified by primary kin terms. The composite terms of *bit 'amm* (HZ, literally meaning "daughter of paternal uncle") and *mart kayne* (WBW) are an exception to this rule.

Descendants of Egos's collateral relatives are consistently described by means of composite terms, thereby accentuating the genealogical distance between Ego and these collateral relations. Also the spouses of Ego's paternal und materal "uncles" (*'amm*, FB; *khal* MB) and "great-uncles" (*'amm*, FFB, MFB; *khal* MB, FMB) are always identified by composite terms, such as *mart 'amm* or *mart khal* contrasting with the primary term *azize* (see above) which is used without exception to denote the spouses of Ego's paternal and maternal "aunts" (*'amma*, FZ; *khala*, MZ) and "great-aunts" (*'amma*, FFZ, MFZ; *khala*, MMZ, FMZ).

In conclusion one notices that the AAT features many cases of – if one follows Lounsbury's (1969b: 208) terminology – "skewing". Across the generations one finds for instance a combination of the terms *'amm, 'amma, khal*, and *khala* with those of *mart*, *hmate,* and *azize*.

If one considers the kinship vocabulary as a whole, a hierarchical order comes to the fore giving consanguinal relations priority over affinal ones. If there is the choice to either denote a person by a consanguinal or an affinal term (irrespective of whether the term is a primary or a composite one), as a rule, the consanguinal term will be preferred as it implies a closer degree of relatedness. This also comes to the fore in the above mentioned situations when affinal relatives – if present in the room – should be designated by a consanguinal and not by their affinal term.

On the whole, the AAT is characterised by a clear differentiation between the sexes and the accentuation and strict separation of lineal segments in ascending and descending generations. This is also important for articulating *'a`ile*-relations. As mentioned before, among the Alawi the *'a`ile* is the core social and ritual group consisting of an agnatic line (F; [sometimes FeB], Sb, S, D) stretching over not more than three generations (see Prager 2010a). However, the AAT as a whole only partially covers the inherent agnatic tendency; in fact the structuring logic of this system of classification only comes to the fore from the perspective of one generation to the other. Thus as an Ego one has to take over the perspective of the successive generations in order to acknowledge the differences made between the different patrilines.

The AAT is determined by a strict differentiation of sex in all generations allowing for a clear separation of laterality.

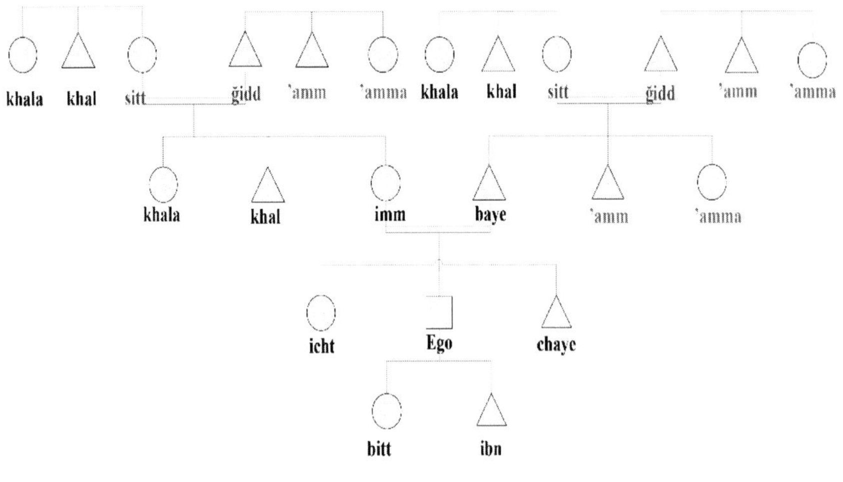

Fig. 3: Change of perspective in the AAT

Accordingly, Ego's relations in G 0 for instance to an *'amma* (MFZ) in G 2 is not self-explanatory, as the term *'amma* in its first meaning denotes the relation of Ego to a FZ, thus a paternal genealogical relation. Thus Ego has a "real" *'amma* (FZ) in G 1, mediated through the father, whereas the other *'amma*-relations are mediated either through the FF or the MF. Accordingly,

Ego in G 0 would designate the MFZ as *'ammte*[4] *la-immi* = the FZ of my M. This implies that "skewing" always results from a "change of perspective", i.e. Ego deducing specific relations from the perspective of another Ego. The apparent "asymmetry" in the relationship terminology which emerges from the fact that Ego has *khal-*, *khala-*, *'amm-* and *'amma*-relations both on the maternal and paternal side in consecutive generations, can only be explained if one takes into account that the AAT has to be "read" from the perspective of one generation at a time. This is further corroborated if one turns to the affinal kin terms. As already discussed, a variety of affinal kin terms in G 0 and G 1 indicate the ideal of FBD-FZD-marriage. However no comparable terminological equivalences are occurring in G -1 and G -2. Thus, the AAT does not show a consistent pattern which would hint to the existence of a marriage prescription as understood by Needham (1973).

The Alawi-Turkish Terminologies

Apart from the AAT – as already mentioned above – three other kinship terminologies are in usage among the Alawi of Turkey. I shall designate these systems of classification as "Alawi-Turkish Terminologies" as these vocabularies all have been influenced – although to a different degree – by the Turkish language and its concomitant kinship denotations. Nevertheless, the Alawi-Turkish terminologies constitute genuine forms of kinship classification as they display significant differences to other Turkish kinship terminologies.[5] The three variants of the Alawi-Turkish terminology reflect – as will become evident in the following discussion – a gradual transformation from a traditional "Arabic" system of kin classification to a modern Turkish form of denoting kinship relations, particularly in urban contexts.

The first variant of these terminologies (Alawi-Turkish-Terminology I, henceforth ATT 1) still shares some similarities with the AAT. It is mainly used by Alawi work migrants (irrespective of age) who originally came from Hatay and have migrated to the Adana-Mersin region some 40 years ago. Usually, these migrants are still upholding close contacts with the Hatay-region through marriage arrangements, economic and ceremonial exchanges, and by a network of relatives and friends.

The second variant of the Alawi-Turkish-Terminology (henceforth ATT 2) is characteristic for the older generations of urbanised Alawi in Adana city, whereas the third variant (henceforth ATT 3) is typically used by the younger generations in this city. Although the Alawi from the rural and urban Çukurova region are familiar with all of these variants, individual speakers usually

4 *'Ammte* is built from the term *'amma* with the possessive suffix for the first person singular.

5 Other Turkish terminologies have been described by Stirling (1965), Cuisenier (1967), Casson (1973), Delaney (1991), Gokalp (1994), and others.

use only one of these terminologies, depending on the origin of the speaker, his/her age, and the influence that Turkish culture had on his/her upbringing.

In the following description and analysis, I shall focus on the ATT 1, comparing it with the other two variants (ATT 2 and ATT 3) only in those cases where significant differences are coming to the fore.

In all ATTs the primary kin terms for consanguinal relations are as follows:

F	baba	M	anne
B	kardeş; ki-kardeş	Z	kiz-kardeş; baci
Be; FFBS; FFZS; FMBS; MFBS; MFZS; etc.	abi	Ze	abla
S	oğul, erkek	D	kiz
FB; FFB; MFB;	amca	FZ; FFZ; MFZ;	hala
MB; MMB; FMB	dayi	MZ; MMZ; FMZ;	teyze
BS; BD; ZS; ZD	yeğen		
SS; SD; DS; DD; SSS; SDS SDD; SSD; DSS;DDS;	torun		

Fig. 4: The primary kin terms for consanguinal relations in ATTs

The terms *baba* (F) and *anne* (M) mark the parents of Ego. For Ego's siblings four terms are used, as in contrast to the AAT in the ATTs a difference is made between "elder" (*abi*: eB and *abla*: eZ) and "younger" siblings ((*ki-)kardeş*: yB*;* (*kiz-*)*kardeş* or *baci:* yZ). A male Ego calls his younger brother *kardeş* (B, Sb) or uses the composite term *ki-kardeş* (B), if it is necessary to mark the sex of his younger sibling. The sex of a younger sister, however, will always be accentuated by a male speaking Ego as the term *kiz* ("girl", "daughter") will be used as a prefix to *kardeş* (= *kiz-kardeş).* In some cases a younger sister can also be designated by the alternative term *baci* (Z). A female Ego also uses the terms *kiz-kardeş* (yZ), *ki-kardeş* (yB) or *kardeş* (yB) to denote her younger siblings.

Self-reciprocal terms such as *kardeş* or *kiz-kardeş* are only employed if two same sex younger siblings have an elder sibling of the same sex. The eldest female sibling would be called *abla* by her younger siblings, whereas the eldest male sibling would be designated as *abi* (see fig. 5). The differential denotation of the siblings indicates that there is an asymmetry, male sex being valued higher than the female one. While it is sufficient to use the term *kardeş* to identify a younger brother, a younger sister is designated by the composite term *kiz-kardeş* (or *baci*). This implies that the term *kardeş* ("sibling") primarily has a male connotation.

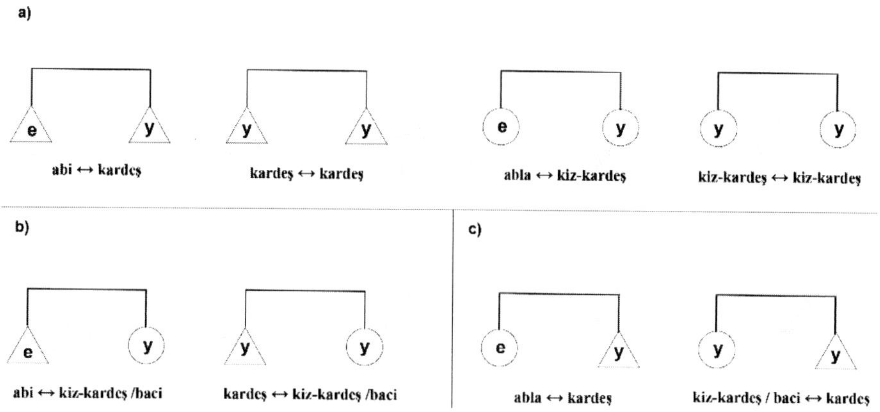

Fig. 5: Sibling-relations in ATTs
a: terms used between same sex siblings;
b: terms used between siblings of opposite sex, seen from the perspective of a male Ego;
c: terms used between siblings of opposite sex, seen from the perspective of a female Ego

The term *kiz* also denotes Ego's daughter, whereas Ego's son is named *oğul* or *erkek*. While a sex differentiation is made for Ego's children, no such differentiation occurs among the children of Ego's siblings (irrespective of their relative age), they are all classified by the categorical term *yeğen* (BS, BD, ZS, ZD). In G -2 and G -3 all lineal descendants of Ego or those of Ego's siblings are denoted as *torun* (SS, SD, DS, DD, SSS, SDS, SDD, SSD, DSS, DDS, ZDD, ZDS, BDD, BDS, BSD, BSS, etc.), thus no longer distinguishing between relative age, sex, or laterality. In G 2 the terms *babaanne* (FM), *banneanne* (MM), and *büyükbaba* (FF, MF) to some extent still hint to the importance of linearity, whereas in G 3 with the terms *büyükbaba* (FFF, MFF, MMF, FMF) and *büyükanne* (FFM, FMM, MFM, MMM) no distinction between patri- and matrilateral relations is made at all, as only the sex is being distinguished.

The last four primary terms designate the siblings of Ego's parents and those of the grandparents. In G 1 the term *amca* denotes FB and in the ascending generations (G 2, G 3, etc.) it covers various types of FB-relations (FFB, MFB, MMFB, etc.). This skewing rule is mirrored by the terms *hala* (FZ, FFZ, MFZ, etc.), *dayi* (MB, FMB, FFMB, etc.), and *teyze* (MZ, MMZ, FMZ, FFMZ, etc.). However, in contrast to the AAT the children of these collateral relatives are not denoted by descriptive composite terms which would allow for a clear distinction of lineal segments, but are all subsumed under the terms *abi* (for example MFBS) and *abla* (for example FMBD), i.e. the same terms which are also used to denote the elder siblings of Ego. Hence, the ATTs in contrast to the AAT are only partially consistent in the separation of lineal segments, laying instead more stress on relative age.

The primary kin terms for affinal relations in ATT 1 are as follows:

Generation 0			
H	koca, ärkik, eşi, hayatar kadaşi	W	hayatar kadaşi, eşi, kari, kadi
HB, WB	kayin	HZ	görümçe
		WZ	baldiz
WZH	bacanak	HBW	elti
ZH, HZH,	enişte	BW; WBW;	yenge
Generation –1			
DH	damat	SW	gelin
BDH, ZDH	enişte	BSW, ZSW	yenge

Fig. 6. Primary kin terms for affines (ATT 1)

The majority of the affinal terms are to be found in G 0 and in G -1. The terms used to denote a "husband" are *koca*, *ärkik*, *eşi* and *hayatar kadaşi*. The latter two can also be used to refer to a „wife". Therefore, both *eşi* and *hayatar kadaşi* are self-reciprocal terms, the sex differentiation only coming to the fore if one knows the speaker's sex. Apart from these "termes sexe relatif" (Barraud 2001), there are other terms which may be used for "wife", such as *kari* and *kadi,* both, however, being rarely employed in practice.

Other primary affinal terms in G 0 are *kayin* (HB, WB), *görümçe* (HZ), *baldiz* (WZ), *bacanak* (WZH), and *elti* (HBW). *Enişte* (ZH, HZH, FZH, BDH, etc.) and *yenge* (BW, WBW, FBW, BSW, etc.) are classificatory terms which are denoting in-law relations with persons who are not considered to be close relatives. This idea is further corroborated by the fact that the conceptual pair *enişte/yenge* in figurative speech is employed to refer to the "other" or to "foreigners".

As already noted, *bacanak* and *elti* are denoting the spouses of the same-sex-siblings of husband and wife. Many of my Alawi informants – of both sexes – who are using this terminology explained that: "She/He [*elti/bacanak*] is like my own sister/brother." Therefore, a male Ego and his *bacanak* (WZH) consider themselves as identical and employ this self-reciprocal term. The same holds true for a female Ego and her *elti*-counterpart.

Finally, in G -1 one finds two additional primary kin terms for affines: *damat* (DH) and *gelin* (SW). While in ATT 2 and ATT 3 all other affinal relations in this generation are identified by the terms *enişte* and *yenge*, in ATT 1 a range of composite terms is used, clearly indicating that ATT 1 is closer to AAT than the other two variants. Such composite terms are, for example, *abi enişte* (eBDH), *kardeş enişte* (yBDH), *abi gelini* (eBSW) or *kiz kardeş yenge* (yZSW). In addition to G -1, the individual variants of the ATTs also show differences in the way in which affines at the level of G 1, G 2, and G 3 are denoted. Thus, in ATT 2 and ATT 3 *yenge* is employed for all female affinal collateral relations, whereas in ATT 1 composite terms are used, such as *amca hanimi* (FBW), *amca karisi* (FBW), *amca eşi* (FBW), *dayi hanimi* (MBW), etc.

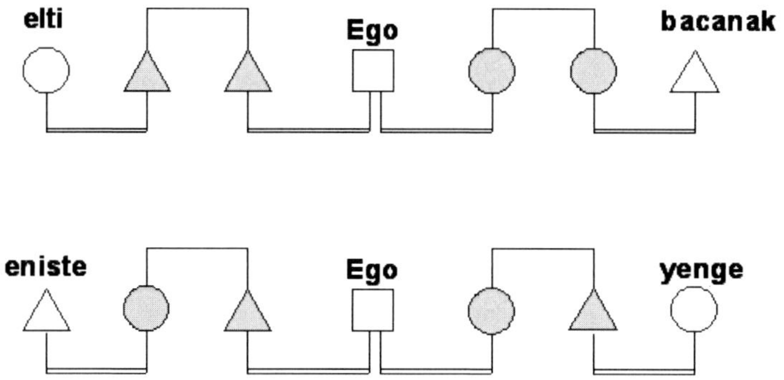

Fig. 7: *elti-bacanak-* and *enişte-yenge-*relations

The composite terms used for affinal relations in ATT 1 – as described above – thus imply firstly a distinction between lineal segments; secondly, they indicate the importance of relative age, the latter feature being completely absent from the AAT and the other two Turkish variants. However, in all three ATTs e*nişte* is used to denote the husbands of the siblings of Ego's parents, grandparents, and *great grand-parents* (FZH, MZH, FFZH, etc.). Hence, while in ATT 1 the female affinal collateral relations from G 1 upwards are distinguished by means of the composite terms, the male affinal counterparts are merged, thereby revealing a noticeable asymmetry within this classificatory system.

As already mentioned, the terms designating the siblings of Ego's parents are subjected to a *skewing* rule in consecutive generations which also applies to their spouses. Moreover, even the spouses of the children descending from these relations (PPSbCh and PPPSbCh, etc.) are classified in accordance with this skewing rule. In ATT 1 and ATT 2 the PSbCh in G 1 are denoted by composite terms, such as *amca oğul* (MBS) or *teyze kiz* (MZD), their spouses being respectively called *amca gelin* (MBSW) or *teyze enişte* (MZDH). While the composite terms for the PSbCh are constructed by referring firstly to Ego's relation to his/her PSb and then deriving the relation to their children, the connection of Ego to the latter's wives is modelled on a quite different principle of classification, in that the PSbChW are not derived from their husbands (i.e. the composite term does not include *yenge*), but are designated by composite terms which all contain the word *gelin,* meaning "daughter-in-law".

The situation is quite different for the male affinal relations, as the composite terms include *enişte*, and not the category of "son-in-law" (*damat*), for example *amca enişte* (FBDH), *teyze enişte* (MZDH), or *dayi enişte* (FZDH). By applying the term *enişte*, these affines are equated with relations to "foreigners", clearly contrasting with the *gelin*-relations described above. In the widest sense *enişte*-relations are thus relations where "consanguinally" related women are married out, whereas *gelin* are women who have married in.

However, in ATT 2 the above mentioned distinctions between male and female spouses as belonging either to the in-laws or the category of foreigners (PSbChSp; PPSbChSp; eSbChSp, ySbChSp) are not accentuated anymore. Moreover, the distinction between the paternal and maternal side as well as the emphasis on relative age are no longer an issue, as only *enişte* and *yenge* are employed to indicate affinal relations.

An even more radical change occurs in ATT 3 where not only the terms *enişte* and *yenge* are employed to denote PPSbChSp, eSbChSp, ySbChSp, etc. (as in ATT 2), but all PSbCh are identified by the uni-sex kin term of *kuzen*, deriving from the West-European and Anglo-American "cousin". In G 0 the non-differentiation between sex, laterality, and relative age also applies to the PSbChSp who are named *kuzen eşi*. This "merging" of all PSbCh in ATT 3 reflects the idea that kinship relations are increasingly conceptualised in terms of bilateral descent or cognation.

Further composite terms for affinal relations are *kayinbaba/kayinpeder* (HF, WF) and *kaynana/kayinvalide* (HM, WM) designating the SpF and SpM respectively. These composite terms are made up by the primary affinal kin term *kayin* (HB, WB) and the primary consanguinal kin term *baba* (F) or *anne* (M). Thus in contrast to the AAT, where the relation to the SpP is described in consanguinal terms (FB) indicating the ideal of FBD-FBS-marriage, in the ATTs the relation to the SpP is clearly conceptualised as a purely affinal one. The same holds true for the relation to a SpB. This implies that FBD-FBS-marriage is no longer indicated in the ATTs, and indeed, in social practice this type of marriage has almost disappeared among those Alawi who are using these "Turkish terminologies" (cf. Prager 2010b, 2015).

From the Arabic to the Turkish classification

As stated at the beginning of this article, the four relationship terminologies used by the Alawi represent different stages within a process of gradual adaptation to the Turkish language and its concomitant system of kin classification. In this process of terminological change, the AAT used by the Alawi from Hatay historically constitutes the oldest variant involving hardly any borrowings from the Turkish kin vocabulary.

The central characteristics of the AAT can be summarised as follows:

1. a clear differentiation of sex
2. the distinction of lineal segments

3. a preference for patrilateral parallel cousin (FBD-FBS)-marriage (indicated in G 0)
4. consanguinity is valued higher than affinity

Some of these features are retained in the Alawi Turkish terminologies, while others disappear or apply only to specific generations.

The ATT 1 used by the rural Alawi from the rural Mersin region still shows a high degree of continuity with the AAT. In practice, most of those persons using the ATT 1 are still familiar with the AAT. Therefore, if they are meeting with Arabic speaking Alawi from Hatay, they can easily switch to AAT, though misunderstandings can occur. In such cases it is becoming evident that the ATT 1 is not merely a Turkish translation of the AAT, but that in the ideological universe of the speakers using the ATT 1 fundamental changes have occurred, impacting on the way in which different categories of kin relations are conceptualised. The ATT 1 differs from the AAT particularly with regard to the following aspects:

1. sex differentiation is decreasing
2. the strict separation of lineal segments is only uphold for consanguinal relatives in ascending generations and their respective affinal relations
3. importance of relative age
4. the affinal terms do not indicate any preference/prescription for a particular type of marriage

The differences vis-à-vis the AAT also clearly come to the fore when taking a closer look at the term *torun* (SDS, DSS, DDD). With *torun* – as already stated before – Ego's consanguinal female and male relatives in G -2 and G -3 are designated. The differentiation of laterality thus disappears in these generations. Nevertheless one observes that linearity is still upheld to a certain extent among the ascendant generations. Comparable with the skewing rules of the AAT, also in the ATT 1 a differentiation between the patri- and matrilateral siblings of Ego's parents and grand-parents occurs in consecutive generations, allowing for the identification of lineal segments.[6]

ATT 2 shows a higher degree of adaptation to the Turkish terminologies (c.f. Cuisenier 1967) than ATT 1. ATT 2 is mainly used by urbanised Alawi in Adana who have settled in this city from the 19th century onwards (Prager 2010b: 23). Undoubtedly, the close cohabitation with the Turkish and Kurdish majority in Adana has led to a more accentuated form of assimilation of

6 However, this distinction only applies for the following ascendants (*amca* FB, FFB, MFB, MMFB; *hala* FZ, FFZ, etc.; *daye* MB, FMB, etc.; *teyze* MZ, MMZ, etc.; *banneanne* MM, *babaanne* FM), whereas in contrast to the AAT the children of the classificatory *amca*, *hala*, *daye* or *teyze* in G 1 and G 2 are subsumed under the terms *abi* and *abla*, thus not being distinguished by the criterion of laterality, but solely by sex and relative age.

"Turkish" categories. In fact, ATT 2 shows various significant differences in comparison to AAT and ATT 1:

1. a further decrease of laterality
2. categories of affinity are less differentiated
3. potential „cousin marriages" are not indicated at all

The further decrease of the importance of laterality manifests itself by the fact that affinal terms across the generations only feature sex differentiation. Hence, in all generations skewing-rules are at work, according to which all male spouses of collateral relatives are referred to as *enişte* and all female spouses as *yenge* respectively. Ultimately, this leads to a progressive dissolution of differentiations among affinal relations, which is also accompanied by the absence of terminologically indicated marriage preferences. The children of Ego's parent's siblings are still denoted by composite terms (such as *amca kiz* (FBD), *teyze oğul* (MZS)), thus allowing for the differentiation of their relative genealogical positions and identifying them as marriageable categories (as in ATT 1), whereas the classification of the latter's spouses is determined by the above mentioned terms of *enişte* and *yenge*. This implies that the spouses of the PSbCh and PPSbCh, etc. are no longer considered being identical with Ego or his/her brothers and sisters.

This development from a classificatory system in which potential marriage partners are designated by consanguinal terms – as in the case of the AAT – to a vocabulary where marriage with close relatives is terminologically more or less excluded – finds its completion with ATT 3. This terminological system which is mainly employed by the younger generations of urbanised Alawi in Adana is marked by the incorporation of the "Western-European" term *kuzen* (cousin) which is applied to all PSbCh.

Although the distinction of linearity persists in ATT 3 for the ascendants in G 1, G 2, and G 3 (again differentiating among PSb, PPSb, and PPPSb), in G 0 any indication of linearity and laterality has disappeared altogether when all PSbCh are designated by the same uni-sex term *kuzen*. In G 0 the paternal and maternal sides are no longer distinguished, which is a major difference to the AAT and the other variants of the ATT. With the merging of the PSbCh the terminology has acquired a clear cognatic orientation where the distinction between relatives from the mother's and father's side in G 0 is "blurred". Moreover, sex differentiations are also less pronounced in comparison with the other Alawi terminologies.

Generally one observes that the transition from AAT to ATT 3 is accompanied by a continuous decrease of the number of kin terms used both for consanguinal and affinal relatives.

Concluding Remarks

In conclusion, it can be noticed that the Alawi relationship terminologies have been subject to a fundamental change initiated by the adoption of – and adaptation to – the Turkish form of kin classification. If one would follow Murdock's (1949) classification, the development from AAT to ATT 3 could be described as a transition from a descriptive "Sudanese" classification system (AAT, ATT 1, and ATT 2) to an "Eskimo"-type of terminology (ATT 3). This raises the question whether these transformations within the classificatory scheme can be attributed to changes in the social structure and practice.

Indeed, the migration of the Alawi to the Mersin and Çukurova regions has entailed various changes in the social structure, in that – for example – larger patrilineal groups could not be upheld in the same way as in Hatay. Only singular families had migrated to these regions and as time went by the nuclear family became the focal social group. The same holds true for the postmarital residence rule of viri-patrilocality which could hardly be applied in a highly urbanised environment. Neolocality thus emerged as the standard mode of postmarital residence. Moreover, from the 1980s onwards, the Alawi living in Adana and the Mersin region were increasingly influenced by state-sponsored discourses on the alleged genetic risks going along with "consanguinal" or "cousin marriages". During the same time, the Turkish state had begun to launch educational and medical counseling programmes in order to prevent the spread of hereditary diseases resulting from such "dangerous liaisons" (c.f. Prager 2015).

Since for the Alawi in the Adana and Mersin region Turkish is the basic language, these programmes had a much stronger impact on them compared with their counterparts in Hatay where only the younger generation uses Turkish on a daily basis. In the Adana and Mersin region the Turkish campaign against cousin marriages on the long run gave rise to changing conceptualisations of marriage, kinship, and family. In my study on the consequences resulting from these state financed programmes against hereditary diseases, I could show that marriage choices among the Alawi in Turkey have been subject to fundamental changes over the last generations. Particularly for Adana city one notices a clear statistical trend showing that marriages with cousins have almost disappeared (Prager 2010b: 131-157, 179-194, Prager 2015). Finally mention should be made that the Alawi from Adana only constitute a small minority among the Turkish and Kurdish majority population. Thus unlike their Hatay and (rural) Mersin counterparts, the Alawi from Adana have been under strong pressure to adapt to Turkish culture. The transformation of the Alawi relationship terminologies – as I would argue – to a large extent reflects these changing social and historical circumstances.

Nowadays, research on the historical change of relationship terminologies as initiated by various scholars (Eggan 1937; Spoehr 1947; Friedrich 1966; Service 1985; Mesthrie 1990) has gained a new relevance, particularly in a time when millions of people are on the move due to diaspora, work migration, natural disasters, and war. Having settled in new surroundings, many of these migrants and refugees are confronted with the situation that they have to adapt to different linguistic and cultural contexts, thereby often using two or more languages simultaneously and adopting new forms of (kin) classification.

The study of relationship terminologies, which often has been considered as an outdated and dispensable tool of kinship analysis, thus acquires a new topicality, as it is precisely through the comparison of such kinship vocabularies that one can assess the various degrees of adaptation and acculturation of social groups to new ideological horizons.

References

AZZI, J. (2002) *Les Noussairites-Alaouites (histoire, doctrine et coutumes).* Paris: Clamecy.

BARRAUD, C. (2001) De la distinction de sexe dans les sociétés une présentation. In ALÈS, C., BARRAUD, C. (eds.), *Sexe relatif ou sexe absolu?,* pp. 23-99. Paris: Éditions des Sciences de l'Homme.

BRAWER, M. (1988) *Atlas of the Middle East.* New York: Macmillan Publishing Company.

CASSON, R.W. (1973) Paired Polarity Relations in the Formal Analysis of a Turkish Kinship Terminology. *Ethnology* 12(3): 275-297.

CUISENIER, J. (1967) Systèmes de succession et de dotation en Yougoslavie et en Turquie. *L'Homme* 7(3): 25-47.

CUISENIER, J. and MIQUEL, A. (1965) La terminologie arabe de la parenté. Analyse sémantique et analyse componentielle. *L'Homme* 5(3-4): 17-59.

DELANEY, C. (1991) *The Seed and the Soil: Gender and Cosmology in Turkish Village Society.* Berkeley et al.: University of California Press.

EGGAN, F. (1937) Historical Changes in the Choctaw Kinship System. *American Anthropologist* 39: 34-52.

EICKELMAN, D.F. (2002) *The Middle East and Central Asia: An Anthropological Approach.* Upper Saddle River, NJ: Prentice Hall.

ENGIN, I. and FRANZ, E. (eds.) (2000) *Aleviler/Alewiten.* Hamburg: Deutsches Orient-Institut.

FLICHE, B. (ed.) (2008): Thematic Issue: The Social Practices of Kinship. A Comparative Perspective. *European Journal of Turkish Studies* 4.

FRIEDRICH, P. (1966) The Linguistic Reflex of Social Change: From Tsarist to Soviet Russian Kinship. *Sociological Inquiry* 36(2): 159-185.

GOKALP, A. (1994): "Mariage de parents": Entre l'échange généralisé et le mariage parallèle. Le cas de la Turquie. In BONTE, P. (ed.), *Épouser au plus proche – Inceste, prohibitions et stratégies matrimoniales autour de la Méditerranée,* pp. 439-452. Paris: Éditions de l'EHESS.

HALM, H. (1982) *Die islamische Gnosis. Die Extreme Schia und die 'Alawiten.* Zurich and Munich: Artemis Verlag.

KHURI, F.I. (1991) The Alawis of Syria: Religious Ideology and Organization. In ANTOUN, R.T. and QUARTAERT, D. (eds.), *Syria: Society, Culture, and Polity*, pp.49-61. New York: University of New York Press.

LOUNSBURY, F. (1969b) The Structural Analysis of Kinship Semantics. In S.A. Tyler (eds.), *Cognitive Anthropology*, pp. 193-212. New York: Holt, Rinehart and Winston.

MERWIN, S. (2007) Des Nosayris aux Ja'farites: le processus de "chiitisation" des Alaouites. In DUPRET, B. et al. (eds), *La Syrie au Présent: Reflets d'une Société*, pp. 359-364. Beirut: Actes Sud.

MESHRIE, R. (1990) The Linguistic Reflex of Social Change: Caste and Kinship Terms among People of Indian Descent in Natal, South Africa. *Anthropological Linguistics* 32(3/4): 335-353.

MORGAN, L.H. (1871) *Systems of Consanguinity and Affinity of the Human Family.* Washington, DC: Smithsonian Institution.

MURDOCK, G.P. (1949) *Social Structure.* New York: Macmillan.

NEEDHAM, R. (1973) Prescription. *Oceania* 42:166-181.

NIMIER, A. (1987) *Les Alawites*. Paris: Édition Asfar.

PRAGER, L. (2010a) Âmes sexuées et idées de procréation chez les Alawites/Nousairites (en Turquie). *Anthropology of the Middle East* 5(2): 77-99.

PRAGER, L. (2010b) *Die 'Gemeinschaft des Hauses': Religion, Heiratsstrategien und trans-nationale Identität türkischer Alawi/Nusairi-Migranten in Deutschland.* Münster: LIT.

PRAGER, L. (2013) Die Zeichen der Wiedergeburt: Körper, Stigmata und Seelenwanderung bei den Alawiten/Nusairiern der Südosttürkei, *Paideuma* 59: 237-260.

PRAGER, L. (2015) 'Dangerous Liaisons': Modern Bio-Medical Discourses and Changing Practices of Cousin Marriage in Southeastern Turkey. In SHAW, A. and RAZ, A. (eds.), *Cousin Marriages Between Tradition, Genetic Risk and Cultural Change*, pp. 88-109. Oxford: Berghahn Books.

PRAGER, L. (2016) The miracle of rebirth. Stigmata, transmigration, and the remembrance of former lives in Alawi religion. In KURZ, S., PRECKEL, C. and REICHMUTH, S. (eds.), *Muslim Bodies: Body, Sexuality and Medicine in Muslim Societies*, pp. 273-302, Münster: LIT.

SERVICE, E.R. (1985) *A Century of Controversy. Ethnological Issues from 1860 to 1960.* London: Academic Press.

SPOEHR, A. (1947) Changing Kinship Systems: A Study in the Acculturation of the Creeks, Cherokee and Choctaw. *Field Museum of Natural History Anthropological Studies* 33(4): 151-235.

STIRLING, P. (1965) *Turkish Village.* Liverpool/London: Charles Birchall.

Exchange

The Sickening Gift[1]

Helene Basu

In an article entitled the "The Healing Gift", Jos Platenkamp argues that Tobelo healers acquire medical knowledge from diverse sources, such as "strangers" or "non-human beings", which they reciprocate with specific things given as gifts (Platenkamp 1996). He emphasises, however, that the "efficacy of the gift" does not reside in the object, but in "the act of transfer as such" (1994: 330). Drawing on this insight, I shall explore concepts and transfers of occult illnesses (madness) in Muslim India. The ethnography is derived from fieldwork at the shrine of the Sufi saint Mira Datar in North Gujarat, which is an important regional center of Muslim and multireligious pilgrimage (Hindus, Jains and others).[2] Mira Datar's sanctity is closely associated with black magic or sorcery, as well as with Islamic and Hindu classes of spirits (*jinn, bhut-pret*). The shrine shelters a "family of saints" whom pilgrims approach for the power (*karamat*) to heal disorders of mind and body and social suffering afflicted by sorcery.

Islam accommodates *jinn*, sorcery (*as'sihr*) and black magic (*kala jadu*) in a hierarchical network of beings created by God (humans made from earth/*jinn* made from fire); religious and medical epistemologies (Dols 1992; Haq 2010) and ritual practices of counter-sorcery or healing. In South Asian Islam, Sufi shrines engaging in the latter often lack the honorable genealogy of medieval Arabian orders and are associated with Islamic learning and institutionalised transmissions of charisma (Green 2012). In Gujarat, too, such shrines are accorded a relatively marginal status by Sufi leaders, whose claims to higher status often involve distancing themselves from healing occult illnesses, which become spectacularly manifest in madness and possession trance.

Anthropologists repeatedly drew attention to the "syncretic" character of Sufi practices, in which Islamic and Hindu forms of worship and ritual performances become merged. Such practices, it has been argued, build on a shared

---

1 The research informing this paper was made possible by a generous grant from the Deutsche Forschungsgemeinschaft to the Cluster of Excellence "Religion and Politics in Pre-Modern and Modern Cultures" (2007-2012) at Muenster University. Earlier versions of this paper were presented at the Anthropology Department Seminar Oxford, Nov 29th, 2011 and during the Workshop "Psychotic Experiences, Religion, and Spirituality" organised by CERMES3 at Paris, May 14-16th, 2012. I wish to thank the participants of both seminars for fruitful comments and discussions. My gratitude goes especially to Rebecca Walsh for her thoughtful editing of this paper.

2 The article is based on intermittent fieldwork carried out between 2008 and 2012.

'South Asian grammar' which transcends boundaries between religious 'systems' (e.g. Werbner and Basu 1998; Bellamy 2011; Quack f.c.). If it can be stated that there is a shared grammar at all, it may stem from the Indian logic of gift exchange, which in the case I am looking at informs equally conceptions of madness resulting from sorcery and ritual practices of healing or counter-sorcery.

Sorcery and witchcraft are subjects of a large body of anthropological literature, spanning debates on rationality, modernity, postcolonial politics, and medical anthropology. Here, I attend to the performativity of black magic in the context of kinship and gender as discursively constructed by the institution of a Sufi healing shrine (*dargah*) in post-liberalisation India. With a rising urban Indian middle class that is wedded to culturally inflected ideologies of modernisation (Jaffrelot and Veer 2008; Donner 2011), ritual healing practices have come under the critical scrutiny of mental health professionals, psychiatry and the state. I have dealt elsewhere with this phenomenon (Basu 2014a; 2014b). A major issue of contention arises from the mind/body dichotomy, which is crucial in psychiatry but absent in Hindu and Muslim practices of ritual healing (cf. Naraindas et al. 2014). In this article, I inquire into the relationships between healing, gift exchange, and undivided bodies and minds. More specifically, I am interested in the ways in which the realities of sorcery and good and bad divinities are produced as ontological beings, and how these realities, drawing on the ideas of Bruno Latour, persist in the face of intensified "practices of purification" (Latour 1991; for details of this approach, cf. Sax and Basu 2015).

A major trope evoked in the discourse institutionalised at Mira Datar depicts mind-body disorders as 'sent illnesses' (cf. also Callan 2012). Black magic works by 'sending' what I call a 'sickening gift' to a vulnerable recipient. 'Sickening gift' seems to be an apt translation for the term employed at the shrine *bala*; it fuses matter with the donor, in this case disembodied agents (human and non-human) and is charged with the Maussian "spirit of the gift": in this case 'bad' emotions (especially envy), ill-wishing, and the aggressive desires of the donor or sender. The sickening gift of black magic, I argue, resembles the Hindu gift of *dana*. *Dana* has been the focus of much anthropological attention in India and beyond because it is a non-reciprocal gift (Parry 1985; Gregory 1997). It is a gift that transfers the sins, inauspiciousness and impurities of the donor to funeral Brahmans, sweepers or wife-taking affines (Parry 1985; Raheja 1988). The category of *dana* seemingly contradicts the Maussian dictum that the morality of gift exchange (in distinction to market transactions) resides in the obligation to return the gift at a later point in time (Parry 1985; Mauss 1990). Unfolding its effects according to the logic of *dana*, the sickening gift introduces non-reciprocity into the domestic life-worlds shaped by reciprocal transactions in the domain of "nurture kinship"

(Schneider 1980; Holland 2004). Nurture kinship is created through the informal, everyday exchanges of food, speech, emotions, touch, or, more generally, embodied transactions of coded substances (Marriott 1976). The sickening gift (*bala*) reverses the life-giving effects of transactions facilitating social reproduction and turns them into destructive, fragmenting and deadly forces.

The Logic of the Sickening Gift

The shrine of Mira Datar orchestrates a complex discourse of black magic. Magic, as Marcel Mauss famously proposed, may pervade religion (or medicine, check intro), but it is still distinct from it because of its "association with evil" (Mauss 1972: 27). According to Mauss, ideals and moral values represent one pole of 'magic religion' and the other pole is constituted of 'evil magic' (*ibid.*). Between these poles "we have a confused mass of activities whose specific nature is not immediately apparent" (*ibid.*). The institution of the shrine of Mira Datar appears as just one such muddled site. According to Mauss, religious acts articulate moral values through the obligations arising from gift exchange; but he saw no morality attached to magical rites since they seemed to lack reciprocal gift giving. Magical rites, Mauss wrote, are "always considered unauthorised, abnormal and, at the very least, not highly estimable" (*ibid.*: 29). The institution of Mira Datar, however, provides an arena for ritual and narrative performances, which stress transacting gifts equally in regard to 'religion' (saints) as in regard to 'magic' (*kala jadu*).

Black magic, as is well known, is denounced in both religious reformist and modernisation discourses as 'superstitious' and 'false belief'. Against modernist notions of 'belief' Latour coined the oxymoron of 'factish' entities or 'divinities' (Latour 2010), which emerge from networked interactions of 'making' or 'fabricating' them, so that they become experiential realities. 'Factish' in the sense of Latour thus refers to something or somebody moving through a passage which links 'fabrication' with 'reality' (*ibid.*: 35). In such a process, black magic and spirits become tangible realities, too. In the journey of the sickening gift, this passage is marked by the traces it leaves in narrative and material and bodily substances before unfolding as a reality that is open to inspection. In the case of the shrine of Mira Datar it is the reality of an institution which – following Mary Douglas – does the classifying and labelling of persons, things and events and thus brings into being what it describes (Douglas 1987). This process is mediated by transactions that link multiple human and non-human agencies, which are visible in the body and behavior of pilgrims suffering from occult afflictions. The sickening gift becomes known by its effects on the mind and body of those who got hit. Through behavioral enactments, the agencies playing havoc with the mind and body of the afflicted become tangible. The reality of black magic is manifested in the mad individual.

The model of black magic fabricated at this shrine establishes a passage of *bala*: how the sickening gift moves through the hands of multiple actors and agencies. It starts from a 'bad' emotion, which develops into envy (*irsa*) (my interlocutors associated envy with selfishness), then malevolence and eventually a perverse satisfaction in seeing other people suffer rather than flourish. In order to achieve this kind of satisfaction, the envious person seeks help from a magician (*jadukar*), who may be a Muslim *faqir*, a tantric Hindu ascetic or an astrologer. These categories of ritual specialists share ambiguous knowledge and can inflict curses, spells and ritual attacks, as well as their remedies. Some of them advertise a host of services in vernacular newspapers – from magical love spells to solving 'problems' with an enemy. The person harboring envy is said to approach a magician if he/she wants help harming an enemy. 'Infecting' an enemy often occurs indirectly, i.e. the sickening gift might not be sent to harm the person resented and envied but to destroy a desired object, which can be human (the sanity of a loved one) or non-human (e.g. a flourishing business).

Practitioners of black magic (*jadukar*) are paid money for their services. Magicians transgress the line between human and non-human worlds and know how to manipulate natural and supernatural phenomena, including ghosts (*bhut-pret*) and spirits (*jinn*). Some of the latter are under the command of a practitioner who can order them to transmit the sickening gift to a targeted subject. The passage is made of transactions: money (magician's payment) is converted into a gift which brings harm and disorder to a self and his or her family. The most common form the sickening gift takes is food. Like all gifts, the sickening gift embodies the donor, but in this case it does not mediate sociality but anti-sociality. The sickening gift of black magic is loaded with the donor's evil intention to destroy or kill the object of desire. The targeted subject who unknowingly ingests the sickening gift becomes host to the transformation of food into weapons. Inside the stomach, the food contaminated with black magic is 'cooked' and turns into indigestible things, such as razor blades or glass from broken bangles, or it may even become a snake. The victim experiences not only extreme pain in the stomach, but also an utter confusion of the mind (*dimag*). Madness is equally spread throughout the mind and the body, making the 'dividual' Indian person inseparable from the social body of the kinship group or family (*khutumb*) (Marriott 1976). A mad person behaves in ways which violate the normative frameworks and moral values of the collectivity. The sickening gift of black magic and ensuing madness individualises the recipient and fragments the wholeness of the family (*khutumb*).

I encountered one young man (whom I call Mohammad) who was separated from his family and had been at the shrine for the past eight months with his mother. Mohammad's story ran like this: About two years ago he began to experience trouble. He said he suddenly could not think properly any more, he began to do terrible things like beat up his mother and he felt extreme pain in his stomach. "My brain (*dimag*) was out of order," Mohammad said, "but now I am half well again." His father's sister, i.e. his mother's sister-in-law (*narand*), had fed him biscuits and milk when he was a young child. But these entirely trivial items, which are given routinely to children by the women of a household, were contaminated with feces in Mohammed's case. Feces is a prime substance in black magic; it carried the aunt's envious emotions and bad intentions to harm Mohammad's mother. "My sister-in-law," Mohammad's mother interjected, "envied me because I got a son while she had remained childless. She sent the *bala* because she wanted to kill my son." The effect of the sickening gift did not unfold immediately but only later when Mohammad was expected to earn money, get married and gradually assume the duties of a son who looks after his parents in old age. Mohammad could not assume any of these duties. He failed to reciprocate the care he received as a child. Because of the confusion clouding his mind, he lost his job as a worker in a factory. His aggressive behavior towards his mother and other members of the household caused tensions within the family, which were further intensified by his enormous appetite. Though he had no income, Mohammad began consuming large amounts of food, which actually weakened him more than giving him strength. "It's because nothing of the food remained in my body, it all went to her, the *bala* [here: the spirit of the sending aunt]," Mohammad explained, "the more we eat, the more the *bala* strives." Although Mohammad ate more than anybody else in the household, his body remained almost skeletal.

Although the pilgrims I met with came from diverse social, religious and geographical backgrounds, their narratives were strikingly similar. Individual stories were constructed along a similar pattern: A sister-in-law, neighbor or another relative was unhappy about how much money another woman's husband made, or how successful her son was in school, or how promising her beautiful daughter's prospects of finding a good husband were. Her envy caused her to turn to a sorcerer, telling him about her complaints and asking him to solve the problem; the son, daughter, husband or father of the envied person was then "fed something", which carried the harmful emotions and intentions of the donor/sender. For my interlocutors, black magic was at the root of experiences of all kinds of loss: poverty, failure, illness, madness, death of loved ones, etc. Relationships between women of a household related by affinal ties play a central role in the fabrication of black magic.

Guardians of the shrine proposed a gender theory of aggression and black magic. They claimed that men are prone to physical violence, whereas women resort to the hidden, secret forms of 'spiritual' violence realised in practices of sorcery. Accusations of sorcery are not confined, though, to intimate relationships between women but may arise in various types of relationships strained by competition, rivalry, and politics (cf. Kapferer 1997). Still, the template institutionalised at the shrine locates the main incentive for doing black magic in the emotions of envy and hatred straining relationships between women.

The logic of the sickening gift thus underscores a built-in conflict in the structure of the Indian extended or joint family which arises from the incorporation of strangers, i.e. women who marry into patrilineal domestic units. Along with ethnographic accounts, popular Indian media such as comics, Bollywood films and TV-serials untiringly address this conflict, preferably in the guise of the relationship between a power hungry, harsh mother-in-law and the subordinate young wife of her son. In-marrying wives not only threaten the privileged emotional bond between mothers and sons, so the argument goes, but also the solidarity of brothers and other male members of a lineage. By convincing their husbands to set up a nuclear household, wives threaten to tear apart the joint family. This account stresses the superiority of the values of the patriarchal family and lineage over the interests of women. The discourse of sorcery articulates women's anxieties arising from competitive relationships between women that are mediated by men. It focusses on everyday concerns such as fertility and childbirth, the economic dependence of women and children on the husbands' earnings and the hope invested in sons and daughters to improve the social, economic and symbolic capital of the family-unit in the future through professional success and gainful marriage matches. While the ideal of the Indian family emphasises cooperation between the wives of brothers and the subordination of female subjective emotions and desires to the common good, everyday realties of women related by ties of affinity (HZ, HBWe, HBWy) are often strained by quarrels over domestic issues, status rivalry based on the number of sons/children, and competition in regard to their husbands' material gains. Accusations of sorcery give a voice to women's subjective feelings of injustice within the order of kinship.

Married women, moreover, take a great share in the performance of "nurture kinship" (Schneider 1980; Holland 2004). Indian cultural theories of relatedness, as Marriott pointed out, fuse actor and action, morality and biology in coded substances (Marriott 1976). Gifts, whether 'good' or 'bad', transport particles of embodied coded substances. Transactions of coded substances constitute social processes which occur simultaneously within and between persons (dividuals) – whether in the realm of kinship, ritual, or the division of labor between castes (Raheja 1988). In addition to the dyadic transactions of coded substances which are reciprocal, Indian history and culture has produced a host of asymmetrical forms of reciprocity (pure and impure, high and low castes). Indian regulations of give and take, especially those enacted in the realms of eating/commensality and marriage, introduce social boundaries and hierarchies. Nurture kinship, although gendered, supersedes hierarchical relationships. It arises from the mediation of affective care, nourishment and mutual protection transacted between permeable mind-body-selves (dividuals). Nurture kinship aims at the reproduction and growth of the family unit as a whole (or lineage segments) through reciprocal transactions between elder and younger family members across life-cycles and generations. Acts of black magic invert the practices of nurture kinship, such as feeding, by using them as a means to harm or attack another. The sickening gift thus interrupts the circulation of care and nurture as the donor transfers his or her ill wishes to a segment of the family, like, for example, Mohammad and his mother. Because of his madness, Mohammad had almost become an 'outcaste' from his family. The sickening gift effected the severance of the son and his mother from participating in everyday domestic life.

Mohammad said that he had become 'half good' since entering the shrine. When I asked him what he meant by this, he said that his 'upper body' was better, but that his 'lower body' was still not quite well. He pointed to the lower body, hinting at the 'indigestibility' of the sickening gift – it got stuck in the intestines and caused a kind of constipation. Children and other recipients of an inverted gift of nurture can neither return it nor pass it on. As Brahmans receiving *dana*, those afflicted with sorcery illness are stranded with the sickening gift. How healing practices draw on cultural metaphors of digestion, on cooking to make the sickening gift digestible, i.e. return it to circulation, is the theme of the next section.

The Logic of Healing: Digesting the Sickening Gift

The shrine facility emerges as a network linking family, a law court, and a body sustained by processes of nourishing (cooking) and digestion. For the description of this network, it is useful to draw on Latour's distinction between 'mediators' and 'intermediaries' (Latour 2007). "Mediators", Latour (2007: 39) wrote, "transform, translate, distort and modify the meaning or the elements they are supposed to carry" 'Intermediaries', by contrast, are mere transmitters of meaning. Thus, family and law become linked through the transactional processes between pilgrims and saints mediated by the body and transmitted by *khadims* in ritual services.

The body here emerges as a mediator, as it translates family values into the generalised law of the social. Speaking of the body, Mary Douglas becomes relevant again, as she famously scrutinised the relationship between 'two bodies', the body of the self and the body of society (Douglas 1973). The human body, Douglas argued, in many cultures symbolises social relationships and thus provides a model for representing hierarchical relationships between society as a whole and its limbs or parts. How the social body of society is perceived by members of a community in turn shapes the bodily perceptions of a self.

Healing agency is distributed over a network of actors who have been mobilised to digest the sickening gift (for the notion of distributed agency, cf. Sax 2009). In this case, healing agencies are first of all distributed over members of the family of the saint and their tombs and shrines. The shrine is known by the name of Sayyid Ali Mira Datar but encompasses a whole lineage and a domestic family unit (*khutumb*). There are many titles by which practitioners and pilgrims invoke the saint: *jinnat ka sarkar* (master of spirits), *daktar* (doctor), *judge* (in English), and *daatar* (supreme Provider/Giver of life). Mira Datar shares these qualities or substances of occult power, authority and knowledge of healing with his mother (*Rasti Amma*), father's mother (*Dadima*), and mother's brother (*Mamu Saheb*). Each occupies his or her own 'throne' (*gadi*) within the wider ritual landscape constituted by a network of burial sites.

The narratives attached to these sites evoke memories of a medieval Muslim family partaking in the establishment of Muslim rule in the region. Sayyid Ali Mira Datar's father, buried next to his tomb, was a military commander. Sayyid Ali Mira Datar followed in this profession. On the day his marriage was to take place; he was called to war against a 'sorcerer king'. Mira Datar killed the king, but lost his own life in the course of the battle as well.[3] His mother, Rasti Amma, rests in a separate shrine in another village, where she had taken *samadhi* (or committed suicide). When Sayyid Ali Mira Datar was six months old, his mother died because the father had married a second wife who was given to him as a tributary gift from a defeated Rajput king. The second wife became Sayyid Ali's 'milk mother'. Her tomb sits in a deep burial crypt within the fort-like structure of the shrine. The tomb of the milk mother's brother, Mamu Saheb ('Lord Mother's Brother'), also a Rajput serving under the command of Mira Datar's father, has its own shrine a short distance from the main *dargah*. The father's mother is objectified in a dome on the upper level of the fort-like structure of Mira Datar's shrine. The dome is called 'Father's mother's grinding stone' *(Dadima ki chakki)*.

The narratives told by *khadims* evoke scenes of suffering, self-sacrifice, and death (transcended by the souls, *ruh*, of the saints) to which pilgrims can relate their own experiences of misery and grief. In the transaction of coded substances, pilgrims become 'bound to' and thus incorporated into the body of the saintly family. Pilgrims of suffering coming for the first time to the shrine offer gifts, such as a piece of cloth, sweets, rose petals, money, etc., to the tomb of the saint and are received by a *khadim*, who then initiates them into the cult with a *dua*-prayer of supplication. He henceforth prays with his clients after the routine rituals performed at dusk and dawn (*loban*). *Khadims* pray with their clients in a cross-legged position over a bowl containing figures of horses made from green cloth. The clients, usually a group of relatives accompanying a sick member of the family, place their hands on top of the horses. One after another, each client (including children) is made to hold one end of a red thread in both hands, while the *khadim* takes the other end. The

3 For details of the mythical narratives, cf. Pfleiderer (1981) & my film Spirits of Envy (2012).

prayer is spoken line by line into the thread by the *khadim* and repeated in the same style by the clients. One line requests the saints "to send away" the agents of adversity, who are carefully invoked by name (*jinn*, ghosts of female suicides, shadows of the dead, etc.). The horses made of cloth represent the vehicles that take away the evil spirits. Through transactions merging touch, sound/words and materials transmitted by *khadims*, clients incorporate particles of the moral code of the saints and deposit the distress effected by the sickening gift upon them in turn. In this process, male and female pilgrims of suffering often develop especially intense affective ties with one of the saints (the son, mother, or father's mother), who visits them in dreams and who is, after pilgrims have returned home, consulted in all sorts of everyday affairs.

The fabrication of the ontology of invisible realities and divinities also works with symbolic mediators. The tombs are covered with bundles of red threads, which represent the incorporation of help seekers into the *body* of the saintly family. Significantly, bodily symbolism attached to the tombs stresses gendered relationships between internal organs and external extremities. Mira Datar, the son, is at the 'heart' – the seat of 'pure' emotions – of the family. The 'womb' of the tombs – the shelter of fertility – consists of the tomb of the milk-mother, which can be reached by climbing down into a deep, narrow staircase. The 'head' – seat of the mind (*man*) – of the domestic family unit is represented by the father's mother (Dadima), whose dome is built on a terrace at the highest level of the shrine. Below and at the 'feet' of Mira Datar (south of his tomb) is the place of the defeated sorcerer king.

The symbolism of the tombs as a 'whole' social body merges values of hierarchy with personal sanity, both of which are attacked by the sickening gift of black magic. According to my interlocutors, their capacity to reason and sanity depended on the fluid relationship between the heart and the mind. The sickening gift interrupts the flow of morally valued emotions (love, compassion, surrender) from the heart to the mind. This gift also brings fertility as the source of life to a halt. Fertility is transformed into social and physical barrenness or death. The sorcerer king is an icon of the forces and agencies acting upon destructive emotions and unsettling human minds. Caught "under the feet" of the saint, these forces are brought under control.

While hierarchical relationships between men are articulated in terms of the morally superior and inferior magical powers (derived from God/spirits) that are symbolised in the relationship between the position of the 'head' and 'feet' of the son, the head of the domestic family unit is female.

A common tool in Indian kitchens is a grinding stone. The 'grinding stone' of Mira Datar's father's mother metonymically refers to the process of cooking. It serves as a mediator which links cooking with the process of digestion (cooking understood as the pre-digesting of food by fire, e.g. Cantlie 1981). In ritual healing practiced at the shrine, body-centered processes, especially cooking and digestion, were brought together with proceedings at law courts.

Digestion and trials have in common the separation and re-configuration of matter. Physiological processes of digestion separate nourishing elements of food from excreta. Court trials process evidence to separate law-abiding acts from crimes. Both dimensions converge in healing black magic and the discourse of counter-sorcery, which declares the sickening gift as indigestible and transactions of sorcery as a crime against social morality.

Processes of digestion and the law become merged in the epithet of the shrine as an 'open court' (*khulo darbar*). The *dua*-prayer begins with the line: "Bismillah-Ir-Rahim, Sayyid Ali Mira Datar, Rasti Amma, Dadima, Mamu Saheb please accept my plea in your court [...]." The saints act as 'judges' who bring the agencies of black magic to trial. The tomb of Mira Datar is designated 'Supreme Court', and those of the others are 'High Courts'. The father's mother's grinding stone is a 'police-station' where spirit suspects are 'grinded' until they succumb to confession and spill out who they are, what occult crimes they committed, and how they justify black magic machinations against a vulnerable Other. The shrine's status as a ritual law court is further stressed by the *kacheri* ('office') where people submit written supplications requesting justice to the saints. The designation 'open court' (*khulo darbar*) references the specificity of the work of the body that the saintly family performs. At their open court, the secret, hidden and evil machinations of black magic are brought to light and made publicly known – in a process of 'digestion'. This occurs in performances of trance (*hajri)*.

Hajri literally means 'presence': it refers to the presence of *jinn*, the ghosts and spirits of living senders of black magic (collectively called *bala*), and the saints. In trance, afflicted persons perform dialogical interactions with the saints and with the senders of black magic. *Hajri* thus marks the ontological realisation of invisible, immaterial spirits in the body of afflicted persons, such as the souls of saints (*ruh*), *jinn* and other categories of ghosts, including the hidden, immoral dispositions of living humans (senders of black magic). Healing involves ritual production, a process drawing on the human voice as a medium, which allows disembodied forms of existence to become experiential to humans (Basu f.c.).

The appellation of the *dargah* as an 'open court' establishes a homology between transactions within and between the human and the social body. The shrine surfaces as a gigantic body that digests the impurities of occult aggression. Pilgrims relentlessly deposit feelings of despair, pain, terror and the desire for revenge at the feet of the saints.[4] The embodied particles of the sickening gift are evident in the afflicted person's wailing, screaming, shouting abuses, or vomiting. The sickening gift's non-reciprocity locks it up in the mind and body of the afflicted person, where it causes a kind of 'constipation'. It cannot be passed on in transactional flows constituting dividuals embedded

4 In this sense, it does not seem very different from the 'turbulent centers' described by Kapferer (1997).

in hierarchical networks of kinship. It thus effects the severing of human social relationships. By transferring the sickening gift to the body of the saintly family, it becomes unlocked in a process that merges moral control with the excretion of evil. The latter, all the material and immaterial filth, is channeled into a sewer pit outside the shrine premises in acts of purification performed by *khadims*. In dreams and trance, persons receive orders (*hukm*) and decisions (*faizla*) that determine how the agents of black magic possession are to be separated from the mind-body-units of a self. A recurring saintly order sends the afflicted to bathe in the sewer pit, which is the 'natural' abode of the impurities of black magic and its agencies.

Healing is liberation from the clutches of occult aggression. It is achieved when the sickening gift is freed from its deadlock in the stomach and mind of the afflicted person and returned to circulation. Lemons nailed to a tree embody the free spirits of envy that were ousted or excreted from one person but are ready to enter the next.

Final Remarks

I started this article by suggesting that the sickening gift resembles *dana* in so far as it is a non-reciprocal gift. Muslim discourses have inverted the value of the gift. While the donor of *dana* accumulates religious merit, the sender of *bala* accumulates sins. Amongst Muslims, no social category parallels that of Brahmans, who are caught in the dilemma of being obliged to accept a non-reciprocal gift which carries negative effects for the recipient. Muslims distinguish, however, as Werbner emphasised, between 'upward' gifts (to God) and 'downward' gifts (to the poor) (Werbner 1990). *Nazrana* is the term for the gifts to God that pilgrims give to saints to pass on to Allah. The *khadim*'s status is not endangered by the pilgrims' gifts, because he is merely transmitting them to the saints. The latter are thereby empowered to act as protective patrons. Their dependents are human beings in crisis. Although the circulation of *nazrana* creates asymmetrical bonds between hierarchical subjects, the gifts given as *nazrana* are not considered dangerous. The perils of asymmetrical transactions and relationships are entirely shifted to the borderland of occult realities.

I have analysed processes of the transferal and re-direction of a gift considered dangerous because of its non-reciprocity. A further step in the analysis would reveal how asymmetrical, reciprocal transactions of upward gifts pass into reciprocal transactions of downward gifts. The result of these transactions – an institution which offers remedies and shelter for have and have-nots alike – oddly contradicts two middle class stereotypes in India: that ritual healers or healing institutions economically exploit people in distress rather than helping those who turn to them for help; and that ritual practices are fake – completely beyond the cure 'really' needed for someone who suffers from a severe mental disorder. The focus on exchange has revealed how the gift as a "total social fact" (Mauss) connects domains severed in modern thought – religion, magic, law, possession, and kinship. At the shrine of Mira Datar, such a network is fabricated through embodied practices of healing the damages caused by the immoral, socially harmful emotions embodied in black magic.

References

BASU, H. 2012. *Spirits of Envy. Ethnographic Film.* Cluster of Excellence: Muenster University.

BASU, H. (2014a) Dava and Dua. Negotiating Psychiatry and Ritual Healing of Madness. In NARAINDAS, H., QUACK, J. AND SAX, W.S. (eds.), *Asymmetrical Conversations. Contestations, Circumventions, and the Blurring of Therapeutic Boundaries*, pp. 162-199. Oxford: Berghahn.

BASU, H. (2014b) Listening to Disembodied Voices: Anthropological and Psychiatric Challenges. *Anthropology & Medicine* 21(3): 325-342.

BASU, H. (f.c.) Voices of Possession – Dialogical Selves in India. In WAGNER-EGELHAAF (ed.), *Voices from Beyond.*

BELLAMY, C. (2011) *The Powerful Ephemeral.* Berkeley: University of California Press.

CALLAN, A. (2012) Patients and Agents: Mental Illness, Modernity and Islam in Sylhet. Bangladesh. New York: Berghahn.

DOLS, M. (1992) *Majnun: The Madman in Medieval Islamic Society.* Oxford: Clarendon Press.

DONNER, H. (ed.) (2011) *Being Middle-Class in India: A Way of Life.* London and New York: Routledge.

DOUGLAS, M. (1973) *Natural Symbols: Explorations in Cosmology.* New York: Random House.

DOUGLAS, M. (1987) *How Institutions Think.* London: Routledge and Kegan Paul.

GREEN, N. (2012) *Sufism: A Global history.* Chichester, West Sussex; Malden, MA: Wiley-Blackwell.

GREGORY, C.A. (1997) *Savage Money: The Anthropology and Politics of Commodity Exchange.* Amsterdam: Harwood Academic Press.

HAQ, S. N. (2010) Occult Sciences and Medicine. In COOK, M. (ed.), *The New Cambridge History of Islam*, p. 640. Cambridge: Cambridge University Press.

HOLLAND, M.P. (2004) *Social Bonding and Nurture Kinship: Compatibility Between Cultural and Biological Approaches.* University of London.

JAFFRELOT, C. and VEER, P.V.D. (2008) *Patterns of Middle Class Consumption in India and China.* Los Angeles: SAGE.

KAPFERER, B. (1997) *The Feast of the Sorcerer: Practices of Consciousness and Power.* Chicago: University of Chicago Press.

LATOUR, B. (1991) *We have never been modern.* Cambridge: Harvard University Press.

LATOUR, B. (2007) *Reassembling the Social. An Introduction to Actor Network Theory.* Oxford: Oxford University Press.

LATOUR, B. (2010) *On the Modern Cult of Factish Gods.* Durham: Duke University Press.

MARRIOTT, M. (1976) Hindu Transactions: Diversity without Dualism. In KAPFERER, B. (ed.), *Transactional Meaning. Directions in the Anthropology of Exchange and Symbolic Behaviour.* Philadelphia: Philadelphia Press.

MAUSS, M. (1972) *A General Theory of Magic*. London: Routledge.

MAUSS, M. (1990) *Die Gabe*. Frankfurt am Main: Suhrkamp.

NARAINDAS, H., QUACK, J. and SAX, W.S. (eds.) (2014) *Asymmetrical Conversations. Contestations, Circumventions, and the Blurring of Therapeutic Boundaries*. Oxford and New York: Berghahn.

PARRY, J. (1985) The Gift, The Indian Gift, and the Indian Gift. *Man (N.S.)* 21: 453-473.

PFLEIDERER, B. (1981) Mira Datar: The Psychiatry of a Muslim Shrine. In AHMAD, I. (ed.), *Ritual and Religion among Muslims in India*, pp. 195-234. Delhi: Manohar.

PLATENKAMP, J.D.M. (1996) The Healing Gift. In HOWELL, S. (ed.), *For the Sake of our Future. Sacrificing in Eastern Indonesia*, pp. 318-336. Leiden: Centre of North-Western Studies.

QUACK, J. (f.c.) Leading and Misleading Religious Boundaries: Lessons from (mental) health seeking practices in India. In BASU, H., LITTLEWOOD, R. AND STEINFORTH, A (eds.).

RAHEJA, G.G. (1988) *The Poison in the Gift. Ritual, Prestation, and the Dominant Caste in a North Indian Village*. Chicago: University of Chicago Press.

SAX, W.S. (2009) *God of Justice. Ritual Healing and Social Justice in the Central Himalayas*. New York: Oxford University Press.

SAX, W.S. and BASU, H. (eds.) (2015) *The Law of Possession – Ritual, Healing, and the Secular State*. New York: Oxford University Press.

SCHNEIDER, D. (1980) *American Kinship. A Cultural Account*. Chicago: University of Chicago Press.

WERBNER, P. (1990) Economic Rationality and Hierarchical Gift Economies: Value and Ranking among British Pakistanis. *Man (N.S.)* 25: 266-285.

WERBNER, P. and BASU, H. (eds.) (1998) *Embodying Charisma. Modernity, Locality and the Performance of Emotion in Sufi Cults*. London: Routledge.

Healing Rituals and Sorcery among the Lanten of Laos

Joseba Estevez

Josephus Platenkamp's works "Rituals and the comparison of societies" (1990) and "The healing gift" (1996), among others, deeply influenced my understanding of health, healers and healing rituals in the Southeast Asian socio-cosmological context – which are the subject of my on-going research among the Lanten.[1]

Known in China as Lan Dian Yao (蓝靛瑶), meaning Indigo [dye] Yao, and in Vietnam as Dao Làn Tiển, the Lanten call themselves Mun, which is their pronunciation of the Chinese character 人 (pinyin: *rén*), which means "human", "person". In Laos, the Lanten are known in Lao language as Lao Huay, meaning the "Lao of the streams". They are one of the 39 ethnic minorities living in Luang Namtha province in Laos.

Lanten religious traditions are composed of particular representations of animist, Confucian and Daoist provenance,[2] cults addressed to ancestors and to particular deities. Their understanding of the underworld and afterlife originated from Chinese Buddhism. Within this composite religious setting the Lanten person, as a whole, is comprised of three different types of constituents:

- the physical body, created and initially nourished by the parents;
- the spiritual body, composed by *hon*, "spiritual parts", "souls"; three *hon* constitute what I label a person's "persona" while the others "animate" the physical body. *Hon* are granted by the ancestors through the intervention of the female deity *Fa Man*;
- the social body, consisting of the progressively expanding sets of socio-cosmological relationships developed in the course of a person's life.

The cyclic existence of a Lanten individual starts at the *Mo Kang*, the House of the Ancestors in the invisible realm, where souls are conceptualised as

1 Research conducted from 2010 onwards in Luang Namtha, Lao PDR, was financially supported by the German Research Council and supervised by Prof. Josephus Platenkamp. I am grateful to Prof. David A. Palmer, Prof. John Lagerwey and Prof. Guido Sprenger for the many discussions and insightful comments on this contribution.

2 These influences include the imperial bureaucratic system, Daoist manuscripts, some of which date from the 15th Century, and Chinese script. For further information about the Lanten manuscripts see our on-going project with the British Library's Endangered Archives Programme: http://eap.bl.uk/database/overview_project.a4d?projID=EAP791 [01/11/2015].

"flowers"[3] waiting to be reborn into the world of the living. As part of particular ceremonies serving to honour the ancestors, an application is sent to the *Wa Kang*, the third – and most powerful – generation of ancestors who are responsible for granting the "birth certificates". *Fa Man*[4] who takes care of the "flowers" mediates during the birth process. She also protects the newborn children along with their ancestors until boys pass *jai sai,* Lanten 'ordination' and girls get married. Those children who die before getting married do not qualify to be reborn again.[5] The newborn children are considered as "spirit-babies" until the naming ceremony takes place, between the third and seventh day after birth. Only after a social identity is granted do they become *Mun,* human.

Lanten *hon* share characteristics with both Chinese and Southeast Asian views[6] of the soul and body. As mentioned above, *hon* vary in number[7]: there are ten in male bodies and twenty-seven in female bodies. Three of these *hon* constitute the "persona" while the rest "animate" the body. The "persona" is located on the left side in the male, and on the right side in the female. *Hon* are perceived as "fluids" that move around their side of the body and can be ritually reinforced or debilitated and individually detached, which makes the individual "stronger" or "weaker" respectively. To assure that *hon* remain attached to the body the Lanten wear silver jewelry[8] that effectuate the protection of their ancestors.

Adulthood manifests itself as the competence to interact with the spiritual world, particularly to communicate with the House of the Ancestors. In males this competence is developed by becoming an expert of ritual. This expertise comprises of two ranks with differentiated roles: *Tao Kong*, 'priest', and *Tai*

---

3 Red for the girls, white for the boys.

4 A goddess who resides in the forest and is aided by some other inferior female deities, see Doré (1987: 1-6). The ancestors alone cannot articulate the reproduction process, they require the aim of a divinity beyond the social space (i.e. "forest") to create descendants. See Barraud and Platenkamp (1990) for a discussion about the distinction between the internal order of the society and the domains beyond it in Southeast Asia.

5 Individuals who do not contribute to the reproduction of the socio-cosmological order are usually expelled from it. Those who have attained divine status constitute for instance an exception.

6 For Chinese views about the soul (Chinese: *shen/hun* and *kwei/po)* see de Groot (1901/4) and Schipper (1993). For Southeast Asian views of the soul (Lao: *khuan*) see Platenkamp (1996, 2010).

7 Since Lanten ritual experts are ascribed to different traditions in terms of instruction with different textual materials and lineages of masters, both the number and location of the *hon* may differ. This account is comparable to that of the Lao ethnicity that counts 32 *kwan,* 'souls'. Thus, the number of the spiritual component parts should be analysed in the context of ethnic identity and ethnic differentiation. See Platenkamp (2010:199).

8 Such as the silver torques wear around the neck by children; women wear earrings and men bracelets for the same purpose.

Kong, 'master'. As part of *jai sai*, a double 'ordination', the candidates become mediators between their ancestors and deities and their community. This newly attained ability is asserted by channelling features attributed to the ancestors such as health, wealth, good fortune, wisdom, etc. to the community. These features are transmitted into sweetened sticky rice cakes, which are then considered as embodiments of the *jong feng ji nu* 'milk of the [all ancestral] mother[s]'.

Another important religious rank is that of the *Kwa Mun*, 'shamanistic healer'[9]. *Kwa Mun* use 'spirit horses' that they receive from their ancestors in *Ho Kwa,* an initiation ceremony in which they learn the basic techniques of shamanistic healing. The 'spirit horse' enables its owner to 'jump along the stars' in order to find the lost *hon*, the loss of which leads to illness and ultimately, death.

When an individual dies, the persona is sent in the course of the primary mortuary ritual to the *Di Ñok*, the Underworld's Ten Courts of Hell in order to be judged, punished (if required) and to learn how to become an ancestor. Most of the animating parts of *hon* remain in the cemetery with the buried body while some others return to the Sky. At least three years later, a secondary mortuary ritual takes place. The ritual experts assault the gates of the Underworld Hells and liberate the 'persona'. They also create a physical body made out of bamboo sticks and bamboo paper that, according to the Lanten, reproduces the thirty-six parts of a human body. In the course of this ritual the 'persona' is invited to take possession of this body and to attain a new status as *Pibu*, ancestor. Afterwards the ancestor is relocated[10] along with a spirit house, spirit servants and joss offerings at the *Mo Kang*, the House of the Ancestors, in 'Heaven'. Therefore, individuals are re-constructed and re-socialised after death: the physical body is shaped as an archetype of eternal youth and beauty, the 'persona' is purified in the Underworld and the social body qualifies as the most valorised, one that is a *Pibu*, an ancestor. On rare occasions, a posthumous deification ritual[11] may be conducted to transform remarkable ancestors[12] into deities who will join the local pantheon. Ancestors remain in their new house for at least five generations, taking care of their descendants. After this period of time, they can become 'flowers' and be reborn again or they can settle in the 'Heaven' for good.

9 Also called *Kwa Tai*, 'master of healing'. Although not usual, during the course of this research several cases of Lanten female healers have been documented.

10 In order to transfer any object to the Otherworld it has to be burnt. The smoke symbolises the transition.

11 A very rare ceremony nowadays because of its high costs and the current lack of ritual expertise. See Puett (2002) for further analysis about this ritual in ancient China.

12 For example, founders of the village are incorporated as *Pun Keng*, Deities of the [Ritual] Territory.

Lanten Conceptualisations of Health, Illness and Healing

Life in the Northern region of Laos was until recently[13] a harsh and challenging experience. The forest is still perceived as a source of sickness and accidents: it was only forty years ago that most Lanten villages suffered attacks from tigers at night. Illness and misfortune could only be counteracted by using herbal medicines, respecting taboos and performing sacrifices that provide protection against a dangerous environment.

Health among the Lanten is perceived as an ideal state in which an individual is in balance within his particular socio-cosmological realms of life. Thus, imbalance results in illness and death when not treated. There are three main categories of illness:

- minor cases without 'supernatural' inference. The condition either does not require particular therapies or it responds quickly to traditional or modern medicines;
- cases in which the condition originates in socio-cosmological domains
- cases related to conditions attributed to sorcery.

Hence, illness is mostly considered as an indicator of a failing spiritual condition. Conditions that qualify within this category include, for example 'abnormal' maladies that do not respond to established treatments, ideas of 'bad luck', 'accidents' and 'natural disasters' such as floods destroying the rice fields. Modern medicine is sought to treat the symptoms of illness, that is to say, to heal the physical body.[14] However, to re-establish the necessary balance of life in the socio-cosmological sense, ritual experts must mediate between patients and those actors causing the disease.

The incorporation of Daoist and Buddhist repertoires and practices into the Lanten animistic–shamanistic beliefs also entailed superimposing the Chinese imperial bureaucratic system onto the Lanten cosmovision and socio-cosmological space.[15] By doing so, the Otherworld became a replica of the Imperial Government and accordingly all space was credited to deities claiming the ownership of the universe, divided into Sky, Earth and River. These deities

---

13 Electricity has arrived to most of the Lanten settlements in the course of the last decade. For most villagers the only source of drinking water is that of the surrounding river and streams.

14 Ritual experts are particularly adept to modern – Western – medicine. They employ the newest locally available medicaments for treating infections, gout, and muscular and joint pains. As a matter of fact it is them, as heads of their households, who mediate the incorporation of modern medicine into their communities.

15 This 'bureaucratic system' has been only partially incorporated into the Lanten cosmology due to the limited understanding of both the language (Classical Chinese) and the cultural system (Imperial China) from which it originated. Moreover, this idiosyncratic incorporation has been affected by centuries of isolation from its origins in China.

have a hierarchical organisation, which is part of a divine administration the higher representatives of which are the Three Old Sages. These three well-known Chinese gods are identified among the Lanten as Boung Wung (Chinese: *Pangu*, first living creature and creator of the universe), Lui Wung (Chinese: *Lei Gong*, God of Thunder) and Yui Wung (Chinese: *Yú Huang*, the Jade Emperor, the highest power). They are *The Three Pure Ones*[16] and together they rule the Lanten universe.

This system creates a hierarchy in which every part of the Lanten cosmos is under the authority of named and individualised deities and ancestors. The ritual experts play the role of high-ranking officials,[17] even godlike agents, within this divine bureaucracy.[18]

Lanten deities and ancestors share negative features ascribed to the Chinese imperial bureaucrats such as fallibility, bias, greed and corruption. From the Lanten point of view, they are in a permanent state of famine, which manifests itself in their predisposition to steal the *hon* of the living in order to 'consume' it, to call attention to their needs or to punish improper behaviour. Since life stems from these deities and humans continually 'invade' their domains, the living are in a constant state of debt and the only way the deities can remind them of their obligations is to recall the precious gift they benefit from, i.e. their *hon*.

Illness caused by loss of one or more souls

Physical illness follows in most of the cases upon the loss of *hon*. One or more parts may have been kidnapped, stolen, reclaimed, scared away or expelled from the body. In particular cases, possession may also take place when deities or ancestors occupy the body in order to express their needs. Sorcery results in conditions that also entail the loss of *hon*.

Since the Lanten cosmos is organised in domains subject to divine agents, every interaction requires a proper relationship with the divine bureaucracy. Thus, cutting, burning or even urinating near a big tree may cause the deity that inhabits the spot to retaliate. Going hunting without offering sacrifices to

---

16 Several other deities are former humans, sometimes with well-known biographies, who became gods and occupied a position within the 'divine bureaucratic system'.

17 According to *Liji*, the Book of Rites, "he who rules the world sacrifices to seven generations of ancestors; he who rules a state sacrifices to five generations; he who rules a territory of five chariots sacrifices to three generations; he who rules a territory of three chariots sacrifices to two generations. He who eats by the labour of his hands is not permitted to set up an ancestral temple" (in Lagerwey 2010: 4). Hence, by sacrificing to five generations, Lanten ritual experts locate themselves as rulers of a state (without the state), right after the Emperor.

18 For instance, the four bronze seals that identify 'masters' and 'priests' as high officers, not only are used to stamp official documents but also to imprint the body of people. This act constitutes a healing ritual of its own since it subordinates persons and objects to the authority of the 'divine administration' (Strickmann 2002: 131).

the deities of the mountains or refusing to share game with the hunting deities will anger them. Falling into a river, being scared by a noise in the forest, having an accident while driving or collecting wood in the forest, all may result in the loss of *hon*.

Ancestors and descendants have comparable needs: they require food and clothes, goods and servants, money for betting and paying bribes. All these needs must be covered by their offspring by means of sacrifices to the five generations of ancestors. If the ancestors feel neglected, their primary response is to withhold their protection and then to retrieve their primary gift of *hon*. Therefore, failing to sacrifice to the ancestors weakens both the *hon* and their attachment to the body. The recently deceased constitute a particular menace. If their graves are unattended or their liberation from the Ten Courts of Hell takes more time than expected, they can perform retaliation acts to fulfil their needs. Sometimes they simply do so because of the inexperience of their newly assumed condition[19].

Immoral behaviour not only results in the withdrawal of the protection of deities and ancestors but also in punishment. In a similar way, rituals that have been improperly performed or taboos that have been broken will invite misfortune and disease.

Restoring the balance, recalling the souls

When supernatural agents are suspected of causing illness, the head of the house requests a 'healer' to perform *Ai Kwa*, a shamanistic healing séance. In this ceremony the 'healer' tries to detect the origin of the illness. The diagnosis involves a dialogue between the healer and the patient who is considered as part of his household. Hence, symptoms, context and personal history of misdeeds of the all household's members (including the recently deceased and ancestors) are examined. The healer then "rides his spirit horse" to the stars, the domains of the deities, in order to locate the lost *hon*. Once the socio-cosmological origin of the illness has been identified, the *Kwa Mun* prescribe the appropriate treatment.

Healing rituals belong to sets of hierarchically structured ceremonies. The primary prescription consists of the basic sacrifice serving to honour particular deities. If the basic sacrifice fails to redress the distorted relationship, a more complex ceremony will be performed entailing more performers and higher-ranking offerings. *Sit Ñung* written ritual contracts may be filled with vows to perform higher sacrifices or particular ceremonies in a future date if the household lacks the required resources (i.e. appropriated sacrifices, manpower or cash for ritual payments) at the moment.

19 The dead can take the *hon* of the members of their household in order to force their beloved ones to share their destiny –death – or to punish those they hate.

If the full set of sacrifices to particular deities fails to redress the imbalance, a major ceremony devoted to fully reconstruct the socio-cosmological relations of the patient must be performed. Thus, the patient has to be reconnected to the Lanten universe, represented by the ancestors and the *Three Pure Ones*, heads of the Divine Administration. There are two stages of this major healing ceremony which physically embodies the re-connection with the two socio-cosmological realms of life:

- Consuming a gift from the ancestors. The ritual experts articulate the channelling of spiritual parts from the ancestors into matter that can be consumed by their descendants, such as grains of raw rice served in glasses of water covered by bamboo paper. These grains are labelled as 'fishes' and must be consumed by the household's members.
- Tying a thread of cotton around the wrist. All men attending the thanksgiving banquet that finalises the ceremony consecrate a cotton thread. This act conveys verbal wishes for protection and blessings from the divine bureaucracy that they represent. The thread is tied to the wrist of the patient and serves to re-impose the authority of the *Three Pure Ones* upon the person so that the *hon* will remain attached to the body.

Partaking in sacrifices performed by other households or by the community (for instance, in the case of the Lanten New Year) serves as prophylaxis and as a healing method since *hon* from the ancestors of the community can be transmitted to any participant. In these cases, rice cakes receive and convey these transmissions.

Ritual experts are always remunerated. Payments consist of meat (e.g. one chicken or the head and/or leg of a pig), basic needs like food, alcohol, cigarettes and opium covered during the ritual performance, and the final banquet. Particular ceremonies require the payment of silver coins.[20] If the ritual payments are not transferred, the ritual effectiveness will decrease. These ritual payments must be understood as a part of an on-going exchange that originates in the ritual payments made by the apprentices to their masters in exchange of ritual knowledge. Lanten ritual knowledge is constituted of three elements:

- Physical content, in the form of a corpus of manuscripts that contains procedures to engage the deities in the Lanten ritual life;
- Transitive content, which encompasses the performance of the ritual;

20 Nowadays, these are converted into standardised payments of e.g. 50.000 Kip (5€) per coin.

- Spiritual content, the connection with the lineage of the 'original masters' that enables ritual effectivity.

All these are provided by the masters to their apprentices who respond with ritual payments in the form of sacrifices, labour and cash. Hence, the payments from patients to the experts (i.e. former apprentices) for performing the rituals constitute the final authorisation of the healing process as a whole (see Platenkamp 1996: 328).

Conceptualisations of Sorcery among the Lanten

The Lanten belief system replicates ideals of justice of this world in the Otherworld, where all wrongdoings are judged at the *Di Ñok*, the *Ten Courts of Hell* – colloquially, the "prison[21]" – and penalties are administered. By doing so, the ritual and legal discourse and praxis are intertwined so as to create the conviction that the universe works like a flawless system that allows no immoral activity to pass. This conceptualisation of justice is integrated into what Katz (2009:47) defines in the context of the Chinese legal culture as "judicial continuum". However, when the legal system does not provide a satisfactory outcome and the Underworld's justice may not be reliable or takes too long, sorcery provides tools for those who want to take the law into their own hands.[22] There are three major types of 'sorcery' among the Lanten: (1) ritual indictments, i.e. 'cursing'; (2) malevolent use of ritual knowledge; and (3) malevolent sorcery.

Eo Man, reporting to the deities (cursing)

Eo Man, to report [somebody's misdeeds] to the deities, is a form of ritual indictment common among the Lanten.[23] These reports are perceived as immoral but in some way legitimate since they are a justified means of self-defense[24]. Although the particular formulae differ from case to case, the ritual, performed in vernacular Lanten language, follows these two steps: The case is reported to the major deities and a sacrifice is performed to them.[25] *Lui*

---

21 In the last decades notions of the Laotian penal system such as modern prisons, re-education centres and rehabilitation camps have permeated the Lanten popular imaginary of the *Ten Courts of Hell*, which are based on ideas original to the Imperial China.

22 When these methods fail, physical violence may become a self-perceived legitimate option.

23 Although it is difficult to establish how often 'cursing' takes place due to its usually secretive character, the counter-acting ceremonies are public and are performed regularly.

24 In ritual indictments the identity of the offenders is not requisite: the deities will find them.

25 From lower to higher orders, sacrifices include: breaking a tree leaf with the hands; hammering a stone into pieces; cutting down a tree with a machete; taking off the head of chicken; hitting with a stick until death – or throat cutting – a chicken, a dog, a pig, a goat,

Wung (Chinese: Lei Gong, God of Thunder) is in charge of applying the punishments, however, specific deities may be addressed depending on the claimant's particular cosmological relationships and the requested penalty. Although a virtuous individual should have nothing to fear from the divine administration, due to the venal nature of the Lanten deities, a large sacrifice may provoke a biased arbitration.

The primary result of a successful ritual indictment is a disconnection of the offender from the realm of the deities. Serious transgressions invite severe punishments and even lead to the offender's 'persona' being judged in the Ten Courts of Hell, which amounts to a death penalty. These outcomes affect not only the offender but also his household. As a result relatives, livestock and crops may suffer the consequences of individual misdemeanors. To counteract a curse, *Ai Yun*, a ceremony aimed at resolving ritual indictments, must be performed. In this ceremony, deities are not only appeased with sacrifices but are also bribed into changing their verdict so the claimant – instead of the offender – may get harmed. If *Ai Yun* is effective, the claimant will have to perform it in turn, increasing the sacrifices. The cumulative economic costs of partaking in these mutually provoked ceremonies are known to have eventually ruined households.

Malevolent use of ritual knowledge

The malevolent use of ritual knowledge is a potential weapon of ritual experts engaged in disputes with other ritual experts. Two types of rituals can be corrupted in this way: secondary mortuary rituals and exorcisms. In both cases, by means of targeting the living with ritual practices usually aimed at the dead, the *hon* of a particular individual can be sent to the domains of death, "Out of the World" or the Ten Courts of Hell. If the ceremony is effective, the persona leaves the body, which will die soon after. The victim may still be able to eat, sleep and work, but is considered as a living-dead. A single ritual expert performs selected parts of these major rituals in absolute secrecy. The ceremony only takes around two hours and requires personal information about the victim, such as the ritual name(s) and the date of birth. In some cases, personal items (e.g. hair) are hidden within the sacrifices sent to the Otherworld so that the persona will follow them. The effect of these rituals is considered as almost impossible to undo since the same ritual expert or at least the same ritual manuscript must be used in the counteracting ritual necessary to bring back the persona.

---

or a buffalo. The animal must be owned by the claimant and cannot be consumed afterwards. It has to be left to rot. Higher sacrifices entail rougher and more rapidly served punishments.

Ai Deh, performing malevolent sorcery

Performing *Ai Deh* sorcery enables a specialist in this art to introduce foreign objects into the victim's body. These objects include small fragments of buffalo skin, cloth or glass that are implanted into the stomach or near vital organs. The ritual knowledge involved is highly secretive. As such, the identity and the procedurals employed by the practitioners are subject to local gossip. The transmission of this knowledge takes place by means of dreams. It can be also whispered to somebody who is sleeping. The sources of expertise are either senior practitioners or one of the deities that inhabit big trees. Gender plays no role in the acquisition of the ritual expertise, and no manuscripts are known to be required, so that any individual can qualify to learn *Ai Deh*. However, practitioners of malevolent sorcery are socially isolated when they are openly identified and risk death in the hands of the victim's relatives.

To counteract *Ai Deh*, a *Kwa Mun* must consume (swallow) the foreign object(s) inserted into the patient. Sometimes in the course of the ceremony the healer spits out pieces of cloth, glass or buffalo skin. Sacrifices to ancestors and major deities to secure their protection of the victim conclude this ritual.

In these three cases the victims are engaged in the rituals against their will and usually, unknowingly. Therefore, no sacrifices or ritual payments are transferred to the expert. Lanten ritual experts justify this anomalous situation by conceptualising sorcery as a payback, as part of an on-going exchange. As said, sorcery seeks to adjust perceived deficiencies in this cosmo-judicial system, and so, the previous misconduct of the victim is perceived as a payment "in advance". Most rituals require a formal petition to the ritual master accompanied by a token-gift (such as a cigarette) and ritual payments for the service afterwards.[26] Moreover, during their ordinations, priests and masters as earthly representatives of the divine administration vow to follow a strict code of moral behaviour in which their position and knowledge cannot be employed for obtaining personal profit or to harm others. Thus, a perceived misdeed matches the roles of petitions, token-gifts and payments[27] and enables – compels – the ritual expert as high official to use any ritual means to deliver an appropriate sentence. As a result sorcery constitutes a key feature of the cosmo-judicial continuum entailing both regulation and social control.

Healing and Sorcery as Two Sides of the Same Coin

Although the procedural approaches may differ, the ritual contexts of healing and sorcery are similar. Both of them are articulated in terms of equivalent

26 While ritual experts cannot charge fees for their services, ritual payments are clearly stipulated. Requested rituals cannot take place without a token-gift and a formal petition and without the ritual payment it will not be effective.

27 A negative payment.

formulations that pertain to universal but opposing domains. Whereas healing rituals aim at establishing balance by (re)connecting the patient with his particular socio-cosmological realms of life, sorcery aims at enforcing an imbalance by means of disconnecting the victim from these realms and by creating analogous connections with the socio-cosmological realms of death. When these connections are effective, features ascribed to those universal domains are transmitted onto the individuals: from the domains of life the patient receives health, from the domains of death the victim receives illness and death.

Moreover, by applying parts from those universal domains to individuals the transmission of these features can also be accomplished. Therefore, grains of rice, cakes, silver jewellery and cotton threads that are symbolically related and represent ancestors or deities may transmit features ascribed to the domains of life. Correspondingly, parts of a slaughtered animal and inert matter, such as glass or cloth may transmit features related to the domains of death. In addition, since individuals are part of a household these features affect his relatives, livestock and crops. Health and illness affect the domesticated realm as a whole.

In this context, the *hon* are part of an on-going circulation between these universal cosmological domains and the Lanten social domains. This circulation process is grounded in a particular model of society as part of a socio-cosmological order. The reproduction of that order is dependent on the socially sanctioned appropriate moral behaviour defined by active participation in the ritual life of the household and the community and in the generation of progeny. The immoral behaviour is sanctioned by ancestors, deities and members of the community.

The ritual experts play a key role in this model of society by mediating between their communities, the deities and the ancestors. This role entails the articulation of the exchange of persons, objects and spiritual parts and securing the transmission of ritual knowledge. This knowledge involves not only expertise regarding ritual performance; it also comprises of the Lanten cosmovision and of a traceable connection with its own origins, the original masters. This connection constitutes a necessary condition for the effectiveness of every ritual. As such, its transmission involves the exchange of gifts between masters and apprentices, which in the long term is also reciprocated in the context of healing rituals by ritual payments from patients to experts in order to ultimately authorise the rituals.

Health among the Lanten is thus a spiritual condition that pertains not only to particular individuals but to social milieus. These milieus can be understood in terms of households, since all the members share *blood and bones*, or of higher organisational levels such as the village, where individuals failing to respect communal norms (e.g. taboos) invite misfortune to the entire community. Since health is a spiritual and social condition, healing rituals must address not only individuals but complete socio-cosmological milieus. The Lanten ritual experts, as healers, show themselves as pivotal figures within their communities by assuring not only health but the reproduction of this particular society.

References

BARRAUD, C. and PLATENKAMP, J.D.M. (1990) Rituals and the Comparison of Societies. In BARRAUD, C. and PLATENKAMP, J.D.M. (eds.), Rituals and Socio-Cosmic Order in Eastern, Indonesian Societies, Part II: Maluku, special anthropological Issue. *Bijdragen tot de Taal, Land-, en Volkenkunde* 146(1): 103-123.

DE GROOT, J.J.M. (1901) *The Religious System of China: Its Ancient Forms, Evolution, History and Present Aspect, Manners, Customs and Social Institutions Connected Therewith Vol. IV*. Leiden: E.J. Brill.

DORÉ, H. (1987) *Chinese customs.* Singapore: G. Brash.

KATZ, P. R. (2009) *Divine Justice: Religion and the Development of Chinese Legal Culture.* New York: Routledge.

LAGERWEY, J. (2010) *China: A Religious State.* Hong Kong: University of Hong Kong Press.

PLATENKAMP, J.D.M. (1996) The Healing Gift. In HOWELL, S. (ed.), *For the Sake of our Future. Sacrificing in Eastern Indonesia*, pp. 318-336, Leiden: Centre of Non-Western Studies.

PLATENKAMP, J.D.M. (2010) Becoming a Lao Person: Rituals of Birth and Socialisation in Luang Prabang, Laos. In BERGER, P. et al. (eds.), *The Anthropology of Values. Essays in Honour of Georg Pfeffer*, pp. 180-200. Delhi: Pearson.

PUETT, M. J. (2002) *To Become a God. Cosmology, Sacrifice, and Self-Divination in Early China*. Cambridge: Harvard University Press.

SCHIPPER, K. (1993) *The Taoist Body.* Berkeley: University of California Press.

STRICKMANN, M. (2002) *Chinese Magical Medicine*. Stanford: Stanford University Press.

Tax Morality and the Redistributive Exchange of Gifts: Ethnological Annotations

ANDREAS HARTMANN

Taxes generate, represent and perpetuate social relations. These relations both couple the individual tax-payer to the state and ally the tax-payers to one another. Taxation does not establish an equivalent tax ratio and, in this respect, also no reciprocal relationships between tax-payer and tax collector, but rather involves dues which, by way of asymmetric redistribution, benefit the general public. Furthermore, the model of vertical and horizontal fiscal justice contributes to organising the totality of tax-payers as a community founded on the solidarity principle. Both systems of relations (of tax-payers to state as well as the community of tax-payers) are connected with one another and morally valuated. These valuations manifest themselves in such categories as trust, responsibility and honesty, as are frequently and necessarily observed in discourses on taxation. Though connected, the two systems of relations are also understood as separate, which, as it were, leads to a split both in the latter moral categories and, for example, to a lack of tax honesty both as (more or less pardonable) fraud towards the state, as well as theft of fellow citizens. To the lay public, *the state*, above all, signifies an anonymous counterpart of the tax-payer, comprising an opaque world of composite administrative and executive institutions at local, state and federal levels, whereas what is meant when referring to *the tax-payer*, is a community of affected persons appearing in the form of a collective subject.

But what kind of community is this? In so far as all its members are subject to taxation, it is both economically founded and economically real by way of the flow of capital. However, this is founded indirectly, since the interrelations between members is not directly given, but mediated as a whole via the transfer and redistributive provisions of the state. By way of state mediation and taxation of reciprocal provisions, monetary payments are transformed from the market (directly and indirectly) into relations of gift exchange and, thus, rather than establishing short-term exchange cycles characteristic of the exchange of goods, instead establish long-term and, no less, confidence-building exchange cycles which are the general characteristics of systems of gift exchange relations. In the former case – market exchange relations – this ideally involves equivalent acts of exchange, whereby the value determines the price of goods exchanged. At the same time, however, due to the unequal distribution of resources, those who are participating in the market emerge as a group of unequals. In the latter case, long-term relations of gift exchange in the fiscal context, the reverse is true. Unequal resource distribution among taxpayers establishes a non-equivalent exchange. The tax-payers pay an ideally equally – or more cautiously expressed: equationally – distributed, asset-related and respectively varying compensatory "price". The group of economic unequals, thus, transforms itself into a community of exchange partners who are equal before the exchange partner of the state. Accordingly, tax transactions transform competing, non-egalitarian economic subjects into partners who are loyal and egalitarian towards the state.

Naturally, with respect to the tax-payers, this involves an imagined community since its members are connected solely by means of the common, albeit not communicated, awareness of being tax-paying members of the same state. This common orientation towards the same state generates a latent coherence of the community of tax-payers. To this end, face-to-face relations are irrelevant. The totality of tax-payers sustains the state – at least in a financial respect. And yet it is *not* the state, and does not represent it, but rather the state is far more its exchange partner and counterpart. Neither does it fully represent the nation, although it does indeed display affinity to an emphatic concept of a state-framed, fateful interdependency. Hence, tax payments are not genuine national commitments, since neither the institution of indirect taxes (value added tax, petrol tax, tobacco tax, etc.) nor direct taxes (income tax, tax on assets, etc.) represent voluntary payments, and are thus not, following Ernest Renan, to be understood as a plebiscite occurring daily. Nevertheless, this famous adage is not to be conflated with a resolve renewed on a day-to-day basis. Hence, if one instead understands by the term "plebiscite" a mental structure and habitual disposition, then – presupposing the accuracy of the observation that in the outpourings of media indignation on the subject of tax-sinners whereby the honest masses are frequently stylised as voluntary and intentional confessors – tax payments do not fall that short of the mark.

Occasionally, well-heeled and thoroughly respectable paragons of virtue make an appearance in front of media cameras zealously declaring their desire to accept increased taxation for the benefit of public good . Such exemplars stage-manage their tax payments as a kind of voluntary act of renunciation addressing less the state as institutional superego, but far more society in general and the nation in particular, of a "we-society" founded on relations of solidarity. Why not simply society, but the nation as well? Because this is meant to underscore the fact that in this "we-society" – as distinguished from the state, though mediated and generated through the national tax culture – the social first reveals itself, not in its general, but its specific national character.

In its more recent guise, the concept of tax culture, which first appeared in the 1920s, refers explicitly to the historically emerging national culture, which is to be understood as its subsystem. In the context of national culture, tax culture defines itself as the totality of elements of the country-specific system of taxation, its institutional constitution, the administrative and legal practices governing its application and enforcement, the commensurate interaction between various actors, the thus established social relations and duties, as well as the habits of thought connected with this total complex[1]. One aspect of the national tax culture is the so-called tax mentality, which Günter Schmölders, the astute scholar of social-psychological economics and finance, understands as the dominating attitude or attitude "towards tax and taxation that differs from person to person and age to age".[2] According to Schmölders, tax mentality and, more specifically, tax morality, understood as the evaluation and attitude towards tax fraud or tax evasion, characterises ethnic types. In the manner stereotypical of *Völkerpsychologie*, he wrote:

> "Peoples whose members are predestined, as it were, for a negative tax mentality by virtue of specific character traits, as observed, for instance, in the individualism of the Romanies or the Asians who, due to their cautious and sceptical reserve, frequently slip into lax morality when it comes to taxation – something which, above all, expresses itself in a particularly lax attitude with respect to tax offences. This phenomenon not infrequently evolves as widespread degeneration, such that the tax-payer often sees outsmarting tax authorities as a kind of fair, sporting contest – which is understandable only when considered from his own primitive view that the state would otherwise outsmart him."[3]

Today, we are similarly inclined to consider tax morality, and along with it the relationship of loyalty to the state, as an expression of a country-specific national character. On 17 October, 2010, the Süddeutsche Zeitung conducted a survey of such character types in a ten-country review bearing the title "*Steuermoral – global. Die Welt der Steuerflüchtlinge. Wer besonders dreist ist – und wer noch halbwegs ehrlich* [Tax Morality – Global. The World of Tax Fugitives. Who's Especially Outrageous – and Who's Partly Honest]". When, in the spring of 2012, IMF chief Christine Lagarde reproached the Greeks for having a disastrous tax morality, compounding her reprimand by remarking that she felt greater sympathy for the victims of poverty in sub-Saharan Africa than she felt for the crisis-ridden Greeks, a cross-party wave of outrage rippled through Greece, which felt its national honour and pride cut to the quick.[4] Naturally, there are factors which can be extrapolated that influence tax morality or the readiness of the total population for resistance to taxation; and yet – as elaborated by Schmölders[5] – it has been empirically proven[6]

1 Cf. Nerré, Birger: Steuerkultur – Ein vernachlässigter Faktor der Steuerreformpolitik. In: Blümle, Georg et al. (ed.): Perspektiven einer kulturellen Ökonomik. Münster 2004, pp. 271-286, here p. 278f.

2 Schmölders, Günter: Finanzpolitik, p. 325.

3 Ibid., p. 330.

4 Cf. The Guardian, 28 May, 2012.

5 Schmölders, loc. cit., p. 332.

6 Cf. Franzen, Wolfgang: Steuermoral und Steuerhinterziehung. Eine empirische Studie im Auftrag der Margarete und Johann Ley-Stiftung. September 2008. [Forschungsstelle für empirische Sozialökonomik e.V. , www.fores-koeln.de].

that a plethora of social parameters leads to domestic social differentiation, a state which does not, without further qualification, permit one to speak of a unified or homogenous national tax morality. In this context, the ethnic type of a national tax mentality also invariably involves an ethnic stereotype.

While, on the one hand, a country-specific arithmetical mean value of tax morality does not represent a compelling national collective characteristic, the field of activity of tax is suited – particularly if this is understood as a gift for the benefit of the common good and, thus, as a kind of secular oblation – at least in part, to construe the nation as a living self-determined community of fate. Let us suppose that a German national, a so-called celebrity of German extraction – however this may be defined – commits an act of tax evasion, such as by relocating to a tax haven: while this celebrity will neither forfeit his citizenship nor nationality, the treason with which the irate public censure him – as witnessed on Internet forums – he commits first and foremost towards the nation in the sense of this self-determined community of fate as this is founded on solidarity and joint responsibility. His action would be perceived as a one-sided termination of a morally valuated, reciprocal social obligation; as an illegitimate reconversion of relations founded on effective and coherence-endowing relations of gift exchange into egoistically motivated, market- or trade-oriented monetary calculation. As mentioned, the field of action concerning taxation can only partially construct the nation; by way of a counter-example, this is illustrated by the immigrant of Turkish origin with German nationality, who, as a local employee, regularly and correctly pays tax. By doing so, he unquestionably becomes a contributor to the state and member of the society, and owing to the exchange relations invested by tax-payments, becomes a part of the nation. Naturally, full inclusion into the collective of the nation – "nation" understood as a comprehensive cognitive construct – is thereby not fulfilled, since the national self-allocation thereof is not essentially affected by this.

A sheer inexhaustible body of literature has been published on the subject of tax culture, tax mentality, tax behaviour and, above all, on attitudes towards tax, and the question as to the factors of tax honesty has a long research tradition. In the past, standard economic models mostly presupposed rational actors who, intent on personal monetary gain, entered into a game of risk-assessment against the background of control and the threat of punishment[7]. Homo oeconomicus was the godparent of such models. Homo reciprocans as well as homo culturalis have subsequently been introduced and, in addition to economic explanations, there are, meanwhile, individual and socio-psychological ones which, taken as a whole, have led to partial disempowerment of rational choice theory. Based on the above-mentioned sense of burden as elaborated

7 Cf. Kirchler, Erich; Maciejovsky, Boris: Steuermoral und Steuerhinterziehung. In: Frey, Dieter; von Rosensiel, Lutz: Enzyklopädie der Psychologie. Vol. 6: Wirtschaftspsychologie. Göttingen et al. 2007, p. (Online pp. 5-8).

by Schmölders, there has been a multiplicity of surveys, psychological tests and statistical analyses focusing on the subjectively and inter-subjectively shared perceptions, interpretations, evaluations, justifications and practices, etc., of tax-paying citizens which – following so-called prospect theory – thus identify the psychic mechanisms of tax honesty.

International comparative surveys, along with several individual studies, have been able to show, for example, "that tax honesty in an anticipated refunding by the fiscal authorities was significantly higher than had been the case with expected back payments".[8] The former is connected with a subjective winner-experience, whereby the goal is the securing of profits and the effect is a lower propensity for risk-taking and greater tax honesty. The latter relates to a subjective experience of loss, whereby the aim is the avoidance of loss and the effect is a higher risk-propensity and lower tax honesty.[9] However, such findings, as revealing as they may be for the analysis of stimulus-response phenomena, are only partially convincing from an anthropological comparative perspective since they are based on the premise of universally valid, deterministic behavioural mechanisms. In the process, culture-specific traditions and differences of evaluation in exchange relations between individual and state, of loyalties of moral categories, such as honesty and fraud, are thus overlooked.

However, for a cultural-anthropological review of the topic of tax behaviour and tax morality, economic and psychological approaches are fundamental in so far as the first account for the exchange relations under consideration here, while the latter account for the affective and perceptive forces. On review, we see the way in which system and subject interact with one another, and what incongruities and conflict areas arise as a consequence. System and subject, policy and habitus, structure and consciousness also delineate the poles between which ethnological research oscillates when conducting research, not least in its attempt to disclose recurring, stable cultural patterns behind the relations under observation that would otherwise remain blurred. To date, ethnological research has given little attention to taxes. The premise on which I approach the topic is, firstly, that the totality of institutions, the administrative and legal complex of a tax system represents a cultural object in the sense of an architectural construction, the function of which represents state-regulated reproduction and perpetuation of society, and, secondly, that the totality of these practices, attitudes, ideas, narratives, etc. of this tax complex is to be seen as an explicit cultural practice.

These premises lead initially to a fundamental question which has hitherto received comparatively little attention in the literature under review. It stands to reason that if the architecture of the tax system aims at the perpetuation of

8 Ibid., p. 18.
9 Cf. ibid.

state and society and is, to this extent, directed at the sustainability of established exchange relations, then the inclination will be to naturalise and dehistoricise tax loads in whatever form these may take. One may take as given that for their sustained implementation, their temporal and occasion-specific grounds must evaporate by way of a politics of forgetting[10] and be, consequently, converted into ontologically and normatively binding facts. As is well-known, the above-mentioned sense of burden is historically and interculturally extremely elastic and adaptable. Compared to contemporary tax rates, for example, the tithe, which was formerly experienced as a burden, is fantastically low. Today, any claim that would merely hint at such reversion would be simply dismissed as unrealistic: in fact, if enforced, the sustainability and "naturalness" of the contemporary state could no longer be guaranteed.

Hence, the first fundamental question runs: which role does a tax system play within the framework of the idea that state, society and the nation are eternal; what contribution does it make to the social cosmology of a national, state-constituted commonwealth? In that it supplies the state with those funds by way of the administrative functions of the fiscal authorities,[11] the budgetary application of which is decided in parliamentary procedures at federal, state and communal levels, it represents a powerful, order-endowing and regulating medium for society as a whole. It, thus, consolidates the state order and authority as a whole and, in doing so, legitimises it. Accordingly, fiscal offences are valuated as misdemeanours against the overall order. This comprehensive order designates a social cosmology, namely, the social consensus thereof, such that this order – as well as the fundamental ideas of order as inscribed in the tax system – should, in principle, be ordered as it, in fact, is. In this connection, finance-psychological proposals for the improvement of tax morality aim not at revision, but at increased plausibility. Repeated reference is made to more direct democracy (referendums on the uses of tax revenues based on the Swiss model), greater transparency and information relating to decision-making processes, increased decentralisation and localisation, as well as advertising campaigns aimed at strengthening public spirit with respect to civic taxation.

Whereas tax evaders infringe on public order, tax objectors and tax rebels reject the given order either in part or *in toto*. In its various forms of justification, demonstrative tax objection – equated in this country (Germany) with tax fraud – not infrequently connotes a critical (in both sense of the word), fundamentally impaired and subverted relationship to the state. Counting among

10 An often-cited example of this is the financing of the imperial fleet by the introduction of duties on sparkling wine in 1902; this has yet to be abolished.

11 For the fiscal year 2015, a sum estimated in excess of 95 percent of the federal budget earned by way of the general administration of revenues (Federal Ministry of Finance 2015 online).

these various justifications are, for example, reasons of conscience, self-defence or resistance against repression and arbitrary acts of the state, alleged absence of legal legitimation for tax collection and the fundamental rejection of the state as an apparently illegitimate or criminal organisation. Whether moderate or radical, whether partial or department-related – such as in the refusal to co-finance military spending with taxes – or fundamentally as a matter of principle, as in the refusal to participate in a system conceived as unjust, what remains true for all variants is that they brusquely dismiss moral valuation and oppose it with a counter morality which, for its part, reclaims legitimacy for itself. By this means, it relativises, subverts or negates that which has here been referred to as social cosmology, and, to some extent, underscores, *ex negativo*, the findings that a legally and administratively given system of taxation is carried by a general fundamental understanding and credo that resembles a tacit confession of faith. Thus, in so far as this consent formulates decidedly culturally – ethical, moral, social, economic, etc. – occupied values and valuations,[12] cultural anthropology sees itself challenged to introduce to the forefront of the discussion on tax morality and trust the question as to the socially binding force, as well as the (potentially carrying) position of these values and evaluations within the architecture of the tax system and, therefore, to the state regulation of social reproduction and continuity.

Some remarks on tax evasion are called for to avoid potential misunderstandings before turning to my conclusory discussion on the question of the social binding force and the cultural practices it generates. Tax evasion, tax opposition, tax rebellion and tax revolt all denote strategies of moral/political protest, and yet the motives behind moral/political justifications and, accordingly, the reactions to them are so inconsistent that any generalisation would be misleading. Exemplarily, the motive of tax objection for reasons of conscience – especially that of an objection to qua tax payment for indirectly rendered military service[13] – enjoys comparatively high social acceptance and is certainly capable of discourse.

Citizens of North America, who, so as to elude co-financing military operations in the Middle East and Maghreb, purposely cap their income to remain below the tax threshold may be credited with ethically impeccable behaviour. Now, whoever for the same reasons passes over from tax evasion to tax objection also stands a good chance of being judged as peace-motivated – particularly where the refusal is free from the suspicion of monetary self-interest. The "*Netzwerk Friedenssteuer* [Peace Tax Network]", for example, bearer of

12 These evaluations refer to all fields of fiscal expenditure, from education through social security systems to child care or road construction, etc.

13 This motive has a long history. Activists draw on this regularly and construe a line of ancestors which ennobles their acts according to the principle of descent. Occasionally, chronological lists of tax opposition and tax rebellion read like the heroic history of a democratic tradition.

the *Aachener Friedenspreis* [Aachen Peace Prize], understands itself as a central platform for the peace movement, advocating at least partial decision-making autonomy by the tax-paying citizen with respect to the use of their tax revenues. It advocates that everyone has the "legal possibility to decide between the use of his taxation either solely for civil purposes, or additionally for military purposes".[14] This vision of a right to tax opposition with respect to military expenditure finds support in clerical as well as parliamentary-political circles; it is both the object of seriously discussed political manifestos as well as legal discourses and dissertations.

The degree of tolerance looks entirely different with respect to movements that dispute the state's right to collect taxes as a matter of in principle. North American actor Wesley Snipes, accused of tax fraud in 2008 and later sentenced to three years imprisonment and ordered to pay seventeen million dollars back payment, was suspected of being affiliated to a group of radical so-called tax protesters. Such an affiliation can clearly result in severer sentencing, since according to the charge, this offence could have been punishable with a total of sixteen years detention – namely, on grounds of conspiracy. In 2007, Ed and Elaine Brown – the former a retired pest controller, the latter a dentist – emerged as the figureheads of militant tax resistance. Obsessed by crude conspiracy theories, anti-Semitic resentment, legal myths and pseudo-religious phantasms, the latter initially refused to pay tax, before becoming more radicalised by increasingly rejecting all state authority: heavily armed, they entrenched themselves on their property, held out in the stronghold for several months in a siege situation, and thereby demonstrated their media effectiveness. They recruited droves of supporters, in some cases armed, before being arrested without bloodshed and finally sentenced to 35 and 37 years imprisonment respectively – not only for tax fraud, but also on charges of incitement and conspiracy.

Conspiracy theories connected with ethnic-racist sentiment and phantasies of greater German power also pervade the 76-page invocation to tax resistance published by Roland Bohlinger in 2007 bearing the title "*Steuerverweigerung. Das Widerstandsrecht des unterdrückten Volkes* [Tax Resistance. The Right to Resistance by the Oppressed People.]" Ideologically, the author invoked the pioneer of the '*völkische*' movement and fervent anti-Semite, Mathilde Ludendorff, thereby demonising the state ("the so-called BRD") as an illegitimate crime syndicate. Held up as counter-examples, excrescences of this kind again throw light on the binding or unbinding forces that connect the individual to the state and, thus, to the community of tax-payers.

Tax honesty is redolent of loyalty, whereas tax opposition suggests hostility towards the state and subversion. Tax frustration, which creates a breathing space primarily by way of standardised expressions of discontent – indeed,

14 Cf. online appearance of the Netzwerk Friedenssteuer [Peace Tax Network].

precisely because they are permitted as a vent even in a compensatory sense – is an immanent bearer of the system. As one may also occasionally witness in other relations, the home fires are not burning as they should be. The cause of this troubled domestic tranquillity is, for example, the *Bund der Steuerzahler* [Association of Tax-payers] with its black book "*Die öffentliche Verschwendung* [Public Wastage]". Countless entries in the relevant Internet forums suggest that stories of wasted taxes erode tax morality, since they narrate breaches of trust by the state, of betrayal towards those relations of redistributive exchange based on responsibility for the maintenance of which each individual tax-payer is legally and morally obligated. Who then, one may ask, is accountable to whom and for what? Is the tax-payer accountable to the state or, conversely, is the state accountable to the tax-payer? A telling comparison concerning this question is to be found in a forum article on tax morality in "Spiegel Online" dating from 6 February, 2014:

> "As long as tax revenues continue to be wasted and the state makes little attempt to act with moderation and deliberation when disposing over tax revenues from which half of the tax-payers' hard-earned earnings are deducted month by month, then tax opposition and fraud is, so-to-speak, a civic duty! The situation is much the same with children: when they spend their pocket money on rubbish, then their parents accordingly reduce it."

The essence of this piquant image is: if one believes that the people are sovereign not the state, then the question as to accountability and consent is reversed. The burden of proof of maturity and reliability lies on the shoulders of the state and not the citizen. Should it fail in this respect, then "Uncle Sam" ['Vater Staat'] regresses to the status of an immature child, whereby, for all intents and purposes, the mature tax-payer is then obliged to have recourse to his parental function by issuing disciplinary measures.

The frequently misused expression "tax-citizen" appears in administrative parlance no less often than does the self-coined epithet "we tax-payers" in everyday language. In this connection, the term denotes more than the mere aspect of tax liability. It also implies civic self-confidence, which, as the occasion arises, also raises objections, complains of deficits or makes demands. Alongside the fundamental agreement with respect to the exchange relations outlined above, this reciprocity of designation and self-designation also demonstrates the fundamental agreement about the moral obligations inscribed therein. This begs the question as to the procedures that lead to this agreement. From an economic standpoint, one is a tax-payer if one has personal[15] tax obligations. From a cultural-anthropological perspective, something else is added, namely, the internalisation of moral values as pertains to

15 A child who pays value added tax when purchasing sweets, hardly has the status of a tax-payer.

taxation as a result of socialisation processes. As is well-known, threats of punishment are the most effective educational means in this field.

Daily practices and routines as part of tax declarations are no less a part of the tax-payer's socialisation. Day by day and year by year, we contemporaries collect countless millions of receipts, invoices, bills, tickets, mileages, bank statements, etc., and store these in envelopes, boxes, files and other containers in order to meet our tax obligations and ensure our right to tax remissions. This large-scale work of collecting and tinkering is not a collective parlour game, but a massive and, for the most part, mutely performed gymnastic exercise, which trains not the body but the conscience. Training aids in instructions for the efficient ordering and filing abound as well as countless guidebooks on the subject of saving taxes.[16] Here, tax-saving models can develop into decisive guides for life, and reach far into the emotional economy of domestic life and intimate relations. For a while it even became common practice to cite the "Inland Revenue" as a reason for getting married, indeed, this reason for marrying was considered the least suspect.

Thus, if the enculturation of the tax-payer proceeds largely as part of a tax-saving habit, then this might well collide with the commandment of tax morality, according to which the obligation is conceived not merely as an inescapable contribution to the community, but also as a social and desirable contribution to it. Accordingly, the cultural logic of the tax system favours the civic virtue of thrift, and hence, counters the mortal sin of *avaritia*, greed and acquisitiveness. And yet, it is less the news on tax scandals than the grey zones – the small and big so-called tax-loopholes – which suggest that civic virtue and Christian mortal sin are not archenemies, but illegitimate, though related siblings. How does one learn about tax morality and how does one practice it? To this end, it may, perhaps, be helpful to shift one's focus more towards the model of redistributive relations of gift exchange – where, at least to some extent, gifts are perceived as involuntary donations or offerings – as a general option for the organisation of social relations, and to familiarise individuals with this option long before they have attained the status of tax-payers.

References

KIRCHLER, E. and MACIEJOVSKY, B. (2007) Steuermoral und Steuerhinterziehung. In FREY, D., ROSENSIEL L. VON (eds.), *Enzyklopädie der Psychologie. Vol. 6: Wirtschaftspsychologie*, pp. 5-8 (online).

NERRÉ, B. (2004) Steuerkultur – Ein vernachlässigter Faktor der Steuerreformpolitik. In BLÜMLE, G. ET AL. (ed.), *Perspektiven einer kulturellen Ökonomik*, pp. 271-286. Münster: LIT.

SCHMÖLDERS, G. (1970) *Finanzpolitik.* New York: Springer.

16 Google Search records approx. 3.4 million entries under the keyword tax.

"On the Safe Side of Life": Cultural Appropriations of Funeral Insurances in Namibia

SABINE KLOCKE-DAFFA

Preliminary Remarks

This article is based on field research in Namibia between 2010 und 2014. In some way it dates back to the year 1992/1993 when I conducted my very first fieldwork among the Berseba Nama of Namibia as part of my PhD thesis at the University of Muenster. It was Jos Platenkamp as my supervisor who inspired me to look into the cultural foundations of social security systems – a topic which turned out to be fascinating enough that it keeps my busy up to this day. I am grateful to Jos for the many fruitful discussions we had, his professional encouragement on my way through university, and invaluable advice whenever a new challenge had to be faced.

Security – Social and Individual Challenges

No society can afford to leave the contingencies of life to be dealt with exclusively by the individual. This task always requires social solutions, necessitates a minimum of organisation and must be based on trust. This is where the state usually comes in, provided there is a functioning state, and one of the uppermost responsibilities of a state is to guarantee the safety of its citizens. In fact, nation states and their governments derive part of their legitimacy from the degree of security and support they manage to secure, all the more as social security was declared a human right in the Declaration of Human Rights accepted by the United Nations in 1948 (Van Langendonck 2007). If the state does not provide what is needed or is unable to fulfil this task, other institutions might step in as a complementary source of security. This study concentrates on one of these, namely private insurances. In fact, most people prefer to have recourse to several security provisions – formal as well as informal, private as well as state insurances – so that they can use them as alternative means to serve the same ends, shifting between the different spheres.

Based on my research in Namibia over the past years, I formulate two hypotheses: The first is that the appropriation[1] of modern social security systems necessitates certain preconditions:

1 I use the term *appropriation* in the sense of integration of an external item or idea into the own cultural system; see Hahn 2005: 102-104.

- they must be *accessible* – potential beneficiaries must have a chance to get in contact with them and use them,
- they must be *affordable* – as far as the price is concerned,
- they must be *reliable* – regarding the safeguarding of risks, and
- they must be *adjustable* – respecting the prevailing cultural norms and values serving as a social resource.[2]

I will mainly focus on the last point, arguing that new forms of social security must above all be adjustable to the existing value system in order to be accepted by local actors.

My second hypothesis tackles the preconceived notion of a dichotomy of traditional and modern social security systems: is traditional to modern as informal is to formal? I challenge this assumption as it must be realised that *traditional institutions* can be quite modern and *new formal institutions* can be used according to traditions and based on culture-specific values (Klocke-Daffa 2008).

Namibia is indeed a fine example of the blending of different social security systems, originating in the diverse African and European cultural traditions which shaped the society of this country.

Research Questions and Theoretical Approach

The predominant mode of providing security for the native (black) population[3] in Namibia is the informal system (e.g. Klocke-Daffa 2001, 2009, 2011; Greiner 2011; Pauli 2013). However, there is also a booming business in formal insurances sold by multinational companies operating on a global scale: Old Mutual runs offices in at least six African countries with headquarters in London/UK[4]; Metropolitan Life is a US-operated company with headquarters in New York[5], Sanlam is a South African financial services group with its main office in Cape Town and offices in many parts of Africa, Europe, India, the USA, Australia and South East Asia.[6] Namibians have long been used to private insurance companies operating in the country – for decades available

2 The term *resource* signifies "a base of or means to create, maintain or transform social units and identities on the background of culturally shaped beliefs and practices" (Bartelheim et al. 2015: 39) according to the analytical concept developed by the Collaborative Research Center 1070 'ResourceCultures' of the University of Tübingen where I have been a member since the CRC started in 2013.

3 The terms 'white' and 'black' population are used according to emic distinctions in Namibia where the so-called 'colour bar' between ethnic groups up to date determines access to job opportunities or residential areas.

4 Old Mutual International: http://www.oldmutualinternational.com/en/europe/consumer.

5 Metropolitan Life: http://global.metlife.com.

6 Sanlam: http://www.sanlam.com.

only to the white population – but one would expect that life and funeral insurances are the social protection measures least demanded in a country with a large percentage of the population – between 27 to 40% – considered to be poor or even severely deprived.[7] However, Namibia's life insurance market is the fourth largest in all of Africa and outnumbers by far the non-life insurance segment. Life insurances (including funeral insurances) account for almost 70% of all premiums (KPMG 2014: 6). The volume of premiums per capita is the second highest in Africa after South Africa, amounting to almost one billion US $ (*ibid.*), even though the population is one of the smallest on the continent.

Among the Namibian Khoekhoen (also called Nama), one of the Khoisan peoples of Southern Africa, as well as among other Khoisan of the country such as the Damara, many individuals below the age of 60 have arranged for a formal insurance or are part of somebody else's contract. The word 'insurance policy' (in the Afrikaans language 'poolus') has even become synonymous for funeral insurances. Anyone asked about a 'poolus' will most certainly answer: "Yes, I got a funeral insurance." However, this does not mean that modern social insurances have replaced traditional forms of social support but are used instead as extra benefits.

My research questions were based on the following: What impact do informal security systems have on the implementation of modern security institutions? How do private social insurances match with the existing informal social security system? And why did funeral insurances become so popular? To interpret the empirical data collected during two field trips in 2005 and 2010[8] I will use a twofold approach: a structural analysis of exchange relations combined with models drawn from cultural economics (following Parry and Bloch 1989 and Gudeman 2012). This does not seem to be too daring a venture since all authors have published extensively on exchange, values and the cultural dynamics of economic concepts. It helps to understand why and how modern insurances are accepted by a society whose members describe themselves as tradition-minded.

Namibia and the Namibian Khoekhoen

Namibia became independent no earlier than 1990. Between 1884 and 1918 it was a German colony (called *Deutsch Südwest-Afrika)*. German settlers took possession of large parts of the land and German missionaries sent by the

7 The UN Human Development Report 2013 indicates that about 40% of the Namibian population live in 'multidimensional poverty' (United Nations 2013: 4-5), as against 27% indicated by the Namibian Central Bureau of Statistics (2008: 6).

8 I would like to thank the Fritz-Thyssen-Foundation for generously supporting a research project in 2010 on "Life insurances and social networks" which included research on funeral insurances usually integrated into life insurance policies.

protestant Rheinish Mission Christianised the natives, so that today about 80-90% of the population are Christians.[9] In 1919 the country became a mandate of the League of Nations to South Africa and was renamed *South West Africa*. The political system of apartheid was thus introduced with all the well-known consequences of discrimination against the black population, even though never rigorously enforced such as in South Africa (for a historical overview see Wallace and Kinahan 2011: 131 ff.). Namibia is a vast country with many treasures of the soil such as diamonds, gold and uranium and a relatively small population of slightly over two million inhabitants (Central Bureau of Statistics 2008).

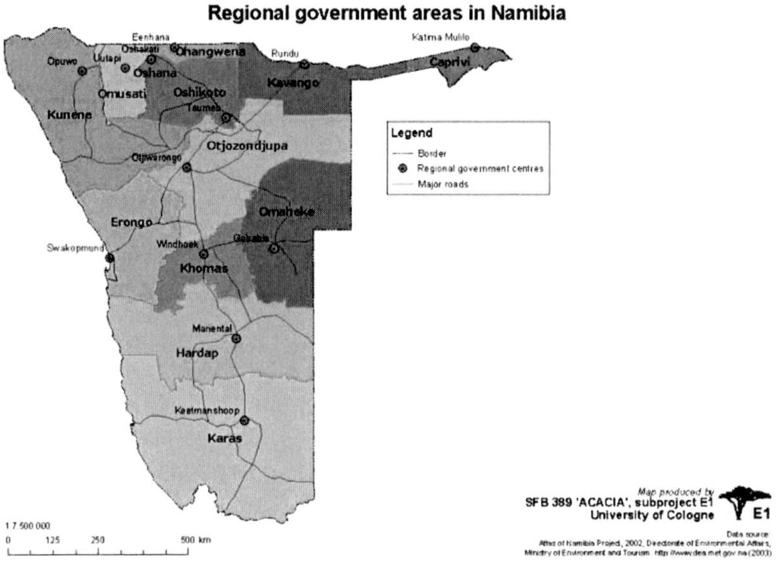

Fig. 1: Regional Government Areas in Namibia
(Source: ACACIA Cologne)

At present many of them continue to dwell in or around the former homelands. In the scarcely populated Southern Communal Areas of Karas Province (see fig.1) jobs are particularly few. As before, young people are forced to leave to work on farms, in mines or in the country`s cities.

Social Security in Namibia – Formal and Informal Systems

So how can there be anything that could be identified as 'social security'? To answer this question it is necessary to examine how *security* is conceptualised

9 According to the ecumenical Council of Churches in Namibia up to 90% of the population are claimed to be Christians, see http://www.ccnnamibia.org/. However, there are growing numbers of Pentecostal churches as well as a still relatively small Muslim community.

and how it is provided. As opposed to the common perception of poor countries lacking social protection for the bulk of the population, Namibia does provide a minimum of social security. In fact, the individual is positioned in three circles – here called circles of security (see fig. 2) – each of them granting a certain amount of support. The population manages to use all three in quite efficient ways:

1. the primary circle, comprised of the informal security system which works most effectively in personal networks of close family members, neighbours and friends;
2. the secondary circle made available by the national welfare system[10], providing cheap medical care, small grants to the severely ill, the veterans, surviving spouses, orphans, the disabled, and women on maternity leave. Most important of all are the state pensions for any person beyond the age of 60 years (currently 1,000 NAM $ = 60 EUR) and a funeral aid (for a person beyond the age of 60) which covers the costs of the cheapest funeral (currently 2,500 NAM $);
3. *the tertiary circle of* private insurances, which covers mainly life and funeral insurances, offered by private insurance companies.

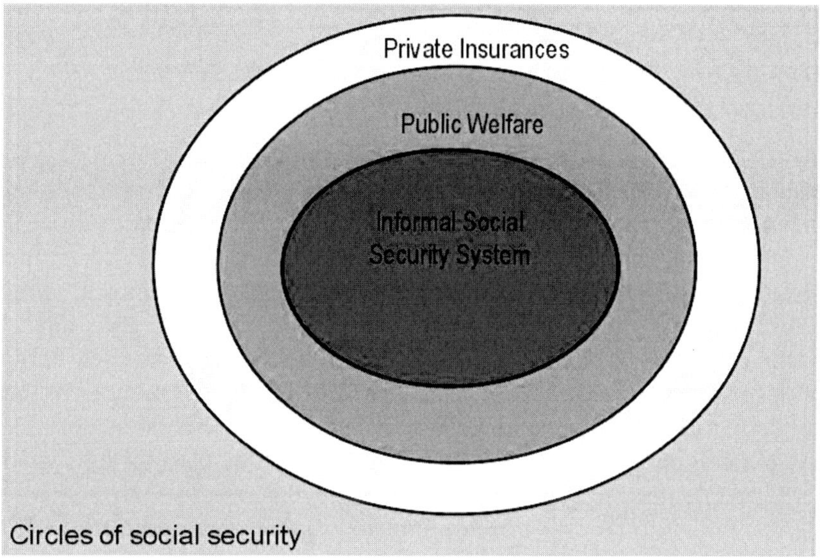

Fig. 2: The Namibian system of social security: positioning the individual in different circles of security

10 For an overview of Namibian social security programs see U.S. Social Security Administration 2013 providing detailed information on all Namibian welfare programs.

For the Namibian Khoekhoen, formal funeral insurances constitute a new form of private security provision, but these appear not to be regarded as a substitute but rather as complementary to existent collective security measures, namely the so-called traditional one.

During the course of my research, I interviewed a number of company representatives selling funeral insurances, as well as private brokers, some of their customers and several of the Namibian undertakers. What is remarkable is that many policy holders do not have a policy only for themselves but manage to place up to 12 or even 18 persons of the extended family on the contract. The companies now demand that the contract does cover the customer personally, as it had appeared previously that some of them excluded themselves which gave rise to problems of responsibility when that particular customer died before any of the beneficiaries and none would be willing to take over payments but all claimed coverage. The insurance companies did, in fact, manage to adjust to different demands and cultural conditions such as accepting unofficial second wives, illegitimate children and classificatory relatives (sisters and brothers of parents or grandparents) as members of the extended family to be placed on the policies. This raises the question why funerals are so important in the first place and leads back to the primary circle of security – if that is understood we will realise how all three merge together.

The Khoekhoen Exchange System as a Social Resource and "Cultural Base"

Funerals are part of extensive exchange relations which Marcel Mauss depicted as "total social phenomena" (Mauss 1954: 76-77). Every sphere of the social system is characterised by an ongoing exchange of goods, services, and persons. Whenever somebody is in need of something, he or she is entitled to demand it. The Khoekhoen would say: "You can go and ask for it", because the rule states: If you have something to give and somebody asks for it, you must give it. It can be anything one might need such as food, clothes, furniture, money or services (see Klocke-Daffa 2001: 219 ff.). If the groceries or the bill from Pep Store cannot be paid for, ask somebody else to foot the costs. Similarly, if a cell phone needs to be uploaded, this can be paid for by somebody else. If people need transport to a funeral, they can ask to share somebody's car. Assistance in the care of farm animals can be requested just as help for the household, encouragement for the ill or solace for a person in grief, since time is something that can also be given. Even persons can be asked for: the elderly might request that a youngster stays with them in the house for a while and a couple without their own children might let the family know that they would like to be given a child (Klocke-Daffa 2015).

All these transactions are considered to be gifts which have to be given freely, without complaints or insults of the recipient. Gifts should not be reciprocated right away or even of the same value, but are expected to be compensated whenever there is a need. The donors do have the right to receive also, since the recipients are obliged to give in return at some point in time. If nothing can be given on a material basis, people can always "help with their hands" i.e. provide services. This kind of generalised reciprocity creates a system of open bills which keeps people in continuous social relationships and helps to create extended networks serving as social capital. The ongoing giving and sharing is further topped by obligations arising on ritual occasions. In a Christian community these are mainly life cycle rituals such as baptisms, confirmations or communions, weddings, and funerals. Observing the performance of these events it quickly becomes evident that they exceed the financial possibilities of individuals and households. In fact, they can only be enacted if all the participants share the costs. Funerals are the most important of all and there is no excuse for absence, because the entire social network is present only on such occasions. A funeral is the last opportunity to see the deceased person and is an act of honouring the family as well as the community.

In this regard, funerals are the acid test of the system: no other event demands so many personal sacrifices and has so many positive effects on the (re)-construction of community. In times of HIV/AIDS when funerals became an almost regular weekend event[11], it could have been expected that the costs be reduced for reasons of mere necessity and financial shortages. But the opposite is true: funerals are celebrated ostentatiously, with luxurious coffins, with up to a hundred and more guests to be fed and expensive marble grave stones set up (Klocke-Daffa 2009: 33). Cemeteries shine from all the new graves. Even a poor person is entitled to a big funeral. The differences between the funeral of a chief and an ordinary person are gradual rather than substantial. Among the Namibian undertakers the Khoekhoen are much appreciated customers said to be "the people who always want expensive funerals".

When it comes to the funeral benefits granted by the Namibian state originally meant to reduce the financial burdens of AIDS-stricken families, it is evident that the Khoekhoen do not use these grants to get rid of their social obligations but instead use the money to arrange an even more lavish funeral to make sure that the guests will remember it as a "really nice funeral" (see fig. 3).

11 During the years 2000-2010 the number of HIV infections was rising continuously. AIDS became a constant threat with an almost national dimension. With the introduction of sex education in schools and churches as well as the provision of antiretroviral medication and therapy (ARVs) available to the population for relatively low costs, the AIDS prevalence rate dropped from around 24% in the mid-2000s to currently 14.3% (UNAIDS 2013).

Fig. 3. Grand funeral on a farm in the Southern Communal Areas, 2010.
Most of the costs were covered by funeral insurances
(© Klocke-Daffa)

The tradition of giving and sharing can only be understood if the value system that backs it up is examined. Asked why people so willingly submit to norms that strain them, they would answer "I do it for the LORD". Since GOD is conceptualised as the ultimate giver and original source of gifts it is HE who demands a counter gift – which is given to a third person. So there is a triadic structure and it is indeed the LORD who keeps the system going. The Khoekhoen are convinced that this pays off in the long run and with a triple effect:

- reciprocals gift can be expected,
- social reputation is granted to generous givers by the community, and
- divine blessings are guaranteed due to the fact that "the LORD loves the givers".

Wealth derives from giving not from keeping. Giving and sharing thus is much more than a functional support mechanism but part of a socio-cosmological model which creates and sustains an elaborate exchange system. Obligations can temporarily be halted but in the long run it is all worthwhile as people want to be good givers in the confidence that the LORD will keep the system going.

Ideally, exchanges within social networks do provide a relatively high degree of security because they grant economic, social and emotional support whenever needed and even before being asked for. Those whose income allows for a decent living are expected to give freely to show their gratitude to the LORD and possibly even before being asked. Those in need can always go and ask. If demands are becoming too much, obligations can be delegated temporarily to other persons to prevent overexploitation. Both sides should change positions now and then in order to make sure that those who give generously for most of the time will be helped in case they should ever be in need, and those who are notorious recipients of gifts do get a chance to reciprocate. This way a minimum of social security can be provided even without national welfare programs though on a rather low economic level. It is hence not only the availability of means that ensures the social security of society but the *exchange* itself is conceptualised as a social resource and with it the circulation of items, goods, services and persons.

The Cultural Appropriation of Funeral Insurances

Then why are private insurances so heavily sought after, with two systems already providing support? One of the challenges the Khoekhoen face today are the challenges of individualism. Young people expect to get what independent Namibia is promising them rather than give away their last bit: decent houses, good education for the children, medical care, economic wellbeing, and social advancement are compelling prospects. Modern Namibia provides all this and, unlike the days of apartheid, the black population is no longer legally excluded from expensive consumer goods. But individual wellbeing has its price. The Namibian state guarantees free access to markets but does not guarantee equal wealth for all its citizens. On the other hand, social welfare programmes cover only part of what is needed. Hence the pressure on the traditional exchange system is increasing and on those expected to give freely, who in turn seize any opportunity to escape excessive demand sharing. A value comparison shows that everybody in need should indeed receive his/her share, but *in reality* not everybody does as too many opt out. That is where private insurances come in.

Insurance companies promise to provide security where individuals or society fail to do so – and are used in a very culture-specific way. Even though insurance policies must be taken out by *individuals*, they are conceptualised as a measure to safeguard *social obligations* and *collective responsibilities*. Khoekhoen customers will use their contracts to fulfil social obligations as well as individual needs. As the funeral is the one event that counts in life every individual is anxious not to fall short of expectations. Whatever hardship life may bring, it is calming to know that funeral costs will be taken care of, including one`s own, thus helping descendants to fulfil their obligations.

Effecting an insurance payment then becomes an individualised way of coping with social obligations under changed conditions.

What makes private insurances so attractive is the fact that they satisfy all necessary preconditions to become acceptable in a society heavily relying on social networks:

- *accessibility*: insurances are easily accessible through offices and agents all over the country;
- *affordability*: anybody with a minimum of income even in rural areas can afford a risk insurance, recently made available even to farm and domestic workers with only minimum wages[12];
- *reliability*: insurance companies are reliable when it comes to payout. There have been few cases of corruption and the money is paid out in cash within 24-48 hours after it is due;
- *adjustability*: most importantly, these private insurances fit perfectly into the existing exchange system: up to 12 or even 18 persons can be covered by one policy. The obligation to pay the monthly contributions can be transferred to another person and, once paid out, the money is fed back into the social network.

The structural characteristics of social security transactions related to funeral insurances show that two different *modes of transactions* (following Parry and Bloch 1989: 23-28) can be distinguished: in a *short-term cycle* (fig. 4) security is conceptualised as a commodity sold to the individual (A) by an insurance company (IC) within the highly capitalised and individualised market of insurances. It is money taken out of the *long-term cycle* of social exchanges (with A > B, C, D). These persons in fact expect to be given, supported and looked after whenever in need as they do amongst each other and towards (A) in case of emergency. Indeed (A) might end up paying for all – and in this case for several successive funerals – but might well receive more than what was given (B, C, D > A).

Taking financial resources out of this cycle makes insurances appear somewhat dubious at first glance as they are conceived of as 'extra money' originally belonging to the social sphere. If funeral insurances would be finalised only for the individual policy holder (including at best the spouse/partner as is the standing rule in most Western countries) this kind of extra money put aside for individual purposes would be highly contested. However, what happens is that funeral benefits are re-transformed into a social resource at death,

12 In 2011, the minimum monthly deposit for a funeral policy at Old Mutual was 10 NAM $ (= approx. 0.90 EUR) with a guaranteed payout of 2,500 NAM $, after six months of payments at the earliest. Most insurance companies lowered the minimum monthly income to approx. 500 NAM $ (50 EUR) as a response to the demands of potential customers from the lowest income group of domestic and farm workers who had been excluded before.

thus becoming morally acceptable. This happens precisely via the funeral ritual when used to cover all or part of the costs incurred.

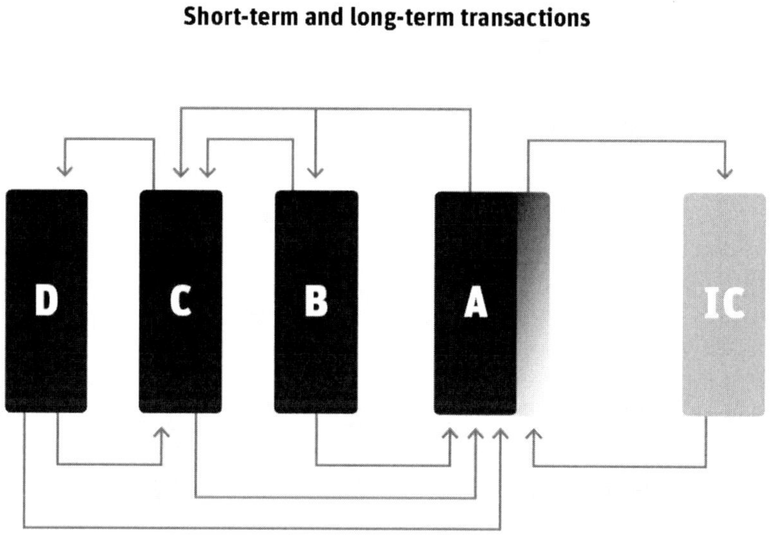

Fig. 4: Short-term transaction (A > < IC) as opposed to long-term transactions (A-B-C-D-A) of resources used for social security

It explains why the Khoekhoen would consider any personal use of funeral money not only as a misuse of what belongs to the community but as an act of morally reprehensible wrongdoing because the individual use of it violates the rules of exchange. Suspecting policy holders of using this money for new clothes, furniture or cars range among the many stories that are communicated at the events of a death and are a source of collective indignation since this kind of behaviour – though occurring frequently – is considered to be a threat to social security and, worse, an offence against the dead. The Khoekhoen are convinced that death benefits "belong to the deceased" who will distribute it to the community of the living at the time of the funeral as a last gift – the gift of life.

However, as the economic and social situation in Namibia forces many people into poverty or odd jobs with a shaky foundation, funeral insurances though much appreciated must sometimes be waived for lack of financial resources. There is one possibility to avoid the disagreeable consequences: many policy holders do run the risk of having their policy dissolved (usually after three months of unsettled accounts) but would then ask members of the extended family to accept some or all of the extra persons covered onto their contracts. Stories abound about phone calls comprising unexpected requests to take over family members on one's own funeral insurance contract ("You must put our mother on your policy"). The person to be asked first is usually the one with the presumably highest income in the family but if that person is not willing or capable to take over the oldest sister (*aussis*) or oldest brother (*kai dab*) will certainly be reminded of their obligations. The only risk connected to this kind of transmission is the postponement of payouts for any additional person for the following six months – but people would be willing to accept the risk and trust the LORD that no death will occur and no funeral will accrue within this period of time. However burdensome, it became such common practice to shift family members and to exchange contractual obligations that many insurance companies nowadays include an insurance waiver in the first place in order to reduce administrative costs. To make sure there will always be someone to take care of some pending funeral, many elderly persons find themselves posted on several policies taken out by their own children or former foster care children, classificatory children, grandchildren or other members of the extended family. With five or more policy holders collecting the payout for a deceased grandfather, a grand funeral will be secured. We then find a *merging of short and long term transactions* (fig. 5) where the insurance company is incorporated into the existing social network and very much treated accordingly: there is always someone to sign a contract provided that the company does not fail to reciprocate (i.e. be a reliable provider, pay out in due time, and ensure transparency concerning administrative structures). To make sure that customers are satisfied, insurance brokers never omit to point to the punctual payout of funeral benefits and some undertakers – who usually will be the direct beneficiaries of the contract money – would even offer in advance of the funeral ceremony to pay for the upcoming grocery bills on behalf of the bereaved family, showing their concern as business people as well as liable members of the community.

Both circles are closely linked and dependent upon each other: the individualised short-term mode of exchanges and the socially much higher valued long-term exchanges ensuring the symbolic reproduction of the socio-cosmological order. If customers of private insurances consent to pay out funeral benefits to their exchange partners within the social network or even make some of them take over part of concomitant obligations, then both orders merge into one coherent mode of exchange – linking social norms with individual agency much to what Itéanu calls "une coherence dynamique" (Itéanu 2010: 369).

Funeral rituals operate as a sort of hinge or transformational focus to turn individually saved money into a socially acceptable resource such as the 'cooking money' among Malaysian fishermen (Carsten 1989) but in a somewhat more enhanced way as there are other options of transformation. At the core of both orders is money as the medium of exchange, serving different ends in different spheres and value systems. Taking into consideration that it not only takes *material accumulations* through the productive use of finances (savings) but also *cognitive skills* and *knowledge* of just how much can justifiably be set aside without being considered an offence against ethical values and *practices* of reasserting one's own position as a moral person, exchange truly constitutes a culture specific 'base' in the sense that "persons in a community are connected to one another through and in relation to the base that lends them an identity" (Gudeman 2012: 98).

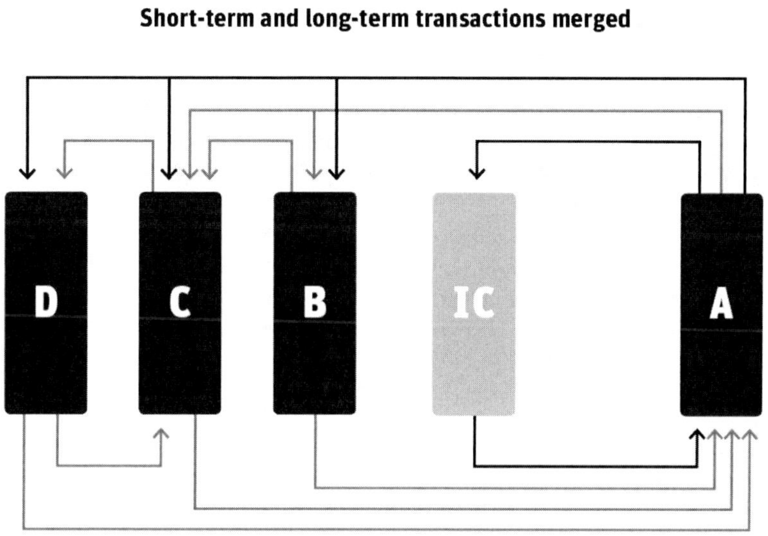

Fig. 5: Short-term and long-term transactions merged into one coherent mode of exchange by integrating private insurance contracts into the social network.

What is then apparent is a blurring of socialised *gifts* and individualised *commodities* or, in other words, the socialisation of individual consumer products (insurances) and the individualisation of social obligations (gifts). Whether this is simply a way of adjusting to new needs or a specific clash of a newly established consumer culture with a traditional gift economy would need further investigation. Arguing with Marilyn Strathern (1997: 307) "that the double orientation of gifts in consumer culture presupposes two kinds of relationships: an individual person's interpersonal relations with others and an individual person's relation with society" it would be interesting to know what kind of person the 'Khoekhoen gift' reproduces under contemporary conditions.

Conclusions

Returning to the research questions, it must be realised that informal support systems and cultural patterns are closely interwoven and should be considered a firm base of modern social security provisions. Tradition and modernity must not be incompatible. It also appeared that the state (and in many cases its local representatives) plays a crucial role in the realisation of security provisions if only to make sure that certain preconditions are fulfilled. In this case it was the independent Namibian state that improved access to private insurances, negotiated low prices for the labourers (thus making insurances affordable), ensured the reliability of insurance companies (by certain control mechanisms), and initiated new welfare programs thus adjusting to existing demands. So when it comes to local state formations and different forms of social assistance, a careful examination is needed of the structures of existing security systems and the extent to which local states manage to take these into consideration when designing public welfare programs and allocating national resources.

The ethnographic example of formal funeral insurances shows that they can be complementary to existing social support provisions. Provided that new market products are accessible, affordable, reliable, and adjustable there are chances that they will enhance security rather than threaten social relations by tying up monetary assets. In Namibia, consumers as well as insurance providers have shown a striking flexibility in the use of funeral benefits: as a social resource they allow the giver to take money out of the social exchange without sanctions, leaving the door open to feeding it back into the system when needed. In the end, the customer might get out more than was invested, keep the opportunity to fulfil social obligations deemed imperative and, above all, gain the reputation of being a responsible social person. This is what makes funeral insurances so attractive. To put it in the words of one of the insurance brokers I spoke to: "You really should get a funeral insurance. Then you are on the safe side of life, even if it is your death."

References

BARTELHEIM, M. et al. (2015) 'ResourceCultures' – A Concept for Investigating the Use of Resources in Different Societies. In DANIELISOVÀ, A. and FERNÁNDEZ-GÖTZ, M. (eds.), *Persistent Economic Ways of Living. Production, Distribution, and Consumption in Late Prehistory and Early History*, pp. 39-49. Budapest: Archaeolingua.

CARSTEN, J. (1989) Cooking Money: Gender and the Symbolic Transformation of Means of Exchange in a Malay Fishing Community. In PARRY, J. and BLOCH, M. (eds.), *Money and the Morality of Exchange*, pp. 117-141. Cambridge: Cambridge University Press.

CENTRAL BUREAU OF STATISTICS/NATIONAL PLANNING COMMISSION (2008) A Review of Poverty and Inequality in Namibia. Windhoek: National Planning Commission. [Online] Available from: http://www.nsa.org.na/files/downloads/8d9_A%20review%20of%20Poverty%20and%20Inequality%20in%20Namibia%202008.pdf [Accessed: 3rd January 2015].

GREINER, C. (2011) Migration, Translocal Networks and Socio-Economic Stratification in Namibia. *Africa* 81(4): 606-627.

GUDEMAN, S. (2012) Community and Economy: Economy`s Base. In CARRIER, J.A. (ed.), *A Handbook of Economic Anthropology*, pp. 94-108. 1st ed. Cheltenham etc.: Elgar.

HAHN, H.-P. (2005) *Materielle Kultur. Eine Einführung*. Berlin: Dietrich Reimer Verlag.

ITÉANU, A. (2010) Rituel et organisation sociale. In ITÉANU, A. (ed.), *La cohérence des sociétés. Mélanges en hommage à Daniel de Coppet*. Paris: Edition de la Fonfation de la Maison des Sciences de l'Homme.

KLOCKE-DAFFA, S. (2001) „Wenn du hast, mußt du geben". Soziale Sicherheit im Ritus und im Alltag bei den Khoekhoen von Berseba/Namibia. Münster: LIT.

KLOCKE-DAFFA, S. (2008) The Modernity of Traditionalists. Culture Change, Identity and the Impact of the State among the Namibian Khoekhoen. In BIESELE, M. and LIMPRICHT, C. (eds.), *Heritage and Cultures in Modern Namibia: In-Depth Views of the Country*, pp. 174-182. Windhoek: Hess Publishers.

KLOCKE-DAFFA, S. (2009) Namibia: Sicherheit in unsicheren Zeiten? Ein soziales System auf dem Prüfstand. In GIELER, W. and BELLERS, J. (eds.), *Fremdes Verstehen. Entwicklungspolitische und ethnologische Beiträge*, pp. 11-40. Baden-Baden: Nomos.

KLOCKE-DAFFA, S. (2011) Is BIG big enough? Basic Income Grant in Namibia. An Anthropological Inquiry. [Online] Available from: https://publikationen.uni-tuebingen.de/xmlui/bitstream/handle/10900/46965/pdf/Rport_2012_Is_BIG_big_enough_end.pdf?sequence=1&isAllowed=y.

KLOCKE-DAFFA, S. (2015) Hauskinder – Großmachkinder – Logierkinder: Die Aufnahme von Kindern als Ressource im Kontext sozialer Sicherung. In FINK, M. and GRONEMEYER, R. (eds.), *Afrikanische Kindheiten. Soziale Elternschaft und Waisenhilfe in der Subsahara*, pp. 59-80. Bielefeld: Transcript.

KPMG (2014) Insurance in Africa. Sector Report. [Online] http://www.kpmg.com/Africa/en/IssuesAndInsights/ArticlesPublications/Documents/Insurance%20in%20Africa.pdf. [Accessed: 23rd March 2015].

MAUSS, M. (1954) The Gift. The Form and Reason for Exchange in Archaic Societies. London: Cohen.

METROPOLITAN (2015) MetLife. [Online] http://global.metlife.com/. [Accessed: 23rd March 2015].

NAMIBIA STATISTICS AGENCY (2012) Namibia 2011 Population and Housing Census. Main Report. [Online] http://www.nsa.org.na/files/downloads/Namibia%202011%20Population%20and%20Housing%20Census%20Main%20Report.pdf. [Accessed: 24rd March 2015].

OLD MUTUAL (2015) Old Mutual International. [Online] Available from: http://www.oldmutualinternational.com/en/europe/consumer [Accessed: 23rd March 2015].

PARRY, J. and BLOCH, M. (eds.) (1989) *Money and the Morality of Exchange*. Cambridge: Cambridge University Press.

UNITED NATIONS (2013) United Nations Development Program: Namibia. HDI values and rank changes in the 2013 Human Development Report. [Online] Available from: http://www.na.undp.org/content/namibia/en/home/countryinfo/ and http://hdr.undp.org/sites/default/files/Country-Profiles/NAM.pdf.

PAULI, J. (2013) Sharing Made Us Sisters. Sisterhood, Migration, and Household Dynamics in Mexico and Namibia. In ALBER, E. (ed.), *The Anthropology of Sibling Relations. Shared Parentage, Experience, and Exchange*, pp. 29-50. New York: Palgrave Macmillan.

SANLAM (2015) Sanlam Global. [Online] http://www.sanlam.com/. [Accessed: 23rd March 2015].

STRATHERN, M. (1997) Partners and Consumers. In SCHRIFT, A. D. (ed.), *The Logic of the Gift. Towards an Ethic of Generosity*, pp. 292-311. London and New York: Routledge.

UNAIDS (2013). Namibia. HIV and AIDS estimates. [Online] http://www.unaids.org/en/regionscountries/countries/namibia/. [Accessed: 23rd March 2015].

U.S. SOCIAL SECURITY ADMINISTRATION – OFFICE OF RETIREMENT AND DISABILITY POLICY (2013) Social Social Security Programs Throughout the World: Africa: [Online] http://ssa.gov/policy/docs/progdesc/ssptw/2012-2013/africa/namibia.html. [Accessed: 23rd March 2015].

UNITED NATIONS (2013) Human Development Report. Namibia. [Online] http://hdr.undp.org/sites/default/files/Country-Profiles/NAM.pdf. [Accessed: 23rd March 2015].

VAN LANGENDONCK, J. (2007) The Meaning of the Right to Social Security. In VAN LANGENDONCK, J. (ed.), *The Right to Social Security*, pp. 3-12. Antwerpen/Oxford: Intersentia.

WALLACE, M. and KINAHAN, J. (2011) *A History of Namibia. From the Beginning to 1990*. London: Hurst.

What Do Ceremonies Do? Ceremonial Exchanges as Operational Processes Elaborating Relations[1] (Kanaky, New Caledonia)

Denis Monnerie

Jos Platenkamp once wrote that "filming rituals provides the other society with a radically new means to surbordinate *us*, who are *their* 'other', to their value system" (2006: 97). He himself and the ethnocinematographer Dirk Nijland were the filmmakers, their subject was a Tobelo marriage ceremony in 1982 (Eastern Indonesia). The context of Jos Platenkamp's observation was, six years later, the screening of the completed film to the Tobelo who – at first had been – "deeply concerned that the marriage ritual be registered on film in a 'complete' form. Only in this way might its efficacy in the multiplication of people and rice be anticipated" (*ibid*: 97). The point I want to make here has to do with this crucial importance of the complete, step by step, performance of ceremonies to achieve their aims.

The emphasis on the importance of the complete performance of ceremonies, I have also found from 1992 onwards in Hoot ma Whaap, a region covering the far north of Grande Terre, in Kanaky New Caledonia, (Melanesia, Oceania). In anthropological parlance, these are labelled *ceremonies of exchange.* Diseases and deaths can be attributed to incompletely or badly performed sequences of ceremonies, expressing the importance of the correct unfolding of their operational process. This chapter shows how to adjust this locally perceived importance of a correct performance of all the sequences of the ceremonies with the classic epistemological emphasis on exchange, by looking at these ideas and acts in a different way. My analysis focuses on northern Hoot ma Whaap. However, the ethnographic data I am going to discuss point to more general theoretical issues concerned with exchanges (cf. Monnerie 2014b) and the ceremonial elaboration of relations and I will suggest that some of my proposals may apply to other ceremonies of Oceania or the world in general.

Exchanges and Beyond

The idea of the crucial importance of exchanges for the societies of Melanesia has become consensual (cf. Thomas 2006 [1995]), leading from Richard

1 Thanks are due to Sophie Chave-Dartoen for her comments on a previous version of this paper and to Chloe Le Mouël for digitalising my diagrams.

Thurnwald's and Bronislaw Malinowski's pioneer analyses of their ethnographies to Daniel de Coppet's (1968) and Roy Wagner's (1974) radical appreciations of the place of ceremonial exchanges in Melanesian societies in the late 1960s and early 1970s. That Polynesia should be integrated into this perspective deeming exchanges central to the understanding of Oceanian (among others) societies is evident from Marcel Mauss' early emphasis on Maori 'gift ' relations in his key text of exchange theory, the *Essai sur le don* (1980 [1923-4]).

In this chapter, I would like to make critical and complementary proposals, which although inspired by these debates, will expand on them, both in my use of concepts to deal with the ethnography and in my anthropological perspective. I will concentrate on the description and analysis of procedures which anthropology usually calls ceremonial exchanges. The term 'exchange', however, can be misleading. Having become more and more imbued with implicit meanings, it has to be carefully defined – as debates about reciprocity have shown.[2] Exchange is also imprecise, which is why some time ago, when describing Kanak ceremonial procedures, I deemed it more relevant to use 'transfer' and 'circulation', i.e. concepts that have frequently been used in anthropology, but seldomly brought together. The term 'transfer' has the descriptive and theoretical advantage of only expressing 'reciprocity' when explicitly specified, as for instance in a transfer followed by a quantitatively and qualitatively equal counter transfer. 'Circulation' on the other hand has often been favoured by ERASME[3] researchers – under the influence of John-Erik Elmberg (1968) – thus underscoring the flow of prestations beyond sequences of exchanges, however never giving up the simultaneous use of the term 'exchange'. Along with different ways of sharing, the concepts of 'transfer' and 'circulation' are relevant tools for the description and analysis of ethnographic data, circulation describing the wider processual movements created by linked successive sequences of transfers. Arnold Van Gennep's *Les rites de passage* (1909) offered the first theory of ritual to highlight the processual dimensions of rituals, linking it with the transformation of social relations.

In northern Hoot ma Whaap where I have worked since 1992, I have had the opportunity to observe and participate in numerous ceremonies of various scope organised and performed by the autochtonous Kanak people of this French territory. At the core of all these ceremonies are transfers and circulations of goods, foods, speeches and people. In Kanak and Caledonian French

2 Barraud et al. (1984) and Weiner (1992), a.o. clearly emphasise and argue the importance of this point.

3 ERASME was an EHESS Paris based CNRS research team inspired by Daniel de Coppet's and Louis Dumont's work with which the CASA research team in Leiden collaborated. Jos Platenkamp belonged to both teams.

these ceremonies are called 'coutumes', a polysemic word extending to everything deemed to be specifically kanak. However, in local Kanak languages they have specific names: for example *thiam* are the prestigious ceremonies of regional relations, *thîîlin uvi*[4] is the annual ceremony for the first yams, *nyang maac* are funerals, and *nyang maôlep* is the name given to designate other life cycle ceremonies.

Hundreds of Kanak men, women and youths may be involved in large ceremonies – some of them working for their preparation and performances more or less permanently for days or weeks. In the case of life cycle ceremonies, other current activities or salaried occupations are reduced, suspended – sometimes even prohibited in the context of funerals. I wish to contribute to the understanding of these collective social activities with the apparently simple question: What do these ceremonies do that is so important that they take priority over many other tasks, actively involving so many people – with such intensity – for long periods of time, particularly during the central phases which I call the focal ceremony?

In order to do so I have to do away with several dichotomies grounded in Western ideological traditions, which tend to hinder or warp the progress of anthropological work, especially in Oceania and probably elsewhere. They strongly oppose spirit and matter, the verbal and the non-verbal, 'symbols' and actions, and relations and substances (Monnerie f.c.). Also, I do not make a distinction between ceremonies and ritual, and will not deal with many theoretical debates on ritual[5] for the obvious reasons of space and also because my perspective aims at a level which underlies most theories of ritual.

The point I want to emphasise throughout this paper is that ceremonies elaborate socio-cosmic relations through precisely defined operational processes. Social relations include not only relations between living humans, but also relations with the world at large, including non-humans: places, ancestors, plants, animals, etc. This world, conceptualised specifically by the societies and cultures of Hoot ma Whaap, I define as a cosmos. Hence my use of the term socio-cosmic relations, or in short: relations. These are elaborated with the active involvement of different configurations of the social, especially kin groups called *yameevu* (and *phwâmeevu*[6]), localised configurations including hamlets called houses making up local societies called great

4 The term *thiam* (and several other shared concepts of regional relations) is used throughout the Hoot ma Whaap region even though ten different Austronesian languages are spoken. But *thîîlin uvi*, *nyang maôlep*, and the other local words and concepts quoted in this text properly belong to *nyêlâyu*, the language of Arama, the local society on which my fieldwork focuses.

5 For more ethnographic details and discussions of kanak ceremonies in relation to ritual theories see Monnerie 2002, 2003a & b, 2005, 2010a, 2014a, f.c..

6 *Phwâmeevu* are groupings of *yameevu*. Belonging to the latter involves personal names, transmitted through a complex mix in which patrilineal relations tend to prevail but where adoptions are frequent (Monnerie 2012b).

houses[7], and especially the pervasive flexible ceremonial dyads I shall dwell upon shortly. By elaboration I mean the fabrication, maintenance, renewal, transformation, expression, performance, display, ordering, acknowledgment, and authentification of relations. The operational processes put forward here consist in highly valued and carefully organised and cognised non-verbal and verbal acts. These aim at the elaboration of socio-cosmic relations through precisely ordered series of named transfers of goods, foods, speeches, and people, constituting more global dynamics of circulation.

I derive the concept of operational process from André Leroi-Gourhan's (1965) anthropological concept of *chaîne opératoire* which should not be restricted – as it often has – to the understanding of technology in its material dimensions.[8] I extend this direction of research to the processual elaboration of relations. A similar inspiration is manifest in de Coppet's (1981) use of 'transformation chains' to describe key dynamics of 'Are'are rituals (Solomon Islands). Ludovic Coupaye's groundbreaking work on Abelam ceremonial yams (Papua New-Guinea) has demonstrated how "relationships between people and things appeared as created through processes of making and ways of doing" (2009: 451; see also 2013). Leroi-Gourhan defined three levels of *chaînes opératoires.*[9] At the fist level there are body attitudes, alimentation, and sexual distinction, the second level involves mechanical *chaînes opératoires*, and the third level implies the confrontation of these former two with the symbols of language. Leroi-Gourhan insists on the fact that these three levels are arbitrary "as are all cuts within continuous processes" (1965: 27). Kanak ceremonies as highly sophisticated operating processes have their own distinctions, some of which will be dealt with here. In ceremonies, they interweave them, achieving densities of interactions and expressions resonating with much of Kanak experience (Monnerie 2014a).

Transfers and Dyads as the Building Blocks of Kanak Ceremonial Operating Processes

The simplest form of Kanak ceremonies is the formal meeting of two men, marking the arrival of one of them in a place and group which he does not belong to and his reception by the other one, a dignitary who belongs locally. It thus is a process elaborating a relation with an *other*. The ceremony unfolds as follows. The two men face each other. In a first sequence of transfer, the

7 In Arama hamlets are called houses (*mwâ*), the 'great house' (*mweemwâ*) is the local concept for the local society. Throughout the Kanak world there are concepts which refer to local societies (or hamlets, or social groups) as 'houses' (or compounds thereof, such as 'great house'). They are designated by words which are often close to the nyêlâyu *mwâ*. This can be materialised by the construction of the characteristic Kanak round houses which 'figures' these social configurations (Monnerie 2010b).

8 Thus I do not use the term 'operational' in Roy Rappaport's sense (1968: 237).

9 These have often been overlooked in the French tradition of *technologie culturelle*.

arriving person shows a composite prestation: money, often wrapped in a cloth, held in one hand, and food, displayed on the ground – ideally containing at least one yam. This double display – held in the hand and grounded – is accompanied by a formal speech expressing the context and relations of the encounter. The prestation is then handed over to the host who solemnly takes it in his hands after having respectfully listened to the arriving man's speech. The recipient, in turn, is holding a similar prestation, and gives a speech relating to the context and relations of the meeting, often linking the relation with past and future ones. He then transfers his prestation to the arriving man and invites him to share food. The same process applies when two groups or parties of persons are involved, each represented by one man.

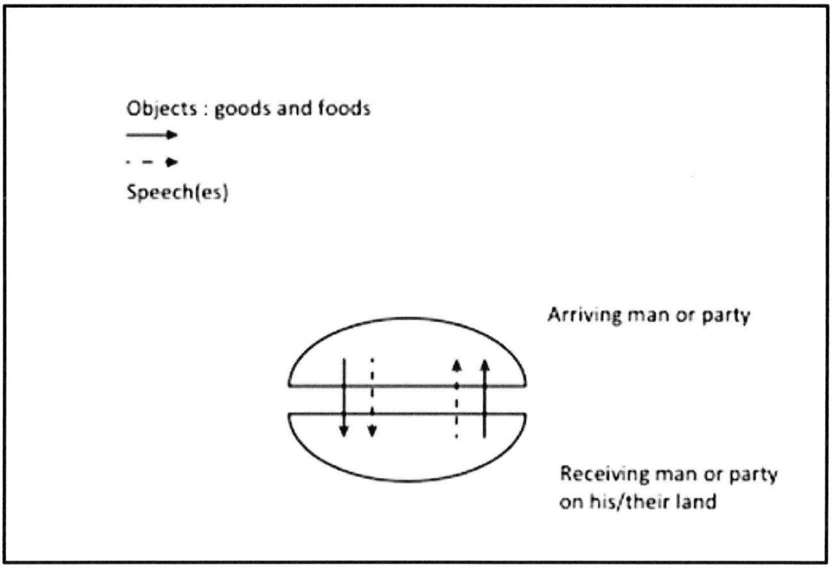

Fig. 1: The building block of Kanak ceremonial operating processes

Through this ceremony, the arriving man will have a great deal of autonomy in the place where he has been received and with its inhabitants. A party may also be received in a similar fashion. Thus such ceremonies simultaneously elaborate relations between people, groups, and with the land. These relations are composites made up of various intertwined strands (Stasch 2009), constituted by valorised tangible acts, objects, foods and speeches, thus conveying verbal and non-verbal meanings. These verbal and non-verbal relational transfers and counter-transfers within dyads are the dynamic building blocks of Kanak ceremonies and socio-cosmic relations. Dyads are ambivalent in that they are at the same time social configurations of separation and contact

(Stasch 2009). Here, note the co-occurence of (i) the dyadic form – two ceremonial sides facing each other, the receiving one locally anchored, and (ii) the dynamic acts of relational transfers combining verbal and non-verbal components. These dyads are made up of people who are others – strangers to each other – but only relatively so as they share this specific idiom of the dyad and relational transfers as the locus of the elaboration of relations.

These dynamic building blocks of the social have characteristic flexibility and I now want to outline how this minimal form is developed and expanded into (sometimes hugely) complex large ceremonies.

Kanak Ceremonies: Some Principles

For an important ceremony, dozens of sequences of transfers and counter transfers based on principles similar to the one just described are performed during several ceremonial stages taking place in various locations and times, all leading to the focal ceremony.

Fig. 2: A (medium to small) preparatory ceremony. The arriving side is laying down some of its food prestations (before transferring them) for a marriage ceremony which will be held in Gomen on September 24
(© Monnerie, Arama September 11, 2010)

Whereas in each stage the transfers and counter-transfers tend to be qualitatively similar, the counter-transfers are quantitatively limited. Such double transfers elaborate the dyadic relation in both ways and at the same time the transfers contribute to the next ceremonial stage. Stage after stage, the quantities of goods, foods, people, and the length of the speeches increase each ceremonial stage contributing towards the main ceremonial circulation. Thus series of dyadic transfers build up ever larger contributions leading to the main – and largest – ceremonial dyad, that of the focal ceremony. Each step of the operational process has a name which is a compound of *ba-*, which means "in order to do" thus verbalising its operating aim.

In regional *thiam,* the focal ceremony is repeated with each arriving delegation of local societies received by that of the inviting society. In the Arama yam ceremony, delegations of each of the hamlets making up the local society are successively received by that of the focal hamlet. In life cycle ceremonies, the delegation of the 'maternal side' (*ajalu*) of the focal person is received by that of the 'paternal side' (*yameevu*). In all cases, the receiving side of the ceremonial dyad is and acts on its land. Both sides of ceremonial dyads are made up of delegations constituted around core relational components. For the regional and local ceremonies, they include the dignitaries of the great houses or hamlet/houses. For life cycle ceremonies, they include (i) on the maternal side, the close kin and *yameevu* of the focal person's mother – or of the bride for marriages; (ii) on the paternal side, the *yameevu* of the focal person – or of the bridegroom for marriages. In ceremonies people participate around these core persons, groups and relations. They do so according to highly complex principles involving *yameevu* relations, maternal ones, residence relations, and individual preferences. Thus they may assemble along interpersonal relations, expressing relations of residential proximity and/or of friendship: one transfers prestations to neighbours or friends belonging to a local *yameevu* linked to the ceremony's core relations. Finally, it is important to underline both the flexibility of the recruitments of all sides of all the ceremonial dyads and the fact that they change from ceremony to ceremony.

An important point distinguishing life cycle ceremonies is that the main circulation of foods and goods is oriented in a one-to-three proportion in favour of the maternal side, called *ajalu*: "those who receive [more]". This has been prepared throughout the ceremonial stages and is counted, apportioned, performed, and displayed in the focal ceremony. This orientation of the ceremonial circulation is linked to life giving attributes of social entities related to the mother of the focal person – her and her mother's *yameevu*, etc. – and, in weddings, the bride's. Two points are worth mentioning here: First, the repetition of ceremonial processes throughout the 'life cycle' expresses that marriages are followed by the birth of children who become the focal persons of future life cycle ceremonies. The links of past, present and future relations are one of the main themes of ceremonial speeches, a verbal emphasis coordinated with this aspect of the operating process. Second, the ceremonial operational process opens several quantitative possibilities for the main circulation: the life cycle 'one-to-three' orientation as well as the (roughly) 'one-to-one' configuration common in other ceremonies. This is important for anthropology – and economics – in that it shows that quantitative aspects of 'exchanges' are actually subordinated to the operational processes underlying them – and their purpose: elaborating relations.

One crucial aspect is underscored in myth (*vajama*): goods and foods must be "held with the hands" (*cuva*) and shown for everyone to "see" (*kiia*), acts and perceptions which are strongly valued aspect of the operating process, expressing the truth (*paer*) of relations (Monnerie 2012a; fc.). During the ceremonial stages, the goods and foods are thus carefully handled, counted, unfolded, exhibited, and transferred. They are then apportioned, rearranged into new prestations and so on. Always carefully synchronised with the ever increasing transfers and counter-transfers, the speeches themselves expand, become more sophisticated. The whole process leading to the focal ceremony is thus a repetitive and progressive one of accretion multiplying, expanding, and complexifying the dyadic relational building block described above. In it, one contributing group after the other gather prestations and transfer them towards the core component of their ceremonial side in ever expanding ceremonial stages performed in different places. This entails transferring more and more goods, foods and people, making longer and more beautiful speeches, preparing larger meals and ceremonies, and living the process more intensely – thus elaborating more and more complex and enduring relations through the many verbal and non-verbal relational strands circulated in the process. These time and energy consuming collective activities take place on both sides (fig. 2, see top and bottom) and contribute to the main ceremonial circulation as the focal ceremony approaches.

When both sides are ready to perform the focal ceremony, this culminating stage is held soon after, lasting one or two days. The receiving side, on its land, has prepared the focal ceremony with the help of many men and women participating in the work: erecting buildings, preparing the food, organising the meals, etc. For funerals, marriages and *thiam*, dozens, sometimes hundreds, of people gather into two sides facing each other. On each side, men who have a special status and/or ceremonial competence organise, with the cooperation of many others, the transfers of the considerable amounts of goods and foods previously gathered by their ceremonial side.

During the focal ceremony, the main circulation crosses from one side of the focal dyad to the other. The main relations are elaborated: mostly transformed or reaffirmed. Again, just as was the case during all the stages of the whole preparatory process, but with even more flourish, during the focal ceremony, the circulating objects and foods will be handled, one by one, displayed, and piled in several sequences of transfers and counter-transfers synchronised with long, often magnificent, poetic speeches made by the best orators available. All participants respectfully observe and listen. Thus the focal ceremony is at once a spectacular performance and a crucial moment of the operating process, condensing the numerous acts, objects, foods, speeches, places and times of preparatory ceremonial stages, as well as all the composite relations. This is the great locus of the elaboration of the main relations of the ceremony.

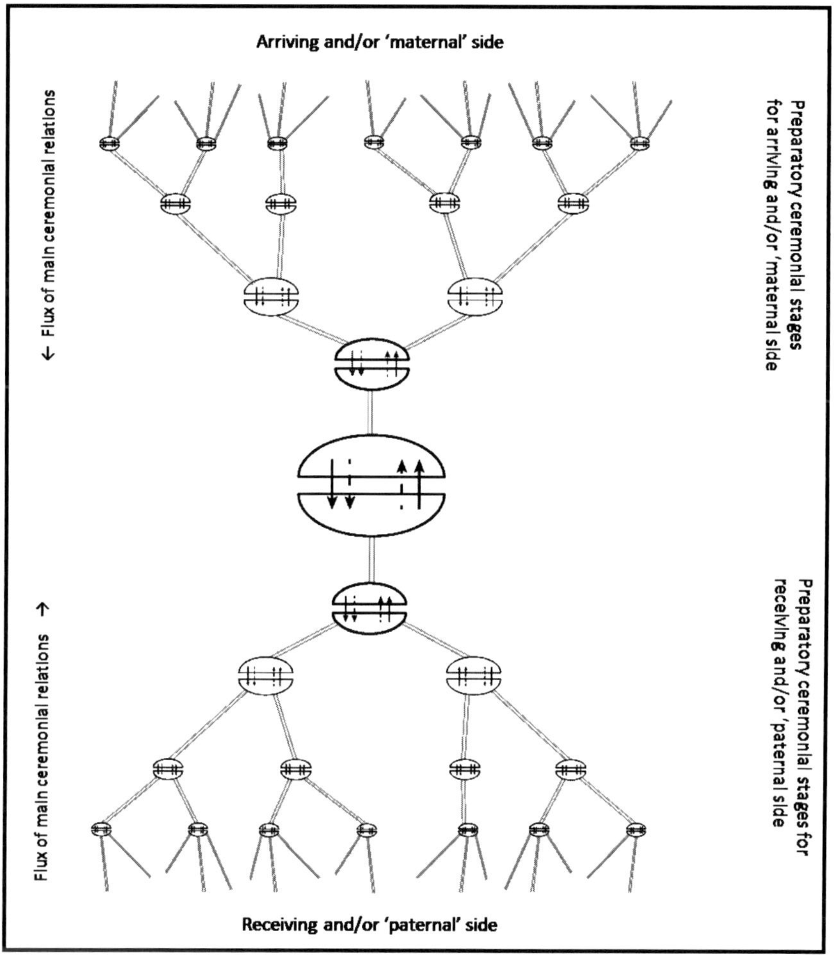

Fig. 3 Some basic principles of transfers and circulations leading towards a large focal ceremony

Not all the people who have taken part in the previous ceremonial stages leading to the focal ceremony come for that grand event. But their representatives have brought all the objects, meanings, and relations they have contributed to the focal ceremony where speeches express the gist of all these combined relations. Thus, every participant in the whole process is represented – made present – through their ceremonial acts, objects and foods, and the relations and meanings expressed in them and in speeches.

Focal ceremonies are scenes where everyone participates to varying degrees in different roles: children, adults, men, women, ancestors and other elements of the cosmos, including places, vegetable and animal foods, building up an especially intensely shared and meaningful performance and gustative experience with a systematic condensed association of intellectual and pluriperceptual dimensions. This plays a great part in the authentification, transmission, and rememberance of the ceremony, especially of the relations it has recalled and those it has elaborated (Monnerie 2014a). After participating, some will talk about strong feelings of well-being. Conversely, ancestral entities are said to deliver sanctions when ceremonies are not performed in the prescribed way: deaths, physical and mental illnesses, getting lost at sea or in a forest, etc..

Fig. 4: One sequence of the ceremonial transfers of the focal ceremony for a marriage. The *yameevu* (husband's) side having just completed the laying out of its prestations, the ceremonial speech of its orator begins.
(© Monnerie, Tiari July 19, 2012).

When the goods and foods transferred on both sides have been gathered and counted, all these strands of the main relations are divided, the goods and remaining food then being apportioned again to be displayed in a line of bundles laid out in a precise order, each destined for a named contributing ceremonial group.

They are distributed with short speeches. Each bundle is now taken by the assigned group, without counter-transfers. The general circulation having crossed from one side of the focal dyad to the other, from the point of view of the participants on each side, it is now reversed. They all will receive return prestations, redistributed along both sides in the days that follow. This reverses the long process which has led to the focal ceremony. Dignitaries of the ceremonial groups will again divide and apportion the resulting composite prestations. They will redistribute these prestations to all the groups and persons who have contributed to the whole process. Thus these return transfers will be received by all participants on each side of the ceremonial dyad. Now, ideally, this should be done quickly, within a few days if possible. The recently ceremonially elaborated relations thus irrigate all former relations expressed by all the contributors.

This sketch of the operational process of Kanak ceremonies in the north of Hoot ma Whaap is an anthropological model developed for analytical and comparative purposes. It has relied on the systemic stability, coherence, family resemblance and flexibility of ceremonies in adaptating to many sorts of social circumstances and relations.[10]

What Do Ceremonies Do to Relations?

Up to the focal ceremony, each stage elaborates 'thicker' and more complex composite relations contributing to the main ceremonial relations which are thus maintained or transformed. Yam ceremonies usually maintain local relations, reaffirming the order of precedence of the hamlets/houses, but may authentify decisive local changes in social configurations (for another region see Pillon 2002). *Thiam*, while maintaining regional relations renew some of their strands, for instance when new dignitaries replace former ones in their roles and the ceremonial pathways (*daan*) between them have to be reaffirmed. Life cycle ceremonies' main relations – for marriages, deaths, adoptions, etc. – are elaborated during the focal ceremony around the focal person(s). First of all they reconfigure the relations between paternal *yameevu* and maternal kin. They can also give relations with neighbours and friends more visibility, strength, and authenticity.

One of the achievements of all this ceremonial process is the creation of robust composite relations authentified through their intensified, condensed, and explicit elaboration. Within the time span of the preparation of the focal ceremony, its performance and redistributive stages, it strongly imparts actualised relations on all the actors/participants. It does so in specific places of the land tightly linked to the existence of such basic social groups as ceremonial dyads, *yameevu, phwâmeevu,* hamlets/houses, or great houses. This elaborates relations 'concretely' as well as 'abstractly', non-verbally and verbally, through objects, and through intense embodied and intellectual experiences,

10 For the highly complex practices that characterise actual Kanak ceremonies see Bretteville (2002) for funerals, for *thiam* see Monnerie (2005).

the latter related to the oratory poetic *tours de force* of the best orators. It produces communication about them including their meanings, transmission, and remembrance. During these ceremonies, crucial Kanak relations are elaborated which will then prevail in current, everyday life until new ceremonies elaborate new relations.

A New Perspective on Kanak Ceremonial Exchanges

Classically, in Oceania, such ceremonies have been interpreted in terms of exchanges and/or in terms of communication. My argument is that what lies beneath exchanges or communication are culturally/socially specific operational processes elaborating meaningful relations.

The exchange interpretation falls short of accounting for what Kanak ceremonies actually do. Consider some of their crucial aspects in relation to debates about reciprocity: up to the focal ceremony and its redistributive sequence, transfers mostly imply counter-transfers in which qualitative equivalence tends to predominate. Quantitatively, ceremonies can circulate goods and foods either in a one-to-one or in a one-to-three proportion. The latter quantitative difference is underscored by the explicitly valued qualitative life giving character of the maternal side, which receives three times more than what it has brought. Some of the difficulties we have in accounting for such practices – roughly reciprocal qualitatively in both cases, however not quantitatively in the second – can be reduced, if we consider these transfers, counter-transfers, apportioning, and their combined sequences as creating relational strands and circulations within operational processes. Here qualitative reciprocity being a localised artefact of this shared operating idiom. The omnipresent, flexible dyads are poles of transfers which constitute intricate dynamics circulating relational strands in composite acts, intertwining subjects and objects in the aim of elaborating relations.

As for the communication angle, it combines meanings with embodiment (Lakoff 1987). Ceremonies associate non-verbal procedures and verbal language, they are food for thought as well as for satiation and all the senses. They share a common idiom with a shared 'vocabulary': the various and numerous actors, goods, foods, gestures, words and their meanings. This is organised by a common 'syntax' and 'rhetoric': displaying acts of handling, apportioning, and transferring within the canonical dyadic spacetime; developing the build-up and redistribution of the circulation of prestations throughout the ceremonial process; expressing relations and their truth in this process. From this angle ceremonies are, in their vocabulary, syntax and rhetoric, strongly iconic. They also make use of several other social modes of meanings, i.e. symbols, indexes, and 'qualisigns' (Munn 1986), figures (Monnerie 2010b), and through objects akin to 'perissologic resonators' (Lemonnier 2012). Kanak ceremonial speech acts which are great poetical and rhetorical moments, parallel and reinforce what ceremonies 'concretely' aim at, for instance in their use of dyads in rhythm, meaning, and rhetoric (Monnerie 2005: 218-260) which parallels the largely dyadic idiom of ceremonial processes and goods; or in the *thiam* formula with the sequential order of utterance of names paralleling the ceremonial order of precedence of the great house (Monnerie 2003a; 2005: 72-80). Hence much of the communication aspects of ceremonies are outcomes of their combined verbal and non-verbal operational processes, which do more than communicate about relations, they elaborate them while expressing their meanings.

Conclusion

What does the operational process perspective bring that is anthropologically useful? It gives a wider, more relevant view of what Kanak ceremonies do. It shows how the transfers, apportioning and circulations, with their quantitative and qualitative dimensions, the multiplicity of meaningful ceremonial acts, goods, foods, and speeches forcefully elaborate – and express – relations for all to understand and sense. These are concrete/abstract meaningful composites that can be combined and/or divided over and over again. They transform and expand the dimension of otherness inherent in their basic dynamic dyadic form to express meaningful dimensions of affinity and diversity and complementarity about the socio-cosmic world. The larger the process, the more strands, the more stable and wide ranging the relations.

By now it should be clear why I consider the 'exchange' and 'communication' paradigms – for all their past heuristic value – have become too narrow to understand Kanak ceremonies and relations and why I advocate using more relevant concepts such as transfers, circulation, and forms of sharing. This enables both to specify what exchanges really are and take a wider perspective of their role in ceremonies, that of sequences of relational, meaningful operational processes. This fits with local interpretations underscoring tangible meaningful objects. The acts which display and handle them for everyone to see are elements of the operational process which express the truth of relations. The latter being composites which are always simultaneously given verbal meanings in speeches.

Somewhat related ideas about the specificity of rituals or ceremonies have been developed by Roy Rappaport and Joel Robbins. However my emphasis is not on the fact that speech can be false (Rappaport 1999), nor on the perspective of language or even ceremonial ideology (Robbins 2001), nor even on Victor Turner's precious concept of multivocality. The antidote against falsity of speech, the existence of a ceremonial ideology or the multivocal dimension of ceremonial expressions, and other aspects of ceremonies – and ritual –, I argue, build-up on more basic dynamics and features pertaining to ceremonial operational processes. Underlying all these and many other important points is the fact that in northern Hoot ma Whaap at least, ceremonies elaborate socio-cosmic relations through finely defined and organised operational processes, non-verbal and verbal. That this is also the case for other Oceanian societies where ceremonial exchanges are a recognised (or not) 'structuring' feature will be the subject of further research.

From this perspective, the kind of ceremonies – or rituals – which elaborate socio-cosmic relations do not appear *primarily* involved in the sacred (Durkheim 1912) or the religious (Rappaport 1999), but the latter are, I suggest, developments of socio-cosmic relations elaborated by them. Beyond Oceania, this hypothesis puts such ceremonies and rituals at the forefront of the social/cultural life and implies that their culturally/socially specific operational processes elaborating relations have a central importance which should be studied comparatively.

This brings us back to the Tobelo's emphasis on registering ritual in a 'complete' form in order to ensure its 'efficacy'. This emphasis derives, I suggest, firstly from the processual dimension of rituals and ceremonies, and secondly, from its operational potency. Thus Jos Platenkamp's argument that anthropology participates in radically new ways of surbordinating our ideology to other human value systems rightly suggests new approaches and concepts to deal with ritual, ceremonies, and the social at large, including our own society.

References

BARRAUD, C. et al. (1984) Des relations et des morts. Quatre sociétés vues sous l'angle des échanges. In GALEY, J.C. (ed.), *Différences, valeurs, hiérarchies*, pp. 421-520. Paris: Editions de l'Ecole des Hautes Etudes en Sciences Sociales.

BRETTEVILLE, D. (2002) *L'os et le souffle: le système social et cosmique d'une société Kanak de Nouvelle-Calédonie: Les Paimboas* [unpublished PhD thesis]. Paris: Ecole des Hautes Etudes en Sciences Sociales Paris.

COUPAYE, L. (2009) Ways of Enchanting. Chaînes Opératoires and Yam Cultivation in Nyamikum Village, Maprik, Papua New Guinea. *Journal of Material Culture* 14(4): 433-458.

COUPAYE, L. (2013) Growing Artefacts, Displaying Relationships. *Yams, Art and Technology amongst the Abelam of Papua New Guinea.* New York, Oxford: Berghahn Books.

COPPET, D. de (1968) Pour une étude des échanges Cérémoniels en Mélanésie. *L'Homme* 8(4): 45-57.

COPPET, D. de (1981) The Life-Giving Death. In Humphreys, S. and King, H. (eds.), *Mortality and Immortality*, pp. 175-204. London : Academic Press.

DURKHEIM, E. (1912 [1960]) *Les Formes élémentaires de la vie religieuse. Le système totémique en Australie*. Paris: Presses Universitaires de France.

ELMBERG, J.E. (1968) *Balance and Circulation.* Stockholm: The Ethnographical Museum.

GENNEP, A. van (1909 [1981]) *Les Rites de Passage*. Paris: Picard.

LAKOFF, G. (1987) *Women, Fire, and Dangerous Things. What Categories Reveal about the Mind.* Chicago: The University of Chicago Press.

LEMONNIER, P. (2012) *Mundane Objects. Materiality and Non-Verbal Communication.* Walnut Creek, CAL: Left Coast Press.

LEROI-GOURHAN, A. (1965) *Le geste et la parole. Vol. II La mémoire et les rythmes.* Paris: Albin Michel [English translation (1993) *Gesture and Speech.* Cambridge, Massachussets: MIT Press].

MAUSS, M. (1980 [1923-4]) Essai sur le don. In *Sociologie et Anthropologie*, pp.142-279. Paris: Presses Universitaires de France.

MONNERIE, D. (2002) Nouvelle-Calédonie. La terre et les hommes, la culture et la politique: Consensus et Confrontations. *Ethnologie Française* 32(4): 613-627.

MONNERIE, D. (2003a) Social Organisation and the Naming System. A New Caledonian Case: Arama. *Zeitschrift für Ethnologie* 128: 249-268.

MONNERIE, D. (2003b) Résistance au colonialisme, culture, coutume et politique (Arama et région Hoot ma Whaap). Pratiques et représentations historiques et contemporaines. *Journal de la Société des Océanistes* 117: 213-231.

MONNERIE, D. (2005) *La parole de notre Maison. Discours et cérémonies kanak aujourd'hui.* Paris: CNRS Éditions et Éditions de la Maison des Sciences de l'Homme.

MONNERIE, D. (2010a) Following the Pathways. Contemporary Ceremonies, Representations of the Past and Catholicism in Northern New Caledonia. In DOUAIRE-MARSAUDON, F. and WEICHART, G. (eds.), *Religious Dynamics in the Pacific,* pp. 85-11. Marseille: Edition du CREDO.

MONNERIE, D. (2010b) Symboles et figures, deux modes sociaux de signification. L'exemple de la Grande Maison d'Arama (Nouvelle-Calédonie). *Journal de la Société des Océanistes* 130-1: 147-163.

MONNERIE, D. (2012a) Objets cérémoniels, chaînes opératoires et élaboration des relations sociales (Arama et Hoot ma Whaap, Kanaky Nouvelle-Calédonie, Mélanésie) *Techniques et Cultures* 58: 122-141.

MONNERIE, D. (2012b) Significations, valorisations et performativité de la nomination dans l'organisation sociale à Arama (Nouvelle-Calédonie). In CHAVE-DARTOEN, S., LEGUY. C and MONNERIE, D. (eds.), *Système, résilience et créativité nomination et organisation sociale,* pp. 29-165. Paris: Armand Colin.

MONNERIE, D. (2014a) Expériences, performances et compétences dans les cérémonies Kanak (Arama et Hoot ma Whaap, Kanaky Nouvelle-Calédonie). In BORNAND, S. and LEGUY, C. (eds.), *Compétences et performances*, pp. 265-290. Paris: Editions Karthala.

MONNERIE, D. (2014b) Les échanges en Océanie et l'anthropologie. In DOUSSET, L. et al. (eds.), *Les sciences humaines et sociales dans le Pacifique Sud. Terrains, questions et méthodes*, pp. 127-159. Marseille: Pacific-credo Publications.

MONNERIE, D. (2016 f.c.) Relations et substances : l'anthropologie et les enseignements des cérémonies kanak. (Nord de la Grande Terre, Kanaky Nouvelle-Calédonie). In BARRAUD, C., ITEANU, A. and MOYA, I. (eds.), *L'impuissance de la valeur, ouvrage en hommage à Louis Dumont*, pp. 235-260, CNRS éditions.

MUNN, N. (1986) *The Fame of Gawa.* New York: Cambridge University Press.

PILLON, P. (2002) Cérémonies de l'Igname Nouvelle, Intégration Sociale, Modalités Binaires et Ternaires en Pays Mèa (Nouvelle-Calédonie). *Journal de la Société des Océanistes* 114-115: 129-140.

PLATENKAMP, J.D.M. (2006) Visibility and Objectification in Tobelo Ritual. In CRAWFORD, P. and POSTMA, M. (eds.), *Reflecting Visual Ethnography - Using the Camera in Anthropological Fieldwork*, pp. 78-102. Aarhus: Intervention Press & Leiden: CNWS Press.

RAPPAPORT, R. (1968) *Pigs for the Ancestors. Ritual in the Ecology of a New Guinea People*. New Haven and London: Yale University Press.

RAPPAPORT, R. (1999) *Ritual and Religion in the Making of Humanity*. Cambridge: Cambridge University Press.

ROBBINS, J. (2001) Ritual Communication and Linguistic Ideology. A Reading and Partial Reformulation of Rappaport's Theory of Ritual. *Current Anthropology* 42(5): 591-614.

STASCH, R. (2009) *Society of Others. Kinship and Mourning in a West Papuan Place*. Berkeley and Los Angeles: University of California Press.

THOMAS, N. (2006 [1995]) Exchange Systems, Political Dynamics, and Colonial Transformations in Nineteenth Century Oceania. In BELLWOOD, P. et al. (eds.), *The Austronesians. Historical and Comparative Perspectives*, pp. 287-308. Canberra: ANUE Press.

WAGNER, R. (1974) Are there Social Groups in the New Guinea Highlands? In LEAF, M.J. (ed.), *Frontiers of Anthropology: An Introduction to Anthropological Thinking*, pp. 95-122. New York: Van Nostrand Company.

WEINER, A. (1992) *Inalienable Possessions: the Paradox of Keeping while Giving*. Berkeley and Los Angeles: The University of California Press.

Care in a Context of HIV and AIDS: Beyond Exchange and Reciprocity?

ANKE NIEHOF

Introduction: Care and Context

In 2007, together with Monica Karuhanga Berahu, then my PhD student, I visited Jessica, a 52 year-old woman who lived together with her adult son, Kavuma, on a small banana farm in Rakai district, Uganda. Kavuma had AIDS and was on the brink of death. His wife had left him and had taken the children with her. He had gone back to his widowed mother to be cared for by her. Jessica said:

> "The pain of losing three children and the responsibility that has come with it – five orphans, and now waiting for Kavuma to die despite *the five tortuous years of futile care*, is not comparable to anything that I ever experienced" (Karuhanga Berahu 2008: 24, my italics).

In a nutshell this statement captures the plight of many caregivers in rural areas in sub-Saharan Africa that have been struck by the HIV epidemic. In this paper I shall explore the different dimensions of care in this context, using the concepts of exchange, reciprocity, and gift in an attempt at understanding.

To receive and to give care are part of the universal human experience. At the personal level the balance between giving and receiving care shifts according to one's phase in the life course and specific circumstances. Care practices are part of everyday life and are embedded in the moral economy of households that have to make available and allocate the resources needed, also when the care involves non-household members or external caregivers. Households function in political economies, in many of which care work is undervalued, underpaid and mostly left to women (Tronto 1993, 2013; Razavi 2007; Makina 2009). Since households are family-based, care tends to be given to and provided by family members and when extended to non-household members often concerns members of the kinship network. Hence, care is also about family responsibilities and their normative and moral underpinnings (Finch and Mason 1993). Family relationships are the pillars of the household as a social institution. The responsibilities and entitlements these entail are part of the everyday – most of the time taken for granted – experience of the household members. Household, family and kinship constitute overlapping moral contexts within which people acknowledge obligations to give or provide support and act upon these, making the household a context of 'condensed morality' (Pennartz and Niehof 1999: 206).

The above is particularly pertinent to countries in sub-Saharan Africa in which high levels of HIV and AIDS prevail that have given rise to a huge care burden. Even though AIDS impacts manifest themselves in diverse ways (Gillespie, Niehof and Rugalema 2010), the prolonged and erratic care needs generated by AIDS are a common feature. In Africa, apart from the role of medical professionals in antiretroviral treatment or crisis mitigation, care for people living with AIDS (including palliative care) is provided in the home, mainly by women (Akintola 2006; Peacock and Weston 2008; Du Preez 2011; Niehof 2015). Overwhelmingly, women bear the brunt of caring for persons suffering from AIDS, and communities and governments depend on their doing so (Peacock and Weston 2008; Makina 2009). Tronto (1993: 103) has described care as

> "a species activity that includes everything that we do to maintain, continue and repair 'our world' so that we can live in it as well as possible. That world includes our bodies, our selves, and our environment, all of which we seek to interweave in a complex, life-sustaining web."

This broad definition is particularly applicable to home care for persons living with AIDS in sub-Saharan Africa. Without the 'life-sustaining web' provided by their caregivers, these persons would live shorter, suffer more, and die miserably. The life-sustaining web of the many children orphaned by AIDS is woven by the relatives who provide or arrange for foster care (Kasedde et al. 2014).

The picture that emerges from what we know about AIDS care in Africa is a harsh one. Because of the synergistic relationship between HIV infection, AIDS and related afflictions (malaria, TBC), food insecurity, and poverty (Stillwaggon 2006; Gillespie, Niehoff and Rugalema 2010), affected persons and households hardly have the resources to cope with AIDS morbidity and mortality. The caregivers are not only poor, they generally also get little external support, because other people are equally poor (Nombo 2007, 2010).

The plight of AIDS sufferers and their caregivers is aggravated because of stigmatisation. Susan Sontag (1978: 58) once said: "Any important disease whose causality is murky, and for which treatment is ineffectual, tends to be awash in significance." The causality of AIDS is indeed 'murky', as it is difficult to comprehend that one is ill now because of a sexual encounter with an infected person that took place six or seven years (incubation time) earlier. And even when the epidemiological mechanism is more or less understood, it remains hard to accept why after all this time AIDS hits you and not your neighbour, for example. This complexity has given rise to the image of HIV as a secret invader of the body sent by witches (Allan 2007; Niehof and Price 2008). Furthermore, although treatment may mitigate symptoms like weight loss and makes AIDS patients less susceptible to infections, it cannot cure the condition and is – in that sense – ineffectual. Without treatment, however, persons suffering from AIDS are living with an imminent death sentence and because of their thin and weak appearance often look like 'living dead'. Niehaus (2007) attributes the stigma surrounding people suffering from AIDS and their caregivers to these features in particular.

Caregivers are generally overburdened. Care giving has been called a 'labour of love' (Graham 1983). The labour of caring for persons living with AIDS comprises basic nursing care (treating sores, turning bedridden patients); help with feeding, bathing, going to the toilet, dressing and with household chores; taking the patient to a health care facility; assisting with collecting and taking medication; rehabilitative care; and psycho-social and spiritual support (Akintola 2006; Du Preez and Niehof 2008). Apart from care giving in the ways listed above, caregivers often also organise the care arrangement, which may involve moving to another homestead by the caregiver or the person in need of care, or sending children away to be fostered by relatives (Du Preez 2011; Niehof 2015). The persons who provide foster care for children orphaned by AIDS look after the children's welfare, even when they hardly have the resources to sustain themselves (Kasedde et al. 2014).

This begs the question of why people are doing all this. How can we understand this commitment to care as a "labour of love in which the labour must continue even where the love falters" (Graham 1983: 16)? How do we place this phenomenon of resilient care giving in the world of 'exchange', the subject matter of this book section? Should we in this case equate exchange with reciprocity, however asymmetrical, or should we contend with framing care as a gift? These are the questions that I shall discuss and try to answer in the remainder of this paper.

Care and Exchange

Since in sub-Saharan Africa care for persons living with AIDS is provided mainly in the home, it is part of the moral economy of the household. Unlike what happens in political economies, in a moral economy goods and services are not exchanged to gain short-term benefits but to underpin social relationships that are a source of long-term social security. In the moral economy of the household "the actions of individual household members are motivated by a desire to produce socially preferred (i.e. moral) relationships between them which define the domestic division of labour" (Cheal 1989: 14). Care work always involves the use of goods, services, and time. In the moral economy of the household it is neither the goods, services, and time by themselves that counts, nor their exchange value, but what matters is their potential to give expression to moral principles and strengthen social relationships. Seen in this light, care for AIDS sufferers in Africa does definitely not belong to that part of the world of exchange where goods, services, and time are commodities that can be exchanged according to their market value.

However, if we momentarily step out of the African AIDS context and look at the broader picture two disclaimers could be made. The first is that domestic labour, which includes care work, has been made increasingly visible by the work of feminist economists. They calculated the opportunity costs of women's domestic work and, in this way, attached objective economic value to it, irrespective of the social relationships involved (Razavi 2007). The second disclaimer is that in countries like the Netherlands, for example, a large part of domestic care work can be and in fact is outsourced, which is a form of commoditisation. However, at the same time, the periodic surveys conducted by the Dutch Social and Cultural Planning Bureau (SCP) reveal large numbers of unpaid informal caregivers whose work is motivated by the affection and moral feelings attached to the relationship between the caregiver and care receiver concerned (SCP 2009). These findings weaken the second disclaimer and question the justification of postulating care as an exchangeable commodity. Nussbaum (2003: 53) takes a clear stand on this, saying that "care is hardly a commodity, like income and wealth, to be measured by the sheer amount of it citizens have."

In her book, Annemarie Mol (2008) takes the perspective of the person receiving care and contrasts the logic of the market with the logic of care:

> "in one way or another a market requires that the product that changes hands in a transaction be clearly defined. It must have a beginning and an end. In the logic of care, by contrast, care is an interactive, open-ended process that may be shaped and reshaped depending on its results" (Mol 2008: 23).

She is opposed to the increasing use of the term customer instead of patient for a person receiving health care, because the term customer conveys a false image of choice.

Going back to the African AIDS context, we can conclude that the large majority of the affected people does not have the means to outsource AIDS care, that there is no market for it, and that care by the state is minimal. On the stage of AIDS care, given the virtual absence of the actors 'Market' and 'State', informal caregivers occupy prime of place. Mol's logic of care seems very much applicable to AIDS, since according to this logic "even though care is result-oriented, it is not necessarily bad when 'health' and a 'good life' remain out of reach. Some diseases can never be cured, some problems keep on shifting" (Mol 2008: 23).

Care and Reciprocity

Reciprocity is the give-and-take, or, as we should say in the context of care, give-and-receive part of the world of exchange. In discussing the different forms of reciprocity, Marshall Sahlins saw generalised reciprocity as exemplifying solidarity par excellence.

> "The material side of the transaction is repressed by the social: reckoning of debts outstanding cannot be overt and is typically left out of account. This is not to say that handing over things in such form, even to 'loved ones', generates no counter-obligation. But the counter is not stipulated by time, quantity, or quality: *the expectation of reciprocity is indefinite*" (Sahlins 1996: 31, my italics).

In terms of social distance Sahlins (1996: 35) places the enactment of generalised reciprocity in the contexts of the 'house' and the 'lineage sector', in other words in the context of the household and kinship relations, which is the context in which care for people living with AIDS in Africa is performed. So, does applying the concept of generalised reciprocity to AIDS care in Africa contribute to our understanding of the phenomenon?

In answering this question, I shall distinguish between intra- and inter-generational generalised reciprocity. To both applies that in the African AIDS context antiretroviral drugs, which can turn AIDS into a chronic condition instead of the fatal affliction that it is now, are still far from widely accessible (Whitehead and Strauss 2014). This implies that for the caregivers the expectation of reciprocity is not 'indefinite' (see above) but definitely negative. Evidence on the exchange of care and support between persons of the same generation shows that people are aware of this looming end of the reciprocity equation. "No one is willing to assist someone from January to December knowing that this person will not be able to reciprocate", says a woman in a group discussion in Mkamba, Tanzania (Nombo 2010: 73). Poverty plays a role in this as well. For instance, in a village study in Ghana one of the persons interviewed comments on what he calls 'individualism', which he blames on poverty; people want to care for their relatives, but they cannot afford to (Radstake 2000: 29). Two more statements from Mkamba, Tanzania, also relate the lack of reciprocity to poverty (the first statement more explicit than the second):

> "It is not that the sense of brotherhood is dying but the big problem is poor living and economic conditions for many of the families. Life has become very difficult. Even if your relative is sick, you are unable to assist due to the lack of income."

And:

> "Real relatives were there in the past. Nowadays there is no one who can even pay you a visit, very few do that. Your own children are now your relatives."(Nombo and Niehof 2008: 247, 248)

This shows that the network that would sustain horizontal generalised reciprocity is short-circuited by AIDS. This is aptly illustrated by the fact that in Mkamba relatives and community members only give support at the occasion of the funeral once the sick person has died (Nombo 2007). Poverty would be a 'legitimate excuse' (Finch and Mason 1993: 97) for relatives to withdraw themselves from care and support obligations, if they are not already sick themselves or have AIDS in their own family. AIDS also puts a strain on the marital relationship. Women are far more often the caregivers for their ailing husbands than *vice versa* (Peacock and Weston 2008). Cultural scripts that denote women as caregivers and men as providers give men a position of what Tronto (2013: 103) calls 'privileged irresponsibility' with regard to care. Sometimes the partner, being overwhelmed by fear and the daunting care responsibilities, just flees. As Akintola (2006: 241) observed: "Abandonment of partners emerged as a subtle but effective and pervasive way of avoiding caring roles."

The second Mkamba statement above illustrates how AIDS has led to the contraction of care and support networks, leaving only the parent-child relationship as the active context for performing generalised reciprocity. In this regard, the demographic impact of HIV and AIDS has to be taken into account (cf. Niehof 2012). Since particularly adults in their prime of life are afflicted, the care-giving potential of this middle generation vis-à-vis both their children and their elderly parents is reduced. The burden of care giving is shifted to the older generation. Numerous cases have been reported of elderly women who nurse their sick adult children and take care of their orphaned grandchildren (e.g. Akintola 2006; Nombo 2007; Nombo and Niehof 2008; Du Preez and Niehof 2008; Paradza 2010; Du Preez 2011; Kasedde et al. 2014). The fact that elderly parents have to care for their adult children who have AIDS instead of – in their old age – being cared for by their children, upsets the perceived natural order of things. A study on the social dynamics of AIDS in Buhaya, Tanzania, quotes village elders saying that AIDS is "killing the young [...] and leaving the old to carry the burden of farming and child-rearing in their weakened states" (Foster Githinji 2015: 28). In short, the archetypical form of generalised reciprocity of the exchange of care between children and parents across the life course is no longer anchored to the social system.

Kasedde et al. (2014), however, do place the foster care by older people in Uganda in a framework of reciprocity. From a large data set of children in foster care, the research team selected 13 children, their care providers and additional relatives for an in-depth study. Eight children were fostered by (great) grandparents, in six cases by the (great) grandmother alone. In the remaining cases the foster parent was a (great) aunt or an uncle (in one case) (Kasedde et al. 2014: 164). They found that although fostering the children could take a physical toll on the older caregivers and that providing for the children's needs (food, school expenses) was challenging, the children's companionship and their help with daily chores, however, were positively valued. Additionally, they point at the respect the elderly foster parents earned from the other relatives and the community by taking on the role of caregiver. "Their involvement in these social roles and spaces had, in many ways, given many of them not only companionship and support in their households but had helped these older foster-parents find their place in new social networks" (Kasedde et al. 2014: 166).

Using the lens of reciprocity thus yields a mixed image. In the case of relationships between closely related persons of the same generation, AIDS wreaks havoc on reciprocal practice, in which poverty plays an important role. In the case of ageing mothers caring for their adult sick children, the – however – 'indefinite' (see above) expectation of reciprocity the mothers may have had, is annulled. Only when elderly persons are fostering grandchildren, a new reciprocity equation seems to be created. It has to be noted that, in this case, the social space to which it applies is larger than just the grandparent-grandchild relationship. As Kassede et al. (2014) show, the elderly caregivers are respected for their life experience and moral baggage, which earns them the support of the community. The positive moral identity or 'reputation' (Finch and Mason 1993: 129) they built up during their life time is rewarded in the form of support by relatives and community members at the time they need it.

Care as Gift

In the concluding part of his seminal work *Essai sur le don* Marcel Mauss says:

> "It is our good fortune that all is not yet couched in terms of purchase and sale. Things have values which are emotional as well as material; indeed in some cases the values are entirely emotional" (Mauss 1969: 63).

Having concluded that the denominator of generalised reciprocity does not well fit the case of the care of African mothers for their adult children with AIDS, and that this care cannot be purchased, should we then interpret this care giving as gift giving?

In the vision of Mauss, gift giving is part of a total social system. Gifts have and create social history. As put by Mary Douglas (2011: ix): "The whole idea of a free gift is based on a misunderstanding." In the case of mother caregivers, the mother-child relationship captures its social history. It is also a relationship with a heavy moral loading. Parental care is a core element in the web of family responsibilities, and there are few legitimate excuses for parents to opt out. In many societies, parental obligations of care for minors are underpinned by legal sanctions. But there are no such sanctions to enforce the care by mothers for adult children. Poverty and physical frailty could be used as legitimate excuses by poor and ageing African mothers to decline taking their adult sick sons or daughters into their home to care for them until they die,but the literature shows that this rarely happens.

Aafke Komter (1996) analysed women's gift giving from the perspective of gender and power relations. In Western societies, women are more frequent gift givers (and receivers) than men, because of women's embeddedness in networks "of particular social expectations, norms and rules regarding their societal rights and duties and their position within the family" (Komter 1996:

120). Of the four models that Komter developed to explain women's gift giving in relation to power, the "model of asymmetrical reciprocity in favour of men" seems particularly applicable to the care of mothers for their adult children in a context of AIDS. In the following citation we could easily substitute 'gift' by 'care':

> "Women's greater share in giving giftsis [...] related to their position within the family and their traditional responsibilities for keeping social contacts alive [...]. Women can barely escape their gift giving duties, whether they like them or not" (Komter 1996: 125).

The care provided by African mothers to their adult sick sons and daughters could thus be seen as a gift, but it is a gift of a special nature. It is based on personal social history and on gendered frameworks of duties and obligations, but it does not entail expectations towards the receiver. Mary Douglas (2011) concluded in her discussion of the work of Mauss that there are no free gifts, but the care of the African mothers does not create new social histories. The inevitable death of the receiver marks the end of the social and of the gift giving relationship.

Conclusion

From the discussion above, we can conclude that caring for people living with AIDS in the African context has certain specific characteristics. It is definitely not a commodity that can be exchanged on the market, it tips the balance of relationships characterised by generalised reciprocity to the point of (almost) destroying these relationships, and as a gift it has a finality that precludes its potential to create social history. Instead, this type of care is a case of refuted reciprocity and is a gift with no strings attached. It is inspired by a combination of emotion (love) and perceived (and ascribed) moral duty. Additionally, the (in majority female) caregivers often lack the proper capabilities and means to do their care work and are inadequately, if at all, supported (Niehof 2015). Nussbaum (2003) envisions capabilities as fundamental entitlements for social justice. Seen in this light, the care by African women for their AIDS-afflicted adult (grand)children constitutes a case of grave social injustice. In her article Nussbaum lists ten central human capabilities, but notes that "we would need to add other capabilities as well to the list, in areas such as health care, work conditions, and emotional well-being" (Nussbaum 2003: 55). Indeed, these are the type of capabilities that the African women caring for their AIDS-afflicted children and spouses would desperately need.

Improved and sustained access to antiretroviral therapy (ART) would lighten the care burden and would alleviate many of the problems involved. In the words of Whitehead and Strauss (2014: 103): "Today people on treatment can expect to live long, healthy and productive lives." That would ease the plight of caregivers, also because AIDS patients on treatment do no longer resemble 'living dead' (Niehaus 2007) and stigmatisation can be expected to decline. Not only would that lift the social confinement of AIDS patients and their caregivers, it could also restore reciprocal support relations between caregivers and care receivers. Unfortunately, as Whitehead and Strauss (2014) note, the drugs are expensive and require a commitment on the side of the patients to take them for the rest of their lives and of governments and funding agencies to provide for them. At present, even though antiretroviral therapy (ART) coverage has substantially increased in many countries in sub-Saharan Africa, there is no overall 'epidemiological transition' of HIV and AIDS, meaning that in most parts of the continent the numbers of new infections still exceed the numbers of people entering treatment (Whitehead and Strauss 2014: 105). To conclude, for the caregivers there is light at the end of the tunnel, but the light is weak and the end of the tunnel is still far away. For now, these caregivers have to continue their labour of love, for which they may gain only moral authority in return, if gaining anything at all.

References

AKINTOLA, O. (2006) Gendered Home-Based Care in South Africa: More Trouble for the Troubled. *African Journal of AIDS Research* 5(3): 237-247.

ALLEN, T. (2007) Witchcraft, Sexuality and HIV/AIDS among the Azande of Sudan. *Journal of Eastern African Studies* 1: 359-396.

BARNES, M. et al. (eds.) *Ethics of Care: Critical Advances in International Perspective*. Bristol UK, Chicago: Policy Press.

CHEAL, D. (1989) Strategies of Resource Management in Household Economies: Moral Economy or Political Economy? In WILK, R.R. (ed.), *The Household Economy. Reconsidering the Domestic Mode of Production*, pp.11-23. Boulder, San Francisco, and London: Westview Press.

DOUGLAS, M. (2011[1990]) Foreword: No Free Gifts. In MAUSS, M. *The Gift: The Form and Reason for Exchange in Archaic Societies*, IX-XXIII. London and New York: Routledge Classics.

DU PREEZ, C.J. (2011) *Living and Care Arrangements of Non-Urban Households in KwaZulu-Natal, South Africa, in the Context of HIV and AIDS*, PhD Thesis Wageningen University, the Netherlands.

DU PREEZ, C.J. and NIEHOF, A. (2008) Caring for People living with AIDS: A Labour of Love. *Medische Antropologie* 20(1): 87-105.

DU PREEZ, C.J. and NIEHOF, A. (2010) Impacts of AIDS-related Morbidity and Mortality on Non-Urban Households in KwaZulu-Natal, South Africa. In NIEHOF, A., RUGALEMA, G. and GILLESPIE, S. (eds.), *AIDS and Rural Livelihoods: Dynamics and Diversity in Sub-Saharan Africa*, pp. 43-61. London: Earthscan.

FINCH, J. and MASON, J. (1993) *Negotiating Family Responsibilities*. London and New York: Tavistock/Routledge.

FOSTER GITHINJI, V.E. (2015) *Everyday Social Dynamics and Cultural Drivers of Women's Experiences with HIV/AIDS: Voices from Buhaya, Tanzania*. PhD Thesis Wageningen University, the Netherlands.

GILLESPIE, S., NIEHOF, A. and RUGALEMA, G. (2010) AIDS in Africa: Dynamics and Diversity of Impacts and Response. In NIEHOF, A., RUGALEMA, G. and GILLESPIE, S. (eds.), *AIDS and Rural Livelihoods: Dynamics and Diversity in Sub-Saharan Africa*, pp. 10-18. London: Earthscan.

GRAHAM, H. (1983) Caring: a Labour of Love. In FINCH, J. and GROVES, D. (eds.), *A Labour of Love: Women, Work and Caring*, pp. 13-28. London: Routledge & Kegan Paul.

KARUHANGA BERAHU, M. (2008) *Living with AIDS in Uganda: Impacts on Banana-Farming Households in Two Districts*, AWLAE Series 6, Wageningen: Wageningen Academic Publishers.

KASEDDE, S. et al. (2014) They are not always a burden: Older People and Child Fostering in Uganda during the HIV Epidemic. *Social Science & Medicine* 113: 161-168.

KOMTER, A. (1996) Women, Gifts and Power. In KOMTER, A.E. (ed.), *The Gift: An Interdisciplinary Perspective*, pp. 119-131. Amsterdam: Amsterdam University Press.

MAKINA, A. (2009) Caring for People with HIV: State Policies and their Dependence on Women's Unpaid Work. *Gender and Development* 17(2): 309-319.

MAUSS, M. (1969[1954]) *The Gift: Forms and Functions of Exchange in Archaic Societies*. London: Routledge & Kegan Paul.

MOL, A. (2008) *The Logic of Care: Health and the Problem of Patient Choice*. London and New York: Routledge.

NIEHAUS, I. (2007) Death before Dying: Understanding AIDS Stigma in the South African Lowveld. *Journal of Southern African Studies* 33(4): 845-860.

NIEHOF, A. (2012) HIV/AIDS in Sub-Saharan Africa: Impacts and Social Change. *Afriche e Orienti, Special issue on AIDS 2012* [Bologna]: 36-46.

NIEHOF, A. (2015) Contours of Matriarchy in Care for People Living with AIDS. In BARNES, M. et al (eds.), *Ethics of Care: Critical Advances in International Perspective*, pp. 139-151Bristol UK, Chicago: Policy Press.

NIEHOF, A. and PRICE, L.L. (2008) Etic and Emic Perspectives on HIV/AIDS Impacts on Rural Livelihoods and Agricultural Practice in Sub-Saharan Africa. *NJAS Wageningen Journal of Life Sciences* 56(3): 139-153.

NOMBO, C.I. (2007) *When AIDS meets Poverty: Implications for Social Capital in a Village in Tanzania*, AWLAE Series 5. Wageningen: Wageningen Academic Publishers.

NOMBO, C.I. (2010) Sweet Cane, Bitter Realities: The Complex Realities of AIDS in Mkamba, Kilombero District, Tanzania. In NIEHOF, A., RUGALEMA, G. and GILLESPIE, S. (eds.), *AIDS and Rural Livelihoods: Dynamics and Diversity in Sub-Saharan Africa*, pp. 61-76. London: Earthscan.

NOMBO, C.I. and NIEHOF, A. (2008) Resilience of HIV/AIDS-affected Households in a Village in Tanzania: Does Social Capital Help? *Medische Antropologie* 20(2): 241-259.

NUSSBAUM, M.C. (2003) Capabilities as Fundamental Entitlements: Sen and Social Justice. *Feminist Economics* 9(2-3): 33-59.

PARADZA, G.G. (2010) Single Women's Experiences of Livelihood Conditions, HIV and AIDS in the Rural Areas of Zimbabwe. In NIEHOF, A., RUGALEMA, G. and GILLESPIE, S. (eds.), *AIDS and Rural Livelihoods: Dynamics and Diversity in Sub-Saharan Africa,* pp.77-95. London: Earthscan.

PEACOCK, D. and WESTON, M. (2008) Men and Care in the Context of HIV and AIDS: Structure, Political Will and Greater Male Involvement. New York: United Nations, Division for the Advancement of Women.

PENNARTZ, P. and NIEHOF, A. (1999) *The Domestic Domain: Chances, Choices and Strategies of Family Households.* Aldershot, Brookfield USA et al.: Ashgate.

RADSTAKE, M. (2000) *Secrecy and Ambiguity: Home Care for People Living with HIV/AIDS in Ghana*. Leiden: African Studies Centre.
RAZAVI, S. (2007) The Political and Social Economy of Care in a Development Context: Conceptual Issues, Research Questions and Policy Options. *Gender and Development Programme Paper* no.3, Geneva: UNRISD.
SAHLINS, M.D. (1996[1979]) On the Sociology of Primitive Exchange. In Komter, A.E. (ed.), *The Gift: An Interdisciplinary Perspective*, pp. 26-38. Amsterdam: Amsterdam University Press.
SCP (2009) *Mantelzorg: Een overzicht van de steun van en aan mantelzorgers in 2007* [Informal care: A review of the support by and to informal caregivers in 2007], The Hague: Sociaal en Cultureel Planbureau.
SONTAG, S. (1978) *Illness as Metaphor*. New York: Farrar, Strauss and Giroux.
STILLWAGGON, E. (2006) *AIDS and the Ecology of Poverty*. Oxford: Oxford University Press.
TRONTO, J.C. (1993) *Moral Boundaries: A Political Argument for an Ethic of Care*. New York and London: Routledge.
TRONTO, J.C. (2013) *Caring Democracy: Markets, Equality and Justice*. New York and London: New York University Press.
WHITEHEAD, A. and STRAUSS, M. (2014) The End of AIDS: Possibility or Pipe Dream? A Tale of Transitions. *African Journal of AIDS Research* 13(2): 101-108.

Production is Exchange: Gift Giving between Humans and Non-Humans

GUIDO SPRENGER

"But is this really new?" – I have heard this question, rendered in an inquisitive, curious voice, repeatedly from Jos Platenkamp when he was confronted with concepts heralded as significant theoretical advances. For him, quite a number of ideas that have been put forward in recent decades as innovations actually belonged to anthropology's standard fare. This would involve the inclusion of non-human beings as social persons into society. In his writings, Jos Platenkamp never bothered to explain ancestors or spirits in terms of symbolism, psychological needs, or the expressive aspects of culture. For him, these are beings to be considered in terms of their sociality, as part of exchange systems that encompass more than living human beings. He never cut through the socio-cosmic wholes that were the object of his analysis by separating humans from non-humans. In his vision, they were all integrated as equally existing beings through their constitutive relationships.

From this point of view, the ontological turn, promoted by scholars like Bruno Latour, Philippe Descola, Eduardo Viveiros de Castro and others, would, in many respects, look not so much like the revolution it is often said to be, but rather appears as a reinstatement of what anthropology has achieved for some time already. In that sense, I am afraid, the following reflections are not really new.

Production and Exchange

The separation of production and transfer as two different types of relationships axiomatically underlies much of economic anthropology. The classical types of production, hunting-gathering, herding, agriculture and industrialism, describe relationships between humans and objects – tools, animals, plants etc. – while transfer covers relationships among human beings: sharing, giving, gift exchange, trade, etc.. Such models suggest that humans first have to engage in production, i.e. in human-resource relationships, before they can turn to transferring the yields of their productive activities. This indicates a naturalistic point of view: Relationships with nature provide the foundation for society.

However, the revision of the relationship between humans and their non-human environment in the past two decades has given rise to the possibility of considering modes of production as systems of social relationships not just among humans. With 'social', I mean relationships based on communication and meaning among beings that, in a given context, count as persons. These beings are not always humans, and it is with this in mind that I choose to use the term 'society' instead of the currently common 'collective'. A collective includes beings that are not necessarily persons, but still appear as actants, as beings with an agentive impact on relationships (Latour 2008; Descola 2011). A society, in my understanding, focuses on the production and recognition of persons – however, with no given distinction between human and non-humans, and with the possibility to differentiate degrees of personhood.

Here, I follow Luhmann (1984, 1995), whose focus on communication does not exclude non-humans, as Teubner (2006) has argued. Persons appear in societies first as senders and receivers of communication. From this point, they might accrue other indicators of sociality: intentionality, accountability, kinship relations, etc.. The communications between humans and non-humans thus potentially establish both parties as persons. At the same time, personhood might not be fully accomplished or recognised in any given context, but remain emergent and graded.

In the following, I will take these recent approaches to the social into the fold of gift exchange theory. As Platenkamp has argued, the particular mode of communication that establishes beings as persons in the first place is exchange (e.g. Platenkamp 2004, 2010). What appears as a relationship between humans and objects from a naturalist point of view, is for many other societies often a relationship of exchange between human and non-human beings (Gregory 1982: 91-93). With each gift, as Mauss (1990 [1924]) has observed, one gives something of oneself. Following this, one can consider the gift as a transfer of aspects of personhood. The social efficacy of the gift lies in the inalienable relationship it retains with its giver or origin. Otherwise, the relationship it entails would be ephemeral, as Bloch and Parry (1991) have argued. If persons are made up of relationships, as Strathern (1988) has proposed for Melanesia, then the beings that constitute or embody such relationships, the gifts themselves, are aspects of personhood – and therefore may be neither passive object nor complete person. Thus, gift exchange is particularly important for the production of hybrids or quasi-objects, as Latour (2008) calls those entities that cannot be neatly classified as human or non-human, natural or cultural (Caillé 2010: 183).

However, while current approaches like that of Latour tend to stress the symmetrical nature of human-non-human relationships, my aim is to draw attention to the hierarchies at work in them. Value hierarchies are a central factor of gift exchange, as Barraud et al. (1994) have demonstrated, but are often overlooked when the equality of exchange partners is assumed (see Descola 2011: 568). I will thus highlight the asymmetries gifts entail for the exchange partners.

In the following, I will thus propose a model of gift exchange that enables me to make an argument about asymmetric and hierarchical communication within human-non-human societies. Next, I will go through a series of examples from different modes of production that appear as ways of exchange. Finally, I will explore some consequences of the acknowledgment of hierarchies.

Gift Exchange

In this section, I will briefly sketch a particular concept of the gift which I find useful for the argument that follows.

1) The concept of gift exchange is often reduced to reciprocity between two parties (e.g. Laidlaw 2000; Karatani 2014). This position, however, is unable to account for the integrative character of the gift that Mauss envisioned. Therefore, I follow Barraud e.a. who define exchange as "*all activities in the course of which something is seen as circulating*" (1994: 5, italics in original). Mere reciprocity between two parties reduces gift exchange to its contractual, individualist features. Circulation, on the other hand, defines socio-cosmic wholes (Platenkamp n.d.). Any act of exchange is inconceivable without other exchanges it is related to.

2) Gift exchange generates difference. Other than the exchange of commodities, gift exchange is not so much based on the idea of equivalence, but of an asymmetry of values (Foster 1990, Platenkamp n.d., Strathern 1997). Gifts are representations or aspects of their givers or origins, conceived as values. They are thus qualitatively different from each other in a way that cannot be reduced to any type of quantification. Therefore, the act of exchanging them accomplishes two things: It acknowledges the representative nature of the gifts, while at the same time linking their value to a different value. In the moment of exchange, the two values are claimed to be a virtual whole, the realisation of a mutually constitutive relationship between two qualitatively different, albeit complementary sources. Still, equivalence is not inconceivable in gift exchange. People involved in gift exchange may pursue equality between partners, but this demands elaboration and attention to the sameness of the gifts (Foster 1990).

3) Therefore, gift exchange provides an "elementary experience of otherness" (Moebius 2009, my transl.; see also Därmann 2010). Alterity is at least realised in the difference between the giver and the taker of a particular item – even when the exchanging persons are familiar with each other, the mere act of assigning them the positions of giver and taker of a particular gift may turn them into representatives of differing, complementary values. The difference is even more pronounced when the giver and taker belong to different categories of people, ethnicities, or beings.

At the same time, the item that represents the giver may become part of the relational being of the taker, thus integrating the other into the self as a constitutive element. Exchange constitutes the persons it connects as beings with an internal split, a self-other dichotomy. The complexity of persons emerges as a series of (often internalised relations to) others. Birth rituals in gift-oriented societies illustrate this (Platenkamp 2010; Sprenger 2016). Becoming other for good is thus a perennial risk of gift exchange. Alterity appears as a condition for the integration of parts and wholes in a gift exchange system. This is also true for societies composed of humans and non-humans.

4) Gifts are exchanged as representations or embodiments of such values. Value relations are in any given case, when they are enacted as decisions, judgments, etc., by necessity hierarchical, as one value needs to be superior to the other (Dumont 1991). Thus, if gift exchange means relating two qualitative values, this implies establishing a hierarchy (Barraud et. al. 1994). This asymmetry is sometimes realised as social hierarchy between persons. Hierarchies of this kind have three important aspects – their stability, their range and their reversibility.

Stability refers to the question for how long the hierarchy is acknowledged and if and how it is reversed. Range refers to the field of communication it structures. Some hierarchies only apply to the moment of a particular transaction and involve two persons, while others structure permanent relations of dependency between large parties. In this sense, hierarchy is not identical with power in the usual (non-Foucauldian) sense of the term. Hierarchy of this kind does not necessarily imply oppression, control of action or chains of command (Iteanu 2009).

As Dumont (1991) has argued, value hierarchies may be reversed according to context or level of the ideology. This also applies to the relationship between exchange partners, insofar they represent the values of their respective gifts. Consequently, the hierarchies between humans and non-humans in many non-modern ontologies are reversible.

5) The connectivity of a gift, that is, its potential to establish and maintain relationships, is partially constrained by its physical qualities – if it can be stored or transferred for several generations, if it is bulky or light, if it can be transformed, be hidden or is exposed to plain sight etc. In particular the question what endures and what is being consumed is crucial for the consideration of production as exchange.

6) Gift exchange involves non-human beings and at the same time generates personhood. As gifts transfer aspects of personhood, they acquire person-like traits themselves in order to serve this purpose. Therefore, some major exchange items are considered to be animate, in particular high-ranking ones, like the Northwest Coast copper plates (Mauss 1990) or Southeast Asian bronze drums (Lundström and Tayanin 1981). Their rank is connected to persons and domains of origin that are themselves ranking high in the value structure of the exchange system. Thus, their relationships tend to become more complex and durable than low-ranking items. Personhood or animation of gift

items may indicate degrees of complexity in the making of gifts and persons through their circulation.

For the same reason, the personhood of the beings involved in gift exchange might shift. Beings that are treated respectfully as persons in one context might be treated as things in another. Personhood might thus be temporary, graded, and contextual, according to the transfers a being is involved in and its current position in them (see Remme 2016).

There are thus three possible positions for persons in gift exchanges – as givers and receivers of a gift and as the gift itself. Some beings even move between these positions – the aforementioned Bronze drums are also receivers of offerings. If marriages are gift exchanges, with a spouse as a gift, as Lévi-Strauss (1993) has argued, then the spouse is the highest-ranking exchange item because he or she is a potential giver or receiver him/herself.

Production is Exchange

All human societies encompass non-humans, like animals, plants, deities etc. Societies are thus human societies only insofar as humans consider themselves to be those beings that link the others together. Societies may also draw beings into the range of their sociality that are not permanent members, like spirits of the environment or human foreigners. When the integration of non-humans within society or at its boundaries involves exchange, their difference from human members of society appears as one of the differences maintained and valorised by the exchange. If, for example, animals are engaged in gift exchanges with humans, this does not make them human in the same way that, e.g. behavioral scientists or some animal rights activists stress similarities between humans and animals. The latter tend to account for the similarity of humans and animals in terms of graded resemblance, that is, of identity. Animals appear as human-like because they share with humans features like intelligence, tool use or ability to communicate. In some non-naturalist ontologies, however, such similarities may be acknowledged but are not the main reason why people engage in exchange relations with animals. Their exchanges demand, first of all, difference. Exchange with non-humans makes sense because they are different persons – even if the difference is formulated under the condition of what Descola (2011) has termed the identity of interiority. In Amazonian ontologies, humans, jaguars or peccaries may consider members of their own species as human, sharing identical social insitutions. In these cases, the difference is maintained through the body as a mediator (Viveiros de Castro 2004). The bodies of different species make them relate to their environment in a different way. If these differences are valorised, the relationship will be conceived as hierarchical – however, often this hierarchy demands negotiation in the course of human-animal interaction (Lima 1999, see below).

I will now give a brief overview of various types of exchanging with non-humans. The classification is preliminary and, rather unresourcefully, follows the traditional classification of production types. More research may produce a different classification that is better suited to account for the variations in human-non-human relationships.[1] Thus, for the time being, I merely demonstrate that within each of the traditional categories of production modes there are examples of exchange between humans and non-humans. I am not implying that the categories in their entirety are characterised by the type of exchange, nor that the kind of relationships I am pointing out are exclusive to them. However, insofar particular production types imply the intense heeding of certain relationships with non-humans, the exchanges these relationships might involve are characteristic, even prominent, in these types. In other modes of production, similar relations might exist, although in a somewhat subordinate manner. In any case, a classification of exchange types cannot be achieved by classifying societies in their entirety, but rather according to relationships (e.g. Kishigami 2004).

Hunting

I begin with hunting, not because it is particularly primordial, but because my argument has been elaborated best for venatic ideologies, systems of ideas applied in hunting. These may occur even in societies that practice horticulture or other forms of production. In these ideologies, game animals appear either as property of spirit lords or as a kind of human being in their own right, and sometimes as both. A successful hunt thus results from a negotiation with the spirit owners or the animals themselves – the observation of ritual rules, seduction, making demands, etc. That this relationship is one of gift exchange has been elaborated most explicitly by Nadasdy (2007).

A typical case is hunting in northern Asia. Regarding hunting shamanism in Siberia, Roberte Hamayon (1994) argues for a system of reciprocal exchanges between human communities and the spirit owners of the animals. The shaman marries the daughter of a lord of the animals and therefore acquires the right to demand gifts. These are presented to his community in the form of prey. However, the transfer needs to be balanced, and thus the spirits demand similar gifts from humans, appearing as human death and illness. It is the shaman's task to keep the balance to the humans' favor.

---

1 In this respect, Kojin Karatani's project to move from modes of production to modes of exchange is highly stimulating (Karatani 2014). Instead of Marxist modes of production – shared ownership, agricultural society, slave society etc. – he proposes three modes of exchange: reciprocity (gift exchange), protection and control (the state) and commodity exchange (capitalism). These modes always exist at the same time, but in most cases, one is dominant and subordinates the others. Although Karatani does consider relations between human beings and 'nature' as a type of exchange, his model primarily focuses on relationships among living humans and therefore has a different focus than my present effort.

Rane Willerslev has opposed this view from the perspective of demand sharing. Demand sharing, not reciprocity, is the most important form of transfer among Yukaghir in Siberia, and there is no sense of balancing out a gift. Those who are in need have the right to demand from those who have. The same applies to relationships with animals. Humans communicate their neediness to the spirit owners and therefore gain access to the animals which are persons in their own right. However, a strong depletion of animals by hunters might push the spirits into neediness themselves and therefore into a position to demand human lives in turn (Willerslev 2007: 45-49). With this model, Willerslev elegantly explains the coexistence of hunting taboos with the killing of more animals than needed, the latter being gifts which one should not refuse. At the same time, mutual but unbalanced demand sharing establishes society as a whole. The readiness to share decreases when social distance grows (*ibid.*, 41). This indicates that the relationship with game animals is rather close.

While several authors have stressed the difference between gift exchange and sharing (e.g. Kishigami 2004), there is no pronounced difference between them from the perspective of gift exchange theory. Balancing is not a necessary feature of Mauss' overall model. In the most integrative version of gift exchange, he argues, two moieties are linked in a constant process of giving to each other – a phenomenon which, more than any other, inspired Mauss to claim that gift exchanges produce societies as wholes (Mauss 1991). If each moiety transfers its marriageable children as spouses to the other or performs mortuary rites for the other, there is no point in asking for balance. One cannot reckon that two deaths need to be balanced by two others in the other moiety. The difference between sharing and reciprocal (or, in Sahlins' (1972) terms, balanced) exchange might be an important analytical distinction, but for the present argument both still fall within a general theory of gift exchange (for a strong counterargument, see Widlok (2013)). It is not the *do ut des*, but the actual asymmetry of giving that creates societies as wholes.

In this case, the values of need and abundance are variously attached to both parties of the exchange. Abundance obliges. The social hierarchies between parties are therefore quite negotiable and might be reversed at any point in time. The perishable character of the gifts transferred – the bodies of the animals, which are often seen as replaceable aspects of their person – indicates the flexibility and reversibility of the relationship. Like the gifts, the hierarchy is not permanent. Among Yukaghir, the relationship takes the form of seduction. Hunters put on their most beautiful dress and abstain from intercourse with their wives before pursuing a spirit or the elks, which appear to them as beautiful women. It is, however, crucial to keep the relationship unbalanced. If hunter and prey were to fall for each other equally, the result would be highly destructive for the human partner – he would become other altogether and turn into a victim or an animal himself (Willerslev 2007: 105, 113). In

Amazonia, in contrast, the reversibility is one of predation and warfare. The hunters see themselves as hunting animals. For the peccaries, however, hunters appear as enemy warriors. These two values, experienced as perspectives, are in permanent conflict. If peccaries manage to kill a hunter, this amounts to their perspective becoming superior for the hunter, who joins the peccaries as a prisoner of war, being dead only for those who remain human (Lima 1999; Viveiros de Castro 2004).

Herding

While hunting provides examples of almost symmetric reversibility of human-animal relationships, herding rather points to the graded way personhood might be achieved by non-human exchange partners. Pastoralism demands an intense and intimate collaboration between humans and animals, often in a way that seems to have few benefits for the animals. The relation is thus based on a rather stable hierarchy. Earlier attempts to frame it in terms of sociality have likened it to 'feudal' types of clientelism and asymmetry (e.g. Tapper 1988, see also Ingold 2000, chapter 4), thereby foregrounding its hierarchical features. In Karatani's (2014) scheme of exchange types, it would count as 'protection', not 'reciprocity'. Even though the relation between humans and animals is more intimate here than in hunting, the idea that herding animals see each other as humans and full persons, seems much less pronounced. This relationship between intimacy, hierarchy and restricted personhood demands further research. Yet, in certain forms of reindeer herding, relationships are not very dissimilar to hunting ideologies. In these cases, the spirit owners of the reindeer receive animals as sacrifice, while in hunting, humans appear as the receivers of gifts of prey from the spirits (Beach and Stammler 2006: 12-13). As in many cases of hunting, the exchange partner, as representative of a species or category of animals, is a being that is at once real and evolving from the exchanges.

This kind of negotiated personhood is also apparent in cases where herding is less informed by animist concepts. Here, the long-term relations with animals leads to what Beach and Stammler (2006: 8) call 'symbiotic domestication', a mutual socialisation in which animal and human agency complement each other. Charles Stépanoff's (2012) account of Tozhu reindeer herders in Siberia highlights this mutual dependence of humans and animals, without denying its hierarchical character. Stépanoff argues that the only way to herd reindeer in this region is to keep them half-wild, and herders valorise their animals' freedom highly. Still, they nurture the dependence of their animals by little services and by getting them used to consuming more salt than they need. Especially the latter is reminiscent of the tendency of gift exchange to focus on luxury or prestige items not needed for survival. This stabilises the hierarchy, but animal independence remains to be a value, one, however, that is rarely realised to its full. Humans live in permanent fear of this, as sometimes wild herds will drive the herded ones away from their human counterparts, leaving them to starvation (*ibid.*, 289, 292). This reversal of hierarchy, even if rare, establishes animal independence as the higher, more powerful value. Thus, while reindeer do not appear as full persons, their own agency is highly valorised in a manner that is effectively social.

Once again, such exchanges are not so much defined by balanced give and take, but by a continuous and quotidian flow of services, objects and substances that bind the transacting parties together. Such communication patterns lead to a familiarity of herders and animals that even makes the latter accepting their human counterparts as riders (*ibid.*, 296-7). From the side of the herders, this demands a strong attention toward the individual animal's actions and moods, an attention that makes them approach personhood, without realising it fully.

The hierarchies, like the gifts being transferred, are highly stabilised. Two substances stick out as important, salt, which the reindeer cannot obtain in great amounts without humans, and milk, where the opposite holds true. There is thus a clear asymmetry in the gifts that marks the difference between the exchanging parties. This contrasts with hunting ideologies, where the relationships can always be reversed, as hunters and animals join possibly symmetric relationships – as love interest, warriors, prey. The gifts in the herding relationship, however, are not symmetric and cannot be reversed. But because they are not durable, the herders constantly fear that their herd will abandon them.

Stépanoff highlights the asymmetry in his term 'joint commitment' – herders and reindeer pursue entirely different aims within their relationship. The stabilisation of the hierarchy also might lead to a more pronounced identification of the reindeer as animals, while venatic ideologies consider game animals as complete persons. Still, as has been observed among reindeer herders in meat-processing plants, a sense of social obligation pertains even towards dead animals. The notion that nothing of the body should be wasted leads herders to process even those parts whose use is uneconomic by market standards (Beach and Stammler 2006: 16). Just as the personhood of animals is not fully realised, but always emergent, herders behave like recipients of a gift that should not be wasted without calling it so.

Agriculture

Many accounts of agricultural ritual suggest that the land or the crops themselves are person-like, communicative beings. Consuming crops may provide relationships with the ancestors or other cosmological domains (e.g. Geirnaert

2002). Stable crops are not simply fuel for the body, but provide life as a relationship with potency and fertility.[2]

Rice cultivation in Southeast Asia is an example. Numerous oral traditions hold that rice was initially given to humanity as a gift that would satisfy their hunger in abundance for hardly any work. In these stories, rice moves on its own into the granaries. Only later, due to human misbehavior, the initially huge rice grains shrunk and their production became more laborious (Terwiel 1994). Furthermore, rice rituals parallel the human reproductive cycle, e.g. when the ripening rice is likened to the growth of girls who ultimately come to the village as marriageable maidens, as among the Iban of Borneo (Sather 1994).

Agricultural ritual among Rmeet, swidden farmers in northern Laos, establishes relationships with two types of spirits as a precondition for cultivating rice (Sprenger 2006). The first are the male spirits of the earth that are propitiated with bloody sacrifices by the male head of the household that is tilling the field or a male relative of his. These spirits are dangerous, and the gifts given to them secure their nonintervention into human affairs. Not including them into human sociality by sharing food with them would lead to spirit attacks causing illness. However, this inclusion only occurs at the margins of human social space, on the rice fields and not inside houses or field huts. It establishes the spirits as different from humans but still renders the relationship as one of respect.

The other type of spirit is much less bounded and personalised. The spirit of the rice is female and propitiated by women, principally by the mother of the household. The gifts given to the spirits in both cases represent the human potential to process food and turn it into social and luxurious value. Fermented tea leaves, dried and cut tobacco and, above all, the real lubricant of any social relationships, rice processed into whiskey, are presented to the spirit as an enticement to satisfy human demands. However, the rice spirit is also lured into the field hut by means of 'rice medicine', a collection of curio ranging from oddly shaped stones to metal weights or sea shells. The durability of these items, heirlooms of specific households, corresponds to the desired stable relationship with the rice spirit.

The first rice to be harvested is called 'rice child', and the house mother who carries it to the field hut is the 'rice mother'. Thus, human beings offer not just village sociality to the spirit but also kinship, creating a relationship of mutual nurturance. The rice spirit makes the harvest plentiful and the stored rice last long. Impermanent relationships with the spirit quicken the consumption of the rice in the granaries. Bloody sacrifices are performed at the granaries, thus, like those for the field spirits in the fields, indicating the desired place of the spirit.

2 For a stark contrast of the naturalist and a personalized approach to rice growing, see Mueggler (2002: 11-13).

As in the case of seduction in Siberian hunting, otherness is coded in terms of gender – in both cases, the nurturing non-humans, prey or rice, appear as women. But importantly, the relationship with the rice spirit does not highlight sexual pleasure but procreation, and it is not man and woman which are paired in the exchange, but two men for the (male) field spirit and two women for the (female) rice spirit. Thus, the household as represented by a couple relates to the field as the coexistence – although not in marriage – of field and rice spirit.

The great respect paid to both spirits indicates the superiority of the values they represent: the fertile soil and the nurturing rice. However, both sets of ritual serve to impose human (village) sociality upon them – in case of the field spirits, they are made to respect the boundary between their and human sociality, while the rice spirit is bounded and introduced into the granaries, endowing her with an increase of personhood. On the level of the imposition of order, village sociality is superior to the world of the spirits.

However, in particular the field spirits might exercise power over humans. They might abduct human souls and, in return, insert illness-inducing objects from their domain into human bodies – small pieces of stone, earth or wood. This amounts to a reverse socialisation which renders humans sick, their personhood incomplete. Some dangerous spirits outside of the village are thought to live in their own villages, and it is the task of ritual to keep these socialities apart. Rmeet value highly the potential of their own villages and houses to accommodate select external forces – including the rice spirit. A symmetric power inhers to the spirits' sociality, this time as a destructive force for humans.

Industrial Modernity

This is obviously the most difficult type, as dominant modern-naturalist cosmologies seem to make absolute distinctions between the domains of nature and culture. In this ideology, personhood is exclusively assigned to the domain of culture. The question is then, are there contexts of modernity in which non-humans appear as exchange partners? Can we detect traces of emergent personhood in this respect?

First, the cosmological domains of 'nature' and 'culture' are hierarchised in a way that is also reversible according to context. In modern naturalist ideology, human beings are superior to objects and resources due to their ability to manipulate non-human beings to their own ends and through an understanding of their very nature, which is not accessible to the non-humans themselves (Descola 2011). Humans know how to domesticate horses and pigs in ways that cannot be replicated symmetrically by horses and pigs. In a different context, however, nature as a reified general principle is superior to humans as they are subject to natural laws – to a degree that allows for theories that consider human behavior and societies as mere extensions of nature. Thus, we can distinguish two levels, one individual and one cosmological. On the individual level, any non-human being is subordinate to any human being, but the cosmological principle that governs the reproduction of both humans and non-humans, 'nature', is superior to both. This is part of what Latour has called the 'modern constitution' (Latour 2008).

This ideological constellation also extends to the economy. Production as a relation between human agents and nonhuman non-agents and exchange as a relation between human persons are separate activities, distinguished by the ontological chasm between nature and society. Their separation is what Latour has called purification (2008) and what denies non-humans an active status in gift exchange. But the personification disallowed to non-humans in the context of purification bears better chances in another assembly of modern practices which Latour identifies, the creation of networks. Networks produce hybrids and here, non-humans attain aspects of personhood. These range from pet animals to electronic agents whose possibility to enter contracts has been acknowledged in a number of jurisdictions (Teubner 2006). This endows modernity with substantial ambiguity regarding the alterity that is integrated into the relationship. Humans are internally split between nature and culture, and, as Latour points out, so are numerous non-humans. The tension between purification and ambiguity actually supports the production of hybrids.

However, gift exchange as a principle of livelihood seems largely excluded in these contexts. Rather the opposite seems to be the case. As Wilkie's (2010) study of farmers and slaughterhouse workers has shown, personalisation increases with sustained relationships, e.g. of raising, while it is nonexistent among those who only meet an animal for killing it. Yet it is unclear if the sociality of farmers and their cattle can be described as gift exchange.

We thus have to look at the other level on which production could be understood as gift exchange in modern societies, cosmology. The exchange partner on this level needs to complement human action in the type of long-cycle transactions that Parry and Bloch (1989) identify as constitutive of socio-cosmic relationships. This draws, once again, our attention to 'nature', but this time not as an engine governed by natural, impersonal laws, but as an encompassing, reified quasi-person – contextually emergent like some of the spirits or animal persons mentioned above. While certain ecological philosophies, like the Gaia concept (in a popularised version that differs from Margulis' and Lovelock's initial, impersonal one) or, even more so, Mother Earth cults, make this step towards a unified, almost person-like entity quite explicit (critically, see Latour 2011), recent years have seen more mainstream attempts to cast nature into the role of an exchange partner. In these cases, even when models of exchange and personhood are applied to nature, commodification and the human-non-human divide still structure the relationship, thus replicating the basic value-ideas of industrialism in this context.

Since 2008, the constitution of Ecuador recognises nature as a person with protectable rights. The reasoning behind this, however, is not so much the establishment of a new form of communication with 'nature', but an easier way for civil society organisations to stop environmental damage. Nature as a person is thus a jural side effect and not so much a being to communicate with – even though the constitution explicitly links it to the Inca earth goddess, Pacha Mama (Fitz-Henry 2012).

A comparable process can be observed in 'ecosystems services'. This notion establishes a view of nature as a company that provides humans and other beings with services that, at least for a number of its supporters, can be quantified and valorised, often in monetary terms – which suggests a degree of personhood on the side of the giver. However, as critics have pointed out, the ideology behind this rather objectifies nature (e.g. Sullivan 2010). The move to consider it as a service provider helps to integrate ecology into quantitative economic calculation, for whom 'nature' used to be a frontier with immeasurable capacities. The ecosystems services discourse turns nature into an exchange partner, but on the base of equivalence models that do not provide lasting moral relationships. Instead, the value of nature becomes identified with the value of the financial products derived from it on a volatile, unpredictable market (*ibid.*: 114-115; for balance, see Schröter et al. 2014). It thus fails to turn the relationships of humans with nature into a moral project.

Here, Mauss' primary model, as mentioned above, helps to frame the exchange. In this model, two moieties that exhaust society (and maybe even the cosmos) stand in a context of permanent mutual transfer. In western modernity, these two constitutive halves are presumably called nature and culture. Like in some hunting-gathering societies, their relationship is informed by demand sharing. People take what they need from nature, but do so with an increasingly bad conscience: What if nature takes its toll on humans? This idea is framed in different narratives of the consequences of the exploitation of the ecology – if we take too much, we will be deprived of our livelihood very soon. The plausibility of such narratives of mutual depredation, leading to catastrophe, does not only rest on exact scientific research and calculation. The awareness of the imbalance of giving and taking might even be the ideological backdrop that informs some scientific research, although this needs much more analysis. These menacing scenarios indicate nature's potential personhood even more that the Ecuadorian constitution, albeit in a subdued manner – on the base of how a social relationship should play out.

The identification of production as exchange in modernity comes at a prize. First, on the levels of immediate relations, there seems little in terms of exchange with non-humans, although there are clues to the acknowledgement of their personhood. If these relationships are cyclical or otherwise similar to exchange is as yet an open question. We thus have to look at the level of reified value-ideas like nature and society to find features of gift exchange relations.

Secondly, even on this level, the exchange partners can be identified only implicitly and through a number of indicators, like fears of the 'revenge of nature'.

The asymmetry of the exchange here extends into the ontological domain. On the one hand, nature gives generously but holds the stronger position, and humanity has to play by its rules – a hierarchy one would expect of a gift exchange relationship. However, humanity is free to explore the rules of nature and turn them to their own advantage, thus making human 'individuals' the masters of nature (see Dumont 1991). However, this potential reversibility of the hierarchy is fraught with anxieties, i.e. on cloning, genetically engineered food and climate change – the result of generosity betrayed and corrupted, leading to sickness and corrupted personhood. Such fears resemble macroscopic versions of the dangerous reciprocities envisioned by some of the above-mentioned ideologies. However, the exaggeration of this notion in modernity, that nature's backlash may be mankind's ultimate demise, can be interpreted as the consequence of suppressing the acknowledgement that this exchange comes with obligations.

Thus, modernity seems stuck in a dynamism of contradictory and yet mutually reinforcing ideas. On the one hand, nature is dominant but passive. On the other, there is a nagging fear that nature might behave like an exchange partner whose generosity has been stretched to its limits. Attempts to acknowledge reciprocity by instating nature as a business partner in a global market of services paradoxically lead to its further objectification.

Would Latour's 'parliament of things' (2008) offer a way out – a site where all beings are represented equally, with any hierarchy suspended for the period of negotiation? The proposal seems suggestive. However, the idea of such a parliament seems rather Eurocentric (see Kapferer 2010: 204-205). It denies the fact that any equality comes at the price of hierarchy. It seems to me that the idea of the natural equality of humans, the foundation of democracy, individualism or parliamentarism, rests on the equally naturalised divide between human beings and things. In modernity, ontological equality in the human domain depends upon the inequality between humans and non-humans. Thus, any attempt to dissolve the ontological hierarchy between humans and non-humans in the name of equality tendentially runs the danger to create a disturbing paradox – the ontological equality among humans might be dissolved along the way as well, creating humans 'of lesser degree', as it were. This, in turn, might question the very project of equality itself.

Conclusion

Hierarchy is an ongoing issue in exchange systems that involve humans and non-humans as recognisable persons. These hierarchies are not necessarily ontological – or rather, they are subject to a relational ontology (Bird-David

1999). In these exchange systems, beings sometimes appear as persons or person-like and sometimes as non-persons, as objects. These shifts are often shifts in hierarchy. In an animist ontology, for example, the status as an object might be nothing but a being's current position within a relationally defined hierarchy – the result of a curse that prevents the object from turning into a person (e.g. Karim 2004). The status of object or person often depends on the question if communication is recognised as such and considered socially relevant. This applies to relations with non-humans in general and to those that are involved in human projects of livelihood in particular.

In this respect, the recognition of hierarchies, especially value hierarchies in relationship to power, is crucial. This applies to gift exchange. Gift exchange is a relationship between persons, human and non-human, that is itself productive of persons. In that sense, it is a variation of what recent theorists in economic anthropology have called 'human economy' (Graeber 2012). A human economy is not defined by the production and consumption of things, but by the reproduction of human beings. As such, the concept still builds upon a humanist separation of objects and humans. A human economy that includes non-humans while leaving the exact hierarchies between them and human beings reversible would be an economy of persons. In these systems, aspects of personhood are unevenly distributed across a field of human and non-human agents, their allocution to beings appearing as a function of the organisation of values and actors into hierarchies. These hierarchies are flexible and reversible, depending on the values and conditions of the exchange.

The distinction between humans and non-humans in these systems is not necessarily denied. In fact, the alterity at the core of gift exchange may create them as persons of different kinds. But these economies of persons differ from naturalist ones not only by the relationality of their ontologies. Gift exchanges do not only create different persons, but also create them through their entanglement, through the transfer of their different aspects. This makes human-non human relationships a social and moral project of mutual obligation. Gift exchange thus appears as the focal process by which persons and socio-cosmic wholes, encompassing humans and non-humans, are being created.

But is this really new?

References

BARRAUD, C. et al. (1994) *Of Relations and the Dead: Four Societies Viewed from the Angle of their Exchanges*. Oxford/Providence: Berg.

BEACH, H. and STAMMLER, F. (2006) Human-Animal Relations in Pastoralism. *Nomadic Peoples NS* 10(2): 6-30.

BIRD-DAVID, N. (1999) 'Animism' Revisited: On Personhood, Environment and Relational epistemology. *Current Anthropology* 40(1): 67-91.

DÄRMANN, I. 2010 *Theorien der Gabe: zur Einführung*. Hamburg: Junius.

DUMONT, L. (1991) [1983] *Individualismus: Zur Ideologie der Moderne*. Frankfurt/New York: Campus (First: *Essais sur l'individualisme*. Paris: Seuil).

FITZ-HENRY, E. (2012) The Natural Contract: From Lévi-Strauss to the Ecuadorian Constitutional Court. *Oceania* 82: 264-277.

GEIRNAERT, D.C. (2002) Witnessing the Creation of Ancestors in Laboya (West Sumba, Eastern Indonesia). In CHAMBERT-LOIR, H. and REID, A. (eds.), *The Potent Dead: Ancestors, Saints and Heroes in Contemporary Indonesia*, pp. 32-47. Honolulu: University of Hawai'i Press.

GRAEBER, D. (2012) [2011] *Schulden: Die ersten 5.000 Jahre*. München: Beck. (*Debt: The first 5,000 Years*. New York: Melville).

HAMAYON, R. (1994) Shamanism in Siberia: From Partnership in Supernature to Counterpower in Society. In THOMAS, N. and HUMPHREY, C. (eds.), *Shamanism, History and State,* pp. 76-89. London: Routledge & Kegan Paul.

INGOLD, T. (2000) *The Perception of the Environment: Essays on Livelihod, Dwelling and Skill*. London: Routledge.

ITEANU, A. (2009) Hierarchy and Power: A Comparative Attempt under Asymmetrical Lines. In RIO, K.M. and SMEDAL, O.H. (eds.), *Hierarchy: Persistence and Transformation in Social Formations*, pp. 331-348. New York and Oxford: Berghahn.

KAPFERER, B. (2010) Louis Dumont and Holist Anthropology. In BUBANDT, N. and OTTO, T. (eds.), *Experiments in Holism: Theory and Practice in Contemporary Anthropology*, pp. 187-208, Oxford: Wiley Blackwell.

KARATANI, K. (2014) *The Structure of World History: From Modes of Production to Modes of Exchange*. Translated by M.K. Bourdagh. Durham and London: Duke University Press.

KARIM, W.-J. (2004) [1981] *Ma' Betisék Concepts of Living Things*. Oxford and New York: Berg.

KISHIGAMI, N. (2004) A New Typology of Food Sharing Practices among Hunter-Gatherers, with a Special Focus on Inuit Examples. *Journal of anthropological research* 60(3): 341-358.

LAIDLAW, J. (2000) A Free Gift Makes no Friends. *Journal of the Royal Anthropological Institute N.S.* 6: 617-634.

LATOUR, B. (2008) [1991] *Wir sind nie modern gewesen: Versuch einer symmetrischen Anthropologie*. Frankfurt a.M.: Suhrkamp [First: *Nous n'avons jamais été modernes. Paris*: La Decouverte].

LATOUR, B. (2011) Waiting for Gaia: Composing the Common World through Arts and Politics. http://www.bruno-latour.fr/node/446. [Accessed: 13rd July 2015].

LÉVI-STRAUSS, C. (1993) [1949/1967] *Die elementaren Strukturen der Verwandtschaft*. Frankfurt a.M.: Suhrkamp (First: *Les structures élémentaires de la parenté*. Paris: Mouton).

LIMA, T.S. (1999) The Two and its Many: Reflections on Perspectivism in a Tupi Cosmology. *Ethnos* 64(1): 107-131.

LUNDSTRÖM, H. and TAYANIN, D. (1981) Kammu Gongs and Drums 1: The Kettlegong, Gongs and Cymbals. *Asian Folklore Studies* 40(1): 65-86.

MAUSS, M. (1990) [1926] *Die Gabe. Form und Funktion des Austauschs in archaischen Gesellschaften*. Frankfurt a.M.: Suhrkamp (First: *Essai sur le don*).

MOEBIUS, S. (2009) Die elementare (Fremd-)Erfahrung der Gabe. *Berliner Journal für Soziologie* 19(1): 104-126.

MUEGGLER, E. (2002) Dancing Fools: Politics of Culture and Place in a „Traditional Nationality Festival". *Modern China Studies* 28(1): 3-38.

NADASDY, P. (2007) The Gift in the Animal: The Ontology of Hunting and Human-Animal Sociality. *American Ethnologist* 34(1): 24-43.
PARRY, J. and BLOCH, M. (1991) [1989] Introduction: Money and the Morality of Exchange. In PARRY, J. and BLOCH, M. (eds.), *Money and the Morality of Exchange*, pp. 1-32. Cambridge et al: Cambridge University Press.
PLATENKAMP, J.D.M. (2004) From Partial Persons to Completed Societies. *Zeitschrift für Ethnologie* 129: 1-28.
PLATENKAMP, J.D.M (2010) Becoming a Lao Person: Rituals of Birth and Socialisation in Luang Prabang, Laos. In BERGER, P. et al. (eds.), *The Anthropology of Values. Essays in Honour of Georg Pfeffer*, pp. 180-200. Delhi: Pearson.
PLATENKAMP, J.D.M (n.d.) The 'Gift' as Model: A Social-Anthropological Perspective on the State and the Market. Paper presented at the workshop *Modern Gift Exchange*, Heidelberg, 12-13 July 2013.
REMME, J.Z. (2016) Actualizing Spirits: Ifugao Animism as Onto-Praxis. In ÅRHEM, K. and SPRENGER, G. (eds.), *Animism in Southeast Asia*, pp. 138-153. London and New York: Routledge.
SAHLINS, M. (1972) On the Sociology of Primitive Exchange. In SAHLINS, M., *Stone Age Economics*, pp. 185-276. Chicago and New York: Aldine-Atherton.
SATHER, C. (1994) The One-Sided One: Iban Rice Myths, Agricultural Ritual and Notions of Ancestry. In WALKER, A.R. (ed.), *Rice in Southeast Asian Myth and Ritual*, pp. 119-150. Columbus, OH: Ohio State Univ., Dept. of Anthropology.
SCHRÖTER, M. et al. (2014) Ecosystem Services as a Contested Concept: A Synthesis of Critique and Counter-Arguments. *Conservation Letters* 7: 514-523.
SPRENGER, G. (2006) *Die Männer, die den Geldbaum fällten: Konzepte von Austausch und Gesellschaft bei den Rmeet von Takheung, Laos*. Münster: Lit.
SPRENGER, G. (2014) Kosmologische Zoogamie: Zur Ehe zwischen Menschen und Tieren. *Paideuma* 60: 25-44.
SPRENGER, G. (2016) The Birth of Strangers: Exo- and Endosociality in Rmeet Birth Rituals, Laos. In ALEX, G. and POLIT, K. (eds.), *Childbirth and Its Accompanying Rituals*, pp. 16-36. Heidelberg: Draupadi.
STÉPANOFF, C. (2012) Human-Animal 'Joint Commitment' in a Reindeer Herding System. *HAU* 2(2): 287-312.
SULLIVAN, S. (2010) 'Ecosystem Service Commodities' - A New Imperial Ecology? Implications for Animist Immanent Ecologies, with Deleuze and Guattari. *New Formations* 69(6): 111-128.
TAPPER, R. (1988) Animality, Humanity, Morality, Society. In INGOLD, T. (ed.), *What is an Animal?*, pp. 47-62. London: Unwin Hyman.
TERWIEL, B. J. (1994) Rice Legends in Mainland Southeast Asia: History and Ethnography in the Study of Myths of Origin. *Contributions to Southeast Asian Ethnography* 10: 5-36.
TEUBNER, G. (2006) Rights of Non-Humans? Electronic Agents and Animals as New Actors in Politics and Law. *Journal of Law and Society* 33(4): 497-521.
VIVEIROS DE CASTRO, E. (2004) Exchanging Perspectives: The Transformation of Objects into Subjects in Amerindian Ontologies. *Common Knowledge* 10(3): 463-484.
WIDLOK, T. (2013) Sharing: Allowing Others to Take what is Valued. *HAU* 3(2): 11-31.
WILKIE, R.M. (2010) *Livestock/Deadstock: Working with Farm Animals from Birth to Death*. Philadelphia: Temple University Press
WILLERSLEV, R. (2007) *Soul Hunters: Hunting, Animism, and Personhood among the Siberian Yukaghirs*. Berkeley et al: University of California Press.

Labour and Land in an Eastern Indian Steel Town: Historical Transformations of Gift and Commodity Values

CHRISTIAN STRÜMPELL

Steel-Making and State-Making in Independent India

Until the 1950s Rourkela had been a small settlement in the hills of the eastern Indian state Odisha, far off the urban centres of the subcontinent and considered in the latter's elite circles a backward periphery. The region was also rich in minerals and Rourkela had already been connected by rail to Kolkata and Mumbai, so that it became the obvious choice for establishing independent India's first integrated steel works. Steel is a key commodity that India needed to produce herself in order to gain economic autarchy from the advanced industrial powers of the West. India's first prime minster Jawaharlal Nehru was convinced that only this would maintain the country's recently hard-won political independence (cf. Parry 1999; Nayar 2004[2001]: 50-85). Moreover, steel works such as the one at Rourkela were supposed to make a fundamental contribution to nation-building in a far broader sense. They were to integrate India's internal peripheries closer into the nation-state, and at the same time provide a model for the integration of the nation as a whole. They would attract workers from all over the country who by their common labour in the steel plant and by their common living in the township that the state established nearby would transcend their many distinctions of caste, ethnicity and religion and turn into a model citizenry.

The foundation of the steel plant in the 1950s as well as its operation ever since rest on a series of exchanges of knowledge, capital, land and labour: in 1954 the public-sector company engaged a consortium of West German steel companies – later on joined by the West German government – to provide industrial expertise and equipment required to build the steel plant-cum-township as well as some capital for it (cf. Röh 1967); the state government acquired the land for the urban-industrial complex from local villagers and handed it over to the RSP and the latter employs tens of thousands of workers to run the steel plant and the attached facilities. The crucial point is that all three exchanges exhibit features of commodity exchanges as well as features akin to gift exchanges. Thus, the RSP paid the West German companies for their services at market rates, but the West German government backing the project with federal guarantees framed it as developmental aid (ibid.); the land for the urban-industrial complex was compulsorily acquired from local villagers by the Odisha state government which fixed the monetary compensation at the land's market value, but also offered alternative fields and housing plots as well as permanent employment with RSP; finally, the RSP offers its permanent workforce well-paid and secure government employment, but since the 1970s also increasingly relies on an ill-paid, temporary workforce supplied by private labour contractors depending on the market situation. My aim in this article is to scrutinise the gift and commodity values inherent in the exchanges involving land and labour. During the time of my long-term field research in Rourkela,[1] they were of major concern for the people in Rourkela, not least because their values had been significantly transformed over the last decades. In this endeavour I will draw in particular on the work of Josephus Platenkamp for whom a historically-informed and comparative approach to exchange, inclusive of modern industrial societies, is a long-standing concern.

Gift and Commodity Exchange

In a suggestive paper on the relationship between gift and commodity exchanges, Platenkamp (n.d.) draws on the thesis of the co-existence of different transactional cycles that Maurice Bloch and Jonathan Parry develop in *Money and the Morality of Exchange* (1989). Here, Bloch and Parry (ibid.: 23ff) distinguish between, on the one hand, the short-term cycle of exchange in which competitive, individual strife for material gain is legitimate, and on the other the long-term cycle of exchange in which the reproduction of the socio-cosmic order is at stake and in which such behaviour is considered a social threat. Moreover, as the contributors to the volume show, in various non-western societies both transactional spheres are hierarchically interrelated, the short-term cycle being subordinated to the long-term one. Whether the protagonists are Andean peasants (Harris 1989) or Malay fishermen (Carsten 1989), they all evaluate self-interested acquisitions as legitimate or even laudable only if part of them is fed into long-term exchanges with deities, ancestors or households; whereas transformations in the reverse direction are highly condemned. As Bloch and Parry (1989: 29-30) further suggest, capitalism transformed this hierarchy, not necessarily by conflating both spheres into the Mandevillean proposition that public benefit derives from private vice, but by perceiving both cycles so distinct that it appears impossible to relate them. In contrast to that, Platenkamp (n.d.) argues that a hierarchy between different transactional spheres prevails also in capitalist societies, that also here commodity exchanges continue to be subordinate to gift exchanges.

As a prominent example Platenkamp (n.d.: 9ff) refers to money and taxation. First, unlike the prices paid for commodities in the market, taxes are not

1 My field research in Rourkela extended over more than 30 months between 2004 and 2014. I gratefully acknowledge the support of the Deutsche Forschungsgemeinschaft and the Max Planck Institute for Social Anthropology. I thank Sandra Bärnreuther and Guido Sprenger for their comments on an earlier draft of this paper. I also remain deeply indebted to Rajat Singh and Ganesh Hembram for their invaluable research assistance.

paid in proportion to the services the taxpayer receives from the state. Second, unlike – in capitalist ideology – anonymous, socially independent actors on a market, it is as a nation's citizens, or as social persons, that people pay taxes. Third, unlike – in capitalist ideology – the symmetrical relationships between sellers and buyers of commodities, tax payments are representative of the asymmetrical relationship between the tax-giving citizen as a part to the tax-receiving nation-state as a whole. Fourth, unlike the alienability of commodities in the market, taxation rests on the idea of the inalienability of money from the state, or its central bank, as its original owner. This finds expression in the immaterial value embodied in the currency's iconography depicting the nation's 'glorious history' and its 'transcendent origins'.[2] Thus, Platenkamp concludes, though gained in commodity exchanges, money is transformed through taxation into a gift (ibid.). Furthermore, "money-valued-as-commodity" also remains subordinated to "money-valued-as-gift". This is revealed by the simple fact that the inalienability of money from its original state owner is in fact the precondition for market actors to trust its stability and to accept it in exchange for commodities (ibid.: 11). Hence, consumers and producers both handle money as commodity only because it is at the same time also valued as gift.

However, despite the fact that the use of money does not necessarily imply its evaluation as commodity, as the contributions to Bloch and Parry (1989) show in great detail, not even in capitalist societies, as Platenkamp argues, the assumption that it does remains deeply ingrained in modern, western thought. Thus, Platenkamp (2016) criticises that time and again archaeologists, for instance, over-hastily interpret deposits of coins merely as indicators of a society's integration into markets and trade networks. In another paper (2012), he shows how Dutch missionaries in the colonial North Moluccas considered transactions of 'living money' from wife-takers to wife-givers a sale of women and aimed to prevent it. In fact, however, 'living money' was transferred as an acknowledgement of the inalienable relationship between the fertility of the bride and the ancestral owners of the domain, as an acknowledgement of the latter's cosmological sovereignty, not as an equivalent to an alienable commodity. The Dutch, with their universal understanding of state sovereignty and market economics, were bound to misinterpret this exchange.

My focus in this article, too, lies on conflictive understandings of an exchange relationship. In this case, however, the disagreement is between the exchanging parties themselves: the postcolonial state and its administrative elite, on the one hand, and the displaced people at Rourkela on the other. This disagreement has intensified since the 1990s when India embarked upon a project of 'economic liberalisation' that integrated her economy more closely into global markets. As I will show in the following sections, this entailed the

2 Thus, currencies depict famous monarchs, scientists or artists and religious or other ideological symbols.

massive reduction of the regularly employed RSP workforce or the transformation of labour-valued-as-gift into labour-valued-as-commodity.

Labour as Gift and Commodity

Platenkamp (n.d.) defines the relations established through gift exchanges by a series of attributes that all apply to Nehru's vision for the Rourkela Steel Plant. First, the Government of India created the RSP, just like other public sector companies, to build a model citizenry, and through it the nation. These companies are thus also about the production of social relationships, not only commodities; they create social value, apart from utility value. Second, the Government of India, and through it the people of India as the sovereign, remains the owner of these public-sector companies. That means ownership of the RSP remains inalienable, though it is in possession of the public-sector holding company, the Steel Authority of India Ltd., (SAIL).[3] Third, SAIL, as the possessor of the RSP, acts as a custodian of its original owner, the Government – and through it, the people – of India. This is true also for the RSP itself and for its workforce. They all act in the interest of, and hence are guardians of, the nation as a whole. Fourth, the relationship between the original owner and the various mediating possessors of the gift is thus coined by transitivity. Fifth, the steel plant, the township and the employment RSP offers is the material form that embodies the immaterial value of a secular, 'socialistic' industrial modernity, which, sixth, originates from the Government of India as its original owner. Seventh, and lastly, the secular, 'socialistic' industrial modernity embodied in the gift is only a part of that value as a whole belonging to the state as its original owner. Thereby, all receivers of that gift turn into a community of people sharing a common relationship to the original owner.

The state's gift of secular, socialist industrial modernity by the state is reciprocated by the gift of labour. As public-sector workers, RSP workers enjoy protection by labour laws as well as, relatively speaking, excellent wages and fringe benefits (Parry 2009: 179). These are negotiated between the public-sector holding company and trade union representatives from all SAIL units (ibid: 189) with the state acting as the compulsory arbitrator in cases of conflict (ibid: 180). The privileged treatment of this section of India's industrial workforce goes back to a pragmatic attitude of the nationalist leadership towards workers in the country's key industries, most of which were in the public sector (ibid; Breman 1999: 31).[4] However, it also derives from the conviction that the state and its public-sector companies should act as benevolent model employers who do not treat 'labour' as a commodity (Parry 2009: 179-

3 Until 1974 the holding company was Hindustan Steel Ltd.

4 By treating the 'labour power' of public-sector workforces as a 'counter-gift', not a commodity, the state also aims to be gifted with recognition by its citizens as legitimate power.

80; cf. Rudolph and Rudolph 1998[1987]: 284). Indeed, in significant ways their labour does not appear alienated: it is a public-sector company that appropriates the product of their labour as national property, not a capitalist as private property. Since time and work discipline is not very stringent their working time does not belong exclusively to their employer; and since they have lifetime jobs with lifetime workmates to whom they are also neighbours in the company town, RSP workers are also not alienated from each other. Furthermore, another crucial attribute of gift exchange in Platenkamp's terms is the distinction between the material substance of the gift object, here the labour of the RSP workforce and its products, and the immaterial value it embodies, its labour power and creative potential, the original owner of which remains the RSP workforce.

Of course, RSP workers (as well as executives) are involved in other gift relationships, too. They pay taxes and make donations to charitable institutions; they conduct or sponsor sacrifices and they give and receive offerings etc. at temples, mosques, churches or other sacred abodes; they establish affinal relationships and make prestations during marital, mourning or other life-cycle rituals, to name just the most prominent examples. The RSP workforce of course also engages in myriads of commodity exchanges. They purchase their essential and white goods, their mobile phones and bikes, etc. on markets; they also purchase the services of housemaids, rickshaw-drivers etc. at market prices; and all those who do not reside in the company town's quarters either pay rents or they have bought or leased homes. In fact, RSP workers are courted as customers and also as donators, patrons, and in-laws, because of their secure, well-paid and 'modern' RSP employment. Thus, a bricklayer with 15 years of work experience earns up to 5,000 rupees (75 euros) per month,[5] if he is employed throughout the month, which is not the rule. In contrast, a RSP worker with 20 years of service and in pay-grade 8 (out of 11 among non-executives) receives a gross salary of monthly 25,000 Rs. He will do so until his retirement at the age of sixty on which he will receive a substantial lump sum and beyond which he and his wife will continue to receive cost-free treatment at the company hospital.

However, since the 1990s the RSP has continuously reduced its manpower, from 37,000 employees in 1989 to 19,500 in 2009. The RSP achieved this reduction through natural attrition, not retrenchments. In the late 1990s, the first generation of RSP workers started retiring and it was only partly replaced with a new one. This drastic policy was itself the outcome of the wider reorientation of India's economic policy in the 1990s, away from the Nehruvian legacy of a protected national economy led by a strong public sector, towards a closer integration into global markets. Among other actions, the so-called "economic liberalisation" entailed in particular the restructuring of India's

5 The numbers refer to the situation in October 2008.

public sector companies, their privatisation or their transformation into undertakings that prioritise the generation of profits, not employment (Nayar 2004 [2001]: 129-155). The effect was that the children of retired RSP workers had to seek jobs in the private sector.

The private sector is, of course, very heterogeneous and encompasses well-paid, secure employment in, to name a prominent example, India's IT industries, but also work in construction, in transport, in smaller, privately-owned steel plants, or also in the RSP, but employed as so-called contract labourers by privately operating contractors, not by the RSP itself. Apart from the IT industries, none of which is settled at Rourkela and of which I therefore have no ethnographic experience, the jobs these industries offer are rarely covered by labour laws, thus leaving workers without minimum wages, sick pay, maternity leave or also compensations for industrial accidents. The people working on these jobs have no doubts that their labour is treated as a commodity, that they sell their labour power to someone using it to make profit and that they offer their labour power in competition with others. However, the question remains whether, even under these circumstances, labour in its commodity value is nevertheless encompassed by its gift value.

Platenkamp argues that in modern capitalist societies the hierarchical encompassment of commodity values by gift values imbues all kinds of exchange objects, not only money. This is so, because the seller of a commodity is economically successful only if he is able to convince buyers that his product relates them to a valued whole beyond the mere commodity. As Platenkamp shows, this is standard advertising practice and as an example he discusses the advertisement of an aftershave that promises its buyers the scent and freshness of the ocean, the material substance of an immaterial value whose origin and owner remains the ocean. In this sense, also the labour bought and sold as a commodity remains encompassed by its gift value. On the one hand, companies, contractors and others in Rourkela (as elsewhere) purchase the material substance 'labour' because it relates them to the immaterial value of the creativeness of human labour power that they seek to turn into a source for profit. Construction workers, on the other, accept money in exchange for their labour only because the political community comprising the Indian state backs it.

However, despite its subordination to the gift value of labour, the evaluation of labour as commodity makes the relationship between employers and labourers on construction sites, in private rolling mills or in iron ore transport businesses in Rourkela fundamentally different from the relationship between the RSP and its regular workforce. With their labour valued as commodity workers are indeed alienated from the products of their labour as well as from their work.[6] Thus, as Polanyi famously argued in *The Great Transformation*

6 Whether they are also alienated from each other remains debatable. Very often workers are recruited through caste and village networks (cf. Parry 1999).

(2001 [1944]: 76-77), although labour – like land and money – is only a 'fictitious commodity' – because none of it is produced for sale – this fiction nevertheless forms the cornerstone for the actual organisation of the utopian 'self-regulating' market. Following Platenkamp, we could argue that all commodities are fictitious, not because none of them are produced for sale, but because all of them are also gifts. Interestingly, the gift value in Platenkamp's example derives from the products ultimate origin in nature, the ocean, which in Polanyi's scheme renders its commodity value fictitious.

Polanyi also argued that the history of 19th century capitalism in Britain is the history of a double movement; the proliferation of markets for genuine commodities, on the one hand and on the other hand, society's commitment to restrict the markets for the fictitious commodities labour, land and money, which threatened it in its very existence (ibid: 79-80). The commodification of exchange objects, however encompassed by their gift values, is a contingent historical process. In Rourkela, India's economic liberalisation and its concomitant manpower reduction of the RSP has furthered the evaluation of labour as commodity and restricted its evaluation as gift. As Polanyi would have foreseen, this triggered attempts to avert this process. Thus, in the late 1980s contract labourers working on the premises of the RSP under privately operating contractors, not the RSP, demanded their regular employment by the RSP, or, in other words, the decommodification of their labour; and after a Supreme Court judgement in 1995 half of the RSP's total contract labour force or 4,500 workers were indeed regularised (cf. Strümpell 2012: 211ff).

Land as Gift and Commodity

Around the same time, the 'displaced people', i.e. the people whose land the state government had compulsory purchased for the establishment of the whole urban-industrial complex, also raised claims to RSP employment. The people living in and around Rourkela before its industrialisation belonged in their majority to several so-called scheduled tribes, but some also to a variety of castes, and they were Muslims, Christians, Hindu or followed a local religion. Despite these differences socio-economic inequality was much less pronounced than elsewhere in India. There were only few large landowners and landless, and the vast majority consisted of smallholders who seasonally also worked in nearby mines and quarries.[7]

Like the contract labourers, the displaced people voiced their demands in courts and on the streets, in the latter case often in the face of repressive state action. They demanded – and still do so today – regular RSP employment, because of the inadequate compensation they, or their forefathers, had received for their land. Civil servants as well as RSP managers, bank officers or

7 Cf. Government of Orissa's Report on Rourkela 1959.

Rourkela's middle class in general claim the opposite, that compensation far exceeded constitutional dues.

The colonial Land Acquisition Act, 1894,[8] together with the Orissa Act XVIII of 1948[9] that applied for the land acquisition at Rourkela granted the government the power to acquire all land required for public purposes and the affected landowners compensation in form of the market value of their land as on the date of the preliminary notification announcing the acquisition. In addition to that, in Rourkela the people displaced by the RSP, its township and allied industries[10] where granted house plots in resettlement colonies adjacent to Rourkela, alternative land in reclamation camps elsewhere in the administrative district as well as employment of the 'able-bodied' among them, first, for the construction of the RSP and, later on as regular workers. These additions were of course in the spirit of Nehru's vision of the RSP as a 'mini India' that also levels regional disparities. However, they were also the outcome of early local protest against land acquisition and subsequent negotiations of the terms of compensation (Strümpell 2014).

In contrast, the displaced people complain that the compensation rates were meagre compared to what the land reaps now; that the reclamation camps were far off, that the land there was often barren if not already under occupation; and that the RSP had employed only few of them. In 2004, when I began ethnographic research in Rourkela and witnessed the then vigorous protest of the displaced people and their descendants, their main concern was the provision of RSP employment and the return of 'surplus land', i.e. land acquired for the RSP, but not under its occupation, to its original owners, i.e. to them. The demand for land was a result of rising land prices and the desire of the RSP, the government, real estate developers and many others including the displaced people to cash in on it. Among the latter it was also an attempt to ward off the eviction threats they regularly faced in case they settled on surplus land, which many did, but until returned into their possession only as illegal encroachers. The demand for RSP employment was a response to the manpower reduction discussed above, which had hit the displaced people particularly hard. Most of them had grown up as children of RSP workers, but in either the resettlement colonies or in the slums into which the village sites on surplus land had transmuted, not in the RSP company township. Here, outside the township with its well-equipped schools, they had not gained the education that was now required to either apply for the reduced number of regular RSP jobs or to aspire to comparatively well-paid jobs elsewhere. Hence, it was

8 This act remained valid until 2013.

9 The full form of that act is 'Orissa Development of Industries, Irrigation, Agriculture, Capital Construction & Re-settlement of Displaced Persons (Land Acquisition) Act, 1948'.

10 The Orissa Act XVIII amends the 'public purpose' clause to include allied and ancillary industries regardless of whether they are in the private or public sector.

people badly positioned on the local labour market who struggled to escape that market by (re)gaining a gift relationship with the RSP.

Most of the second generation of displaced people grew up as children of regular RSP workers, who had now retired. However, they argue that this doesn't contradict their claims for compensatory RSP employment. Their fathers had been employed, because the RSP needed workers, especially for the 'hot shops' such as coke ovens, blast furnaces and steel melting shops, where working conditions were considered particularly tough and for which they as tribals were – in the eyes of non-tribal personnel managers – particularly suited for (Strümpell 2014). Furthermore, as the displaced people never tired of saying, their land was a means of livelihood for which regular RSP employment for one person is not an adequate compensation. Unlike their public-sector jobs, their land was inheritable and only if their jobs will be, too, will they be adequately compensated.[11] In other words, their relation to their land and the livelihood it provides was permanent and only a permanent relation to the RSP through permanent employment across generations would render their compensation of equivalent value to what the state has taken from them. In this view there is no price for their land, because the price would have to cover the livelihood costs of their descendants ad infinitum. They thus value their land as a gift linking transitively all generations as receivers to their ancestors as original owners, and they demand a compensation that would substitute for this gift in this relationship, i.e. that they can pass on their RSP employment to future generations. This conceptualisation resembles what Platenkamp (n.d.: 4) calls an equivalence model of exchange of the substitute type, i.e. one in which the reciprocal gift is a substitute for the immaterial value embodied in the original gift, in this case livelihood or reproduction embodied in both the gift of land as well as the substitute counter-gift RSP employment.

The displaced people bring their claim for permanent RSP employment to court, but in 1995 the Supreme Court finally rejects it as violating the constitutional equality of the Indian citizenry (cf. Strümpell 2012: 217). In Rourkela itself, the usual response of RSP managers, civil servants, lawyers and other middle-class people including RSP workers and also some unionists is to brush away the displaced people's claims for RSP employment as baseless. For them, the displaced people's current lack of access to RSP or other 'respectable' employment lies in the latter's tribal nature. They never tire of telling that the displaced people mainly live for the moment, drink and dance, but never care about the morrow, let alone about saving money or educating their children. They complain that the state and the RSP gave these people every-

11 This was the claim of the first case of the 1990s the displaced people launched at the Indian Supreme Court. I discuss this as well as some subsequent cases in more detail in Strümpell 2012.

thing, good employment, schools and a whole town, but that they did not develop any sobriety, diligence or thrift; that they drank away their compensation payment and later on their wages, but that they stayed away from duty and thus lost their jobs quicker than they got them. These vituperations are of course, if at all, only partly true. Many among the elder displaced people also told me about how outsiders cheated them in joint business operations; many more told me that though they worked for RSP they all ended up at the bottom of the ladder in the 'hot shops' and that this was the reason why many drank and quite some quit their jobs. However, the above mentioned claims of Rourkela's middle class staffing local state offices and other influential positions reveal that they consider the protesting displaced people as defaulting on the state's and the RSP's gift of industrial modernity as they interpret it. The middle-class allegations suggest that in their eyes the acceptance of the gift of industrial modernity entails the obligation to reciprocate by transforming oneself into a 'modern', decent and diligent industrial worker, i.e. by labouring on oneself, not only on coke ovens and steel converters.[12] Thus, both the displaced people and the local elite claiming to represent the state argue that the respective other side failed to meet their obligations arising from the gift exchange between them.

Asymmetrical Recognition: A Concluding Remark

Whereas with regard to labour the disagreement in Rourkela is on its evaluation as gift or commodity, with regard to land it is on the particular nature of gift relationship its exchange entails. In both cases there is what Chris Gregory (1997: 65f) calls 'asymmetrical recognition' concerning the significance of the exchange. Gregory argues that asymmetrical recognitions become more likely in case the parties involved in an exchange are alien to each other. Indeed, in the eyes of Rourkela's (non-tribal, higher-caste) middle class the displaced people are indeed a 'tribal' Other just like the latter consider the former as *diku* ('foreigners'). However, the trajectory of events in Rourkela reveals that asymmetric recognition is also a historical phenomenon. It was in the 1990s, when India's public-sector industries were restructured into profitable, competitive players on global markets, that conflicting understandings regarding the evaluation of labour and land became acute.

12 Few of the people in Rourkela raising allegations against the displaced people at least nowadays subscribe to the socialist dimension of Nehru's vision, though some more to its secular ones.

References

CARSTEN, J. (1989) Cooking Money: Gender and the Symbolic Transformation of Means of Exchange in a Malay Fishing Community. In BLOCH, M. and PARRY, J. (eds.), *Money and the Morality of Exchange*, pp. 117-141. Cambridge: Cambridge University Press.

GOVERNMENT OF ORISSA. Political and Services (Planning) Department. 1959. *Report on Rourkela*. Cuttack: Orissa Government Press.

GREGORY, C. (1997) *Savage Money: The Anthropology and Politics of Commodity Exchange*. Amsterdam: Harwood Academic Publishers.

HARRIS, O. (1989) The Earth and the State: The Sources and Meanings of Money in Northern Potosí, Bolivia. In BLOCH, M. and PARRY, J. (eds.), *Money and the Morality of Exchange*, pp. 232-268. Cambridge: Cambridge University Press.

NAYAR, B.R. (2004 [2001]) *Globalization and Nationalism. The Changing Balance in India's Economic Policy*, 1950-2000. New Delhi: Sage.

PARRY, J.P. (1999) Lords of Labour: Working and Shirking in Bhilai. In PARRY, J.P., BREMAN, J. and KAPADIA, K. (eds.), *The Worlds of Indian Industrial Labour*, pp. 107-140. New Delhi: Sage Publications.

PARRY, J.P. (2009) "Sociological Marxism" in Central India: Polanyi, Gramsci, and the Case of the Unions. In HANN, C. and HART, K. (eds.), *Market and Society: The Great Transformation Today*, pp. 175-202. Cambridge: Cambridge University Press.

PARRY, J.P. and BLOCH, M. (1989) Introduction: Money and the Morality of Exchange. In PARRY, J.P. & BLOCH, M. (eds.), *Money and the Morality of Exchange*, pp. 1-32. Cambridge: Cambridge University Press.

PLATENKAMP, J.D.M. (2012) Sovereignty in the North Moluccas: Historical Transformations. *History and Anthropology* 23(3): 327-353.

PLATENKAMP, J.D.M. (2016) Money Alive and Money Dead. In HASELGROVE, C. and KRMNICEK, S. (eds.), *The Archaeology of Money*, pp. 161-181. Leicester: Leicester Archaeology Monographs.

PLATENKAMP, J.D.M. (n.d.) *The 'Gift' as Model: A Social-Anthropological Perspective on the State and the Market*. Unpublished conference paper.

POLANYI, K. (2001 [1944]) *The Great Transformation. The Political and Economic Origins of Our Times*. Boston: Beacon Press.

RÖH, K. (1967) *Rourkela als Testfall für die Errichtung von Industrieprojekten in Entwicklungsländern*. Hamburg: Verlag Weltarchiv Gmbh.

RUDOLPH, L.I. and RUDOLPH, S. (1998 [1987]) *In Pursuit of Lakshmi. The Political Economy of the Indian State*. New Delhi: Orient Longman.

STRÜMPELL, C. (2012) Law against Displacement: The Juridification of Tribal Protest in Rourkela, India. In ECKERT, J., DONAHOE, B., STRÜMPELL, C. and BINER, Z.Ö. (eds.), *Law Against the State. Ethnographic Forays into Law's Transformations*, pp. 202-227. Cambridge: Cambridge University Press.

STRÜMPELL, C. (2014) The Politics of Dispossession in an Odishan Steel Town. *Contributions to Indian Sociology (n.s.)* 48(1): 45-72.

Exchange and the Constitution of Persons under a Welfare Regime

TORBEN A. VESTERGAARD

That individualism and Human Rights are the basic values of the modern west is often declared and celebrated. It is also an overstatement. It may be a useful outset for comparative purposes with non-western societies, particularly when it is at the core of political antagonisms as in Jos Platenkamp's paper on values and peacekeeping tasks in Militaire Spectator (Platenkamp 2007). However, there are rifts and cleavages revealing rival or conflicting value systems within the West. Depending on who you are, your short term social transactions may have to articulate with more than one system of values.

The association *EuropeanValues.info* recognises a historical hierarchy of values as the basis of freedom in Europe. With its point of origin in classical Greece, the historical hierarchy develops from the value of humanistic thinking through rationality, secularity, rule of law, democracy to human rights. This partly mythological trail of cultural history, however, leaves out other systems of value and social relations that have not all disappeared.

These European values pertain to the whole or the long term order of transactions: the social order (Parry and Bloch 1989: 23ff). They do so in a contradictory way, since they have the smallest part, the individual as the highest value. This means that the part relates to the whole by being identical with it. You simply have to be human to represent the highest value. In terms of economy a similar identification applies: market economy after Adam Smith identifies the value of the short term self-interested transaction with the long term good. There are, however, other domains of social activity, where the articulation of transactions with human rights is not simply that of identity. And then there are traditional holistic systems of values that the same citizens may be involved in or long for. Some of them are demonised, others just live quietly in the spaces left for them in the human rights based political order.

The individualist human rights values do leave a space for some articulations of traditional holistic values (see also Dumont 1986). Europe houses a number of large religious confessions, particularly Christian and Muslim. The space for religion within the modern system of values is placed in the privacy of the individual. Religious institutions are compartmentalised separately, and particularly from politics. In countries like Denmark, the Lutheran Christian church is not clearly separated from the state. Therefore the political question of separating the two turns up regularly. Not that the Church threatens the State, but as a matter of principle it does offend some political groups. If the connection between religion and politics were elsewhere as harmless as this, we could rejoice.

If holistic systems of value would remain quietly in the compartment of private individual matters or as a separate institutional domain, this could be a model of coexistence between value systems. It does not seem to existentially satisfy the lives of all inhabitants in Europe. Depending on social position and occupation, social life within the modern configuration of values may be a bland and unheroic affair. The quest for honour and image may find more promising fields in nationalist, separatist, religious or armed movements.

In some cases traditional and modern European system of values coexist smoothly, and the transition may be relatively peaceful. In my earlier work on the fishing profession in Denmark, I was puzzled by frequent references to royalty and use of the term estate (*Stand*), both of them representations of society as a traditional whole. Other occupations and professions, too, would include references to the monarch and to society as an organic whole in their self-representation, for instance the army. The combination of democracy and monarchy in a constitutional monarchy may in fact allow for the smooth coexistence of otherwise antagonistic systems of value. It does seem remarkable that the demise of monarchies in Europe in the early 20th Century was followed by quite unfortunate political developments.

The case of the fishermen is very clear cut for the demonstration of combined holistic and individualist value systems. It is a high profiled occupation that constitutes fishermen as individuals and as a group in society. So, the fishermen may here serve as the traditional modernist model.

The Values of Fishermen

The configuration of values in the fishing industry in Denmark from the late 19th Century to the late 20th Century was not simply traditional (Vestergaard 1990, 1999). Fishing was a commercial enterprise and its practitioners celebrated their individual freedom and the open access to marine resources. This has contributed to the image of Danish fishermen as pure capitalist entrepreneurs which they shared with North European fishermen generally. The contrasting case in fisheries literature is that of Mediterranean fishermen and their communitarian *cofradias* and *proud'hommies* (Feral 2004). However, the fishermen in Denmark did, nonetheless, consider themselves as subordinate contributors to society as a whole.

Paradoxically, it was the modernisation of the fishing industry from the 19th Century that placed the fishermen in a position as partners in a holistic system of exchange. In the short term order of transactions fishermen caught fish at sea and sold them for money in land. But the credit structure had other dimensions than the money, goods and services as such. The money value was not the only value to the fishermen. Fishing made them, who they were: contributors of food, *life* for society, for which they received significant symbolic recognition, honour or *image*, to speak with Jos Platenkamp's terms (Platenkamp 1988, 2006).

The credit structure in the fisheries was not just between the fisherman and the fish buyer; it was between the fishing estate (*Stand*) and society, i.e. the Monarch. There are historical as well as structural explanations for the situation (Vestergaard op.cit.). From the late 18th Century the Absolutist government and the men of progress campaigned for the development of fisheries and of any industry that could benefit the state. The fisheries did not take off until later, but the heroising of the fishermen began already then. A widely used patriotic school book glorified fishermen of North Sealand for rescuing shipwrecked English sailors. This again became the theme of a theatre play popular throughout the 19th Century. From the mid-19th Century, formally organised rescue services manned by fishermen consolidated fishermen's image as rescuers of life from the dangers of the sea. And they received medals of honour for their deeds from private as well as royal sponsors. Conscription of fishermen for naval service also underscored their position as an estate in the service of the State against sea-borne threats from the outside.

During the late 19th Century, the fishing industry took a development stride forward in terms of technology, economy and demography under the aegis of the Danish Fisheries Association. The association was initiated by nobility, officers, merchants and senior civil servants. Fishermen were called upon to develop their industry for the sake of society and for their own sake. His Royal Highness Prince Valdemar was the Honorary President of the Association. With his wife, Princess Marie, he organised very substantial fund raising to compensate fishermen for losses of boats in a storm in 1894.

In due course (1913) the Fisheries Association became the interest organisation of the fishermen themselves. Nonetheless, it continued to portray its members as an estate (*Stand*) in an organic whole of labour divided contributors to the country. Visibility or image also played an explicit role. In 1935-36 a two-volume publication, Dansk Fiskeristat, presented the industry, its local associations and biographies of 4800 fishermen including portrait photos of the fishermen in their best clothes or in navy uniforms. The introduction said: "Come and see who we are, and learn about our circumstances" (Strubberg 1935: Forword). Similar biographical calendars were published for many other industries and professions at the time, and it was repeated in a wave of biographical calendars around 1960.

The division of labour in society obviously implies an organic whole, but in the modern configuration of values it is kept neutral, and does not in principle involve a person's standing in other domains of society. But it used to, and it is not really over. The trades and industries that contribute to the national economy are still held in esteem for their contribution to the whole.

For fishermen, however the situation was more extreme for structural reasons. The fishermen operate at sea, suffering very real hardships and danger of losing their lives. In fact, fishermen drown every year. Their work takes place outside the land where, in terms of segmentary logic, they represent not their town, but their country. Furthermore they operate their gear under water in an invisible domain requiring knowledge of things invisible. So, outside of society they extract life from the claws of another element and bring it home to society. Out of the same segmentary logic, it is only fitting that they should draw the grace of a higher representative of society than the local mayor, and indeed the Monarch has responded. Apart from the early sponsorship or protection from the royal house, fishermen received attention in terms of medals, compliments in New Year speeches and royal visits to ports. At the congresses or general meetings of the Danish Fisheries Association a telegram was always sent to the King or Queen. Later in the day a complimentary telegram would arrive from the Monarch and be read aloud for the standing assembly.

So the short term transaction of fish for money was part of a long term, holistic exchange of values between the fishermen and the state. Fishermen provided food, rescue and naval defense service, i.e. life, and received in return recognition, visibility, name, honour, i.e. image. At the same time they were prize representatives of individual free enterprise and technical progress.

Things changed with the advent of EC and EU regulations of the fisheries. From the 1970'ies onwards regulations that deprived them of their entrusted responsibility for marine resources were gradually imposed. Mechanical systems of quotas were imposed on an increasing number of fish species. This prevented fishermen from pursuing the usual flexible strategies of switching between species and waters. Fishermen generally took offence for now being portrayed as self-interested threats to marine resources. Bankruptcy because of incompetence seemed more acceptable than bankruptcy caused by barred strategies. A fisherman wrote a letter to the editor of a paper suggesting that he might have to write a letter to the Queen, since the lower echelons of government imposed impossible regulations (Fiskeritidende 25 April 1996: 2). The Queen was the apparent representative of the appreciated social order of exchange and values.

Since 1977 the number of professional fishermen has dropped with 80 % from about 10.000 to 2000 (Statistisk Årbog 1980: Tabel 103; Fiskeristatistik 2012: Antal beskæftigede). Fish resources at sea have in the meantime been privatised and concentrated on fewer owners. Willingly or not, the remaining fishermen have become capitalist entrepreneurs. In the process, the relationship between producer and consumer interests has been inverted. The Ministry of Fisheries has become of part of the Ministry of Food, the research assignments of the Fisheries Research Institute have shifted towards consumer oriented themes. Civil servants and politicians are no longer the helpful patrons of the producers, but rule sticklers for EU and the consumers. The exchange of life for image between fishermen and the state has deteriorated. The term estate (*Stand*) does not turn up very often any more as a collective identity. The individual distinction of receivers of medals for rescue efforts or contributions to the advancement of fisheries let alone biographical calendars with individual portraits is hardly to be expected either.

The Value of Welfare

Though less pronounced than in the fishing industry, other industries have experienced a similar change from being national projects to being merely economy. Part of the explanation is that the economy is much less a national system of exchange than it used to be. The discourse on social value has shifted to welfare for the consuming citizens rather than obligations of producers. Unlike the economy, the welfare system is still predominantly a national system of exchange. How does the welfare system relate to social value and systems of exchange? Various arguments could be suggested.

Tax money is important for the financing of welfare, but seems not to involve the parties in a gift giving logic. To the apparent regret of some tax payers, they and welfare receivers are with some success kept in separate domains. Welfare has become a right rather than a gift, which liberates the receivers from embarrassment. The perspective lurks in there any way. This would explain the insistence of the Danish Social Democratic party that rights should be accompanied by obligations also for welfare receivers.

Is there any holistic dimension to the welfare exchange of services? Providing or receiving welfare prestations do affect the social identity of the parties involved. The constitution of persons in society could expectedly involve exchange relations with transcendent or long term sources of being and value. Medical doctors, social workers and other welfare providers are definitely recognised as actors in praiseworthy activities. It does give image to provide welfare work. Who are the other partners to the exchange? The case could resemble that of the fishermen. Welfare professions provide society with better lives. This is a service to society that in return grants recognition to the practitioners of these professions. In this scenario the receivers of welfare (social, economic or medical) are left out of the exchange as resource or raw material. A term like welfare mode of production springs to mind.

There is a problem in this logic: if the source of the welfare service is the state, and the recognition of the practitioners is also bestowed by the state, then the state is the source of both life and image. This is not exchange.

An alternative model would be that the exchange is between the state and the welfare professions, on the one hand, and clients/patients, on the other. The doctor or social worker gives better life to the receiver, who in return allows the provider to earn recognition. This would make sense in relation to the inverted positions of producers and consumers. The state and its officials represent the producers of services guaranteed by human rights, and the human rights are represented by the consumers. In such case the state is the provider of life and the citizens provide the state with the image of a good state. The state is the source of life and the citizens are the source of image. Welfare society is held to be particularly developed in Denmark, and this model might offer an explanation for the unfailingly high Danish score in international surveys of happiness. Confirming happiness will at the same time fulfil the ultimate objective of being well and place the citizen in the role of giver to the state.

How does this constitute citizens or consumers as persons in society? The life constituent can be accounted for, but the universal individualism leaves the citizens with an undifferentiated image, faceless. The citizens will be differentiated as welfare recipients in terms of the incompleteness that is to be amended. The professionals can indeed be heard to refer to clients by the name of the person's illness or condition. This would hardly be an image of value in the long-term transactional order. Another avenue of personal image differentiation for consumers could be just that: consumption, i.e. acquisition of commodities that associate the consumer with endless varieties of styles and atmospheres. That, too, would seem to belong in the short term order of transactions, and in a separate social domain.

It is difficult to see individualism and Human Rights as a sufficient set of values for the complete constitution of persons in society. If the short term transactions in social life need to be inscribed in a transcendent long term order to be existentially satisfactory, then traditional holistic systems of value have definite advantages. The working population provides life for themselves and society, and the highest representative of society provides recognition, visibility, and image to persons and groups for their respective kind of contributions to the whole. All parties enjoy life as well as image. And it is not a problem that the highest representative has an undifferentiated image, an office, since the highest is one, anyway.

Inversely, under the welfare individualist system of values the state and its welfare authorities provide life support for itself and its citizens. In turn, when citizens receive benefits they recognise the giver and provide honour and image to the state and its operators for living up to the human rights of its citizens. All parties live, but the receiving end of the exchange is left with an undifferentiated image within this set of exchange relations. The implied aspiration in life is just to be well.

With the individualist system of values an inversion or short-circuiting has taken place. The lowest unit in the hierarchy becomes the highest value. The whole is identical with the parts, which characterises modernity in several domains (nation, market). Striving is unnecessary to connect with the whole as a source of value, if the individual is identical with the whole. But striving is necessary, if the individual desires to be differentiated from the universal image, the unknown human. Striving is also necessary, if the incompleteness that merits receiver status is not enough.

In Durkheimian terms Human Rights qualify as a religion, the sacred in which society worships itself (Durkheim 1971 [1912]: 18, 47, 439f.). It also serves as a political charter particularly esteemed in the West and used against various traditional systems of value and institutions. It is, for instance, used to argue for the expulsion of the other religions from the state. It does, however, seem to suffer from incompleteness as a basis for social exchange and the constitution of persons. So it is hardly surprising when citizens seek sources of honour and recognition in rival systems of value. Dispensing with uncompromising Human Rights agendas for a pragmatic approach might serve the maintenance of social peace. It could leave some leeway for the coexistence of Human Rights based regimes with traditional systems of value and prevent some from seeking recognition and image in less palatable radical movements.

References

DUMONT, L. (1986) *Essays on Individualism. Modern Ideology in Anthropological Perspective*. Chicago and London: The University of Chicago Press.

DURKHEIM, E. (1971 [1912]) *The Elementary Form of the Religious Life*. London: George Allen & Unwin Ltd.

FERAL, F. (2004) Martime Societies, Fisheries Law and Institutions in the Western Mediterranean. Rome: FAO.

FISKERISTATISTIK. *Antal beskæftigede 2008-2012 fordelt på regioner og erhvervsstatus.* [Online] http://naturerhverv.dk/fiskeri/fiskeristatistik/alder-og-beskaef tigelsesstatistik/#c9811 [Accessed 10th December 2015].

PARRY, J. and BLOCH, M. (1989) *Money and the Morality of Exchange.* Cambridge: Cambridge University Press.

PLATENKAMP, J.D.M. (1988) *Tobelo. Ideas and Values of a North Moluccan Society.* Leiden: Repro Psychologie.

PLATENKAMP, J.D.M. (2006) Visibility and Objectification in Tobelo Ritual. In CRAWFORD, P. and POSTMA, M. (eds.), *Reflecting Visual Ethnography – Using the Camera in Anthropological Fieldwork,* pp. 78-102. Aarhus: Intervention Press and Leiden: CNWS Press.

PLATENKAMP, J.D.M. (2007) Cultural values, Human Rights and Peacekeeping Tasks. *Militaire Spectator JRG 176*(3): 96-104.

STATISTISK ÅRBOG (1980) København: Danmarks Statistik 1980.

STRUBBERG, A.C. (Red.) (1935) Dansk Fiskeristat I. København: Danske Erhvervs Forlag.

VESTERGAARD, T.A. (1990) The Fishermen and the Nation: The Identity of a Danish Occupational Group. *Maritime Anthropological Studies – MAST* 3(2): 14-34.

VESTERGAARD, T.A. (1999) Fishing: Industry or Ritual? *EUROPÆA: Journal des Européanistes* 2: 89-102.

Fihavanana: Exchange, Social Cohesion and Social Exclusion in Madagascar

INGO WALLNER

The concept of the *fihavanana* counts as a constant in the discussion about cultural phenomena in Madagascar, since it is viewed as the embodiment of *Malgachitude*. This type of relationship even has constitutional status because of its elementary social value; it is viewed as the foundation of social togetherness in the preamble of the constitution. *Fihavanana* has become an inflated catchword in the speeches of all politicians and, thus, in the media in Madagascar, since the Malagasy political crisis in 2009 at the latest – which led to the removal and exile of the president at that time, President Ravalomanana. The only concept competing with the *fihavanana* in the media is *réconciliation nationale*.

The deep-reaching discord in Madagascar's political landscape,[1] which had been increasing in momentum since its independence in 1960, experienced a climax during the interim government of President Rajoelina. The cyclical sequence of events incited Ralambomahay to speak of a "crise interminable" (Ralambomahay 2011). The prominent political camps point to the severe incompetence of the other political camps in their accusations; envy and intrigue are part of this game. Rajoelina's tough stance – his refusal to allow the exiled former president of state to return – only deepened this internal political schism. A *modus vivendi* could only be reached by means of the South African Development Community (SADC), which helped to circumvent civil war-like relations.[2]

The search for national reconciliation, in which political actors continue to invoke the *fihavanana*, raises the question about the cultural origin and content of this concept. A political situation that appears quite hopeless awaits a resolution by this concept. In fact, without political and societal consensus, no basis for the country's urgent economic, ecological and moral development

1 Madagascar has experienced phases of social democracy, socialism, economic liberalism and "controlled democracy", a "crise interminable" since its independence from France in 1960 (Ralambomahy 2011).

2 The interim President Rajoelina was replaced by President Rajaonarimpianina after the general election in 2014. Rajaonarimpianina also denied the return of the exiled President Ravalomanana (2002-2009). Ravalomanana, however, re-entered illegally in the autumn of 2014 and lives, after a short period of imprisonment, under house arrest. Despite his current situation (Jan. 2015), he is associated with the *réconciliation nationale* in political discussions.

can evolve. No political speech manages without the *fihavanana*. The *fihavanana* has become a fixed star of hope, even if this fixed star is only a fiction. In the political conflict, no one reminds the actors anymore of the actual origin of the social and cosmological relations that characterise the *fihavanana*.

This "fixed star" is attributed a quasi-religious character by Njara (1992). The chain that holds two individuals together with a shared ancestor as kin (*havana*) also includes spouses and can even involve close friends. Included as friends are, above all, the blood brotherhood (*fatidra*), as well as those friends that were made in general social relationships, e.g. through school or work. With the integration of close friendships, the ancestrally conditioned *havana*-relations are extended, and family and friends are connected through the *fihavanana*.

Not only is the world of living kin a component of *fihavanana*, but also the world of the ancestors. When the deceased are seen as the "living dead", their influence on the daily life of their descendants is apparent: no important decision is made before asking for the blessing of the ancestor worshipped. The quasi-religious value of the *fihavanana*, as described by Njara, is evident in the indispensable inclusion of the ancestors in the most important decision-making processes of daily life. Each time the ancestor's signal for correct conduct is disregarded, the descendant is punished with shame and retaliating force (*tsiny* and *tody*). The constant threat of punishment by both the "living dead" and the living members of one's own community is, thus, an important part of life in *fihavanana*.

This extension of immediate kin, which includes solidarity relations with selected friends and beyond, makes the multilevel nature of the *fihavanana* apparent, as Kneitz described: Concentric circles of *fihavanana*-relations are set up around the ego, consisting of parents, relatives, ancestors and friends, followed by a circle including the members of the village community (*fokonolona*), the native region and ultimately the Malagasy nation (Kneitz 2014: 26).

The more distant *fihavanana*-relations are from the ego, the more *fihavanana* looses its kin-like and quasi-religious content; it vanishes as it moves in the direction of a "national" *fihavanana*. The threat of possible punishment by the ancestors disappears altogether. Only an amorphous form of solidarity now remains – an empty word used by politicians in the multi-ethnic society of Madagascar. This aspect will be dealt with in more detail later. For now, the "original" content of the *fihavanana* remains in focus.

The numerous attempts to translate the concept of *fihavanana* are limited to "kin", "friendship" (Bergenholtz 1991: 232) or "solidarity relation". However, none of these translations takes into account its cultural background, neither the obligatory exchange relations in the sense of solidarity within the community nor the cosmological ties to the ancestors of the family.

The ancestors (*razana*) are not only hierarchically above humans, but they themselves are the source of the world order. With their nearness to *Zanahary*, the divine creator of all things, they become divine ancestors, emanating sacred power (*hasina*) (Rakotomalala et al. 2001). The world order is the order of the surrounding world of those living on the ancestor's land (*tanindrazana*), who believe they are, through shared ancestors and the *fihavanana*, connected by social obligation.

The course of life, which includes birth, the circumcision of boys, marriage, fertility and death, the cultivation and harvesting of rice fields, etc., is in the hands of the all-determining ancestors, or the "living dead". The ancestor's blessing is, thus, crucial for the ordered course of life. The latter requires ritual relations to the ancestors by immediate or extended kin. Most importantly, the sacrifice of a zebu cow serves this relation; something is passed on as "part of the family", which Molet rightfully depicts as "le bœuf vaut l'homme" (Molet 1979: 182). The ritual becomes a communion with the ancestors, and the consumption of the meat becomes a communion with all who are connected in the chain of the *fihavanana*. The zebu offering strengthens the close, important connection with the ancestors in the hereafter and with those who are living right now. It also becomes a "symbole du pouvoir et de la société" in the rehabilitation of troubled *fihavanana* relations (Kneitz 2014: 29).

The zebu offering in the "bain royal" (*fandroana*) of Malagasy kings renewed their relations to *Zanahary*, as well as their legitimation of power over the people. This exchange relation is not only occasionally protested against by the democratic state of Madagascar, but also intentionally avoided. The interpretation of this ritual as non-Christian made it almost disappear from the political realm. Nevertheless, regional and local institutions sometimes employ it in order to make use of its social-integrative content, even if the exchange relations between this life and the hereafter contained within the *fihavanana* have lost their original meaning.

Njara concentrates on the categorisation of obligations brought about by the *fihavanana* in the social and economic realms (Njara 1992: 156). In the social realm, the avoidance or mediation of conflict, as well as the permanent search for consensus, play important roles in the *fihavanana*. In the economic realm, reciprocal exchange relations determine life in the solidarity community. Anyone who ignores these obligations is subject to punishment. Discontinuing a *fihavanana* relationship that is based on the *fatidra* is imaginable if one of the two parties seeks to break off the connection, but the sacrality of the blood oaths forbids such a discontinuation. Because the *fatidra* is passed on to the following generations of "contracting parties", the punishment of such a break is payment of a penance or social banishment. Njara reinforces this finality, "Que votre amitié ne soit pas comme une pierre, car si elle se brise, on ne peut plus rejoindre les morceaux; mais qu'elle soit comme le fil de soie: s'il est trop tenu, on peut toujours y ajouter pour le renforcer" (Njara 1992: 157).

This "worldly" consideration of the *fihavanana* relation, which helps us to understand the depth of the solidarity relation, has a cosmological basis. It is the elementary relation to the *aina*, or "life force", that connects all members within a *fihavanana* relation with *Zanahary,* the creator of all life. *Aina* is viewed as the chain among the living and was created by *Zanahary*. *Aina* also creates itself from the earth that was left behind by the ancestors of those living, and the ancestors themselves stand divinely next to *Zanahary*. Respect for this sacred exchange relation becomes the obligation of the living with regard to the living and their "living dead". *Fihavanana* provides the behavioural code" for the maintenance of these relations: the code of the obligatory solidarity community. The Malagasy "I" almost disappears through *aina*, which connects all members of the community into a collective "we". The "I" remains, however, responsible for all its actions in relation to the community (Dubois 2002).

Practicing good manners, respecting the elderly and acknowledging the abilities of others are all components of this behavioural code. The common life and all of its phases, its joy and sorrow, similarly determine how community members deal with each other on a daily basis. In more complex, permanent working conditions, obligatory personal engagement becomes especially noticeable, which the *fihavanana* calls on. Such work requires mutual help, but only in the way that those involved can provide. Reciprocity is, however, always the guiding principle. This mutual assistance is especially apparent when building canals to irrigate the rice fields, transporting stones to build a grave or house, or cultivating a field. If an equivalent return cannot be produced immediately, reciprocity is expressed with the provision of food for the workers. Such a situation of asymmetrical reciprocity is shown when, for instance, a wealthy person cannot immediately return the service expected of him. He expresses, instead, a partial reciprocity, such as a meal. This temporary imbalance leads to vertical exchange relations. Over a longer period of time, the renderer of the service must be compensated to avoid disturbing social relations. To avoid such an outcome, the code must be respected. The recipient of such a service can, for instance, hold himself accountable, and help the renderer of the service if his family is in need, even if a material service cannot be offset by an immaterial one.

The momentary verticality in the exchange relation is, therefore, based on a temporary asymmetry of these relations; the asymmetry must be balanced either materially or immaterially over time. It must be kept in mind, however, that pure balance is not a strict criteria in traditional societies, in contrast to pure market economies.

If the recipient of a service uses the exchange relation for political benefit, the momentary verticality in the exchange relation can, however, turn into a long-term verticality of social relations. Social and political power emerge from out of such economically favourable positions in the traditional Malagasy community, which places importance on this compensation. Thus, economic power in the villages of Madagascar becomes the source of political power, which, consequently, leads to a social hierarchy. Despite the "democratic" procedure of public decision-making processes, the position of the wealthy influences outcomes. The wealthy experience a "surprofit social" (Wallner 2012: 240), even though taking advantage of the "codes of behaviour" of the *fihavanana* is not allowed. In this way, the renderer of a service can, thus, become the "puppet" of local political life. Though it may sound like "buying votes", this process remains a part of the exchange relations that are characterised by mutual dependency. Reciprocal economic dependency can, in such a case, lead to "corrupted" generosity, in which the obtainment of political power might be the final outcome.

Here, evident signs of the dissolution of the *fihavanana* in the traditional community are apparent. Due to the people's growing material needs, this has been going on for many years. In addition, there is a second factor leading to the dissolution of the *fihavanana*: The recipients of the services that are provided within the framework of the *fihavanana* are compensating the renderers of the service with money. The structure of traditional social relations, with a largely reciprocal exchange, is being replaced by market economy relations. This affects not only those who belong to immediate and extended kin, but also all non-related members of the village being part of a political community (*fokonolona*).

The network of social obligations, in which social cohesion is based on the *fihavanana*, only secures social solidarity as long as the *fihavanana* code is respected. However, the material need of the people has increasingly led to the erosion of the socially binding force of the *fihavanana*.

Fihavanana has become, above all, a political catchword. Politicians attempt – still with no success – to use the value of the *fihavanana* to appeal for national reconciliation. What they do not consider is the cultural, cosmological origin, and, thus, the actual social value of *fihavanana*.

The values associated with *fihavanana* have been reduced to a political slogan, which calls for the citizens of Madagascar to practice the same collective and brotherly closeness normally seen in the *fihavanana*. However, how the *fihavanana's* values only pertain to individual groups of kin and can only ever pertain to these, since the ancestral relations are an essential component of the *fihavanana* concept, is altogether overlooked; the exception being the integration of *fatidra* relations into the *fihavanana*. The "other" is excluded from its social intensity and, consequently, the sanctions related to it.

The speaker's rhetorical repertoire includes mentioning the *fihavanana* not only in elections, but also in every political speech in which the closeness of the multi-ethnic nation is the focus. However, the slogan is meaningless; the adoption of such a complex cultural value is not appropriate in this case, as politicians seek to nationalise it. The social integrative function of *fihavanana* is increasingly put to the test by the economic needs of the people, even in rural areas. In highly populated urban areas, it could not affect the development of social unification, since all references to its elementary foundation are missing. The social strength of *fihavanana* is lost due to poverty. Under these conditions, the "Chacun pour soi, et Dieu pour nous tous" determines life, yet this survival strategy completely contradicts the code of the *fihavanana*. Reciprocal social obligations fade away, causing envy to develop even among relatives. The decline of *fihavanana* is manifested in the way that both sides watch and suspect the other (*ibid:* 225). The worsening, increasingly visible erosion of the *fihavanana* is clear in a particularly drastic example, which shows that the erosion itself has made its way into the "ruling social circles" and is dividing society: Minister Ramaroson lost her life in a fatal shipping accident at a river's mouth in the autumn of 2011. Several ministers, military officials and politicians were prohibited from attending the public funeral. The reason for the prohibition was apparently the differing political opinions between her family and the politicians in question. One journalist commented on this prohibition in the following way: "How has it got so bad that someone, due to political divergences, can sink so low and stomp on the *fihavanana*" (*La Vérité*, September 3, 2011: 3, Des comportements bizarres, Antananarivo).[3]

The survival of the individual is, under such conditions, no longer dependent on group affiliation. Collective interests no longer determine the actions of the individual, since the individual no longer anticipates punishment for not following the code of behaviour. Even the worst punishment – being banned from burial in the ancestor's grave – is no longer threatening, as one can easily elude social controls and, thus, possible sanctions. This deep-reaching social transformation, which originated in the highly populated urban centres, is also spreading in the same way to rural regions, as there appears to be no improvement in the economic situation.

Thus, a second kind of *fihavanana* is apparent in urban environments. It is manifested by "ethnic faces", because groups such as the "associations ethniques", "associations claniques" or "groupes ethniques" have organised to protect their "diaspora ethnique".

The historically contingent ethnic diversity in rural Madagascar led to the development of peoples (*foko*) as pre-nation groups. The urge to expand, on the one hand, and the will to assert themselves against enemies, on the other, led to a strengthening of the collective consciousness in these pre-nation groups – all of which began with the shared ancestors of each community. The

3 In any case, the cause of death is still unknown, but some say that the river's mouth was forbidden to anyone belonging to the *Merina* and that violating this ban leads to death.

urge to expand in the Malagasy kingdoms, alongside internal power struggles, led to a number of tensions which did not lead to a unified nation. First, Andrianampoinimerina (r. 1787-1810) laid the foundation for a pre-nation in the country with his conquest, alliance and dominance politics, which ended with the annexation of the land by France in 1896. The historical establishment of ethnic boundaries, which, despite developing into a pre-nation, could not be eliminated until 1896, made the exercise of colonial power quite easy. Its politics of *divide et impera* made it easier to control the land and put in place, with the establishment of 18 ethnicities, the cornerstone for today's ethnic segmentation.

This ethnic segmentation resulting from colonial rule based on historical divisions is integral to the current Malagasy nation. In a way, the Malagasy nation has internalised this segmentation, as political and civil societal life are always conscious of this segmentation. The plurality of ethnicities influences political and civil societal life; it creates tension, but without leading to militant conflicts in the style of a civil war. Nonetheless, there are ethnically contingent, suspenseful situations on all levels of Malagasy society. The political instability of the land is a result of this situation. Another result is the identity consciousness of the Malagasy people, which is focused only in a limited way on an actual shared common nation, since the ethnical identities became prominent. The "associations ethniques", "associations claniques", "groupes ethniques", etc., are a sign of this identity diversity. They are also a sign of "us and them", which is expressed in the original *fihavanana*.

This sign, which is made up of opposites: close social cohesion and social exclusion, has developed in the capital city, as well as in the larger cities that have experienced extensive immigration from the inland.

The social cohesion of *fihavanana* focuses on a limited geographical space, and kin are its point of reference. Traditionally living Malagasy orient their actions and behaviours towards this point of reference. The original local boundaries of the land lent from the ancestors were expanded to new village establishments and to the new construction of graves by patrilineal and patrilocal groups. The personal connection to the land of the ancestors (*tanindrazana*), thus, experienced an expansion. Moreover, there was also an expansion of kin relations. Despite these migrations, the village communities (*fokonolona*) stayed together; even those who definitely immigrated continue to see their social origin and home in their *tanindrazana*.

The "associations ethniques", "associations claniques", "groupes ethniques", etc. in the bigger towns of the country, have become an in-between home in the European Malagasy diaspora. However, the strengths of the "native" *fihavanana* with its profound social and cosmological concepts no longer have all their effects there. The sacrality of the bond no longer serves as the basis of *fihavanana*; it is reduced to the safeguarding of group-specific, ethnic-oriented interests in order to implement the economic and political interests of the group. The activities of these "associations", therefore, perpetuate the ethnic segmentation in the country and neglect the original *fihavanana* values. The concept of *fihavanana* has been transformed from quasi-sacral, exchange-based relations to broad, worldly ones; *fihavanana* has been turned into a "lobby".

The idea of togetherness being practiced by "associations" is most apparent at festivals, in the help they offer to students, including the obtainment of grants, and their support in case of death in the family or return of the deceased. They clearly represent ethnic interest when they campaign for the interests of workers from their own ranks. The "associations" become political actors who exert pressure on political decision-makers in many ways. Their position makes the ethnic boundaries permanent. Social cohesion on the grounds of *fihavanana* also leads in this regard to social exclusion and social separation, as the other is intentionally excluded.

However, the influence of "associations" does not end here, since its ability to pressure political representatives, for instance, is especially pronounced in elections, in decisions about infrastructure measures, etc. They are also driving forces in the attempt to decentralise the country's administration and economy, and limit the centralisation of the economy in the capital city.

With their influence on politicians in the national parliament, whose decisions also involve regional and local levels, ethnicity remains a criterion of political processes. At the same time, ethnicity itself becomes an instrument of politics, as politicians consciously use it in order to obtain or maintain power.

There is also a second segmentation describing reality. The debunking question is, "Where are you from?", and attention is simultaneously directed toward skin colour (light yellow, brown or black), hair texture (frizzy or straight), the way "the others" speak, their birthplace and the location of the graves of their ancestors. All these questions are asked in order to categorise "the other". Such questions lead to the acceptance or rejection of "the other". The outcome of the first encounter with a stranger is, thus, subject to optic, acoustic, genealogical or geographical criteria. They represent a pattern that decides whether "the other" will be accepted or openly or discreetly rejected. Alexandre, the philosopher of religion, sees a pronounced racism in this behaviour, "Le racisme est très fort à Madagascar" (Alexandre 2007: 102).

Located above this detailed pattern is yet a higher ranking one that distinguishes the *Merina* and *Côtiers*. The allegedly light-skinned *Merina*, who literally "carry" their Southeast Asian origin on their faces, stand out in contrast to the *Côtiers* and their African or Arabic background. Skin colour, thus, becomes the most noticeable signal for categorisation. Even if this signal does not lead them to openly distancing themselves from the other, it is noted, and, if necessary, applied subtly. The endeavours of the kings of *Merina* to achieve dominance led in a number of places in the land to the still existing clear rejection of the *Merina* who can be immediately identified as such.

Fihavanana, which originally involved consensus, a balance of interests and peaceful cohabitation, finds its true values are limited. The lines of separation also penetrate life in the universities, where the "associations" also attempt to reach their goals. Again and again, intense conflicts often arise there between different ethnic groups. Their reason for this is daily problems. But there is always the potential of being ethnically exploited and causing a stir.

These developments spurred a Malagasy journalist to say that they would all kill each other because of their deep-seated hate if it were not for the *fihavanana* (*ibid*: 118). However, if this journalist had conducted a sober analysis of the political conflicts and reports in the daily newspaper, he or she would have come to another conclusion. In Ravalomanana's fight against his political opponent Ratsiraka for presidency in 2001-2002, as well as the defence of Ravalomanana's later presidency against his challenger Rajoelina, force of arms was used with deadly outcome. The daily newspapers are filled with news about murder, which contradicts the peaceableness of the *fihavanana*. A further conflict is found in the extensive theft of zebu cows which has taken place in the past few years, above all in the south and west of the country by the *dahalo (thieves of zebu)*. The many murders are, thus, proof of the fact that not even the sacrality of the zebu prevents them from breaking the code of the *fihavanana*.[4]

The idea of a "mysthique d'unité" that proceeds from the *fihavanana* is more or less still upheld in the mostly undisturbed, traditional rural areas. The divine creator of all things, *Zanahary*, and the godlike ancestors – together with the microcosm of humans – form this mystical group-specific unity. *Fihavanana* includes and excludes, and thus, is not an instrument to be used to create an encompassing *unité nationale*. Group-specific unity is based on group-specific solidarity, friendship and reciprocity. The peaceableness that lies in the *fihavanana*, but that is established by social control, is, however, increasingly becoming a myth through everyday occurrences. It entirely achieves the status of a myth on the national level when one considers the differences between ethnicities, and how these differences are based on the historical development of individual groups, different economic levels, the unresolved language problem and the various levels of education.[5]

4 The "conversion" of approximately 4,000 *dahalo* in the south of the land under military occupation in October 2014 was celebrated as a great triumph. After a long stay in hiding, they were once again allowed to enter into their village community (*La Gazette*, October 23, 2014:8, Reddition des dahalo – Sécurité en passe d`être retrouvée dans le Sud. Antananarivo).

5 The national languages of Madagascar are Malagasy and French. Malagasy, as the language of the historically leading *Merina*, continues to be the language of dominance. It, therefore, meets a lot of resentment from most of the members of the groups that count as *Côtiers*,

Even if the *Merina* see themselves as "brothers" of other ethnical groups, it does not mean these groups feel the same way. "L'aveuglement des différences va toujours de celui qui est en position dominante vers le dominé. Le *fihavanana*, s'il se contente de nier les différences, risque de n'être qu'une voile sur les yeux et même une manœuvre de celui qui a le pouvoir, pour prétendre qu'il n'a pas de problème et que tout, en conséquence, peut rester en l'état. Prétendre que l'on est semblable est une excellente stratégie pour garder le pouvoir réel en masquant les écarts pourtant manifestes, il ne suffit pas de s'appeler frères pour le devenir" (*ibid*: 120). With this "blindness" to the reality, Alexandre underlines the differences between the leading *Merina* and the *Côtiers,* and he underlines differences even between some groups of *Côtiers*.

A manifold social exclusion is thus visible. If Madagascar has been able to maintain a relative social cohesion despite social exclusion, it has done so because of the strict hierarchical structure in Malagasy society, which mostly eliminates thinking in competing oppositional categories. The way in which the democratic processes were implemented in this society – a social paradigm shift – will show whether the relative social coherence can be maintained. Alexandre is, therefore, cited here once more because his evaluation of the future of *fihavanana* shows how labile its values have become. He views it as a kind of nostalgia, a dream of the lost past and an unreachable ideal despite its presence in many Malagasy: "Le *fihavanana* demeure présent chez beaucoup de Malgaches au moins comme une nostalgie, le rêve d'un passé perdu, un idéal dont on ne parvient pas à trouver le chemin. Mais pour combien de temps encore, surtout que là n'est pas l'essentiel?" (*ibid:* 129). At the same time, the consequences of social separation are discernible, as the idea of a national society based on the equality of individuals continues to be seen as a goal. "Il appartient à tous de briser la permanence d'une culture de la division en s'attaquant aux sources des points de divergence et à l'idéologie ethno-nationaliste de ceux qui les véhiculent" (Ramandimbilahatra 2010: 17).

As explained above, Malagasy kings used the sacral and social function of the zebu offering in the "bain royal" (*fandroana*) and made it into a symbol of their relations to *Zanahary*, as well as a symbol for their legitimation of power over the people. The democratic state of Madagascar has protested against this exchange relation at public events. The interpretation of this ritual as non-Christian has pushed it into the background. Nevertheless, regional and local institutions occasionally demonstrate for it in order to make use of the social-integrative content of *fihavanana*. But this symbol of *fihavanana* also gives way to "material worldliness". For the people, the satisfaction of hunger takes priority over the social and cosmological coherence expressed by *fihavanana*.

---

even though all their dialect variances are to be seen linguistically as a common language. All attempts to give Malagasy elements from the dialectical variants, to raise the acceptance of Malagasy, have so far failed. A sound language politics remains an unpleasant political challenge, since it, above all, affects the identity consciousness of numerous ethnicities.

References

ALEXANDRE, C. (2007) *Violences Malgaches.* Antananarivo: Foi et Justice.

BERGENHOLTZ, H. (ed.) (1991) *Rakibolana Malagasy-Alemana.* Antananarivo: Leximal.

DUBOIS, R. (2002) L'Identité malgache – La tradition des Ancêtres. Paris: Édition Karthala.

KNEITZ, P. (ed.) (2014) *Fihavanana – La vision d'une société paisible à Madagascar.* Halle: Universitätsverlag Halle-Wittenberg.

NJARA, E. (1992) Le Fihavanana. *Droit et Cultures, Revue semestrielle d'anthropologie,* 152-158.

RAKOTOMALALA, M. et al. (2001) *Madagascar: Les ancêtres au quotidien.* Paris: L`Harmattan.

MOLET, L. (1979*) La conception malgache du monde surnaturel et de l'homme en Imerina.* (Vol. 1). Paris: L'Harmattan.

RALAMBOMAHAY, T. (2011) *Madagascar dans une crise interminable. Etudes africaines.* Paris: Harmattan.

RAMANDIMBILAHATRA, R.A. (2010) *État et Ethnicité à Madagascar.* Antananarivo: Friedrich Ebert Stiftung.

WALLNER, I. (2012) *Madagaskar: Nationale Identität in Sicht?* Saarbrücken: Südwestdeutscher Verlag für Hochschulschriften.

Cosmology

Oral Traditions of the Vagri of South India

GABRIELE ALEX

One of the things that I have learned from Jos Platenkamp is the importance of myths for an understanding of the cultural and social specifics of a given society. His fascination with myths is reflected in various of his articles (Platenkamp 1988, 2004), where he shows in a comparative fashion how the conception of the relationship between the social and the cosmological order is considered as essential for facilitating different forms of reproduction. It is above all his analysis of the depiction and explanation of 'human incompleteness', as manifested in the myth of the half-person, or unilateral figure, (a human being that consists only of a half body), that sheds light on central conceptions of reproduction and exchange. In his comparison of three different mythical traditions, Jos Platenkamp (2004: 2) argues that the

> "analysis of myths about incomplete persons recorded in different societies [...] enables us to acquire a comparative understanding of those specific sets of prototypical relationships, the construction and enactment of which various societies consider as a basic condition for their members to be complete"

He distinguishes between on the one hand social relations between living persons, which constitute the internal order of the village, and on the other cosmological relations, which connect the social order and world of living persons with the domain of the spiritual world and consequently the cosmological order. He argues further that changes in the identity of those beings related to each other in one of the various forms are achieved either by changes in the direction of transfer [...] or by changes of the identity of the beings or things transferred; this latter change he calls conversion, the first one he calls transformation. His analysis of these myths is enlightening for an understanding of how reproduction, life and completeness are conceptualised and decoded in myths and then enacted in rituals.

> "The notion of human incompleteness that is articulated in the myths signifies the idea that it is impossible to multiply by reproduction rice and human beings as long as certain prototypical socio-cosmological relationships have not yet been constructed. As is to be expected when dealing with three distinct societies, these socio-cosmological relationships are conceptualised in different ways, yet they have certain representations in common." (Platenkamp 2004: 24)

In this article I want to follow this inspiration by looking at a creation myth that I encountered during my first fieldwork in Southern India, shortly before I came to work from 2001-2004 with Jos Platenkamp at the Department of Anthropology at the University of Münster. From 1998-2001 I carried out fieldwork in Tamil Nadu with an interest in the anthropology of childhood. One of the communities I then worked with were the Vagri, a South Indian community of North Indian origin. Not only did I stay most of the time during fieldwork with the Vagri, they are also the people with whom I engaged the most intensively. I am taking Jos's approach as a starting point to investigate a particular oral tradition, the central creation myth of the Vagri. My question is how this myth with its three sections organises and explains the relations between different categories of humans as well as between humans and the cosmological sphere. The topic of 'completeness' plays a role in this myth which I shall be discussing, albeit in a slightly different way than proposed by Jos Platenkamp. It is not the corporeal incompleteness that is here at stake, but the social completeness.

I will give a very brief introduction to the Vagri themselves and their oral traditions, and then recount and analyse the creation myth of how the Vagri came into being.

The Vagri

The Vagri, as they call themselves (their Tamil neighbours call them Narikuravar), migrated around 400 years ago from Gujarat southwards to Tamil Nadu. The Vagri speak a distinctive language called Vagriboli, which belongs to the North Indian language families and bears many similarities to Mahrathi and Gujarathi. The Vagri used to follow a peripatetic lifestyle until the end of the twentieth century, but due to the settlement programmes of the Indian state they have become more or less sedentary. After independence, they were given small plots and houses in newly constructed settlements, so-called colonies, following the revision of Indian settlement policies. Today, there are a few hundred Vagri colonies in Tamil Nadu, ranging from 20 to 400 houses in size. Internally, Vagri society is divided into two patrilineal principal clans, who act like moieties. These two clans are the *bexavala* – those who sacrifice buffaloes and the *bodkavala* – those who sacrifice goats. The *bexavala* are headed by the goddess Vairagan or Vexli, the *bodkavala* by the goddess Novkod. Within each of these groups are a number of subgroups – patrilineal exogamous clans. Each clan is headed by an ancestor, called respectfully 'Guru', and his specific goddess. Each family possesses a goddess in the form of a little silver statue, to whom sacrifices are to be made regularly. These clans stand in a fixed hierarchical relationship to each other according to seniority and juniority. The family goddesses are honoured at the annual sacrifice, which constitutes the main ritual event for the Vagri. Whereas Tamil people locate themselves socially and spiritually with reference to their *ur*, the 'earth' or 'soil' they come from (Daniel 1987), the Vagri stress no such connection. Instead, they acquire identity and structural relationships between individuals and groups through their position within the kinship and clan system. This finds its material expression in the goddesses. Internally, Vagri society is divided into a number of hierarchically ordered patrilineal exogamous clans, which stand in a fixed relationship to each other and are unevenly distributed over Tamil Nadu. Inter-colony relations are chiefly established through long-standing perpetuated marriage relations, supplemented by political and economic activity. Geographical proximity, on the other hand, does not necessarily lead to such cooperation or intermarriage. Until the 1970s the Vagri were still peripatetic, and would tour through Tamil Nadu using fixed routes and resting places. Today they are settled in colonies, mostly on the margin of villages and towns. Their mode of subsistence includes hunting (rats from the fields, as well as snakes, foxes, birds, and turtles), and the production and sale of medicines or of bead fallalery. The Vagri are strictly endogamous, marriages or illegitimate love relationships outside the Vagri community are the exception and frowned upon. The Vagri were not part of what is traditionally called the *jajmani system*[1], and they also have no role in the temple rituals of the Hindus. However, they do economically interact with the other communities, e.g. when they catch the rats from the paddy fields for the peasants of the surrounding areas or when they sell their beads and herbal medicines on the markets.

The relationship between the Vagri and the neighbouring communities is ambivalent. By the Tamils the Vagri are seen as being of low status in terms of economic and educational as well as ritual status (Trawick 1991). At the same time their knowledge of plants and animals as well as their proximity to the wilds, which is linked to their healing powers, is well acknowledged and respected. The Vagris' self-perception is different: even though they accept the superiority of the dominant castes in village politics, they consider themselves and their own religion as superior.

Physical wellbeing, or rather its absence, plays a huge role in everyday life and is linked to concepts of reproduction and fertility. The Vagri have their own classification system of diseases based on the causal agents. There are diseases that are considered 'natural', meaning that they come from biological causes such as a virus. Other diseases are a result of a misconduct of members of the community. Misconduct means breaches of norms, such as non-observance of the incest taboo or other kinds of disrespect towards the goddesses. Such a behaviour irrevocably leads to punishment, mostly in the form of diseases, childlessness, or accidents.

1 The Jajmani system is defined by Wiser as "a feudalistic system of prescribed, hereditary obligations of payment and of occupational and ceremonial duties between two or more specific families of different castes in the same locality" (Wiser 1936: 6).

Oral Tradition of the Vagri

Oral narratives are a means to connect different generations with each other and to transmit the experiences, values and concepts of one generation to the next. They further help to constitute "a shared understanding of the world and transmit the worldview to others in the group" (Lawless 1988: 67) and as Yamane has pointed out, oral traditions can best be understood as an instrument, or as she terms it "a primary linguistic vehicle" with which meaning is constituted and communicated (Yamane 2000: 183). Dundes has pointed out how folklore is a means for people to express what they think in their own words in a genre that is different to everyday speech, and therefore allows people to say those things which normally go unsaid, so that "one finds a people's own unselfconscious picture of themselves" (Dundes 1975: xi).

Oral traditions and the practice of telling myths and stories play an important role in the everyday life of the Vagri, despite the wide distribution and enormous popularity of television and radio. The loci and occasions of traditional narrative performance vary according to the genre, but narrative traditions are part of everyday life. One can differentiate broadly three genres of oral traditions:

1. Myths, explaining the origin and the particularities of the Vagri and their ritual and religious practices;
2. Folktales and fairy tales, which are part of a broader folktale corpus in the region;
3. Stories that are based on historical events and which I would accordingly classify under oral history. Some of these tales concern the migration of the Vagri from North

India to Tamil Nadu, others deal with conflicts or breach of norms and jural decisions.

The myths that are concerned with Vagri cosmology and religion are mainly told around the time of the animal sacrifices in August but also at smaller ritual occasions. The buffalo sacrifices take three to four days including the necessary time for the preparation of the sacrifice, and thus offer ample time for the retelling of myths. The climax of the sacrifice is the drinking of the blood: for the Vagri men and the goddesses this is also the moment when the goddesses enter the men through their hair and possess them. The folktales, fairy tales and also to some extent the historical stories are everyday stories told to pass time, and parents narrate them to their children. The stories that deal with conflicts are used as reference for current conflicts or in the discussion of rules and norms.

In the following I want to look at a particular myth which explains the creation of the Vagri by the God Dadaji, the genesis of diseases and the origin of the buffalo sacrifice.

The Creation of Man and Woman and the Origin of the Buffalo Sacrifice

The creation myth, *hanmalna dúUdúuvo ni kahini,* of the Vagri is based on the archetypal mythological theme of the slaying of the buffalo-demon Mahisasura by a goddess, a theme that can be found all over South Asia (Shulman 1976; Werth 1996). The Vagri version is distinct insofar as it combines this specific mythical element with two other parts which refer to the central elements and values of Vagri religion, with on the one hand human reproduction and sexuality and on the other hand the buffalo sacrifice and the moiety structure. The myth recalls the role and responsibilities of the *bexavala* – those who sacrifice buffaloes – and also introduces the second group, the *bodkavala* who sacrifice goats. Dadaji is the ultimate creator of everything, who stands at the top of the divine hierarchy. This is a version of this myth as recorded by Thomas Malten in 1986, told by a Vagri called Lalgudi, aka Shankar:

> "When Dadaji had gotten up he took some earth, mixed it with clay and formed the corpus of a human. After he had shaped the corpus of human he formed the arms, the legs, he formed the finger and the nails, he formed the thick intestines and the long intestines, he formed the liver and the gall bladder, he formed the rips and after that he formed the head and the ears. Then he made the eyes and the nose, he made the tongue and the teeth. When he had put them down the sun was coming down quickly and was sitting. When the sun was sitting [Dadaji thought] 'if I poured life into them now they would be half-finished. I should not pour life into it' he thought. After he had made the figure and put it down he went away. After he had gone the mares saw [what he had done] and came down quickly. And once they had come down and looked around they said: 'once this has become a human he will beat us and take our life. We seven must trample this [figure]'. They trampled the figure and flew away. Dadaji came back and when he came to see, [the figure] was lying there trampled. When he saw this [he thought] 'who has trampled this [figure]?' When he saw this, what did he do, he took again earth and made a lump from it. 'Dadaji, what are you doing' [he asked himself?]. 'What do you know? What do I do with this?' After he had taken up the clay he made the hands, he formed the finger and the feet. Then he put the big intestines inside, the rips, then the waist and he made it all, oh Dadaji. After he had done it all, when he wanted to pour the life inside the figure the sun went down quickly. When he saw the sun setting quickly, [he thought] 'if I pour life inside now, it would be half-finished. If it were half-finished it wouldn't be right.' So he thought and went away. After he had gone the mares, who had come down again from above trampled [the figure] and went away. When they were trampled, what did he do then? He got up the next morning, when he came then and looked, this [figure] was lying there again trampled. 'Who trampled [the figure]? What is to do now? One should not do it like this. If I create quickly another being for chasing away, I will understand his ideas.'"

So Dadaji subsequently creates a cock, a dog and a jackal to protect the figure, but they all fail to do so and the mares continue with the destruction of the human. Next Dadaji creates Hanmal the buffalo to protect them, and the buffalo is able to chase the seven mares away and finally Dadaji can blow life into the clay figure and make the first human come alive. Then he creates a woman from clay, puts Amrita on her tongue and she too comes alive. Dadaji shows the couple sexual intercourse and then returns home where he waits for the man and the woman to come to his place. But the couple engages in sexual intercourse. When after eight days they have not turned up, he sends them diseases, which render them so sick that they come to him begging to take the diseases away from them. This is told in the following manner:

> "Dadaji saw the couple copulating and thought 'I cannot allow this.' Then he said: 'Oh Goddess of cough, o Goddess of cold, Goddess of eczema, you three go and attack them with fever, capture them, eat them' so Dadaji said and sent them on their way. When the goddesses of cough, cold and eczema had attacked them they started to scratch themselves. When the husband and wife were about to die from the scratching they came to Dadaji, more dead than alive and they fell down at his feet saying: 'Oh father, you gave us life, why are we attacked now by fever and else, why did cough and cold attack us? What should we do?'
> Dadaji replied: 'I waited for you one day, I waited for you two days, I waited for you three days. When I waited for you all days, eight days had passed, and since you didn't come for all these days I said you will only come if I send cough and cold and fever and so I sent them. Had I not burdened you with them you would not have come to me.' Then he turned his hand over them and both were good again. Dadaji said: 'Now you go and you create your village and you conceive many children and get a good name and live in a proper way. Wherever you go, don't forget my name, and preserve the memory of Dadaji.'"

This myth then continues explaining the origin of the buffalo sacrifice as follows: Dadaji goes to see Hanmal, the buffalo, but when he approached him, Hanmal looked very angry and tried to dislodge Dadaji. "Who can eat Hanmal?" Dadaji thought. Several goddesses were summoned by Dadaji and he asked one after the other whether they would be able to get rid of Hanmal, but all ran away in fear. At the end the goddess Vairagan (a Vagri form of Kali) came after a long sleep of 24 years and said she would be willing to fulfil the task of driving away Hanmal. She asked Dadaji to bring chalk, betel and a plate and then consumed them all in one go. After this Vairagan went to face Hanmal but before she faced him she transformed herself at first into an old non-Vagri woman and then into a beautiful Vagri girl, so Hanmal would not recognise her. They agreed to fight against each other with their best tricks, but after a period of eight days Vairagan went to the smith and asked for a trident and a knife to be made for her and then she bought hard liquor, turmeric powder, a wooden plate, a coconut, flowers, betel, incense and everything else she needed for the puja. Hanmal tried to destroy Vairagan by casting at her the ocean, the wind and the top of the world, but Vairagan was able to protect herself. She asked the goddess Pahaduudan to grasp Hanmal by the front legs of and to pull him down, which Pahaduudan did. Then she took the tripod and the knife and stabbed Hanmal, but both the knife and the trident broke, since the smith had not made them properly. Vairagan tied Hanmal's legs together, dug a little pit and asked Dadaji for his little golden knife with which she sliced Hanmal's cheek down to his neck until his blood issued. She offered the blood first to Dadaji and then drank from the blood herself. Then the other goddesses came and asked for their share. When the goddess Novkod came she tripped over a basket which Vairagan had put over Hanmal's intestines, and from under the basket appeared a goat. Novkod thus became the goddess of the *bodkavala*.

The myth contains three major parts, which deal with the relationship between living beings and the divine sphere.

The first part describes the creation of man and woman. Dadaji creates the human couple from clay but is hindered in blowing life into them because the mares trample on the figures and destroy them. Only the fourth being which he creates, the buffalo Hanmal, is able to chase the mares away. So in order to create humans it needs more than just the relationship between the couple and Dadaji, Hanmal is the essential part in it, only with his presence is Dadaji able to give life to the human couple.

The second part commences with the couple having come to life: Dadaji has taught them reproduction. Being occupied with each other they forget that they should go and see Dadaji. Not following Dadaji is an act of disobeyance. Having waited for them for eight days Dadaji punishes the human couple with diseases, which make them feel so miserable that they are forced to seek Dadaji in order to be freed from their afflictions.

Here the relationship between humans and Dadaji is expressed and rules are established. Dadaji shows them intercourse, but this should be linked to reproduction. He demands to rank first for men and women. The diseases that he sends them are a clear sign that they forgot their duties to Dadaji, and that they need to show respect. This part establishes the relationship between humans and Dadaji as well as between husband and wife: for a man Dadaji is the highest authority.

The third part of the myth connects to the first part and closes the story. From being benevolent and life-giving, Hanmal has become the buffalo demon and a threat to Dadaji. Dadaji has to search again for someone strong enough to face the threat. This time it is not the human couple that is under threat, but he himself. Only one goddess can kill him, and that is the goddess of the Vagri, Vairagan. It takes her 24 years of sleep to amass her powers for her battle with Hanmal. They then fight for eight days, but to no avail. The goddess is able to withstand the ocean, the wind and the top of the world, none of the worldly powers Hanmal has ordered to fight her are able to destroy her! This makes her the supreme goddess of all, a status that is also stressed today by the Vagri. But she is also the only one able to protect Dadaji. The power of life in the form of protection given by Hanmal to the human couple is returned in a kind of reversed transaction. Dadaji's knife is an important part of this process, no human, 'smithmade' tool is able to strike Hanmal. But before Vairagan can strike Hanmal she needs to transform herself, interestingly first into an old non-Vagri woman, and then into a young Vagri woman. She kills Hanmal in the form of a ritual sacrifice: thus the killing of Hanmal takes the form of a classical *puja*. The first blood from Hanmal is offered to Dadaji, only then the other goddesses are allowed to drink – the rules of respect are clearly articulated.

This third part also clarifies the position of the humans against their creator Dadaji: they return the gift of life by protecting Dadaji with the help of Vairagan against Hanmal. Two remarkable transformations take place: 1. the direction of exchange has turned around and 2. Vairagan transforms from an old non-Vagri woman into a beautiful young Vagri woman. Different issues are articulated here: reciprocity (of) protection, the giving of life, and reproductive powers. Women can reproduce if the rules of reciprocity are observed.

The three parts articulate the rules of honour and respect towards Dadaji, Vairagan and Novgod and clearly place them in a hierarchical relationship with Dadaji at the top and Vairagan as the highest goddess of the Vagri, entrusted with the task and power to protect. Novgod is secondary, as is her position within Vagri social hierarchy when it comes to the relationship between the two groups – i.e. the goat sacrificers and the buffalo sacrificers. Novgod's position in the ritual and her role of sacrificing goats is explained somewhat in passing, but she becomes an integral part of the setup and is ascribed a role in the ritual killing. The mythical elements further lay down the behavioural rules between humans and humans and between humans and the cosmological order.

Through the buffalo sacrifice, which constitutes the religious and ritual climax for the Vagri and where the central values are articulated, it is exactly this reciprocity of protection which is enacted. Completeness of society and of humans is reached by the reciprocal transactions between Dadaji and the humans. Vairagan (and Novgod) are intermediary goddesses, standing between the highest sphere of Dadaji and the human domain, who become essential in the communication between these two spheres.

What is also interesting is the role and kind of diseases that are sent by Dadaji – these are simple, everyday afflictions: skin diseases, coughs, colds. But despite their ordinary character they make life so miserable for the humans that they are forced to seek the help of Dadaji. Diseases signify a sin on the side of the humans, such as a loss of respect or the breach of taboo, and they demand subordination under Dadaji in order to restore the order and consequently human health again. My study on health-seeking behaviour and health-explanatory models (Alex 2010) showed that for the Vagri the breach of a taboo and disrespect to the goddesses are seen as important causative agents for disease, but not as the only ones. The climate, the food, infectious agents such as bacteria or viruses, hereditary conditions or sorcery are also made responsible for diseases.

Conclusion

Jos Platenkamp showed how myths are able to articulate society's need for the establishment of the relationship between the social and the cosmological order, because it is those relations which guarantee social and biological reproduction. Platenkamp is mainly concerned with the incomplete human, who manifests in the myth of the half-person, or unilateral figure, that sheds light on central conceptions of exchange, reciprocity and reproduction. In the myth segment that I have retold here we are not confronted with a half-person, but with a half-finished clay figure, whose body is ready formed from clay but whose spirit is not yet alive, and who thereby remains in a liminal state. In order to get out of the state of 'half-ready-ness' Hanmal needs to be created as a protector. When he turns evil, it is the goddess of the humans, Vairagan, who gives protection to Dadaji and ensures the continuity of the cosmological order. The humans experience protection from Dadaji and then return the act of protection with the help of their goddess. Every year they celebrate and articulate this gift of protection through the buffalo sacrifice. It is this repetition of the reciprocal relationship that keeps the humans out of the liminal, half-ready state and through which fertility, reproduction and health are secured.

References

ALEX, G. (2010) *Medizinische Diversität im postkolonialen Indien. Dynamik und Perzeption von Gesundheitsangeboten in Tamil Nadu*. Berliner Beiträge zur Ethnologie 21. Weißensee Verlag: Berlin.

DANIEL, V. (1987) *Fluid Signs. Being a Person the Tamil Way*. Berkeley: University of California Press.

DUNDES, A. (1975) *Analytic Essays in Foklore*. New York: Mouton Publishers.

LAWLESS, E. (1988) *Handmaidens of the Lord: Pentecostal Women Preachers and Traditional Religion*. Philadelphia: University of Pennsylvania Press.

PLATENKAMP, J. (1988) Myths of Life and Image in Northern Halmahera. In MOYER, D. S. and CLAESSEN, H.J.M. (eds.), *Time Past, Time Present, Time Future. Essays in Honour of P.E. de Josselin de Jong*, pp. 148-167. Dordrecht: Foris Publications.

PLATENKAMP, J. (2004) From Partial Persons to Completed Societies. *Zeitschrift für Ethnologie* 129: 1-28.

SHULMAN, D. (1976) The Murderous Bride: Tamil Versions of the Myth of Devī and the Buffalo-Demon. *History of Religions* 16(2): 120-146.

TRAWICK, M. (1991) Wandering Lost: A Landless Laborer's Sense of Place and Self. In APPADURAI, A., KOROM, F. J. and MILLS, M.A. (eds.), *Gender, Genre, and Power in South Asian Expressive Traditions*, pp 224-266. Motilal: Banarsidass Publishers.

WERTH, L. (1996) *Von Göttinnen und ihren Menschen. Die Vagri, Vaganten Südindiens*. Berlin: Das Arabische Buch.

WISER, W. (1936) *The Hindu Jajmani System*. Lucknow: Lucknow Publishing House.

YAMANE, D. (2000) Narrative and Religious Experience. *Sociology of Religion* 61(2): 171-189.

Ethnographic Notes on Incest and other Socio-Cosmic Prohibitions in the Kei Islands: Exchanges and Relations

CÉCILE BARRAUD

This paper[1] is based on essays written for conferences, and on discussions held with Jos Platenkamp during our never-ending talks on rituals over a period of more than twenty years, including our conversations while jointly editing a special issue of *Bijdragen* (Barraud and Platenkamp 1989/1990). For multiple reasons, until now, these notes and thoughts never have been put together and finalised for publication. As many of the following considerations are directly related to my conversations with Jos, this Festschrift is the appropriate place to reveal parts of them and to put them forward for discussion. The text presents mainly ethnographic data from Tanebar-Evav society in the Kei Archipelago (Southeast Moluccas).[2]

It focuses on incest and other prohibitions, but not on the question of incest as the basic foundation of exchange. The latter theme has been amply discussed in the history of anthropological theory and it is not necessary to go back to Durkheim (1896-1897) or Lévi-Strauss (1967) to postulate, if necessary, the universality of incest.

Why has incest been isolated as being 'the' prohibition and extracted from a system of which it is but an element in each particular society? In looking exclusively to the problems of exogamy and social groups, of kinship and marriage, the anthropological tradition has tended to forget the 'total social fact' aspect of a society and has elevated incest and its prohibition to the rank of a negative expression of a universal law of exchange, the universal principle of reciprocity whereby societies are ordered. The aim of this paper is more pragmatic.

My argument is that beyond the arrangement of social groups, incest, as a prohibition, is only one among a set of prohibitions constituting a system that is part of the overall socio-cosmic order. It participates in a more general arrangement of ideas and values ordering a society and is to be studied as such.

1 I wish to thank the editors of this volume for their patience and their corrections which considerably improved the paper. Thanks are also due to Michael Houseman who so kindly and promptly accepted to make a final revision.

2 Tanebar-Evav is the name of the island, of the village and of the society. Only one village of about 600 inhabitants exists on this tiny island that is situated southwest of the two main islands of the Kei Archipelago. Their main activity was to grow millet, the ritual organisation of which constitutes the basis of the organisation of this 'house' society. Fishing plays also an important role. The building and selling of sailing boats made the Kei people famous all over the Moluccas (Barraud 1979).

This argument does not deny the universality of incest, but tries to put the study of societies and their comparison before the development of more universal principles.

In Kei, this system, summarised by the word *fafelek*, 'prohibited', includes different types of incest, murder, adultery, all sorts of misconduct, marriage outside the rules, and the collecting of flotsam and jetsam washed up on the shore of the island. In all these cases, society as a whole and not only the responsible person, is concerned and has to be 'repaired'. Extensive ritual procedures are called for to undo the damage resulting from such transgressions.

My argument is also grounded in the comparison of rituals. In a former article, published in the special issue of *Bijdragen*, I was struck by the fact that there may exist a single ritual that implicates the whole society, under two different denominations, while generally, in other rituals, one or the other set of values is put forward but not both. These sets of values, which are present in Tanebar-Evav under the names of the society as *lór* and the society as *haratut*[3], have different parts to play in the same ritual and at the same time. The case was demonstrated in a paper on a ritual which is performed when a turtle has been caught while laying eggs in the sand (Barraud 1990a). But there is another case, a ritual held in the village when it has been sullied by incest, a ritual called *sob lór*, in which something similar takes place. As explained by a Tanebar-Evav woman: as members of *haratut*, all villagers in the morning give some money that they will receive back in the evening as members of *lór*.

This case raises the problem of the relationship between *lór* and *haratut* and more specifically, of the movement of beings and things between both entities in a sort of never ending exchange. Focussing on the 'incest' ritual might thus shed additional light on the circulation of beings and things in the Tanebar-Evav socio-cosmic order. I observed the ritual in question only once from its opening to its closure, but I witnessed similar ceremonies performed for 'lesser' forms of transgression on a regular basis.

Inward and Outward Movements

In Tanebar-Evav, two deities, the spirits *mitu* Hukum and Adat (who once migrated to the island from elsewhere) and to a certain degree another one, the spirit *mitu* Aturan, govern all the prohibitions mentioned. It is necessary to

---

3 *Lór* and *haratut* are two names given by the people to evoke their society, Tanebar-Evav. They refer to two sets of ideas and values, the value called *lór*, 'the society in relation to the world beyond' (beyond the island territory proper), and the value *haratut*, 'the society in relation to the land, the territory, the inner organisation' (houses, ancestors, etc.). The relationship between these two values, referring to different ideas, entities, rules and ritual practices, constitutes the overall ideological configuration underlying Tanebar-Evav social life (Barraud 1979).

begin with a general overview of these prohibitions and then to see – by way of their analysis – whether they form part of a system. As a guideline, one can follow a Keiese proverb referring to incest: '(Sea-)water sweeps into the hold of the boat, (rain-)water drips from the ridge-pole of the house'. The association of the sailing boat and of the house comes to the fore here; in addition the marriage ritual must also be taken into account.

That which is connected with the rules of the society as *lór* and the spirit Hukum (and to a lesser extent with the spirits Adat and Aturan) is characterised by specific movements, either inward or outward: beings and things washed up on the island, and beings and things expelled from the island.

In the case of incest, the guilty couple was sentenced to be drowned at sea. If the law of *lór* (Hukum *ni aturan*) is not followed, gardens are destroyed by animals, pests and vermins belonging to *lór* and epidemics will kill the people (this is also the consequence of not making the proper offerings when a turtle is caught laying its eggs in the sand.) A number of rituals are held at times to clean the village from all the misdeeds and to reaffirm the proper order either of particular houses (*mngam lór*) or the society as a whole (*tiva lór, sob lór*).

Other movements can be observed, which are at first sight not directly connected with *lór*: when sailing boats are sold outside and people bring back precious goods to the village, and when the life-principles[4] of beings and things are ritually sent outside in order to bring back pigs, millet and good fortune to the village during the millet ritual. In these cases, the movement is two-directional: going out and coming back. There are also occasions in which diseases are expelled in the name of *haratut.*

It seems that in the case of that which concerns *lór* exclusively, the movement is either in one direction or the other, never both.

However, one observes some differences between these movements: the turtle is caught alive on the sand, then killed, its parts offered both to *lór* and to *haratut*, while all other animals washed up on the beach are not treated in this way. They are offered to the spirit Hukum, i.e. *lór* only, and no special parts are distributed, except one: when the bones of the head and of the jaw (*kakean*) have dried up, they are placed in the ceremonial house of Hukum.

But the turtles that are born on the sand then leave the shore and go to live in the sea, which is a movement outward.

As a rule, everything that is washed ashore belongs to Hukum and before entering the island's territory, an offering of some kind has to be made to this spirit. The turtle is no exception, and the particular parts offered to Hukum testify to the idea of the encompassment of *haratut* by *lór.* Primarily, the turtle belongs to *lór*, but as it has not been washed up on the shore and since it is not

4 Persons are consisting of *itumun* ('body') and *mat-inya*, concepts that I formerly translated as 'souls'. Finally I chose the translation 'life-principles' which is less connoted.

dead, special parts are also offered to the society as *haratut*, that is, to the ancestors, *wadar*. Thus, both *lór* and *haratut* are implicated in this ritual.

In case of incest, there is in addition an explicit idea of 'repairing' what has been damaged (*dad, fadad, fadoar, fanahak dos*). What is damaged is the society as a whole, as stated in the proverb above, referring to the idea of a damaged sailing boat or a non-waterproof house. The repair is done by the whole society through the intermediary of certain houses whose function, in this respect, is named after the different tools used to construct or to repair a sailing boat or a house. These actions made during the ritual are ambiguous for, as we shall see, there is destruction and repair at the same time (cf. the details of the ceremony).

Actions ruled by lór and the spirit Hukum

A number of very different actions are found in this category. Some of them are known as *sa*, others, like adultery or murder are called by specific names; the remaining actions are connected with the collection of flotsam on the shore.

Sa, "wrong, not true, mistake, misconduct", designates all kinds of misconducts and wrongdoings. When followed by kinship terms (*sa uran, sa yanat-ite, sa avan, sa renan, sa yaman-yanan* 'brother-sister', 'child-parent of an upper generation', 'father's sister and mother's brother wife', 'mother', 'father child') it designates sexual intercourse with forbidden relatives, and other related acts, such as quarrels, overt speeches, or incitement to bad behaviour. *Sa* also means 'odd number' (while its contrary 'even' *ken*, also means 'true'). These acts are defined by reference to a non-transitive formal scale of measurements applied to the amount of the payments which are due for their atonement. The unit of these measurements is a fathom, *ref*, defined by the distance between two stretched out arms, a measurement also used in the context of the construction of sailing boats and houses. Unfortunately, the information given about this scale varies, even with the same informant, and although its overall order is clear, with respect to the less heavy sanctions, the type of act and the exact amount to be paid is not absolutely certain. A degree of secrecy surrounds these matters.

- *Ref* 44 is the heaviest payment and is required if sexual intercourse between people belonging to the category of father/mother and children, and brother and sister (*telfuhar:* same father and mother) has taken place. This *ref* is differentiated in *ref ya'an* and *ref warin*, older and younger, corresponding to the different houses receiving the payments. The ceremony held is *sob lór.*
- *Ref* 33 is also heavy, and is paid in the case of sexual relations between classificatory brother and sister standing in a relation of first or second degree cousinhood.

- *Ref* 25 and 70 are still fairly heavy, and have to be paid for sexual intercourse between brother and sister who have the same father but different mothers. The ceremony held is *sob lór kavinin*, 'hidden'.
- *Ref* 88 is a less heavy payment, the type of relationship is not certain.
- *Ref* 17, *sob lór kavinin* again, is the payment for *sa avan*, intercourse with FZ or MBW.
- *Ref* 15, is a light payment and concerns sexual relations among relatives separated by a distance of three to four generations; this payment is also due when one acts as an intermediary, *toar taruk*, between two persons who are not allowed to marry .
- *Ref* 19, is a light payment for intervening between two people who want to marry.
- *Ref* 9, another light payment, concerns relations between relatives of generations far removed, because "the placenta went in the right direction and flows far away, straight, (*beb nakso, den nasvak raro*).

However, if a misdeed considered as light is repeated three times it has to be 'repaired' by the highest payment like the heaviest one.

For each of these measurements a certain amount of precious goods has to be given as payment to the house where the spirit Hukum (associated with *lór*) is honoured, and to two other houses which act as the right and left hand of Hukum. In some cases, a fourth house receives a payment in the name of the spirit Aturan. For instance, in the case of *ref* 44, a canon or a gong[5], a jewel of great value *mas tal tel*, one or two others small jewels, a small amount of money (*mas tal* for the house called 'right hand', a cannon for the house called 'left hand') are due. These precious goods are not intended to replace the body of the guilty couple, as it is the case with the bride price. They are offered to the society as *lór*, that 'eats them'. If no payment is made, *lór* will ask: "where is my share? I will eat your food".

The role of the 'left hand' house is that of the axe that cuts down trees needed for the building of boats and houses. The role of the 'right hand' house is that of a chisel used to carve the planks of boats and houses. The role of the fourth house is that of an iron tool (with parallel tips and two legs of different length) used to mark two planks before carving them, so that they will be adjusted to one another.

Apart from the reference to the sailing boat indicated by the role played by these three houses, another reference is made: that of an animal with its head, trunk and tail, evoking the three parts of the village. This suggests that in this ritual, the village as a whole is concerned and affected.

According to the misdeed's weight, the payment is either made publicly during a long ceremony in which the whole village participates (*sob lór*), or privately, during a small ceremony at dawn involving only the members of the

5 Cannons, gongs, jewels are precious goods always used in exchanges.

concerned house (*rinin lór*). White cloth and white plates are used, because everything has to be clean and because the white colour 'protects' the village society. It is said that people will no longer 'talk' about the misdeeds once the payments have been made.

Other misconducts are considered as *sa* but not measured in terms of *ref*, although they require a compensation:

- sexual intercourse and marriage between persons belonging to different social ranks, i.e. nobles, commoners, dependants. Different types of payments are made whether the woman is noble or not. If she is noble, a jewel is offered to Hukum "to close the front door", another is offered to Adat, "to close the back door", and still another is given to the woman's mother's brother who can pass it on to his own mother's brother's house.
- when a woman is pregnant without being married *marvuan fa ovun*, ("he made his garden, there is grain already"). The payment is *vil fa ob* ('to envelope, close the shame' so nobody talks about the event any longer).
- the breaking off of an engagement, *mem fahar*, ('forbidden name').
- sexual intercourse in the house of Hukum.
- if people take back land that they had formerly given.
- if a woman from the village marries a man from another village and has to leave her own, "the wall is destroyed", *lutur varaha*.
- if a villager converts to a religion.
- murder, *fedan umat* ('to murder people').

The final category of forbidden actions ruled by *lór* is collecting flotsam and jetsam from the shore. One distinguishes between different categories: *lór mas* – 'gold': canoes, bottles, etc.; *lór tomat* – 'humans', or *ye ya'an* – 'shark's elder brother': tortoise, dolphin, whale, sea cow, human being (same idea than *lór tomat*); *lór balanun* – 'poison': stranded waste, garbage, etc. Human beings and edible flotsam (such as big sea animals) are thus clearly differentiated from non edible flotsam such as timber, pieces of wood, bottles, canoes and so on. All these beings and things have to be ceremonially welcomed before entering the village. Parts of the animals are presented in Hukum's house before being eaten by everybody, and in the case of a whale, nine shares are offered to the nine village's ancestors *wadar* and eaten by nine elders. Precious goods such as cloth or money are given to Hukum before using a washed up canoe. For smaller things, an offering of betel leaf and areca nut is made in a special plate for *lór*. When many flotsam arrive and are collected over a short period of time, a special ceremony called *tiva lór* is held in the house of Hukum. The villagers gather, beat drums and sing the songs of *lór* during a whole night.

When flotsam are collected and these offerings are not made, the consequence is the same as in the case of other misconducts: small animals belonging to *lór* will destroy the harvest. To avoid this, a ceremony called *mngam lór* (deriving from *mam*, to disappear; it means that *lór* takes the things and makes them disappear) is held at the house of the guilty person, a bracelet of a specific value is paid to Hukum, and a small pig and a small sack of millet are thrown out of the house through a hole made on purpose in the roof. Drums are beaten.

In the case of the heaviest incest, in the past, apart from the payment of precious goods, the guilty couple was drowned at sea. Another possibility was to send off to sea two different sailing boats, one with the man, the other with the woman; when they were far enough from the coast, the man would get into the woman's sailing boat, and his own sailing boat would come back empty (another version says that the couple goes on board in one boat, precious goods are placed on another one, and once far away from the coast, the goods are thrown into the sea and the couple moves to the other boat; one of the two boats thus comes back empty, and the spirit Hukum thinks that the couple has drowned). When this was the case, no *sob lór* ceremony was performed.

Sob lór, the Ceremony

Nowadays, people are not drowned, only the payments are made and the ceremony is held. This includes the destruction or beating of whatever is encountered when the men cross the village and the forest: people, houses, domesticated animals, poultry, coconut, fruit trees and so on. This is explained as *ta ba fo lór korba nba*, 'we go for *lór* and *korba*', the whale and the buffalo, considered to be the weapons of the spirit Hukum. *Lór*'s share is in the sea; the buffalo's share is inland. Coconut trees are destroyed because of the man's guilt, banana trees because of the woman's. The members of Hukum's 'left hand' house lead the society during this time of the destruction of trees and other things.

In the ceremony that I witnessed in 1985, all the elders of the village gathered in the house of Hukum to discuss the case to be sanctioned and to arrange the ceremony. Each spoke in turn, mainly giving advice to the youngest. "The gardens are growing but the misconduct is there and we have to be afraid of the consequences, that is, the destruction of the harvest by small worms *sasos*". "We do this in order to live quietly, that the gardens and the trees yield fruits; we have to repair for everybody, not for only one person". "This misconduct is poisonous food (*bina balanun*), it does not have consequences for only one person but for the whole society and for generations". "We go for *lór korba nba*, that is, we go in the name of *lór* and of the buffalo". "We pay for the sea and for the inland, drowning is the payment for the sea, destroying is to have silence (*vavear tahat roa/nangan: rlodok, vear roa, hafat, ngoan ven*)". They decided to go to 'beat' all of the houses of the village, proceeding clockwise, and then to destroy trees in the forest. If they were to proceed counterclockwise, they would not be able to 'hold' the case sufficiently firmly so as to resolve it (*taha batil, taha kuk*).

After an invocation in front of the house of Hukum, on a stone called Vat Vunil, they went to the house of the guilty couple were the family was waiting. All the villagers were there and had brought small offerings of money and betel and areca nuts placed on a plate in front of the sitting family. When all the elders had arrived, particularly the elders of the three houses of Hukum, they gathered outside at the right side of the house, where the gong for the payment was placed in front of them, as were then the jewels and the money, the palm-wine, the goat to be killed, and many leaves of betel and areca nuts. The money was worth 44, and there was a payment called *ref warin*, a golden bracelet with a value of two *tal* (a specific size), and another *ref ya'an* of a value of three *tal*. Two small golden jewels were added, one of them for the house of the spirit Aturan, *song sin*, and two other pieces of gold of which one was worth a cannon. The *mas hoan* (litt.: the golden jewel 'spouse') to be offered to Hukum was also placed there. The spirits Adat, Hukum and Aturan received their shares to bring back the life-principles of the guilty people. There was water in a small plate, and a coconut leaf that served to sprinkle water on the people participating in the ceremony to clean them. Afterwards, the goat was killed by driving a stake into its head between the eye and the ear (this is the specific way the first pig is offered to the spirit of his house by a young person who begins to raise animals). By doing so, a small hole is made through which the information about the wrongdoing is passed to the spirit *mitu* of this house, and the *mitu* will know the person. The goat is killed to clean up all the misconducts, and cast them out.

Then the participants assembled in front of the house of Hukum; older and younger men gathered there with their weapons and sticks. The elders made a speech to tell them how to proceed with the destruction and which paths were to be avoided. The women and children were asked to stay behind the men or to stay hidden in order not to be beaten or killed. The destruction lasted for about three quarters of an hour.

At least fifteen coconut trees, thirty fruit trees and fifty banana trees were cut down, twenty chickens and two goats were killed. Houses were beaten, dogs and pigs chased, small huts in the forest destroyed. When they returned, they gathered again in front of the house of Hukum and laid down the killed animals. Most of them were distributed among the villagers to be consumed at home, except one goat which is offered during the *sob mitu* ritual, the final ritual of the ceremony which is devoted to the God Sun-Moon to tell him that everything is over and to ask him to send back the life-principles of the guilty man and woman. People bring food to be offered to the God that later will be distributed among the participants. The small coins of money given in the morning are also distributed and some people throw them down the cliff on which the village is set, in order to cast away the sin. Then a ritual offering of betel leaf and areca nut is made in Hukum's 'left hand' house to tell the *mitu* of that house that everything is over; another such offering is made later at the cemetery to inform the dead.

The next morning, at dawn, an invocation is made in front of a small stone on the right side of Hukum's house by the three officiants, with offerings consisting of a small amount of money, betel leaf, areca nut, and tobacco.

Comparisons

The song of *lór* sung during the night of *tiva lór* explains what goes on during the ceremony. It uses compound words and parts of proverbs and is difficult to translate. It is about a type of wood which prevents the misconduct from doing harm, about a tree placed at the tip of the island to prevent bad things from entering the village from outside (*tal mirin*), about the shark which bites the sailing boat, cracks it, and eats the guilty people. *Lór korba* are the weapons of Hukum at sea and inland. Another of Duad Hukum's weapon is the tornado that sends two stones *vat sibit* to a sailing boat in order to destroy it if people are guilty. What are these stones? According to a myth, two *vat sibit*, black oblong stones, of about ten centimetres, made of a mineral unknown on the island, are said to have fallen from the sky after a violent copulation of the thunderstorm with the earth. They are considered to be the testicles of the tornado that itself is called "the penis of the rain". They are husband and wife. When they fell on earth, they destroyed everything, trees and stones. They are present in two rituals, that of the maiden journey of a sailing boat and that of marriage (Barraud 1990b). In the former, also performed when a sailing boat is going to be sold in far-away islands, one of the stones is guarded and fed by a small girl keeping watch in a specific house in the village during the journey of the sailing boat. She sits with her legs stretched and is not allowed to move or go outside. A little boy keeps watch on board and feeds the second stone. While the stones are considered husband and wife, the two children are considered brother and sister (for more details, see Barraud 1985, Geurtjens 1910). The two children are subjected to a number of food prohibitions of the type prescribed to the elders who lead the millet rituals. The two children cannot move until the villagers have returned to the island with the money of the sale. The crew on board is forbidden to wash and comb their hair, as are the hunters during the ritual hunt in the forest before the millet harvest. When a sailing boat departs to be sold, a second one follows it on the journey and brings back the villagers and the money to the island. On their return, an important ritual is held to bring back their life-principles from abroad. The money brought back – in the old times cannons, gongs and jewellery – is said to be the 'yield' of the sailing boat, as the millet is the 'yield' of the garden and the woman the 'yield' of marriage exchanges.

Using the analysis of marriage and funeral exchanges as a model, I am inclined to argue that here too relationships are primary and allow for the existence of social groups or entities. It is the relationship that gives its meaning to the island society in reference to the undefined far-away world (that of close, known islands as well as that of unknown countries). At the highest level of the ideology, the sailing boat at sea figures the society as a whole (Barraud 1985), implying that on another level, when a sailing boat is sold, something of the society has been given away which is replaced by precious goods that give the society its 'name', i.e. abundance of food, fame, superior strength of the spirits, and so on. The same happens in the case of the millet ritual, when life-principles of beings and things enter the village to make its name famous (after a successful harvest).

One of the two stones *vat sibit*, the male one, also plays a part in the marriage ritual. It is placed in a small basket together with two dried fruits found washed ashore, half a betel leaf and half an areca nut that an old woman from the bride's side hangs around the groom's neck. This is the main act of the wedding ceremony. The groom and the old woman from now on are forbidden to talk to each other until the groom has offered her food from the sea, the type of prestations given by the husband's side (*yan ur*) to the wife's side (*mang oho*) in marriage exchanges. Why only one stone of a pair, why this prohibition of a relationship? The lack of completeness (half a pair of stones, half a leaf and half an areca nut) is as a sign that the totality of the exchange has not yet been achieved but will only become manifest with the birth of the first child who will continue the relationship and honour his *duad-nit*, the ancestors of his mother.

While totality was obtained with the successful return of the sailing boat and the junction of the two *vat sibit*, in the case of the marriage ritual, totality is achieved diachronically through the succession of generations. This connexion with the sailing boat is significant, for the precious goods (cannon, gongs, jewels) of which the bride price is composed and which are said to "replace the body of the bride" are named after the parts of the sailing boat (its keel, planks and oars). People say that in the old times the compensation for the bride consisted in a sailing boat together with one item from the land and one item deriving from the sea. Traditionally a marriage mainly concerns two houses (although the whole village participates), while the ritual for selling sailing boats to distant islands in the Moluccas concerned the whole society and the island (i.e. living human beings, spirits and ancestors). These two rituals enact relationships on different scales and on different levels. As a result of exchanges between affinally related houses, something departs from one house and allows for the very existence of another house, at the same time allowing for the latter's own ancestors to be looked after. In the case of the sailing boat, something connected with the existence of the far-away world departs from the island but at the same time allows for the very existence of the village society, its 'name'.

Of Movements and Relations

Why this long digression about the sailing boat and the marriage ritual? The first reason is connected with the proverb already mentioned which qualifies incest as "(Sea-)water sweeps into the hold of the boat, (rain-)water drips from the ridge-pole of the house". Thus not only the boat and the house are associated with the wrongdoing, but also their destruction that, as we have seen, is effectuated by the *vat sibit*, either the tornado or the thunderstorm. It means that the destruction has already begun when the proper flow of the blood, from the ancestors of the woman' side's (*mang oho*) towards the man's descendants (*yan ur*)[6], is not respected and when the generations are mixed, as in different types of incest. The proper relationship and movement – to depart and to bring something in – is destroyed, which implies the destruction of one's own existence.

The association of all kinds of incest with the collection of flotsam on the shore calls for comparison: flotsam collected without an acknowledgment of its arrival by an offering to the proper spirit, looks like the denial or destruction of the relationship with the far-away world: on the one hand, that of the surrounding archipelagos where boats are sold and whose names are mentioned in the invocations intended to bring to the village the life principles of human beings, of millet, and pigs during the yearly millet ritual, and on the other hand, that of the unknown places from which the spirits Adat, Hukum, and Aturan came. The first one, Adat, is at the same time considered an ancestor and is one of the most important spirits partaking in the millet ritual, while Hukum, his younger brother, plays the most important role in connection with *lór* (incest, murder, adultery and flotsam). Aturan also comes from the outside, but is the King of the Golden Mountain, the spirit to which the entire millet ritual is devoted.

The second aspect calling for comparison is the movement of the two boats with the guilty parties, who in the old times were drowned at sea: one boat goes with the couple, another with the precious goods. The latter are thrown into the sea, the couple climbs on board the second boat, and the first boat comes back empty. When selling a boat, the same movements are involved, two boats depart from the island, one of them is sold and does not return and the other one returns with the money. However, when comparing these two rituals one observes an inversion: in one case, precious goods are thrown away, in the other, they are brought back.

6 As in the case of MBD marriage which is the rule in Tanebar-Evav society.

Another important connection with sailing are the names of the three main officiants of the incest ritual, referring to the specific tools used to build or to repair a boat or a house. As these officiants belong to the three different groups that constitute the village space as a whole, there is strong evidence that it is they who repair the sailing boat, i.e. the society as a whole, while at the same time the particular misdeed of a particular house is erased.

Finally, in the incest ritual as in the ritual for the sale of a sailing boat, the life-principles of the concerned people have to be called back to their right place, i.e. to the territory of the island.

A conclusion can be drawn from these facts. The rituals performed after incest or after the arrival of flotsam are not only supposed to clean the pollution caused by incest. They are meant to re-establish the proper relationships both between houses and between the island society and the far-away world, in this case explicitly called *lór*. This would explain why in the ritual chants reference is made to things coming from the outside which are prevented to enter by timbers, to the destruction of sailing boats by sharks, and so on. It would explain why the villagers themselves destroy and cut down everything instead of waiting for *lór* to do so by means of the *vat sibit* or the tornado, and why they say that: "they go (to destroy everything) in the name of *lór* and the buffalo". They do themselves what *lór* should be doing and thus prevent it from doing so. This would also explain why the payment of a cannon, a gong, and jewels are not meant to replace the guilty persons bodies, while on all other occasions when these items are offered, it is to replace a body. The 'replacement' in question takes place in the context of a relationship, and in the case of incest and flotsam, relationships are no longer acknowledged. The beating of drums during the ceremony *tiva lór* recalls what takes place when the villagers come back from war: it is a welcome acknowledging their return and their success.

It would further explain why the vermins destroy the gardens, because they belong to *lór* and *lór* says: "you do not feed me, thus I eat your gardens". It would finally shed light on the question of why both *lór* and *haratut* as such participate in the ritual, and in such an explicit way: the villagers give money in the morning and receive the same amount in the afternoon (see above, in the introduction). The villagers enact the relationship between *lór* and *haratut* and re-establish it in the internal space of the island and of the village itself in order for this relationship to exist as a whole and in connection with the outside world.

To answer the questions posed in the introduction, one might say that incest actualises at a subordinate level, that of the houses, what the collection of flotsam does at the superior level: it breaks up a totality, that is, a relationship, it creates incompleteness (the odd number, *sa*, as opposed to the even number, *ken*, which refers to the idea of completeness as in the measurements of houses and boats).

References

BARRAUD, C. (1979) *Tanebar-Evav. Une société de maisons tournée vers le large.* Cambridge and Paris: Cambridge University Press & Éditions de la Maison des Sciences de l'Homme, série Atelier d'Anthropologie Sociale, dirigée par Louis Dumont.

BARRAUD, C. (1985) The Sailing-Boat: Circulation and Values in the Kei Islands, Indonesia. In BARNES, R. H., DE COPPET, D. and PARKIN, R.J. (eds.), *Contexts and Levels: Anthropological Essays on Hierarchy*, pp. 117-130. Oxford: JASO (Occasional Papers n°4).

BARRAUD, C. (1990a) A Turtle turned on the Sand in the Kei Islands. Society's Shares and Values. In BARRAUD, C. and PLATENKAMP, J.D.M. (eds.), *Rituals and Socio-Cosmic Order in Eastern Indonesian Societies* (Part II). *Bijdragen tot de Taal-, Land- en Volkenkunde* 146(1): 35-55.

BARRAUD, C. (1990b) Wife-Givers as Ancestors and ultimate Values in the Kei Islands. *Bijdragen tot de Taal-, Land- en Volkenkunde* 146(2/3): 193-225.

BARRAUD, C. and PLATENKAMP, J.D.M. (eds.) (1989/1990) Rituals and Socio-Cosmic Order in Eastern Indonesian societies (Part I and Part II), *Bijdragen tot de Taal-, Land- en Volkenkunde* 145(4) and 146(1).

DURKHEIM, E. (1896-1897) La prohibition de l'inceste et ses origines. *Année Sociologique* 1: 1-70.

DURKHEIM, E. (1969) *Journal Sociologique.* Paris: P.U.F.

GEURTJENS, P.H. (1910) Le cérémonial des voyages aux îles Kei. *Anthropos* 5: 334-358.

LÉVI-STRAUSS, C. (1967) *Les Structures élémentaires de la parenté* (2nd éd). Paris-La Haye: Mouton.

Secularisation or Normalisation of Religion in Indonesia: From Custom to *Pancasila* and Back to *Adat*

STEPHEN C. HEADLEY

Introduction

How could anyone have known that Joko Widodo would defeat Prabowo Subianto in the summer of 2014? There are even more fundamental questions to be asked about contemporary Indonesian society. In the last hundred years Indonesia has moved from *adat* (custom) to *Pancasila* and back to respect for *adat*. Does this imply a promotion, acceptance of a normalisation of extra-legal norms, more explicitly religious ones? After Suharto's three decade long denigration of village rituals as superstition, when cultic diversity as such was considered dangerous, are we seeing the return of a new normalcy, a tolerant one? Are older ideas and practices relating to exchange – the keystone of Javanese village *adat* – not deeply affected by those 'modernising' factors of national politics where the role of Islam has experienced successive new positions? In a country such as Indonesia and especially Java, where private property and the law of eminent domain has been applied for two hundred years are we looking at a healthy new balance in social morphology?

These are basic questions which I am going to address in this paper. After independence in 1949, the standardisation of religion in the Republic of Indonesia was animated by a preoccupation with "unity in diversity". All citizens were to be monotheist for monotheism reflected more perfectly the unity of the new nation state than did the great variety of cosmologies deployed in the animistic cults. Secularisation in Europe (*circa* 1700-1900) initially meant confiscations of church property and only later came to mean lesser attendance to church services. Still it also involved a deeper shift in the epistemological framework. It redefined what it meant to be a person. Anthropology in societies where religion and the state are separate is very different societies where the rulers and the religion agree about man's destiny. For example, the French republic, has no cosmology referring to heaven and earth; their genealogical amnesia concerning the Christian origins of the Merovingian and Carolingian kingdoms is deliberate for the universality of the values of the republic were liberal in the sense of broad. Man, as the measure of all things, must dominate the traditional cathedral-centered view of the *polis*. The person was defined as citizen of the republic.

When the Indonesian Republic was founded as a secular state, this was interpreted to mean a very limited separation of church and state. More specifically this took the form of a shift: monotheism *à la Pancasila* and not Islam would be the basis of citizenship. Naturally Islam seemed to be the common denominator to many inhabitants of the archipelago. The anti-colonial, nationalist movement had already shown that ethnic labels could be removed from the insular origins of Muslim groups when they worked towards "unity in diversity". A disenchanted cosmology provides a secular (i.e., this century) space where nature can be manipulated. In the past Indonesians wouldn't sell their rice fields because it was bestowed on their ancestors by the rice goddess Sri. In the twentieth century this view of ownership was understood to make for poor citizenship (Asad 2003: 27). The domestication of 'religions' to the norms of a republican social morphology required secularising those horizons of transcendence which risked contradicting the pretensions of the new nation to being the greater 'whole'. Cultural diversity, too many cults and customs with their strong public norms and values, challenged the secular structure of the republic which proposes citizenship instead of tribal collective identities.

Only Islam was in a position to oppose such a ground shift, for only Islam, having organised itself into political parties in the first quarter of the twentieth century, was able to operate on the level of statehood. For more than ten years, starting in 1949, S. W. Kartosuwirjo's regional rebellion *Darul Islam* challenged the new republic as would-be founders of an Islamic state. Since the suppression of the Darul Islam other models of re-establishments of Islamic states have been imported from abroad, such as Hizbut Tahrir.[1] Radical groups such as the Solonese branch of this movement give gentile titles to their public meetings in order to obtain permission to hold them. They implicitly recognise that nowadays encouraging violence does not bring in votes.

In 1945, Muslims demanded that *shariah* should be obligatory for all Muslims. The so-called Jakarta charter was premised on the belief that acculturations represented by regional custom or *adat*-based religion presented an unwanted imbeddedness to a local, insular Islam. For different reasons Muslims agreed about this latter diversity, and from 1949 till 1955 the insistence on this *piagem* Jakarta was abandoned. Once the Soeharto regime came to power (1967+), military-imposed unity was strengthened, and openly declared as the 'Raison d'État', and eventually it served as the pretext for suspension of legal procedures that smacked of local *adat*. Regional custom came to be eliminated by a republic in the name of the rule of law, by claiming it did not qualify as religion. Successive Indonesian governments and Islam have seen society through such a prism of uniformity. For *adat* difference was the prism of another kind of social articulation. For Islam, Allah's sovereignty is reflected in

1 In Arabic *Hizb ut-Tahrir* means "*Party of Liberation*" and boasts itself to be the vanguard of Sunni Islam. Its aim is to join all Muslim states into a caliphate ruled by an elected caliph. Its founder, Taqiuddin al-Nabahani, a court judge (*qadi*) from Haifa, founded this movement in 1953; it is now present in some forty countries with around one million adherents, and is especially active in Western Europe.

the mirror of mosque-based prayer, while for the republican government political control and surveillance were activities which the Javanese, and their pre-eminent kingdoms, placed above the laws of God. Such, in brief, is the Indonesian republic's vision of secularisation until recently.

This paper considers the post *Reformasi* (1998+) demand for an application of *shariah* to all Muslims. Such a move could further weaken the diversity of local *adat* which implicitly had been admitted by normal *Pancasila* 'plurality'.[2] This article is less a political or ethnographic study than a general reflection on Javanese *adat*. My stepping off point is the following: Javanese culture presupposes that differences create spaces for communication. Political strategies for accumulating identity hinders the creation of such spaces. When difference is treated as division, as in the Soeharto period (1966-98), the relations created by the differences are destroyed. The 'other' becomes a danger and the different parties can be easily instrumentalised. Difference is 'good' where the cultural and religious groups have kept each other in focus. For that to be so, above and beyond their differences, some common 'higher' values must motivate them to remain contact. Following Louis Dumont, I have called this 'encompassment'[3], a concept he used in his description of Indian castes.

Adat as Javanese village custom was traditionally 'encompassing', but not separated, from the law of the palace. Especially in settling village disputes over land tenure, the authority of the founding fathers and spirits of the village were expressed in local cults' injunctions. *Adat* was an indigenous anthropology, a religious and administrative synthesis for local use. It was rendered redundant by the arrival of direct colonial administration as they converted peasant lands into plantation holdings. I am deliberately opposing state-sponsored *Pancasila* management to spontaneous religious praxis because I consider that Soeharto's approach to this normalisation imposes it as a norm, a substitution of religion by secularisation. This included the obligatory study of the *Ordre Baru's* own version of *Pancasila* ideology[4], the substitution of New Order sovereignty above and beyond Allah. The political regime protected, 'umbrella'-ed (*dipayungaké*) and overshadowed the more traditional, 'normal' wholeness of the Godhead, the *tauhid* of Allah. Paradoxically endemic local insecurity in the *reformasi* period (1998+) encouraged the re-emergence of older notions of sociability which permitted the creation of local micro-holisms, each cultural area possessing its own character.[5] Sometimes,

2 I am referring to *Pancasila's* as propagated under Soeharto's New Order. I am not referring to its 2006 *Pancasila* re-promotion as the indispensable national foundation (*dasar negara*) of the republic by such distinguished figures as Azyumardi Azra.

3 Cf. Louis Dumont (1983).

4 On the debates between 1945 and 1978 on the content and way in which *Pancasila* should be studied cf. Marcel Bonneff et al. (1980).

5 Among the new approaches to kinship, the term sociability is used in different ways. Cf. Janet Carsten (2000).

as on the north coast of Java these outbreaks were orchestrated by outside *dhalang*, but in southern Borneo or central Celebes, the results of tensions emerging from immigration (so-called *transmigrasi*) had produced much deeper roots of local conflict[6]. The return to local *adat*-based administration sometimes reduced the levels of intolerance provoked by the top-to bottom *Pancasila* approaches to local religious diversity. To conceptualise the diversity of *adat* revivals outside of Java, I am introducing more recent data collected on this topic.

Adat in Indonesia Today

The post 1998 re-deployments of local *adat* have been surveyed by Davidson and Henley (*The Revival of Tradition in Indonesian Politics*, 2007). I begin by taking into account the Indonesia-wide results of their research before returning to the Javanese data which is the topic of this paper. Is the revival of different *adat* in Indonesia since 1998 no more than vague appeals to "sedate order and consensus", as Davidson and Henley suggest? Since 1998 has *adat* everywhere come to connote activism, protest and violent conflict? This shift in connotation is certainly true of certain areas with Bugis immigrants, most notably in Kalimantan and Sulawesi, but not everywhere.

The issue of political recuperation of conflict is more general than that of ethnicity, for the social tensions sown by the New Order regime were to be found everywhere. In their first congress in 1999 the Jakarta-based *Aman* (*Aliansi Masyarakat Adat Nusantara*) stated pointedly: "If the state will not acknowledge us then we will not acknowledge the state". After Soeharto's fall in May 1998, this is comprehensible in the context of the "decentralization and dismantling of the authoritarian state". Yet is custom really 'radical conservatism', arming a new-protean politics of *adat*? Although Davidson and Henley admit that "Today much of the way societies are organized, rights allocated, and disputes resolved in Indonesia still has little to do with the state or its law", they nonetheless defined traditional *adat* in two ways: (1) "a complex of rights and obligations […] tying together history, land and law in a specifically Indonesian way" (2007: 3); (2) "a formulation of an ideal society".

Davidson and Henley immediately add: "Neither our concentration […] on power politics nor our use […] of the term 'deployment' should be taken to imply that political manifestation of *adat* necessarily reflect the cynical manipulation of tradition by self-interested actors". But why must one defend *Aman* or any other group from instrumentalisation? If Davidson and Henley (2007: 4) are trying to provide a "nuanced and critical" assessment of the way

6 Over the last few years inter-religious violence has been provoked in Tassikmalaya (*Institut Studi Arus Informasi*, 1998), Situbondo in east Java (Hariyanto 1997), southern Borneo (*Yayasan Lembaga Batuan Hukum Indonesia*, 1997), Aceh, Maluku and Riau (*Tim Peneliti LIPI* 2001).

adat is used to redress basic injustices, the very nature of *adat* sociability, of exchange of values, should be the focus of their evaluation. They share to a certain extent the prejudice of the Soeharto era and its *Pancasila* state-administered normalisation of religion that ignores how local societies reproduce themselves in space and time.

The recent diverse adoption or revivals in Indonesia of local *adat* may be understood as an exercise in empowerment that brings along with it "inherent restrictions and dangers". Having tried to circumvent "Indonesia's notoriously corrupt and ineffective formal legal system" (Davidson and Henley 2007: 4), the adopting of *adat* is not that often a justification for ethnic violence. How often are individual interests really being subordinated to communities dominated by traditional elites? Davidson's and Henley's rush to judgment leads them (2007: 4-5) to the view that as a political ideology, *adat,* is especially subject to political manipulation because it is an informal, un-codified idealisation of stability and order. But this is a comparative political approach, preoccupied by the topics of democratisation, legal reform and ethnic violence and not an anthropological one where in a given cultural setting one would want to ask which values are circulated by customs. A preoccupation with 'power' is a straight-jacket which hinders the understanding of the social morphologies articulated by the revival of *adat.*

Davidson and Henley present four factors to explain the nature and causes of this revival of custom: (1) support from international organisations; (2) oppression of the New Order regime; (3) opportunities of *reformasi*, and (4) the role of *adat* in the political imagination of nationalism. Is it true as the cynical observers of indigenous identity believe (2007: 6), that 'indigenous' is defined not so much by ancestry, culture or marginality and more by wily recuperation in the international discourse on indigenous rights? While these authors realise this is an extreme view, they tend to refuse any coherence to culture because of their horror at its instrumentalisation. What today in our world is not instrumentalised? What of the unending lies which are marshaled to project the practice of democracy? So what if for some 'indigenism' is nothing more than a 'handy tool' in the struggle against globalisation, as long as it serves local interests? Davidson and Henley (2007: 9) show that the second cause of the revival of *adat* is that of the disenfranchisement, the refusal of political rights. Under Soeharto, cultural difference and ethnic identity went unrecognised and were brutally punished if proclaimed in the public sphere. After the fall of Soeharto (1998), the reaction in areas of heavy transmigration was one of renewed self-organisation of the indigenous populations and, more rarely, pogroms towards immigrants by the 'locals'.

Once the top was off the pressure cooker of the Soeharto decades, the *reformasi* era provided positive opportunities (Davidson and Henley 2007: 14) "[...] to find bases for political order other than the bureaucratic hierarchy, economic patronage, state propaganda, and the military force of the New Order". Under Soeharto's New Order, religions were directly administered via *Pancasila* which was touted as promoting 'tolerance'. This in fact meant 'normalisation' or standardisation according to the norms of universal scripture based religions, for those areas 'lacking' real 'religion' (i.e., monotheistic *agama*). The difficulty of creating new democratic institutions at the local level, the prevalence of social injustice, the weakness of the existing legal system, and the threat of violence and disorder inclined people in many places to look to tradition as an alternative source of both consensus and justice. This revival of *adat* in certain areas like Lombok and West Sumatra can be seen to be a secular complement to Islam without any ultimate cosmological reference.

One of the main concerns of Davidson and Henley is to distinguish between protection through, as opposed to domination by *adat* 'politics'. Tania Li, in the final chapter of their book (2007: 365-366), claims that both on the local and the national level appeals to *adat* "tend to privilege elites, especially senior men, who are empowered to speak on behalf of a presumed whole". Worse still, the link between ethnic rights and ethnic cleansing seems to these authors to evoke the *Blut und Boden* of Nazi Germany. Clearly there are limits to the usefulness of an empowerment so locally focused. But local micro-holisms are nonetheless real social groups, and an improvement on the anarchy that the New Order left behind. If everything is treated as a question of empowerment, then there is no need to speak of any hierarchy of values, for power is the only value that characterises social relations.

Davidson and Henley (2007: 31) also defend another thesis, namely that in Indonesia Islam has historically provided the most "effective bridge between local cultures". For if one is looking to create *civility*, "in the Enlightenment sense of toleration and restraint with respect to differences in culture, opinion and faith", then it is possible to claim that Islam "plays a kind of civilising role in Indonesian political life". This is not true everywhere for as the authors write (2007: 32-33): "The current *adat* revival, not coincidentally, is concentrated either in areas where the progress of Islamic conversion is blocked by Christianity or Hinduism, or in areas where Islamic conversion has taken place but pre-Islamic elements remain unusually important in social life [...]".

The authors' ultimate critique arises from their belief that *adat* is almost always an expression of some wishful thinking about the ways things ought to be rather than how they are. Given the weakness of the formulations of *adat* in the last sixty years, this fear is understandable. Nor has *adat*, historically speaking, fulfilled its promise and therefore (Davidson and Henley 2007: 33), "*adat* as a guarantee of peace and harmony is misleading not only as a prescription for the future but also as an interpretation of the past". Yet can not the same thing be said of democracy, that hope-filled renaissance of sociability always profiled on a yet further horizon?

Can *adat* be measured by the yardstick of civil society? What is supposed to transcend class, gender and religion is secularism; failing vertical hierarchies and increasing horizontal solidarities are interpreted as giving people reasons for adopting a secular ethic. Yet the credibility of the concept of 'civil society' has had a short shelf life in many parts of the world. In Eastern Europe (Hann & the "Civil Religion" Group 2006), the cement of nationalism is strongly linked to identifications of forms of eastern and western Christianity. Here we see how the most universal of religions can be provincialised (i.e., Poles are Catholics, Russians are Orthodox, etc.). But then why *a priori* should civility be defined by trust in and the recognition of individual agency in complex societies? If that is the case then the Indonesian *adat masyarakat* communities had better be dissolved in order to attain a more complete reign of the ideology of individualism where a "principled conception of the individual upon which rights and duties and obligations of a civil polity and society" can be brought centre stage. That way we could slowly eliminate "[...] the *adat*-based claim to preferential rights on the basis of historical precedence, cultural identity, and territorial possession, a claim which challenges the principles of equality and legal impartiality on which Indonesian (and most other modern) citizenship, at least in theory, is based" (Davidson and Henley 2007: 35). Having subjected *adat* to such a severe critique, Davidson and Henley finally recognise (2007: 38) that *adat* is a "specifically Indonesian ideological tradition in which land, community and custom [...] provide the normative reference points for political struggles". Secondly that (2007: 41) "Firm legal acknowledgement of customary land rights, together with restitution of already alienated *adat* land [...] are matters of simple justice". But, unless one speaks in the most abstract legal terms, the revival of community trust is no simple matter of justice. If one adopts the point of view of the community, (and the authors recognise this [2007:40]): "[...] national legislation is increasingly irrelevant to local developments) *adat* has been caught up (in) a profusion of local conflict and contestation which for the moment is largely beyond legal and institutional control." So is *adat* just a "tool of ethno-politics" as Li (2007) suggests in the concluding chapter?

The Foundation of Javanese Polities: From Myths to Kingship to State and Back to Adat

Current views of society in Java, while necessarily divergent, can be brought into focus by looking at the partial revival of *adat* as an alternative to the society propagated by the Indonesian republic, as said above. Both Javanese Islam and Javanese *adat* share a preoccupation with promoting non-governmental forms of sociability. *Adat* is good for circulating social values, and as such is a norm immediately beneficial to its local practioners. Not that this is the primary phenomenon in the social landscape, but its use in the revival of village life, both rural and urban, favours local autonomy. This is the best indicator of what the fall of the Soeharto regime has permitted in terms of new forms of sociability.

The word society may not have been part of the Javanese lexicon until recently, but all things have a beginning and even Javanese myths described the origins of their world's order. The initial foundation of Javanese custom and religion by indigenous Javanese rulers is recounted in a myth concerning the first Javanese kingdom (Headley 2004, ch. 2-3). As elsewhere in Southeast Asia (Platenkamp 2008), in Java royal affinity was the foundation of the mythic foundation of the first kingdom. The elder sister and rice goddess, Dewi Srī, abandoned her incestuous relationship on earth with her younger brother Sadana and became a python (*naga*) in the rice fields, guaranteeing their fertility, while Sadana became the first king in Mendhang Kamulan (toponym meaning "origin of the rice shaft"). Shortly afterwards they both leave their new roles and return to heaven when ordered back by Shiva. The elder female sibling is still the rice goddess for the Javanese peasants while her double, the maid (or queen) of the southern ocean (Lara Kidul), becomes the invisible consort of the kings of Surakarta and Yogyakarta, resting on a bed in the heirloom (*pusaka*) room behind the throne. Meeting or copulating with her was the privilege of the reigning king, but the whole palace was mobilised to present her with offerings every Thursday evening, just as the peasant women did at the same moment in Sri's granaries in their village homes. Here elder means not autochthonous but sexually potent and the origin of sovereignty (*daulat*). Kings come and go, but Srī never changes. She is a kingdom's ultimate heirloom (*pusaka*). The primordial myths supposed by *adat* are displayed in her cult on Thursday evening in rituals celebrated by rice cultivators and royalty. *Adat* in this sense encompasses the behavior of female rice farmers and male kings. Srī is both an elder sister and a partner for the king Sadana. Her name, written in Javanese script, is found over the lintels of the doors of the inner rooms of the palace. There it is less a name of address than an epithet; Srī means auspicious; she is the consort of Vishnu (name given to Sadana, once a king). Access from our lower world (*jagad alit*) to the higher world (*jagad gedhé*) passes through elder sister rice goddess. For peasants' access to this macrocosm which blesses their crops and guarantees the appropriateness of the ritual sacrifices is guided by their customs that structure and found her cult. The greater and lesser world need to be brought into correspondence for the prevention of epidemics and coherent sovereignty. The formal properties of that 'well-fittedness', that adequation, are recapitulated in the elder female and younger male incest, which needs to be undone for these are now separated polarities providing trans-generational fertility and intra-generational potent sovereignty. *Adat* remains senior to kingship, providing younger sibling an occasion to encompass what might have remained unapproachable, elder sister. In the myth of Srī and Sadana incest produces a disjunction and then a series of similar 'couples' who inbed the original incest, and then are obliged to abandon it and go their separate way bearing traces of their hierarchical intimacy.

Other myths also describe this period as that of a disjunction or separation of the macrocosm from the microcosm. As the heavens are separated from the earth, the shaping of the first human society is recounted. The human body is taken as a trope of the shape of society. The myth of these primordial times understand a completed society through the engendering, the feeding and the assemblage of incomplete human persons (cf. Headley, *L'Homme* n°174, 2005). In these so-called 'half men' myths, incomplete persons provide an occasion to reflect on what makes society whole and how the incomplete or partial body of a person is a trope of an incomplete society. Here in tabular form are two Javanese 'half man' myths (Headley 2005) showing how both unfinished half-men represent incomplete societies:

CODE	*Half person:* *the Sléwah boy (Jarasanda)*	*Half person:* *young man, Kala*
1. Conception	By fission : a split fruit	by disjunction : sperm fallen into the ocean in the absence of his mother
2. Nutrition	zero degree of marriage: solitary and invincible	Kala seeks human blood for food 'because' he never knew his mother
3. Mariage	by bisection: he can only be killed by being split on a meteor	Kala's is given by his father a cook who is his mother
4. Killing	zero degree of nutrition: Jarasanda becomes a food offering but even the ogress find him uneatable	Kala is weakened and neutralised by the ritualised recounting of his own incomplete conception in a genesis myth

Fig. 1: Two Javanese 'Half Man' Myths

The primary interest of studying these Javanese myths, which today are no longer well known, is to understand how the Javanese used to think a whole human and hence a whole sociability was to be constituted. In the nineteenth century such myths troubled the strictly reformist Javanese Islam but, of course, not the various Javanese 'interiority groups' (*kejawén*; *kebatinan*). Although Java was once a traditional society, in the twenty-first century clearly Java presents hybrid contradictory traits one can label pseudo-holistic (cf. Headley f.c.). Nevertheless, one can say that a person's holist axis still exists behind the multiplicity of Javanese beliefs and religious praxis.

Andrew Beatty (personal communication; cf. 1999) is of the opinion that the notion of an individual spiritual culture is alien to the praxis of social conformity in daily Javanese life. Mainstream mysticism is about group sessions, a hierarchy of anti-individual ethic. Theirs is a universal self, not at all a New Age personal creation. Javanese mysticism is about the relationship between community and cosmos.

Conclusions

At every level of an analysis of Javanese village life, we find both endogenous and exogenous factors constraining the reproduction of local society in space and time. Their simple assignment to one or the other of these categories (local/national; traditional/modern; indigenous/imported, as in Islamisation or *dakwah*) is commonsensical and of itself does not explain anything. To say that the authority of village custom was overwhelmed by the power of the colonial prefectural (*kabupaten*) administrative and police apparatus or, for instance, that villagers were obliged to institute the standard urban Muslim ritual calendar replacing the local ones commemorating local protective spirits (*dhanyang dhusun*), etc., are important descriptive statements, but not yet analytical ones.

Standardisation can take many forms including *pro forma* normalisation, one example where one simply pronounces as a state of fact something which 'seems' to have already taken place, i.e., the recent construction of a mosque 'makes' a village Muslim. This is, in fact, not a statement about the religious praxis of the villagers but about what the builder of the mosque thinks of their 'backward' beliefs. Recall that at the end of the *cultuur stelsel* (1860's), the Dutch colonialists were preoccupied with 'traditional village institutions' protecting an imagined village equality, which they had just done their best to destroy. We could call this contradictory kind of normalisation counter-factual, for the conditions pertaining to its realisation have already disappeared. In the 1980's, a hundred years later, the 'culturalisation' (*dikabudayaké,* in Javanese) of the arts for local or foreign tourism of Javanese *adat* was understood to be the death knoll of village ritual entertainment where its exploitation was organised for by the newly enfranchised middle class (Pemberton 1994). In fact, if 'pure' village rituals are generally all about articulating scales of difference, it is because these are hierarchical. It is thus that they reproduce themselves from generation to generation and they maintain their vitality today. They account for the totality of relations that order society. Clearly the governance deployed by the Republic of Indonesia sought to control villages, yet despite state terror under Soeharto, local hierarchies of values did not always disappear.

What we need to understand is how local society manages to preserve its internal constraints and how on other occasions it is obliged to abandon them definitely faced with 'modernity'. After the long night of the Soeharto New Order regime (1966-1998), what surprised people was the revival of both local Javanese *adat* and, even before Soeharto fall, that of Javanese Islam. This should not have been a surprise, however, because both are *prima facie* addressing the defects of the New Order. In what way do they differ from *Pancasila*? Nominally, in the abstract, the values they express are rather similar. By contrast, the consensus of the New Order ideology was anything but spontaneous. There is less artificiality in the redeployments of Islam (its renaissance (*nadha*) through regeneration (*dakwah*) and in the Javanese (*kejawèn*) efforts to remain traditional (*asli*). A cynical effort to willfully impose a norm unaccompanied by a wide local consensus breeds anomie, the opposite of what the initial '*pancasila*-sation' of religion in the early years of the Republic was supposed to achieve. As Umar Dani, founder of the Solo conservatory of dance and musical arts, said to me in the mid 1980's, *pancasila* is best suited for creating atheism. The authority of custom is not consciously constructed, even if it may be the object of competition. Who really defends it? Momentarily in a village the young may be harangued into action "because this is our *adat*". Debates exist among Muslims concerning whose vision of Islam is the purer one. What secondary customs can be dropped or updated as opposed to primary fundamental traditions? When their normative nature does exclude inspection, when revealed (*wahyu*) texts and the ascetical art of cultivating one's *ressenti* of the divine can be objectified, one can feel pluralism at work.

If what I have said here by way of a conclusion has any general validity, then one should be able to distinguish between the manipulation of a norm and its place in a hierarchy of values which a traditional society has inherited, follows and perpetuates. The second option is clearly foreign to modernity based on a conception of unilinear acceleration, as Hartmut Rosa (2013) has recently tried to show. Since the fashion of the 'modern' democracy has rarely kept its promises, many Javanese find the renewal of stricter Islamic practice, top to bottom sponsorship of Ministry of Religion programs, and sub-prefectural reforms of corporate responsibility less palatable than *adat* such as they remember and practice it. This is especially true when one is dealing, not with building roads or installing high tension pylons, but managing daily life through the help of one's neighbors or marrying one's daughter. In Java, secularisation began with an implicit distinction between local tradition and the foreign and later the secular. For centuries the *santri* managed to avoid being recuperated by the successive *kraton* (palaces). This meant avoiding the authority of the different administrations of religion. Such a ministry was created in 1946 with a Muhammadiyah incumbent, Rasjidi, in 1949. With independence, the NU (Nahdlatul Ulama) began to take over the *Kementrian Agama*, for they still hoped that Islam would be proclaimed the official religion of Indonesia.

Just as the Dutch in the nineteenth century co-opted *adat*'s capacity to mobilise local networks, Soeharto also used local traditions of *sambatan/gotong royong* to impose *corvée* labor. This corresponds to a second period of secularisation: Generally neither effort elicited a defiant refusal to participate on the part of the Javanese peasants, even if there were spates of short-lived peasant rebellions (cf. Kartodirdjo 1973). These passed generally unnoticed, as did the fact that the villagers lost their own socio-cosmic references due to this exploitation. Custom may be stamped out, but this is harder to do than one usually imagines. Normalisation, standardisation of religion, on the other hand, is subject to much superficial acquiescence. Then once the threat that accompanies it is absent, modernity volatises as a consequence of its artificiality. Likewise, the Christian and Muslim efforts failed to discourage the Javanese from praying at the graves of their ancestors on the eve of the New Year. This is just one example of that spontaneity, of that cultural *ressenti* whose roots are deeper than the political imagination of norms. Governments are characterised by evaluative indifference to much of the culture of sociability that guides 'their' citizens' daily lives. Rather than ask *why anyone want to promote or accept a move to normalisation of religion, or consider village rituals superstition while Islam is not,* we should first of all notice how shallow the political imagination can become. When one thinks like a nation state or a bureaucracy, does one forsake the shared experience and social diversity that is unconsciously accepted by the average Javanese with the confidence that such difference actually underpins and sustains a common hierarchy of values? 'Representation' to the average Javanese does not mean the election of presidents but participation in the life-accompanying rituals and in the conviviality their families have always practiced in times of transition and need. To bring these ways of living into practice one has to return to an understanding of the person that leaves a large place for exchange and reciprocity. 'Rational choice' and all such theories imply an ideology of individualism that is not yet vectored by Javanese daily life nor even by Islam. The Javanese living in the same *kampong* still have more to share with each other, no matter what their respective 'religions', and this has been formulated in an *adat* that evolves with each generation, enhancing their sociability.

References

BEATTY, A. (1999) *Varieties of Javanese Religions*. Cambridge: Cambridge University Press.

ASAD, T. (2003) *Formations of the Secular. Christianity, Islam, Modernity*. Stanford: Stanford University Press.

DAVIDSON, J.S. and HENLEY, D. (eds.) (2007) *The Revival of Tradition in Indonesian Politics. The Deployment of adat from Colonialism to Indigenism.* Longon and New York: Routledge Contemporary Southeast Asia Series.

HANN, C. and the « Civil Religion » Group (2006) *The Postsocialist Religious Question. Faith and Power in Central Asia and East-central Europe*. Berlin: LIT.

HEADLEY, S.C. (2004) *Durga's Mosque. Cosmology, Conversion and Community in central Javanese Islam*. Singapore: Institute of Southeast Asian Studies.

HEADLEY, S.C. (2005) Ebauches d'hommes, sociétés et corps inachevés. *L'Homme* 174: 23-44.

HEADLEY, S.C. (2005) Ebauches d'hommes, sociétés et corps inachevés Des hommes incomplets à java. Engendrement, nourritures et assemblages. *Moitiés d'Hommes. L'Homme* 174: 161-202.

HEADLEY, S.C. (f.c.) "Unfinished" Secularization Meets Traditional Hierarchies of Values in Indonesia & India. In *Centenary Conference Louis Dumont*. Paris: CNRS.

HERINGA, R. (2007) Constructing the Whole. Seven Months Pregnancy Ritual in Kerek, East Java. In JANOWSKI, M. and KERLOGUE, F. (eds.), *Kinship and Food in South East Asia. Studies in Asian Topics*, pp. 24-53. Copenhagen: NIAS Press.

KARTODIRDJO, S. (1973) *Protest Movements in Rural Java*. Singapore: Oxford University Press.

LI, T.M. (2007) Adat in Central Sulawesi: Contemporary Deployments. In DAVIDSON, J.S. and HENLEY, D. (eds.), *The Revival of Tradition in Indonesian Politics: The Deployment of Adat from Colonialism to Indigenism*, pp. 337-371. London: Routledge.

PEMBERTON, J. (1994) *On the Subject of "Java"*. Ithaca: Cornell University Press.

PLATENKAMP, J.D.M. (2008) The Canoe Racing Ritual of Luang Prabang. *Social Analysis* 52(3): 1-32.

ROSA, H. (2013) Accéleration: une critique sociale du temps. Paris: La Découverte.

PENELTI LIPI, T. (2000) *Bara dalam Sekam*. Bandung: Mizan.

YAYASAN LEMBAGA BANTUAN HUKUM INDONESIA (1997) *Amuk Benjarmasin*. Jakarta: Yayasan Lembaga Bantuan Hukum Indonesia.

Equality, Religion and Power: A Few Topics for Discussion

ANDRÉ ITEANU

My endeavor is to present a few remarks on the notion of equality, with the expectation to feed an ongoing discussion that Jos and I had over the past many years and which I hope will continue for more years to come.

My starting point evolves from the fact that the anthropologists do not constitute a unified community possessing an unequivocal common language. We originate and multiply in varied cultures in which equality appears under different guises. It would be mistaken to consider these forms as identical, but equally fallacious to claim that they differ completely from one another. Most likely, they are varied renderings of the same set of ideas, which scholars, politicians, activists or artists have elaborated in a dialogue with each other, beyond and across any national or cultural boarders. Therefore, through the years, the distance that separates these different adaptations of the notion of equality was at times reduced by imitation, and at times widened by radicalisation.

A telling example of the diversity of this notion of equality can be found in the contrasting way in which Anglo-Saxon and French politicians and scholars judge the policies that the French authorities apply to their "*banlieues*," the French suburbs where many emigrants are housed. One conspicuous way in which the foreigners question these policies is by calling them assimilationist, meaning to say that they go beyond their scope of promoting equality and force the immigrants to reject their former cultural identities in order to adopt the French modes of living, including religious ones. Whereas the French simply believe that by promoting social integration, they only attempt to achieve equality between poor and rich, indigenous and foreigners and later this "naturally" would bring about cultural transformation. Quite naively, they also think that this task always falls short of being achieved, not because it is unfounded, but because their politicians are dishonest and always betray their promises. Therefore, while the French consider their conception of equality as unproblematic, but lacking practical success, other Europeans ask much deeper questions about what French equality means and why it is never achieved. However, these very same Europeans are quite unable to question their own systems in a similar manner, while the French are at ease to raise crisp reciprocal critiques on their neighbor's policies.

Therefore, if we anthropologists wish to understand each other when we discuss about equality we must firstly decide whether, as researchers, we should adopt a so-called scientific point of view, and devise a notion that resembles none of the variations known in our cultural milieus, or if we must establish a midway vintage view common to all of them, or again, if we must each adopt our own country's notion of equality, a position which may prove difficult to many of us who have quite mixed-up origins, or finally, if we all should position ourselves in a culture in which egalitarianism is not a central value at all.

At first sight, this last alternative may seem outdated as it supposes the existence of cultures ignorant of the idea of equality, thus creating an unfashionable radical difference with the cultures we are dealing with here. However, on second reflection, it appears that two distinct concerns are mixed up here. The first one is the existence of a radical difference between any two cultures, which would mean that one is totally alien in relation to the other. This is probably to be dismissed. The second one is that anthropological comparative practice may require operating a radical distinction, at least as a starting point. This form of comparison does not seem obsolete as it unites the old and the new, the peripheral and the utterly fashionable. The old case is the claim by Louis Dumont in his *From Mandeville to Marx* (1977) to understand the birth of the Western economic ideology from the point of view of the Indians. The new case, many years later, is Viveiros de Castro's (1998, 2003) tactic to compare ontologies from the point of view of the Indians. For sure, I know that in both cases the Indians are not the same, but the methodological continuity is flagrant.

Following up on this idea, my second point is a question: are there societies that do not consider equality as their major value or even do not know of equality at all? For example, are Melanesian societies, as it is often claimed (Jolly 1987), fiercely egalitarian or not? This would probably imply that Melanesians must consider equality as an important (sometimes the most important) value. However, in several languages spoken in these societies, there is neither a word for equality nor one for hierarchy. This sometimes goes even further, as the Orokaiva language that I speak does not even have a dedicated comparative form: one cannot say this is bigger than that, but only, this is big, and this is small. Consequently, the Orokaiva language possesses a way to express identity (this and that thing are identical), but not one for equality. So, for example, when canned food is given in a ritual exchange, the reciprocal gift must comprise identical cans of the same brand and type and not same size cans of another brand or even bigger cans of the same brand. One arrangement is not equal to the other.

This leads us to the important idea that equality is but a specific position on a hierarchical scale: the position in which neither one of a number of elements is bigger with regard with a chosen scale. The absence of a hierarchy in Melanesia should have thus suggested that equality may be absent as well.

However, most anthropologists have taken the exact opposite stand. They have deduced that equality is a Melanesian value on the basis of two observations: on the one hand, they have located there a social characteristic called autonomy, very similar to Durkheim's mechanical solidarity, according to which every social unit or person can perform all indispensable social tasks alone, and on the other, they have insisted on the absence of wealth differences founded on the required equivalence of goods that circulate in the ceremonial exchanges performed. They thus defined inequality by the lack of political subordination and by the inexistence of wealth accumulation. In sum, in these societies, they were able to show the absence of two dimensions systematically associated in Euro-America to power domination, but they have never demonstrated that the notion of equality existed as well in the eyes of those of who they were talking about. Objectively, equality existed. However, was it conceived the way we thought? This remained unsolved.

Fortunately, many years later, Marilyn Strathem (1988: 3) has indirectly answered thisquestion in her famous and provocative statement that Melanesians do not conceive of something similar to the Euro-American notion of society or of the individual, because they do not make use of the idea of totality, which is integral to these notions. If she is right, her assertion then easy extends to what interests us here by adding that, since Melanesians do not use the notion of totality, they probably do not either construct hierarchies based on that notion, and therefore do not employ the notion of equality.

This is so because in the absence of a commonly accepted value, comparison between elements may exibit their difference, or their identity, but these elements cannot be measured against a value-ordered hierarchical scale.

In the end, it appears that equality can, paradoxically, only appear as a local value in social contexts where notions of totality and of hierarchy exist as well.

This is almost a trivial conclusion. However, it may gain much impetus if combined with Dumont's claims that an extreme and probably unique distinction between political power and religion, which here also stands for hierarchy, was gradually fashioned in the course of Indian history. I am, of course, unable to judge the empirical reality of this assertion concerning India. However, taking it at face value, it invites us to think that, for Dumont, in relation to India, all other societies know of a partial indistinction between power and hierarchy as well as between power and religion. This is true of all Euro-American cases where we may think that the mixture remains light, and it climaxes in extreme examples like the well-known figure of the Polynesian "sacred king," who cumulates in his or her own person both the epitome of religion and that of kingly power (Hocart 1970).

History shows that a similar distinction, albeit less radical, has emerged in the West from the Middle Ages on. It gave birth to many contrasting scenarios, only two of which concern us here: in the US, religion was preserved to the 19th century as a very important value, while political power was separated and denied any grasp on it. According to Tocqueville (2000), the function of this very weak political power was only to protect and promote the autonomy and freedom of the individuals. Therefore, equality, including wealth equality, was subordinated to the freedom to achieve personal success. Instead, equality, as the assertion of the high hierarchical value attributed to religion, became the major characteristic of American religious communities that accepted no internal difference of status.

Even today, in the states known as the Bible Belt, characterised by strong religious involvement, political power is scorned and elections show that a majority of voters support parties that do not promote wealth equality, but freedom. This is inverted in the states that vote for social equalising measures, like Obama's social care, as religious involvement is low and decreasing drastically over the past fifty years (Frank 2004).

As opposed to this case, France, as Tocqueville sees it, is massively oriented towards equality. This goes together with the fact that, after the French Revolution, religion was suppressed even in what appears today as minor details: like changing the months' and days' names, mostly by terms used for plants. Political power then became absolutely dominant, with the result that during the few years following the Revolution, thousands of people were beheaded. In the end, equality was finely restrained, to be used only against the privilege of nobility and of the church.

However, this was only the mild form of the revolutionary ideology. In 1796-7, there was a serious conspiracy against the government, *le Directoire*, called the *Conspiracy for Equality* or *the Conspiracy of Babeuf*, by the name of one of its chiefs, that pushed the French Revolutionary logic to its extreme. Buonarroti, one of the few of Babeuf's close companions not to end up on the guillotine, conveyed the story of this movement in a book published later in French in 1828.

> "What passed in France immediately after the creation of the Republic, is, in my view, only the explosion of that discord, which ever exists between the partisans of opulence and distinctions on the one hand, and the friends of equality, or the numerous class of labourers, on the other" (Buonarroti 1836: 8).

And this may not be totally useful for my demonstration, but I cannot resist the pleasure of quoting it:

> "Let us call to mind, that very many writers have made the prosperity of nations to consist in the multiplicity of their wants – in the ever augmenting diversity of their material enjoyments – in an immense industry – in an unlimited commerce – in the rapid circulation of the monied metals – and (as the last step of the analysis) in the *restlessness and insatiable cupidity of the population*. At one time they prefer the heaping together of territorial property, at another, they declare for the multiplication of small proprietorships; and whilst some have believed that the misery and brutalisation of the productive classes are essential to the opulence and tranquillity of the whole, others, by holding forth the unrestricted and unlimited freedom of industry and commercial transactions, as a means of remedying

> the established inequality, have only paved the way for a new system of corruption, and for new inequalities, more extensive and pernicious than what before existed" (ibid, 9).

So in order to make everyone equal, one needs to suppress private property and to make the collectivity the only owner of all things:

> "We are all equal, are we not? This principle remains uncontested, because, without being self-convicted of folly, one cannot seriously say that it is night when it is day.
> Well! We pretend henceforward to live and die equal, as we are born so. We desire real equality or death; behold what we want. And we shall have this real equality, no matter at what price. Woe to them who will interpose themselves between it and us! Woe to him who will offer resistance to so determined a resolve!" (ibid, 314-5).

Here, the project begins to be seriously frightening:

> "We want that equality not merely written in the "Declaration of the Rights of Man and of the Citizen"; we want it in the midst of us – under the roofs of our houses. We consent to everything for it – to become as *pliable wax, in order to have its characters engraven upon us*. Perish, if needs be, all the arts, provided real equality abides with us!
> Legislators and governors, who are as destitute of genius as of honesty – you rich proprietors, without bowels of pity – in vain do you essay to neutralise our holy enterprize [...]" (ibid, 315).

We know what follows because we have seen it happened recently in many different places. To achieve equality, people must firstly be transformed:

> "Bedon and Darthé, who proposed the dictatorship, attached to this word the idea of an extraordinary authority, confided to an individual, to be charged with the double function *of proposing to the people a plan of legislation simple, and suited to ensure to it equality, and the real exercise of its sovereignty; and to dictate provisionally the preparatory measures necessary to dispose the nation to receive it"* (ibid, 105).

The following pages present a long list of recommendations on how all this is to be achieved. Their general goal is to deeply transform the nation's "people." To that end, present materialistic concerns should be renounced to forward an ideal to be later achieved, on a purely immaterial plane. Therefore, the desire for clothing, houses, and so forth, must be extirpated from the French individuals, but also their desire for art, embellishments, religion and knowledge. Individually, the rules prescribed resemble those of religious devotion. Collectivity, there should be only one common goal.

> "In a well-ordered society nothing would escape the searching mind of the legislator; in a veritable republic nothing should run counter to the principle of equality, which is at once its end, bond, and strength; let its citizens only once begin to

> taste the sweets of equality, and soon will every department of civil life be most easily moulded and harmonised upon that principle (ibid, 163-4)"

Just as Marx was going to anticipate it some years later for the Communist state, Babeuf believes that transforming the individuals will cause politics to whither:

> "It appears from the preceding remarks that, in the eyes of the Committee, happiness and liberty depend much more upon the maintenance of equality, and in the attachment of the citizens to the institutions which establish it, than on the distribution of the public administration. There was in their Republic an *Institution of the State and a Constitution of Authority*, and in each are fundamental points which the people itself can neither violate nor modify, because they cannot be touched without instantly dissolving society. Such are, in the first, *rigorous equality*; and in the second, the *popular sovereignty*" (ibid, 196).

For the author of the text, achieving this form of total equality is clearly not a political endeavor, but something that he can only describe by using a mystical language. Dumont's comparative assessment, which I earlier mentioned, previously led us in the same direction as for him equality, which is a hierarchical position, stands alongside religion. Therefore, for Babeuf (or Buonarroti), even the Revolutionary collectivity, which is endowed with the utmost political power, cannot alter the mystical rules of the nation: *rigorous equality* and *popular sovereignty*.

However, on the other hand, politics, which Babeuf calls sovereignty, is only a matter of power, which as Foucault and others have since argued, suffers no limitation, be it the absolute value of equality. The French revolutionaries were thus facing a powerful contradiction: getting rid of religion or hierarchy while in the same time instituting another mystical position by the name of equality, able to stand above politics and to induce its disappearance in time. Because this contradiction was very hard to solve convincingly, many heads have fallen, among which those of Babeuf and Robespierre.

Conclusion

Today, in France, an unusually low percentage of citizens practice religion. However, the latter is still largely considered as radically opposed to politics and to equality. This appears clearly, for example, in the way the notion of *laïcité* (secularism) is used in France both against the Catholics, who have massively opposed gay marriage, and the Muslims, booed for promoting gender inequality. The French government, on the contrary, is still systematically attributed the duty to ensure rigorous equality, both internally and in its foreign policy. As it often fails, it is also constantly judged at fault. In view of the content of Babeuf's ideas, I am tempted to go as far as saying that in France, political power has now captured the hierarchy in the form of equality that was formerly associated with religion. Following this line, I think, that today, France, contrary to India, figures an extreme example of un-distinction between what is left of religion and what is left of power. However, in contrast to the sacred king, its mixture appears to be power-dominated. It is then no wonder that so many French philosophers and social scientists (Foucault, Bourdieu, and so forth) have argued that power and hierarchy constitute one and the same social force, which controls the social world without counterparts.

To sum up, I must say that I understand that the reader could dismiss this paper on the account that it contains several contradictions. I argue, for example, that equality is in fact hierarchy and therefore a religious phenomenon, and then I go on saying that when it mixes with power, it creates a sort of indistinctness. All this could be a simple play on words, on the notion of religion, which is, as I formerly recognised, a bad analytical tool, immoderately wide and prone to encompass any sort of social dimension. My only defense against this critic is that my present argument is very limited and does not claim a general scope. In my view, religion is not present in all forms of equality. For example, in the US, religion was much more preserved than in France from political confinement and politics continues to have an external relation to it, when it does. Therefore, contrary to appearances, in the US, the mixture between both elements is much more limited than in France. Furthermore, my argument was originally meant to apply to extreme cases of egalitarianism, like that of Babeuf, in which equality becomes a universal uncontested value that excludes all others. In my view, this ideology is as well to be found in today's extreme forms of egalitarianism. The conclusion on today's France is just a tentative follow on that I hope will evolve in a discussion with Jos and others.

References

BUONARROTI, F.M. (1836) [1828] *Buonarroti's History of Babeuf's Conspiracy for Equality*. Transl. by Bronterre O'Brien. London: Hetherington.

DUMONT, L. (1970) *Homo Hierarchicus: The Caste System and Its Implications*. Chicago: The University of Chicago Press.

DUMONT, L. (1977) *From Mandeville to Marx: The Genesis and Triumph of Economic Ideology*. Chicago: The University of Chicago Press.

FRANK, T. (2004) *What's the Matter with Kansas? How Conservatives Won the Heart of America*. New York: Holt.

HOCART, A. M. (1970) [1936] *Kings and Councillors: An Essay in the Comparative Anatomy of Human Society*. Chicago: The University of Chicago Press.

JOLLY, M. (1987) The Chimera of Equality in Melanesia. *Mankind* 17(2): 168-183.

STRATHERN, M. (1988) *The Gender of the Gift: Problems with Women and Problems with Society in Melanesia*. Berkeley: University of California Press.

TOCQUEVILLE, A. de (2000) [1835] *Democracy in America*. Transl., ed., and with an introduction by Harvey C. Mansfield and Delba Winthrop. Chicago: The University of Chicago Press.

VIVEIROS DE CASTRO, E. (1998) Cosmological Deixis and Amerindian Perspectivism. *Journal of the Royal Anthropological Institute (N.S.)* 4: 469-488.

VIVEIROS DE CASTRO, E. (2003) *(anthropology)_AND_(science)*. Manchester Papers in Social Anthropology 7.

Conflicts in North Maluku, Indonesia: Refusal to Exchange, Refusal to Establish Relations

Kirsten Jäger

Introduction

After Suharto's resignation in 1998, Indonesia underwent large scale administrative, political, social and economic reforms. New administrative units emerged (Indonesian *pemekaran*) throughout Indonesia, especially on the 'Outer Islands', due to the decentralisation reforms. These new administrative units were established on provincial (Indonesian *propinsi*), district (Indonesian *kabupaten*) and sub-district (Indonesian *kecamatan)* levels (Booth 2011).

North Maluku became a province comprising two districts, Halmahera Tengah ('Central Halmahera') and Maluku Utara ('North Maluku'), and the district town (Indonesian *kotamadya)* Ternate in October 1999. Some months before, in May 1999, migrants from the island Makian managed to obtain an official status for their sub-district called Malifut. These migrants have already lived in the district North Maluku for twenty-four years. In order to establish a new sub-district, neighbouring villages from the sub-districts Jailolo[1] and Kao[2] were supposed to merge with the villages of Makianese residents (PP 42/1999[3]). The people of Kao and Jailolo, however, refused to constitute an administrative unit with the migrants from Makian. Moreover, these inhabitants tried to prevent the establishment of this new sub-district.

Nowadays, the inhabitants of North Maluku along with many contemporary researchers identify this local dispute as the starting point of the violent conflicts that shook North Maluku between 1999 and 2001 (Tomagola 2000; van Klinken 2001; Bubandt 2004; Duncan 2005; Wilson 2005, 2008).

Of course, the violence in North Maluku took different forms in different stages of the conflict and therefore changed people's narratives over – and the local justifications for – the violent conflict (Duncan 2005: 79-80). Hence, it is important to note that this chapter focuses on the early stages of the violent conflict in North Maluku during 1999. At this stage, the conflict occurred

1 Desa Bobane Igo, Desa Tetewang, Desa Akelamo Kao, Desa Gamsungi, Desa Dum-Dum, Desa Pasir Putih (PP 42/1999 Pasal 1c).

2 Desa Tabobo, Desa Balisoang, Desa Sosol, Desa Wangeotak, Desa Gayok (PP 42/1999 Pasal 1b).

3 Peraturan Pemerintah 42/1999, Pembentukan dan Penataan beberapa Kecamatan di Wilayah Kabupaten Daerah Tingkat II Maluku Utara dalam Wilayah Propinsi Daerah Tingkat I Maluku.

mainly between the indigenous inhabitants of Kao[4] and the migrants from Makian in North Halmahera.

Jos Platenkamp and Ronald Lucardie conducted their fieldwork in North Halmahera and Makian respectively, during the 1970s and early 1980s. During this time, the Makianese migrants were resettled to the Kao territory in North Halmahera. As I deem this incident crucial for the understanding of the outbreak of violence in 1999, Lucardie's and Platenkamp's fieldwork have generated most valuable information on the socio-political and socio-cosmological concepts of both, the Kao inhabitants and the migrants from Makian. As I shall demonstrate, the conflict is grounded to a large extent in these differing concepts.

Historical Structures of Social Order

Over one million people reside in North Maluku, Eastern Indonesia, which has a land area of 31982, 50 km². The majority of the inhabitants are Muslims while almost a quarter of the population adheres to the Christian faith. Hindus and Buddhist constitute a small minority of the local population. Twenty-two different languages are spoken in North Maluku, twelve belonging to the Non-Austronesian Papua Phylum and ten to the Austronesian language family (BPS Propinsi Maluku Utara 2011).

There are three basic principles structuring the socio-cosmological and socio-political order in North Halmahera; these are the concepts of *ma dutu*, the *soa* and *Moluku Kie Raha*, i.e. the four Sultanates of North Maluku (van Fraassen 1987; Baker 1988; Platenkamp 1988, 1993; Visser 1989; Bubandt 1991; Teljeur 1994; Jäger 2015). Jos Platenkamp's anthropological and historical research has demonstrated that the societies of Halmahera share some basic ideas about their cosmological order in which all social relations are embedded (Platenkamp 1988, 2013). This concept which is named *ma dutu* encompasses all relations between the people, the land, the water, the ancestors, animals and plants (Platenkamp 1988: 10).

The lexeme *ma dutu* used as a noun, refers to a connection between an owner and an entity which belongs to this owner. In this case, we are able to translate the expression *o tonaka ma dutu* (Tobelo) as 'the owner of the land' (*ibid.*) These ties of ownership gain authority, because all *ma dutu*-relations are hierarchically organised and encompassed by a whole. The top of this totality is represented by the idea of the highest deity (Tobelo *jou ma dutu* 'Lord Owner'). There is no entity in the universe which is able to exist without *ma dutu* relations. The universe is not a compilation of single independent entities, but a system of connected *ma dutu* relations (*ibid.*). If the lexeme is used as

4 The indigenous population of Kao comprises Pagu, Modole, Tobelo Boëng, and Kao Islam (Duncan 2005: 61).

an adjective, *ma dutu* refers to a quality that can be translated as 'true' and 'original' (Visser 1989: 105). In this case, *o tonaka ma dutu* would mean 'the true/original land' (Platenkamp 1988: 10). This concept is of crucial importance connecting the ideas of belonging and identity, because this kind of ideological understanding of ownership indicates that "[...] in the North Moluccas land owns people not vice versa" (Bubandt 1991: 4).

The *soa* is a historical social unit in most parts of North Maluku. Antonio Galvão, a Portuguese Governor-General in North Maluku stated in the sixteenth century that the inhabitants of the Moluccas organise their territory "[...] into quarters called soas. Every family, with its relations and servants has its own [*soa*]" (Jacobs 1971: 105). The social unit is called *soa* in Ternate, which can be translated as 'space', or 'space-in-between' (van Fraassen 1987/I: 140). The terms for these social units in Halmahera, like the *hoana* in Tobelo and the *soan* in Tobaru and Loloda, derives from the lexeme *soa*. These social units consist of 'Houses', idealised descent groups that trace their origin to one ancestor. I refer to the descent group as idealised because it seems more important to connect oneself with the social unit through marriage or adoption and live in the same territory than to descent from a common ancestor (Visser 1989: 100).

Another ideological concept shared by most inhabitants of northern Halmahera stipulates that North Maluku as a whole once was part of a political and cosmological federation named Moloku Kie Raha, a unit of 'four mountains' composed of the Sultanates of Ternate, Tidore, Bacan and Jailolo. Myths tell us that their rulers were brothers born to an immigrant father from Arabia and a mother originating in the heavenly sphere of *kayangan*. The sons divided the land between them, resulting in a constant competition for power (van Fraassen 1987; Andaya 1993). Jailolo's and Bacan's political influence faded in the sixteenth century due to internal warfare between the sultanates and the impact of the colonial powers. Ternate and Tidore expanded their territories as far as Sulawesi and Papua respectively. While exerting power, the sultanates gave these social units in Halmahera an additional meaning as a tribute-paying unit. Thus, the soa should be understood as a category with a twofold meaning. A *soa* is a social group that inhabits a distinct territory and traces its origin to an ancestor. This group establishes relations to this ancestor and the land, and is thus strongly connected to the territory (van Fraassen 1987/I: 140-142; Bubandt 1991: 18). A *soa* is as well a social unit which pays tribute (Indonesian *upeti*) to the sultanate. In this understanding, the *soa* is defined by social status and the function it performs within the sultanate's realm (Bubandt 1991: 30-31).

The sultanates constituted the socio-cosmological and political centre, incorporating their subjects into a broader framework. Therefore, the heterogeneous population was united under the 'umbrella' of the four mountains, the four Sultanates (Andaya 1993: 110). The principles of social order like the *soa*, *ma dutu* and the sultanates are all based on reciprocal relations. *Ma dutu*, even used as an adjective refers always to relations (Platenkamp 1988: 12). The *soa* refers to the descent group as well as to their common territory, and the sultanates as centres establish relations with their peripheries. The sultan constitutes his sovereignty ritually while his subjects receive titles and reputation from him (Platenkamp 2013: 226).

North Maluku's history of creating administrative units reveals, even though the region faced large scale changes due to the constant fighting between the sultanates over territory, the colonial regime, the nation-building and the recent processes of *pemekaran*, that these socio-territorial units preserved much of their indigenous significance (cf. Fox 2011). After a long history of indirect rule and the cooperation between the Dutch colonial power and the sultanates, the latter's influence was restricted in the early twentieth century. The sultanates continued to lose political significance with the rise of nationalism throughout Indonesia. The claims to land made by the sultanates were legally declared invalid by the Basic Agrarian Law issued in 1960 (UU 5/1960[5]). Following this law, the land of the sultans was redistributed, for land not owned by individuals belongs to the State (van Klinken 2007: 150). In spite of their diminishing political influence, the borders of North Maluku's units of administration were still those of the sultanate's historical territories (BPS Propinsi Maluku Utara 2011). Dutch colonial officials mapping administrative units in Maluku largely complied with the existing *soa*-units as determined by the sultanates. These historical borders still constituted the terrain after the Indonesian Independence and even after the implementation of the laws on regional and village (Indonesian *desa*) government (UU 5/1974[6] and UU 5/1979[7]), establishing a uniform administrative structure throughout Indonesia.

Makianese Concepts of Migration and Homeland (Lucardie)

Residents from Makian were evacuated from their island to Halmahera anticipating volcanic outbursts on the island in 1975. The provincial government of Maluku organised the resettlement programme[8] and brought the migrants to the area of Kao in North Halmahera. The migrants are exclusively Muslims whereas the inhabitants of Kao are mainly Christians. When the migrants arrived, the government gave the Makianese land to live on and to cultivate. The government told the Makianese that this land was compensation from the state

5 Undang-Undang 5/1960 Peraturan Dasar Pokok-Pokok Agraria.
6 Undang-Undang 5/1974 Peraturan Dasar Pokok-Pokok Pemerintah Daerah.
7 Undang-Undang 5/1979 tentang Pemerintah Desa.
8 Proyek Bantuan dan Penyantunan Korban Bencana Alam Maluku, Kantor Wilayah Departement Sosial Propinsi Maluku (Lucardie 1983: 333).

for their property they had left behind on Makian. According to the national government, this land was defined as state forest which is property of the state.

Six thousand residents of Makian were resettled to Kao between 1975 and 1980. The evacuation project was aimed to be terminated in 1986. By then, all Makianese should reside in Kao (Lucardie 1983: 334). The government's actual aim, however, was not accomplished; Makianese in Halmahera still returned to Makian and decided to live there despite the threat of volcanic outbursts. About 16,000 Makianese lived in Kao in 1999, while approximately 14,000 inhabitants lived on the island Makian (Wilson 2008: 54). The territory for the migrants was called Malifut, but had no official status as a sub-district (Indonesian *kecamatan*).

Although, the Makianese were perceived as a society with great geographical mobility, many residents refused to partake in the resettlement programme. The main reason for their refusal was the discrepancy between their traditional pattern of migration and the planned governmental programme (Lucardie 1983: 334-335). The Makianese have had no relations to the territory in Kao before, there traditional migratory areas having been situated in Southwest Halmahera and Moti. As Lucardie pointed out, each village had its own traditional immigration area (1983: 335-336). Even after a permanent shift of residence, Makianese usually are still closely connected with their homeland, Makian. As long as they keep these connections, by speaking the Makianese dialect and maintain their Makianese *adat*, the migrants are still perceived as Makianese, even if they never set foot on the island (Lucardie 1983: 336-337).

The Makianese use the island Makian as the centre from which they map out their neighbouring regions based on kinship ties and historical relations, not on topography (Lucardie 1983: 340). "Resettlement of the whole population of Makian in Malifut transmigration project simply would mean the end of the centuries-old migratory traditions with Makian as a pivot" (*ibid.*)

Kao Concepts of Residence and Exchange (Platenkamp)

As the migrants from Makian had started to arrive in the Kao area from 1975 onwards, they were given land for cultivation located in Kao's territory. For the Kao, however, this land was still their ancestors' land and thus did not come under Makianese possession (Wilson 2008: 54). According to Kao perception, one has to establish *ma dutu*-relations to the autochthonous lineages in order to claim access to the land. These ties are established by descent, adoption or marriage.

Those who descend from the same patrilineage (Maluku Malay *fam*) and bear the same last name, share access to land. The sons of a patrilineage continue this last name, whereas the daughters bear the name of the *fam* they marry into (Platenkamp 1988: 66-68). Sons remain in their patrilineage after marriage; daughters move to their husband's residency after marriage and become a member of his patrilineage. Metaphorically, the sons are considered as the 'stem' (Tobelo *utu*) and the daughters are perceived as the 'fruits' (Tobelo *ma hohoko*) (Platenkamp 1988: 62). "Like a fruit/flower [...] is connected to the stem of a tree, so a person is connected to an *utu* composed of other people" (ibid.). Because the sons stay in the *fam* and are perceived as the stem, they are associated with permanence whereas from the daughters' side "[...] the *fam* disappears" (Platenkamp 1988: 69), the "[...] fruit is separated from the stem and 'planted' elsewhere" (Platenkamp 1988: 62). Ideally, the matrilineal relations between the stem of origin and the newly founded stems last four generations (Platenkamp 1988: 72). The autochthonous patrilineage has the same relation to the immigrated patrilineages as the wife-giver has to the wife-taker, the former is superior to the latter (Platenkamp 1988: 95).

According to Kao perception, the only opportunity for migrants to gain access to land is to marry into the autochthonous social group. In this case land would be given from the wife's autochthonous *fam* to the migrant husband. As one stem consists of three generations, a new *fam* would be established after four generations and the migrant husband would have become fully integrated into this society (Platenkamp 1988: 94-96). He would stop being Makianese and become Kao. Membership to a social unit is constructed so that an entity of people could inhabit the same territory. Visser called this principle 'territorial kinship' (1989: 119). The Kao community identifies itself as an entity consisting of the four social sub-units of Kao Islam (includes Toliliko), Modole, Tobelo Boëng and Pagu; they were once connected to the Sultanate of Ternate and still use its title system as part of the *adat*-structure.

Socio-Political and Economic Setting

After Suharto's resignation in 1998, the government under president Habibie passed the new laws on decentralisation No. 22/1999[9] and No. 25/1999[10]. These laws were supposed to increase the provinces' and in particular the districts' autonomy. Law 22/1999 defined the regions' new rights and responsibilities while law 25/1999 regulated the financial aspects of these reforms. As a result, a newly ignited competition for natural, financial and political resources emerged (Schulte Nordholt and van Klinken 2007).

Before North Maluku has become a province, the region was part of the province Maluku, with its capital Ambon. When in October 1999 North Maluku was assigned the status of Province it comprised two districts (Indonesian

9 Undang-Undang 22/1999 tentang Pemerintah Daerah.

10 Undang-Undang 25/1999 tentang Perimbangan Keuangan antara Pemerintah Pusat dan Daerah.

kabupaten), Halmahera Tengah ('Central Halmahera'), Maluku Utara ('North Maluku') and the municipality Ternate (UU 46/1999[11]).

From 2003 onwards, these three administrative units were divided into eight.[12] From North Maluku emerged the districts Halmahera Utara ('North Halmahera'), Halmahera Selatan ('South Halmahera') and Kepulauan Sula ('Sula Archipelago'). The residual territory from the *kabupaten* Maluku Utara was renamed Halmahera Barat ('West Halmahera'). From Central Halmahera emerged the districts Halmahera Timur ('East Halmahera') and *kotamadya* Tidore Kepulaan ('Tidore Archipelago'). The residual territory stays with the name Halmahera Tengah (UU 1/2003[13]). The sub-districts Kao and Malifut are both situated in the district North Halmahera, the island Makian is located in South Halmahera.

Administrative units are often fragmented to gain access to political, financial and natural resources. These processes are further accompanied by significant corruption in the relations between regional and national governments. Nonetheless, the shape of these new administrative units can mainly be recognised as a re-construction of the old socio-territorial units, referring to the four sultanates and the social units of the *soa*.

Malifut: Disputes over the New Sub-District

There were no social relations between the inhabitants of Kao (neither with the Muslim inhabitants of Kao) and the Makianese. Everyday life was rather shaped by mutual stereotypes than actual interaction. Inhabitants of Kao perceived the Makianese as greedy and felt that the migrants had too much influence in the regional government. The Makianese in turn considered the Kao as lazy and backward (Wilson 2005: 73). Both social groups were determined to distinguish between 'them' and 'the other'. Even if villages were located so close to each other that an actual border was not apparent, both groups were specific that Kao were not living in Makianese settlements and vice versa. This dissociation from each other resulted from a fear of subordination, a fear which became very articulated during the disputes over the naming of the new sub-district. The Makianese insisted to call the new sub-district *Makian Daratan Malifut* ('Mainland Makian in Malifut') to underline their connections to the island Makian. The inhabitants of Kao, however, demanded a sub-district name indicating a connection to their territory, like Kao Selatan ('South Kao').

---

11 Undang-Undang 46/1999 tentang Pembentukan Propinsi Maluku Utara, Kabupaten Buru dan Kabupaten Maluku Tenggara.

12 By now, there are ten *kabupaten/kotamadya* in the province. Morotai split from Halmahera Utara (UU53/2008) and Taliabu from the Archipelago Sula (UU 6/2013).

13 Undang-Undang 1/2003 tentang Pembentukan Kabupaten Halmahera Utara, Kabupaten Halmahera Selatan, Kabupaten Kepulauan Sula, Kabupaten Halmahera Timur, dan Kota Tidore Kepulauan di Provinsi Maluku Utara.

The Makianese refused the Kao claim vehemently, because they considered living under this territorial designation "[...] would mean the Makian[ese] had effectively assimilated into the alien culture" (Wilson 2005: 80).

The Makianese migrants managed to obtain an official status for their sub-district called Malifut in May 1999. They usually live in Halmahera in the same socio-territorial constellation, using the same village names as they had in Makian (Lucardie 1983: 333, PP 42/1999 (1a)). In their understanding, they retain their social identity of being Makianese (Indonesian *orang Makian*) as long as they identify themselves as originating from that island; otherwise they would lose their right to bear that name. Thus, the creation of their own sub-district was essential for the maintenance of their distinct identity as Makianese. Naming the new sub-district *Makian Daratan Malifut* ('Mainland Makian in Malifut') would express their intentions to continue their relations to Makian. Moreover, the Makianese perceive themselves as official (local) trans-migrants who had been assigned their own territory by the government after the involuntary resettlement in the Kao territory (Lucardie 1983: 336-341). In order to establish the sub-district Malifut, six villages from the neighbouring sub-district Jailolo and five from Kao were meant to merge with the sixteen villages of the Makianese residents (PP 42/1999). The people of Kao and Jailolo[14], however, refused to constitute such a joint unit with the migrants from Makian. The Kao community stated that the entity of the four sub-units, the Kao Islam, Modole, Pagu and Tobelo Boëng, would be separated, as the border of the new district cut through their territory (Duncan 2005: 63). Furthermore, they perceive the migrants as guests on their ancestral land. Following their concepts of land use and ownership, the immigrants from Makian may receive the title to land only on the condition that they establish *ma dutu*-relations to the autochthonous inhabitants, give up their Makianese identity and become Kao (cf. Platenkamp 1988: 94-96). The administrative unit Malifut would invalidate the Kao's indigenous concept of land ownership and its concomitant identity formation. Hence, the people of Kao refused to be incorporated into the entity named *Makian Daratan Malifut*.

Impacts of Resource-Related Motivations

In addition to such claims based on social identity, resource-related motivations played a key role in the conflict as well.

The local gold resources make the territory between Jailolo and Kao, where Malifut was founded, a highly contested one. The PT Nusa Halmahera Mineral, a joint venture of the Australian Newcrest Mining and the Indonesian PT

14 For further information on the problem of the six villages in Jailolo, see Jäger (forthcoming).

Aneka Tambang, operates between the former sub-districts Jailolo and Kao since the 1990s.

During the decentralisation processes in Indonesia, sultans and *rajas* returned, demanding back 'their' land, subjects and rights (van Klinken 2007; Jäger 2015). Therefore, the Sultan of Ternate came to Halmahera to demand compensation for 'his people' of Kao and Jailolo and for himself, claiming that the mining company operates on 'traditional land' (Indonesian *tanah adat*) that belongs to him. When the mining company commissioned work in June 1999, however, the whole mining area was located in Malifut (Newcrest Mining 2011: 20).

The Makianese were supported by their numerous representatives in the regional government (Wilson 2008: 59). Malifut was a long-term project of many Makianese migrants. Creating Malifut as a sub-district was assumed as "[...] only a first step towards achieving a long-held goal for many connected with Malifut [...] establishing the town as the capital of a district" (Wilson 2008: 58). Such a status would generate new jobs, political influence and public funding.

The Escalation of the Conflict, 1999-2001

After the law on the new sub-district had passed, some quarrels occurred between the Kao community and Makianese. In August 1999 violent disputes between these residents resulted in the destruction of two Kao villages and the death of several people (Duncan 2005: 64). People from Kao were still trying to prevent the establishment of the new sub-district, but as their attempts had failed, several thousand inhabitants of Kao attacked Malifut, destroying all villages and expelling the Makianese from the area. The Makianese fled to Ternate. They had a poor relationship to the Sultan of Ternate, as he was perceived as a supporter of the Kao community. This relationship was getting even worse when the Sultan tried to prevent the arrival of the Makianese refugees. The former mayor of Ternate prevailed against the Sultan and helped the Makianese with their evacuation from Halmahera to Ternate. They found shelter in the southern part of island, where many migrants from Makian and Tidore reside.

Subsequently, clashes between the supporters and opponents of the sultanates as well as between Christian residents from Kao and Muslim Makianese refugees concurred in Ternate (Bubandt 2004). These fuelled religious tensions resulting in assaults against Christians in Ternate and Tidore and against Muslims in Tobelo, North Halmahera.

The local dispute between communities over different socio-cosmic and socio-political concepts in North Halmahera had turned into violent conflicts between mainly religious groups in North Maluku (van Klinken 2001; Duncan 2005; Wilson 2005, 2008).

Conclusion: Reasons for the Conflict

Certainly, there were changes of people's ideas and priorities since the period when the Makianese migrants first arrived in Kao. However, as anthropological research shows, some basic concepts about the socio-cosmological and socio-political order are surprisingly resilient (Fox 2011; Hauser-Schäublin 2013; Platenkamp 2013). The decentralisation reforms were undoubtedly a crucial factor for the conflicts occurring throughout Indonesia in 1999. Nonetheless, every conflict has to be analysed in its own terms "[...] as having its own origins and justifications" (Duncan 2005: 54). Tensions between the Kao inhabitants and the Makianese migrants have grown since the resettlement programme started. The Makianese did not want to move to Kao and the locals were not comfortable with their new neighbours.

Especially Suharto's government completely ignored indigenous concepts stipulating land as a component within a socio-cosmological structure of relations. Therefore, the government defined the land for the resettlement programme as state land and gave it to the Makianese without any consultation with the Kao community.

On first sight, the conflict arose from the misunderstanding between rights to own (Indonesian *hak milik*) and rights to use (Indonesian *hak ulayat*) land. The autochthonous inhabitants, on the one hand, considered the Makianese to only use the land while it was still within the Kao territory. The migrants, on the other hand, perceived the land as their property as the government had given it to them officially.

On a closer look, the perception of land use and possession are embedded within a broader structure of socio-cosmological and socio-political concepts. The Makianese migrants were determined to name their new territory with a Makianese term to keep their Makianese identity. Additionally, it was not necessary to establish relations to the Kao inhabitants because they were given land by the provincial government as compensation. However, their refusal to relate to the Kao inhabitants through marriage violated the foundation of Kao's ideas about acting as social beings living on the same territory. According to Kao perception, the migrants are not able to own land without upholding hierarchical relations to the autochthonous lineages; the migrants should have behaved as "sons in law" (Platenkamp, personal communication).

The reluctance of both, the inhabitants of Kao and the migrants from Makian, to cooperate in the case of the new sub-district is not comprehensible without taking the insights achieved by Platenkamp and Lucardie into account. This reluctance is neither fully explained due to economic and political causes, nor is the problem just situated between migrants versus autochthones, or Muslim versus Christian inhabitants (Duncan 2005: 80).

During the violent conflicts, neither Muslim Javanese trans-migrants living in the same area, nor other migrants from Sangir and Galela were attacked.

On the one hand, the Javanese were official trans-migrants who have had neither any plans (nor support) to establish their own sub-district (Duncan 2005: 56). The migrants from Sangir and Galela, on the other hand, have established marriage relations to the local inhabitants and have become family members (Jäger f.c.).

The reasons for the conflicts between the Makianese and Kao inhabitants are not essentially ethnic or religious identities, but social groups having diametrically opposed concepts about the requirements to keep a distinct social identity and which 'necessary relations' (Oppitz 1975) have to be established in this respect. For the Makianese, these were their relations to their homeland Makian. For the Kao, these were the relations between people living on the same territory who should engage in reciprocal exchange.

References

ANDAYA, L. (1993) *The World of Maluku. Eastern Indonesia in the Early Modern Period.* Honolulu: University of Hawaii Press.

BADAN PUSAT STATISTIK PROPINSI MALUKU UTARA (2011) *Maluku Utara dalam Angka.* Ternate.

BAKER, J.N. (1988) *Descent and Community in Tidore.* Ann Arbor: UMI.

BOOTH, A. (2011) Splitting, Splitting and Splitting again. A Brief History of the Development of Regional Government in Indonesia since Independence. *Bijdragen tot de Taal,- Land- en Volkenkunde* 167(1): 31-59.

BUBANDT, N.O. (1991) *The Soa. The Organisation of Political and Social Space in the North Moluccas.* Unpublished Master Thesis. Melbourne: University of Melbourne.

BUBANDT, N.O. (2004) Towards a New Politics of Tradition. *Antropologi Indonesia* 74: 11-30.

DAVIDSON, J.S. and HENLEY, D. (eds.) (2007) *The Revival of Tradition in Indonesian Politics.* London and New York: Routledge.

DUNCAN, C. (2005) The Other Maluku: Chronologies of Conflict in North Maluku. *Indonesia* 80: 53-80.

FOX, J.J. (2011) Re-considering Eastern Indonesia. *Asia Journal of Social Science* 39: 131-149.

FRAASSEN, C.F. van (1987) *Ternate, de Molukken an de Indonesische Archipel. Van soa-Organisatie en Vierdeling: Een Studie van Traditionele Samenleving en Cultuur in Indonesia (2 Vol.)*, Unpublished Dissertation: Leiden.

HAUSER-SCHÄUBLIN, B. (ed.) (2013) *Adat and Indigenity in Indonesia: Culture and Entitlements between Heteronomy and Self-Ascription.* Göttingen: Universitäts-Verlag Göttingen.

JACOBS, H. TH. TH. M. (1971) *A Treatise on the Moluccas (c. 1544). Probably the preliminary Version of António Galvão's lost História das Moluccas.* Rome: Jesuit Historical Institute.

JÄGER, K. (2015) *Das Sultanat Jailolo. Die Revitalisierung von 'traditionellen' politischen Gemeinwesen in Indonesien.* Ph.D. dissertation. Münster: Westfälische Wilhelms-Universität.

JÄGER, K. (f.c.) Redrawing Borders and Reshaping Marginality in North Maluku. In RÖSSLER, M., HAUG, M. and GRUMBLIES, A.-T. (eds.), *Transforming the Margins. Contesting and Reformulating Power Relations in Indonesia's Outer Islands*. London and New York: Routledge.

KLINKEN, G. van (2001) The Maluku Wars. Bringing Society back in. *Indonesia* 71: 1-26.

KLINKEN, G. van (2007) The Return of the Sultans. In DAVIDSON, J.S. and HENLEY, D. (eds.), *The Revival of Tradition in Indonesian Politics*, pp. 149-169. London and New York: Routledge.

LUCARDIE, G.R.E. (1983) The Geographical Mobility of the Makianese. Migratory Traditions and Resettlement Problems. In MASINAMBOW, E.K.M. (ed.), *Halmahera dan Raja Ampat sebagai kesatuan Majemuk Studi-Studi terhadap Suatu Daerah Transisi* (2 Vol.), pp. 333-346. Jakarta: LEKNAS/LIPI.

MASINAMBOW, E. K. M. (ed.). (1983) *Halmahera dan Raja Ampat sebagai kesatuan Majemuk Studi-Studi terhadap Suatu Daerah Transisi (2 Vol.)*. Jakarta: LEKNAS/LIPI.

NEWCREST MINING. (2011) *Annual Report*. [online] http://www.newcrest.com.au/media/annual_reports/FINAL_NCM_2011_Annual_Report_2011.pdf. [Accessed: 12th December 2014].

OPPITZ, M. (1975) *Notwendige Beziehungen. Abriß der strukturalen Anthropologie*. Berlin: Suhrkamp.

PLATENKAMP, J.D.M. (1988) *Tobelo. Ideas and Values of a North Moluccan Society*. Leiden: Repro Psychologie.

PLATENKAMP, J.D.M. (1993) Tobelo, Moro, Ternate. The Cosmological Valorisation of Historical Events. *Cakalele Special Issue* 4: 61-90.

PLATENKAMP, J.D.M. (2013) Sovereignty in the North Moluccas: Historical Transformations. *History and Anthropology* 24(2): 206-232.

RÖSSLER, M., HAUG, M. and GRUMBLIES, A.-T. (eds.), (f.c.) *Transforming the Margins. Contesting and Reformulating Power Relations in Indonesia's Outer Islands*. London and New York: Routledge.

SCHULTE NORDHOLT, H. and KLINKEN, G. van (eds.), (2007) *Renegotiating Boundaries. Local Politics in Post-Suharto Indonesia*. Leiden: KITLV Press.

TELJEUR, D. (1994) Life-Cycle Rituals among the Gimàn of South Halmahera. In VISSER, L. (ed.), *Halmahera and Beyond. Social Science Research in the Moluccas*, pp. 165-195. Leiden: KITLV Press.

TOMAGOLA, T.A. (2000) *The Bleeding Halmahera of North Moluccas*, Paper presented at the Workshop on Political Violence in Asia, Oslo 5-7 June.

VISSER, L. (1989) *My Rice Field is my Child. Social and Territorial Aspects of Swidden Cultivation in Sahu, Eastern Indonesia*. Dordrecht and Providence: Foris Publications.

VISSER, L. (ed.) (1994) *Halmahera and Beyond. Social Science Research in the Moluccas*. Leiden: KITLV Press.

WILSON, C. (2005) The Ethnic Origins of Religious Conflict in North Maluku Province, Indonesia, 1999-2000. *Indonesia* 79: 69-91.

WILSON, C. (2008) *Ethno-Religious Violence in Indonesia. From Soil to God*. London and New York: Routledge.

The Ritual of Death in Acèh: An Ethnographic Study in Blangporoh Village, West Labuhan Haji, South Acèh, Indonesia

ABDUL MANAN

This article is dedicated to my PhD Supervisor, Prof. Dr. J. D. M. Platenkamp, since his interests lie in on rituals, exchange and personhood. "Ritual pertains to the very relationships that structure society both in its social morphology and in its overall order" (Barraud and Platenkamp, 1990: 121).

Living in this world is only temporary for human beings. Both happiness and sadness are always present in life. In Acèh province, when a person dies, men and women bathe and enshroud the body. It is then prayed for and chanted to before being buried; the dead person's debts must also be settled. All this work, alongside near-continuous hosting, cooking, and cleaning, is morally incumbent, emotionally draining, and physically exhausting. But all this work has a moral purpose and is considered to be the best thing for the bereaved family. This article[1] describes the different phases of the death rituals in Blangporoh village reflecting both Islamic and Aneuk Jamee *adat* rules and ideas. It is concerned with the ritual treatment of the deceased, the actual burial ceremony, and the rituals conducted in the months afterward.

Dying

A death ritual[2] can be said to last from the time of dying till the time of burial both in the form of rites, the performance of certain series of actions, words and gestures, and the mobilisation of particular social and religious relations by the deceased's family, neighbours or others. Before he[3] faces the agony of death, a person who is severely ill but still conscious is strongly urged to recover. He is reminded to always remember Allāh by reciting the "begging for divine pardon" and reciting the two sentences of the confession of faith: "There is no deity except Allāh and Muhammad is His messenger". Meanwhile other people, especially his or her family members, recite from the Qur'ān.

1 I would like to thank Guido Sprenger for his perceptive critical comments and suggestions as well as for his careful correction of the early version of the English text.

2 Three death rituals have been observed. Two were for men and one for a woman. For the woman, much information is obtained from a female informant. Taking pictures during the death rituals is not allowed. Otherwise, the deceased's family will be offended and I would have to leave the village.

3 I use the male form for the dead person.

When agony sets in, the *teungku* (religious leader) or other devout people are called immediately to recite the Qur'ān, especially the *sūrah Yāsīn* (QS 36:1-83), several times as long as the eyes of the dying person are open. After seeing that the person's condition is getting worse, "there is no deity except Allāh" is repeatedly whispered to the dying person in the hope that the dying person is able to repeat the phrase. This act is called *peuintat* (accompanying). For those under the age of puberty who do not completely understand the two sentences of profession of faith, *peuintat* is not performed. Nevertheless, even in these cases reciting the Qur'ān is often performed to provide peace.

Some villagers say that at the time of dying various sufferings are felt. One of them is great thirst. At that moment, "Satan comes and takes the semblance of one of that person's relatives who have passed away. Satan holds a glass of drinking water in front of his or her eyes. If that person renounces his faith by saying, 'there are two gods' or 'god is not Allāh', he will receive the water. However, if his faith is strong, they are not interested in Satan's temptation because he knows that it is only a trick", as one informant explained in contrast, someone who has less faith can easily become Satan's prey.

After the moment of death, the dead person's pulse is checked and his eyes are closed. The person who had stayed with the dying person immediately utters, "we belong to Allāh and return to Him", and this is continuously repeated by others sitting around the body. Next, the *do'a* or "the safety of the deceased", led by a religious leader, is recited. At the end of the *do'a*, all attendants intone *amīn ya Allāh* (please accept oh Allāh).

Besides reciting Qur'ān and *do'a* as described above, there are other rituals taking place, for example, the dead person's head is oriented towards the north. It is said that for someone who had practiced black magic, the body is stretched out on the bed with his head to the east and his feet to the west. Such a person dies in difficulty and can hardly end his life because the *malaikat* (angel) do not press him. Those accompanying the dying, therefore, open the windows, cupboards and even the roof above the dying. The healer then begins to scatter husked and unhusked rice while saying: "One, two, three, four, five, six, seven, get away from here! Your possessor has already gone; there is no need to wait." The dead body is then covered with cloth and his eyes are immediately closed. The hands of the dead are placed together on the chest; the right hand is placed on the left as if he were praying. After that, the body is moved from the mat where he died to a new mat and mattress. During the period of dying, it is forbidden to say or to do something that can disturb the dying person. When he is truly dead, the family of the deceased immediately informs the head of the village and the *teungku*. These people order one of the mosque staff to announce the news of the death to the public using a microphone and loudspeakers. In the past, the announcement was made by beating a tambourine at the mosque or *meunasah* three times.

Bathing and Enshrouding

1. Prior to Bathing

While waiting for the dead person to be bathed, some women sitting near it pray to the Prophet Mu<u>h</u>ammad. If the bathing takes longer, they continue reciting the Qur'ān. People who have just heard the news should say '*inna lillā hī wa inna ilaihī raji'ūn*' (we belong to Allāh and return to Him), then continue with the *do'a* for the safety of the deceased. The members of the deceased's family immediately call their neighbours and distant relatives to prepare and to look for the things needed for bathing the dead person. These are water, soap, lime, camphor, and yellow sandalwood. Besides that, burning incense in a coconut shell is placed next to the dead person's feet to avoid unpleasant smells.

Some men, especially close neighbours, immediately organise the enshrouding and the burial process. They acquire a set of white cotton cloth, a coffin, and a burial site. They also prepare a charity box containing some rice for visitors.[4] The visitors come and give charity to the host by slipping bank notes under the rice. This is done so that their charity is not seen by others and the notes are not blown away by the wind. Those who help with the preparations are not paid and they consider it a moral obligation to help the family of the deceased. The deceased's family just serves them drinks and dried bread.

2. Bathing

When the implements for washing the dead person are ready, the people who have been appointed by the deceased's family for this task carry the body to the washing place in the back room to prevent anyone else from watching. They place it on a raised table, his head pointing towards the east, and cover him with a piece of cloth between his navel and knee. For a woman, the cloth covers the chest to the ankles. The washers pour water over the body from the head to the tip of the feet, using a piece of cloth. They turn the body on its left side to wash the right side. Then they turn it onto its right side to wash the left side. This they do for each wash. The first and the second wash are done with water and soap till the body is clean. The teeth and the nose are cleaned from the outside only. The stomach is pressed gently and whatever comes out is removed. The hair is unbraided, washed and combed. For women, the hair is braided again in more than one braid. All the while, the washers simultaneously utter "in the name of Allāh" and "invocation to the Prophet".

---

4 Visitors here mean the people who come and offer condolences for the dead. In Arabic it is called *ta'ziyyah*, meaning visitation by one group to express condolences to another group.

After finishing the bathing, the body's orifices are plugged up with the pieces of cloth called *ija palam*. Furthermore, the *teungku* performs an ablution and ends with "nine waters", meaning washing the dead body three times on the right side, three times in the centre, and three on the left side respectively, by reciting *do'a*. The Imām recited to me the *do'a* after washing the dead body as follows. During the first three pourings on the right side, he utters: "You are the greatest forgiver, o Allāh", during the second three pourings on the centre he utters: "You are the greatest forgiver, o Merciful Lord", and during the third three pourings on the left side he utters: "You are the greatest forgiver, o Charitable Lord". For this last washing, water is mixed with flowers, sliced lime and scented oil (non-alcoholic) and filtered through white cotton cloth called *ija saréng*. The piece of cloth covering the dead body is then replaced with a dry one after the body is dried with a clean towel and the body's orifices are closed once again with other clean pieces of cloth. After that the body is brought directly to the enshrouding place.

3. Enshrouding

In the enshrouding place, three layers of cloth have already been prepared. The pieces of cloth consist of two layers of long white cotton; one layer of white cotton for the waist and for the dress and a piece of cloth for covering the head which is cut to make it seem like an ordinary dress without stitches. Before laying down the dead body, scented oil, henna leaves, flowers and yellow sandalwood which have been refined are scattered on the shrouds. Some soft cotton is mixed with scented oil and camphor and then placed on the finger joints, elbow, face (eyes, nose, mouth, and ears), knees, armpit, and toes of the dead person. The body is then covered by uttering *bismillāh*; the waist is covered, and the head cloth is put on. The dead person is then covered with long white cotton cloth and next is lifted up first from the middle, then the lower (knees and toes) and upper parts (chest and head). Prior to fastening the head cloth, the face of the dead person is displayed for a while to let the deceased's family, especially his children and other close relatives, see him for the last time. If the person had been bleeding at death, for example, a woman who had died in childbirth or a man or woman who had died due to an accident, at the time of wrapping a broken needle with no hole is placed in the shrouds. In addition, tangled thread and torn cloth are also placed on the body to avoid disturbing other living people in the village. During the enshrouding and bathing process, all speaking and any activities that can disturb the process are forbidden.

4. Praying

Before the pallbearers bring the dead body to the place of prayer, it is placed sideways into the coffin, facing the direction of Mecca. If no coffin is used, the body is placed on a stretcher. While lifting it, the pallbearers utter *bis-millāh* and *seulawet* to the Prophet Mu<u>h</u>ammad. As soon as the body reaches the front door (approximately three meters from the door), an old woman scatters husked and unhusked rice in the hope that the dead person will be safe and happy at the side of Allāh. In addition, a representative of the deceased's family delivers a short speech, asking for forgiveness for the bad deeds of the deceased. He also asks the deceased's debts to be paid for the future peace of the "soul". Furthermore, he admonishes those still alive to keep thinking about Allāh, to be afraid of Him and follow all His commands.

After the speech, the dead person is brought into the mosque of the boarding school. There, a prayer is performed in front of the dead. This prayer is spoken in Arabic, reflecting Islamic belief about the destiny of the deceased. One of the deceased's family members is asked to be the leader of the prayer.

If no one of the deceased's family wants to, the mosque's *imām* will be kindly asked to lead the prayer. The men stand in front and the women behind them. After praying, the *imām* delivers a short speech, asking the followers of *imām*, *ma'mūm* and all attendants whether this person has done good deeds or not, then all of them answer simultaneously: "Yes, he is good and may heaven welcome him". The position of the body varies. A man's head faces to the north in a sideways position in the coffin. Another possible position is the head facing to the south in a sideways position and the third position is the body lying down in a sideways position to the north. A woman is lying on her back on the coffin with her head to the north.

This funeral prayer is somewhat different from a common prayer. There is no bowing and prostration. The *imām* performs it while standing at the waist if the dead person is a woman and at the shoulder if a man, followed by the attendants in odd numbers of rows, in a minimum of three rows. When the *imām* stands up and all followers in rows are ready to pray, he says: "Let us pray collectively". To this, all followers reply: "Please bestow Your mercy on you [the deceased]", and sometimes they also utter: "There is no deity except Allāh and Mu<u>h</u>ammad is His messenger". Furthermore, the *imām* begins the prayer by pronouncing the *takbir* (the act of pronouncing the formula "Allāh is great") which is followed by the assembly with the same thoughts (intention). The intention is as follows: "I intend to perform as prayer on behalf of this deceased four *takbir, fardhu kifayah* (Arabic *fard al-kifāyah*),[5] and to follow the *imām* because of Allāh". Then the prayer is performed.

In the first *takbir,* the *imām* and all followers recite, "the mother of the Qur'ān", that is the *sūra<u>h</u> al-Fāti<u>h</u>a<u>h</u>* (QS 1:1-7), the second *takbir* continuously recites the *seulawet* to Prophet Mu<u>h</u>ammad. The third *takbir* contains *do'a* asking for the forgiveness of the deceased and the fourth *takbir* recites

5 *Fardhu kifayah* is the obligatory funeral prayer for every Muslim adult to perform upon the death of any Muslim; however even when only a few perform it, this alleviates that obligation for everyone else (Federspiel 1995).

do'a asking Allāh's mercy for the living. The prayer ends by saying *salam* or "peace" while the praying turn around, first to the right and then to the left. After the prayer, the *imām* recites the *do'a* for the safety of the deceased and all followers intone *amīn* (please accept o Allāh). Those who do not pray for the deceased wait outside of the mosque to take the body to the funeral site. They are strictly forbidden to speak and or to do anything that can disturb people performing the funeral prayer.

5. Burial

a. The Grave

The public graveyard is situated in the south of the village. Affluent people, however, have their own special burial grounds called *bhom*. The grave is dug by two men who have already been appointed and approved by the village head and been guided by the village *imām*. They first recite *Bismillāh* before digging the hole. The grave is adjusted to the body's length, and it is as deep as a normal men's chest (approximately 120 cm) or as high as an armpit for a woman (approximately 140 cm). It is placed in a north-south direction. Affluent families use a coffin but the poor do not. A dead person without a coffin is placed in a niche (a hole facing the west) called *liang lahat*, dug according to the body's size and then closed with a board. The height of the grave is approximately 30 cm from the ground. At the head and feet, plants such as the 'jarak' are planted.

b. Burial Rites

While lifting the body, the pallbearers recite *bismillāhi wa 'alā millati rasulillāh* (in the name of Allāh and the religion of the apostle of Allāh). They then carry the body, shaded with an umbrella, out of the house with the head first, followed by the deceased's family and other male and female visitors. The *imām* walks in front of the stretcher. The body is covered with a long and valuable cloth with embroideries of the profession of faith: "There is no deity except Allāh, and the Prophet Muḫammad is His messenger". If people and vehicles encounter the funeral procession, they stop for a while to show respect. While carrying the dead person, making any noise is forbidden. After arriving at the burial site, the pallbearers place the body on the west side of the grave while reciting *bismillāh* and *seulawet* to the Prophet. Before lowering the dead into the grave, a *muazzin* calls for prayer. In Lampulo, Banda Acèh, I also observed that he does so when the coffin is half covered with earth. In addition, my informant said that sometimes the *muazzin* calls for prayer before the body is buried. Three people in the grave hold up the coffin following directions given by the *imām*. While lowering the body into the grave, the same *do'a* is recited again by the *imām*. I observed that the *imām* also descends into the grave to place the body exactly in the direction of

Mecca. When finished, five shrouds from the body are loosened, then the deceased is covered with a board (the coffin lid) and the grave is filled in.

The burying starts from the west side. A mat is held up over the coffin onto which soil is thrown till completely covered. Then earth is continuously piled up on the east side so that the coffin cannot shift from its position. Finally, the hole is filled completely and soil is heaped up from the head to the feet as high as 30 cm from the ground surface. The *imām* pours water mixed with flowers and sliced lime into the grave, from the head to the feet. This is repeated by the deceased's close relatives. The pallbearers sit around the grave to listen to the *talkin* (Arabic *talqīn*) recited by the *imām*:

> "Now witness that there is no deity but Allāh and Muḫammad is his Messenger. Death is true. The grave is true. The mercy of the grave and His torture are true. Questions from Munkar are true. The Day of Resurrection is true. That deeds are weighed is true. The bridge of *shirathal-mustaqim* is true. The *syafā'ah* (help) from the Prophet Muḫammad is true. Paradise is true. Hell is true. That you will meet Allāh is true. That the Last Day will come is true. That Allāh will resurrect people from their graves is true. Hai [calling the name of the deceased]! Now you are in the layer of earth, the realm of Barzah [the transitional realm between this world and the world of the final judgment] among the dead. If two angels, Munkar and Nakir come to you, do not be afraid of them! Both of them are the servants of Allāh. If they ask you questions, answer them in a clear voice. The questions will be:
> Who is your God? Answer: My God is Allāh.
> Who is your prophet? Answer: My prophet is Muḫammad.
> What is your *kitāb*? Answer: My *kitāb* is the Qur'ān al-Karīm.
> What is your *kiblat*? Answer: My *kiblat* is the Ka'batullāh [Mecca].
> Who are your *ikhwān*? Answer: All Muslim men and women.
> Answer these questions with a clear voice. In the Judgement Day, if Allāh wills, you will be in the groups of true believers. Allāh remains you with two words of profession of faith as the believers till the day of hereafter. O reassured soul, return to your Lord, be well-pleased and pleasing [to Him], and enter among His [righteous] servants and enter His Paradise." (author's translation from Arabic).

This text is written in Arabic and Indonesian so that it can benefit the living as well as the dead. Some villagers said that the aim of reciting this text is to help teaching the dead not to be afraid of the angels Munkar and Nakir who will come to him. It will enable him to remember the answers to the angel's questions and to utter them fluently in a clear voice. Others said that the text explains what will happen at the grave soon after the *imām* has left and what the deceased should do or say when the two angels come to examine him. But the "modernists" say, "the dead cannot hear, the chanting cannot help the dead".

The *imām* only recites the Arabic version of the text as he sits at the head of the grave. An informant explains that after the burial rites the dead can understand Arabic even when he did not understand it while being alive. He emphases that Arabic is "the language of the realm of Barzah" that is "spoken in Paradise". The prayers refer to the idea that "Allāh returns the 'soul' (Acèhnese *aruwah*; Arabic *arwah̲* plur., sing. *rūh̲*) to the body". Allegedly, the deceased person "is revived for a few moments after burial" so that "his good deeds will respond to the angels' questions" about his moral conduct during his lifetime. If the dead has done many good deeds, he will be able to answer the questions, says the *imām*. If the dead has not done any good deeds, then automatically his *aruwah* cannot answer the angel's questions. If the dead can answer all questions satisfactorily, Munkar will release him in peace till the Day of Resurrection. Otherwise, Nakir will beat the dead till the day of Resurrection. "Neither rationality nor intelligence works in answering Munkar's question; the belief is that the brain is where rationality and intelligence reside and these must also be dead when the body dies. Only beliefs, which are found in the *aruwah*, not in the body, will last forever" (Muhaimin 2004: 55). After the angels' questions about the Islamic catechism have been answered, Allāh will soon provide "mercy in the grave".

Reciting *talkin* takes around fifteen minutes. After that, many people go to the deceased's house again to have drinks and a meal. Here they are informed that the ritual for the *aruwah* will be held again.[6] All tools carried to the burial sites are returned to the deceased's family except the mat used to cover the body; it is given to a place of religious services such as a mosque or *meunasah*.

6. *After Burial*

The extended family of the deceased soon holds a meeting to discuss the recital of *shamadiyah* together on several nights. The *shamadiyah* is held for seven nights at the deceased's house. The recital is performed after sunset prayer. It aims helping the deceased's family feel relief while praying for the deceased. Others say that reciting *shamadiyah* is merely to add to the deceased's good deeds. It is said that the more good deeds are assigned to the deceased, the happier he will be. Below is the order of *shamadiyah* recited from the first till the seventh night at the house of the dead in Blangporoh:

1. Reciting the *sūrah̲ al-Fātih̲ah̲.*
2. Reciting *istighfār*, at least three times, either the short or long version.
3. Reciting seulawet to the Prophet *Muh̲ammad* from three to seven times, either the short or long version.
4. Reciting the *sūrah̲ al-Ikhlās* for about 15 to 50 times or even 100 times. This is the core of *shamadiyah* recitation.
5. Reciting the *sūrah̲ al-Falaq*, once.

6 In a village that is slightly modern or already influenced by "young groups", people go back to their houses directly except for the close relatives of the deceased who go to the deceased's house.

6. Reciting the *sūrah an-Nās*, once.
7. Reciting the *sūrah al-Fātiḥah*, once.
8. Reciting the *sūrah al-Baqarah*, verse one to five, verse 163-255 (verse of '*kursi*') and ending with verse 284 to 286.
9. Reciting *tahlīl* [*lā ilāhā il-lallāh*] between 50 to 100 times.
10. Reciting *do'a* guided by an *imām* and replied *amīn* by all *shamadiyah* followers.

Shamadiyah is recited in the same way from the first till the seventh night respectively. The difference is that the *shamadiyah* on the seventh night is longer as the *sūrah al-Ikhlās* is recited 100 times. This performance is called the *khatam shamadiyah*.[7] In addition, reciting *do'a* for the dead on this night is also longer than during the previous six nights. The *do'a* asks Allāh's forgiveness and guidance for all Muslims. It is specifically aimed at the deceased to be forgiven by Allāh, to have his/her good deeds accepted and to be placed in a respectable place beside Allāh. In addition, the deceased's family members are hoped to remain patient. However, "young modernists" do not agree with the recitation of *shamadiyah.* They say that one person cannot make up for the sins of another.

After finishing the recitation, cakes and drinks are served to the reciters[8]. Especially on the third and the fifth night, the family of the deceased serves them special cakes. On the third night, they are served with *dayak-dayak* spongy cakes, made of glutinous powder mixed with coconut milk, salt and sugar. One informant said that the cake reminds the living that the body in the grave has already become mushy, similar to the *dayak-dayak* cake. On the fifth night, the visitors are served with *apam* cakes. It is said that this reminds the living that the body in the grave has already swollen, similar to the *apam* cakes. Thus these two kinds of cake provide a communication. On the seventh night, a big *khanduri* called *khanduri seunujoh* is held. This *khanduri* is bigger than the previous six nights, it is held from the afternoon of the sixth day until the afternoon of the seventh day to welcome relatives, neighbours and other close friends of the family. The family serves a big meal consisting of rice with various side dishes including curry, after the *shamadiyah* recitation. I also observed especially among the affluent families that soon after the big meal, the group of *qāri'*[9] from various districts in the region of south Acèh arrived and started to recite the Qur'ān for a whole night in Blangporoh, using a microphone and loudspeakers. The more famous the invited *qāri'*, the more

7 One says that if *shamadiyah* is recited till morning for the deceased, Rp 100 to Rp 200 will be given to each reciter.

8 Many *shamadiyah* reciters are people who at the same time offer condolences at the deceased's house and go back after the recitation.

9 *Qāri'* is a male reciter of the Qur'ān who has memorized the Qur'ān for a wide variety of contexts. *qari'ah* means a female reciter of the Qur'ān.

money will be given to him by the deceased's family. In addition, for the rich family, the Qur'ān is also recited at the grave[10] from the first till the seventh day and night. In return, the reciters are also given money for their daily needs as well as meals and cakes. These events constitute a critical point of transformation. They finally release the *aruwah* of the deceased and its ties to the house.

Besides the *shamadiyah* which is held on the first till the seventh night after the burial, Qur'ānic recitations are also held at the house of the dead in which each man present takes turns at reading the Qur'ān till late in the night. One of the participants says that reciting Qur'ān for the dead is a kind of good deed. All the words of the recited Qur'ān contain merit and all merit obtained from its recitation is intended for the *aruwah*. The reciters have a break after an hour of reciting Qur'ān. They are then served with drink and cakes. During this break, the reciters recite *qasidah* (songs in praise of religion) to entertain the deceased family and so the other listeners can take a break. The recitation is then continued until midnight.

In the morning of the seventh day, the close relatives, both men and women, go to the grave to "plant" the tombstones on the grave[11]. The stones are taken from the river; they are as big as a man's head. The *imām* guides the ritual of planting the tombstone. One is planted at the head and another one at the feet while reciting *bismillāh* and *seulawet* to Prophet Muhammad. Then water mixed with flowers and sliced lime is poured over the grave. The relatives sit together to recite *do'a* before leaving the grave.

Tilam Bungong

In Blangporoh, I observed that the deceased's family construct a bed in a room belonging to the deceased, with the deceased's clothes embellished with colourful shawls of his wife and other personal objects, as a charitable act. Pillows with colourful embroidered flowers on their covers are placed at the head of the bed while a bolster is placed on the right side and another one is placed on the left side. The center is left empty assuming that it is "the place where the dead person's *aruwah* sleeps". This bed is called *tilam bungong*. *Tilam* means mattress and *bungong* means flower.[12] The room with the bed may not be used by other persons except for the deceased's core family, his wife and his children. It is not allowed to be left empty and a lamp is always lit there as

10 The Qur'ān reciters who recite Qur'ān at the grave are usually chosen from the *imām* in the district or from the students at the boarding school in Blangporoh.

11 Usually the ritual of "planting" the tombstones is performed on the 44th day of the death. This is related to the situation of the rice plants in the fields.

12 In Nagan Raya, West Acèh, this bed is called *reuhab* as a "station of meeting" with the living family. The local people believe that during 44 days the soul of the deceased has not yet ascended to Heaven? but still roams on the earth. On day 44 the *reuhab* is opened, the ritual of planting the tombstones at the grave is then performed (Husda 1988: 34).

the *aruwah* needs to be accompanied by at least one of his family members. For seven days and nights, white incense is burnt in this room because the deceased's *aruwah* likes its smell. The aim of burning incense is said to summon the *aruwah* so that it will not get lost when returning to its house.

After the seventh day the *tilam bungong* is opened. The clothes and other personal belongings, constituting the objectified body of the deceased, are distributed as gifts among the corpse washers. Different items are given to those who have washed different body parts. In case the dead is a male, his rimless cap is given to the person who had washed the head; his tie to the person who had washed the neck; his watch to the one who had washed the hands; his shirt to the one who had washed the chest; his belt to the one who had washed the hips; his trousers to the one who had washed the legs and his shoes to the one who had washed the feet. The white cotton cloth functioning as a water filter during the washing is given to the *imām* who pours the "nine waters" over the dead body. A woman is treated the same way. The difference is that a woman is only bathed by women and her clothes are shared out accordingly. My informant said that every village in West Labuhan Haji district appoints the male and female *teungku* for bathing a dead person. In return, they receive some gifts from the deceased's family and from the village mosque. In an affluent family, *tilam bungong* is opened up on the fourteenth day. Affluent families do the same as poor families on the fourteenth day of the death. This extension is made because affluent families still hold a *khanduri* for the orphans and the poor that are invited to the deceased's house. The religious merit from holding this *khanduri* is once again transferred to the dead. "The more *khanduri* are held for the *aruwah*, the happier the *aruwah* will be and vice versa the deceased's *aruwah* will be sad if the *khanduri* is not done at his or her house", the *imam* explained.

Giving the clothes to the people who bathe the dead body is very important for the fate of the deceased. It is said that several years ago the previous *imam* of Blangporoh village, Teungku M. Isim, saw a person in the darkness coming back from the Blangporoh mosque to the house of Tgk. Bile Rantau. He was completely sure that the person he saw was the *aruwah* of Teungku Bile Rantau who had passed away seven days earlier. His *aruwah* went back to his house with shorts, a small pail in his hand and a sheet of cloth on his shoulder. Teungku Ahmad said to his children, "this is retaliation from Allāh", because his clothes were not distributed to the corpse washers. Soon after, the children of Teungku Bile Rantau called all washers and gave them their father's clothes. Only after that, the *aruwah* was no longer seen.

Apart from the sharing of the deceased's clothes on the seventh day of the death, the gravediggers are also invited. One of the devout village women immediately comes to talk to the wife of the deceased person:

> "*Alah cukup tu* (it is enough). I know your brother well. He is a good person and he has done many good deeds as well. Moreover, he died because of a motorbike

> accident and therefore he receives the same religious merit as for a martyr's death. His good deeds will become a person whose face is very much like his own but that "person" is better and wears better clothes and will accompany his *aruwah* forever. His *aruwah* is very happy and can sleep well until the day of the resurrection. Vice versa, if a person does bad deeds when he or she was alive, his or her bad deeds will also resemble a person whose face is very much like his or her own but that person's body has much blood on it. His clothes are torn and smell very bad. He comes from the deceased's bad deeds. That person says 'I am like this because none of your deeds are good. You follow your lust and never followed Allāh's command when you were alive. So now I will not leave you. Do not hate looking at me because when you were still alive you were not afraid of doing something wrong. That means you were always looking for me. As you are like your bad deeds, thus I am like you. I will never leave you and accompany you till the day of resurrection.' That person then hugs the *aruwah* even though the *aruwah* refuses that person. The more the *aruwah* hates that person, the stronger that person hugs the *aruwah*. In addition, many other grave's torments will come because of not bringing about the pillar of Islam; prayer, fasting, [...]."

The devout woman emphasised further: "What I tell you now is what I heard from my previous teacher. He tells the story about the person who has done a lot good deeds in his life and receives a reward from Allāh when he dies. So do not be sad." Only after hearing about this moral mirror image of the dead person, the wife of the dead person looks good and no longer cries. The idea here is that one cannot escape from his or her own image. The gravedigger is then given two kilograms of sugar; two packs of cigarettes and half a kilo of coffee by the wife of the deceased person. He also receives some money from the deceased's family and from the village mosque on the following day.

Apart from preparing *tilam bungong*, the deceased's family in Blangporoh also invites a devout person three times a day to eat at the deceased's house from the first till the seventh day (and till the fourteenth day for the affluent families). This is called *pajōh khanduri watèe* ("eating *khanduri* at the appointed time"). The devout person is chosen by the village *imām*. The *imām* said that the *aruwah* is the same as we are. The *aruwah* of the deceased is also "hungry and thirsty". It returns to his or her house and asks for food and water but we cannot hear it. That is why the deceased's relatives, especially his core family has to feed a devout person instead in order to fulfil the deceased's *aruwah*'s needs. The food served to the devout person or *teungku* is that which the deceased liked when he or she was still alive. The *teungku* uses the plates, cups, and spoons, fork and glasses that the deceased used in his life. All crockery used by the *teungku* when he eats at the deceased's house is given to him on the seventh day or on the fourteenth for the affluent families. This is all done with the aim of adding to the deceased's good deeds as well as making the deceased's *aruwah* happy. The *imām* is invited to the dead person's house to eat the "*khanduri* at the appointed time" from the first till seventh day respectively. Empty glasses are placed on the table where he eats, to receive

Allāh's mercy. Fresh water in a bowl on the table symbolises that the *khanduri* for the *aruwah* is held sincerely by the family. In addition, a lamp on the table symbolises that the descendants will always remember and pray for the deceased. Praying for the deceased is said to be like a lamp on the grave. The pious children of the deceased person who invoke Allāh in their wishes may indeed hear their prayer answered by Allāh, and they can be especially effective in transmitting the merit of the recitation to their parents' *aruwah*.

Aruwah

Aruwah (from Indonesian *arwah*) can be translated as "soul or spirit". Nevertheless, in the ritual of *khanduri bu* the villagers say *roh* (Arabic *rūh̲*). Many villagers in Blangporoh whom I interviewed believe that *aruwah* remains in the house and sometimes returns to the grave during the seventh day after death. Therefore, *tilam bungong* is prepared for the *aruwah*. It is said that from the seventh day the *aruwah* of a devout person goes back from its place in *'illiyyūn* (the highest elevation of paradise) to the house and to the grave. The *aruwah* of people who violate Allāh's law go to a place under the earth called *barhud*. Some say their *aruwah* are placed in *sijjīn* (the lowest level of hell). The *aruwah* then stands at the ventilation of the house. Therefore ventilations may not be closed at night. The *aruwah* observes the behaviour of his family members. If the relatives perform the *khanduri*, behave well and never forget to pray, the *aruwah* is happy. However, if the family members violate Allāh's law, the *aruwah* will be sad and then return to *'illiyūn* immediately. The *khanduri* and *shamadiyah*, therefore, have to be held and the *teungku* and other devout people have to be invited to eat and recite *do'a* from the first to seventh night respectively. However, my informant, Mukim Hukom has a different opinion; he argues that the *aruwah* returns and stands at the door. The door has to remain open till late at night, especially on the night when the *aruwah* returns to its house. The *khanduri* has to be held and at least one *teungku* or devout person must be invited to eat instead of "feeding" the *aruwah*. There must be some rice left in the cooking pot when the *aruwah* comes back. If there is no more rice, more must be cooked because the *aruwah* enjoys the smell of the rice. It is said that if there is no *khanduri* held as the *aruwah* arrives, the *aruwah* will not enter the house, thinking: "They no longer care about their deceased brother".

The aim of the *aruwah* returning on the first night is to observe whether his close relatives and close friends who live near his house come to offer condolences as well as how his family members welcome them. The family members must pray for the *aruwah* because the torment during the first night in the grave is said to be very frightening. Therefore, recommended prayers which are called *seumayang hadiyah* ("gift prayer")[13] are recommended. It is said that if this recommended prayer is not performed, one can give charity to the needy instead, with the aim of helping the dead in the grave. "Prayer is like depositing money in a bank. Allāh repays it by sending the 'grave's mercy' to the dead". "The fine for failure to perform prayer" is also paid to the village, as one liter of rice given to the poor through the *teungku*. But this is only done by affluent families who did not perform prayer five times a day when the dead person was still alive.

The aim of the *aruwah* returning on the third night, when the spongy *dayak-dayak* cakes are eaten, is to observe whether his relatives and close friends who live far from his house come to offer condolences for him and how his family welcomes them.

On the fifth night, reciting *shamadiyah* and *do'a*, *khanduri apam* is held to mark the corpse's swollen condition in the grave. The *aruwah* is truly sad and cries when it sees the condition of the body in the grave. The task of the *aruwah* returning on this night is to observe how much attention his family members give to him. Some villagers say that if the debt has not been paid yet, the *aruwah* will be in *awang awang*, a place which is not well defined, somewhere between sky and earth. Some others say that the *aruwah* will be suspended between the sky and earth.

The aim of *aruwah* returning on the seventh night[14] is to observe whether his debt has been paid or not. If not, the *aruwah* will cry and condemn his family members. The body is then said to explode in the grave and be devoured by worms. The *aruwah* is again truly sad and cries when it sees the condition of its former body. From the seventh day onwards, the *aruwah* loosens its ties to the house but still roams around the yard. In addition, it is said to leave the decomposed body. The ritual series called *khanduri seunujoh* reaches its climax. The women work hard in the kitchen to make special sweets and prepare food for the *khanduri*. Relatives come from other villages and stay for the *shamadiyah*, a night time chanting session. A ram is slaughtered for the deceased on the Day of Judgement and guests are served rice with side dishes and drink. Serving rice and drink to the guests is said to equal the religious merit received when one performs *hajj* and *umrah* (the lesser pilgrimage out of the *hajj* season) in Mecca. One *imām* said; "Any guest coming to the deceased's house must be fed. Do not refuse the guests! Refusing them is the same as refusing the Prophet". This indicates how high a value is put on the *khanduri* on the seventh night of death. It is said that although the *khanduri*

---

13 The *teungku* tells me that "gift prayer" can be performed in at least two circles (Arabic. *raka'at*) for the dead. In the first circle, the verse of *qursi* (QS 2:255) is recited once. The second circle, the *sūrah at-Takāthur* (QS 102:1-8) is recited once and the *sūrah al-Ikhlās* (QS 112:1-4) ten times.

14 Nothing was known about the tasks of the *aruwah* on the second, the forth, and the sixth night. In any case, according to many *teungku*, it returns every Friday night.

is not held when the Prophet was still alive, it was held at 'Ali's house[15] for seven days and at Aisyah's[16] house for fourteen days when he died.

The aim of the *aruwah* returning on the fourteenth night is to observe whether his *wasiyyah* (the last minute testament) is carried out by his close relative. If not, the *aruwah* will be very sad and will curse them. Around this time, the body begins decaying, and the *aruwah* is said to start roaming away from the yard. A small *khanduri* with a short *shamadiyah* and *do'a* is held. There is a belief that if the *aruwah* condemns his family, the wealth that he left on the earth will no longer bring 'blessing' and will gradually be lost.

The aim of the *aruwah* returning on the 44th night is to observe whether his extended family, especially *wali hukum* (a group of men from the husband's side who receive a greater share of the inheritance),[17] cares for his children and wife. If not, the *aruwah* will condemn his extended family and return to its abode with sadness. A small *khanduri* with short *shamadiyah* and *do'a* is also held for the *aruwah* at home. The decaying process of the body continues. The *aruwah* is said to gradually leave the house and start roaming in the village. On this day, the deceased's family wraps dried bread that is made by his wife and children. All the dried bread is wrapped in a piece of white cloth and then given to the village *teungku* or other devout people in the village. Traditional cakes such as *leumang*, *leupèk*, *keutupék*, *bada* and *umping* made of *idi* glutinous rice are also made as the proof of his wife and children's love. These various foods are "sent to Allāh" for the deceased through the village *teungku*.

The aim of *aruwah* returning on the one hundredth night is said to remind its close relatives to do good deeds (the living cannot hear this). It reminds them that his death has to be exemplary for his family members and for the living. It says: "The rewards received by those who follow Allāh's command compare to those who do not have already been seen". A small *khanduri* with short *shamadiyah* and *do'a* is also held for the *aruwah* to make it happy. At this time, the decaying body turns from a lying to a sitting position. The flesh has disappeared and the bones gradually dissolve into the soil. The *aruwah* is said to start roaming far away from the village but only sometimes wanders in and out of its former house. In order to have a clearer idea about the tasks performed by the *aruwah*, see the *table one* below:

---

15 'Ali bin Abu Thalib, a cousin of the Prophet Muhammad, an important leader in the early Muslim community and the fourth caliph.

16 'Aisyah binti Abu Bakar, a favourite wife of the Prophet Muhammad in his last years and political personality in the struggle for Muslim community leadership after his death.

17 Also called *wali meukeutam*.

No	Night	The tasks of the *aruwah* returning to its former home
01	1	To observe whether his close relatives and friends living near his house come to offer condolences for him as well as to observe how his family welcomes them
02	3	To observe whether his family and close friends living far from his house come to offer condolences for him as well as to observe how his family welcomes them.
03	5	To observe how much attention his family gives him.
04	7	To observe whether his debt has been paid or not. If not, the *aruwah* will cry and condemn his family.
05	14	To observe whether his *wasiyyah* (last minute testament) is carried out by his close relatives.
06	44	To observe whether his children and wife are looked after by his extended family, especially wali hukum.
07	100	To remind his close relatives to do good deeds as well as to remind them that his death is exemplary for his family and for the living.

Table 1: The tasks of the *aruwah* returning to its former home

Besides these occasions, it is said that the *aruwah* also returns to visit his family on the following dates: 1) on the night of determination; 2) two nights preceding Ramadhan; 3) on the night of power which takes place on the last ten nights of Ramadhan; 4) on the night marking the end of Ramadhan; 5) on the night of the tenth day of the month of Zulhijjah, celebrating the pilgrimage, also known as *'īdul qurban*. Like on the one hundredth night, the *aruwah* again reminds its close relatives to do good deeds as well as that his death is exemplary. On these particular nights *khanduri* is held again and the *do'a* recited. According to many *teungku*, this is like providing a life vest for the *aruwah* (see Table 2 below for a summary of the corpse's condition, the exchanges and the *aruwah*'s location).

Time	Corpse's condition	Exchange	*Aruwah*'s location
1st day	As usual	- *Shamadiyah* + *do'a* for the deceased and food gift	1st to 6th day the *aruwah* remains in house
2nd day	Begins to smell	- *Shamadiyah* + *do'a* for the deceased and food gift	
3rd day	Smells and is mushy	- *Khanduri dayak-dayak* *Shamadiyah* + *do'a* for the deceased and *dayak-dayak*, spongy cakes are distributed	
4th day	Begins swelling	- *Shamadiyah* + *do'a* for the deceased and food gift	
5th day	Swollen	- *Khanduri apam* *Shamadiyah* + *do'a* for the deceased and *apam* cakes are distributed	
6th day	About to explode	- *Shamadiyah* + *do'a* for the deceased and food gift	
7th day	Explodes	- *Khanduri sinujôh* *Khatam shamadiyah* + *do'a* for the deceased and food gift and monetary gift + clothes of the deceased are distributed and his debts are paid	- Unties its bond to the house but still roams around the house
14th day	Begins decaying	- *Khanduri duasén tujôh* *do'a* for the deceased + food gift and monetary gift	- Roams outside of the house or domestic surroundings
44th day	The decaying process continues	- *Khanduri peutproh peut* *do'a* for the deceased + food gift and monetary gift	- Gradually leaves the house and starts roaming in the village
100th day	Turning from a lying to a sitting position. As the flesh has disappeared the bones gradually dissolve into the soil	- *Khanduri sireutoh* + food gift and monetary gift	- Starts roaming far away from the village
Corpse decayed →			Allah

Table 2: Intervals of the deceased's *aruwah*

Conclusion

After a person has died, the body is washed and wrapped in cloth inside the house. Ceremonial prayers are said in Arabic reflecting Islamic beliefs about the destiny of the deceased. On the same day, the dead person is buried in the village burial grounds. This entails a ritual performed by the *imām*, who addresses the deceased in Arabic, the "language of *barzah*" that is "spoken in paradise". The prayers refer to the idea that "Allāh returns the *aruwah* to the body". This signifies that the deceased person is "revived for a few moments after burial" so that "his good deeds will respond to the angels' question" about his moral conduct during his lifetime. Having answered these questions satisfactorily, Allāh rewards the deceased with the "mercy of the grave". Nowadays, however, such ideas are contested, particularly those ideas concerning the collective performance of these ritual actions, as being efficacious in improving the ancestral destiny of an individual person. "Young modernists" in West Labuhan Haji argue, "the dead cannot hear, the chanting cannot help the dead and one person cannot make up for the sins of another". It appears that an individualist understanding of morality is gaining ground against the moral precept of collective responsibility as expressed in these instructions to the deceased. The form and meaning of this ritual is not only a topic for anthropological analysis but also a subject for long-standing and hotly contested religious debates within Indonesia, between the proponents of Islamic modernism and those who would reaffirm the correctness of older, local ritual practices (Bowen 1984: 21).

Meanwhile, the deceased's family has constructed a bed composed of the deceased's clothes and other personal objects. It is between these objects that "the deceased *aruwah* sleeps". Represented by his "soul" and the textile objects, the deceased is surrounded by his spouse and children. For seven days and nights incense is burnt "guiding the *aruwah*" in his domestic surroundings. After seven days the clothes and other personal belongings, constituting the objectified body of the deceased, are distributed among the persons who washed the dead body immediately after death. Different items are given to those who washed different body parts. At the same time, food is distributed among the poor and the orphans of the village, once again generating merit for the deceased. Such ritual injunctions are strongly sanctioned by the expectation that negligence to fulfil them results in the return of the deceased's *aruwah* into the village society. This is seen as "retaliation from Allāh for not distributing the clothes to the people who bathed him on the seventh day of his death."

At certain intervals, the deceased's *aruwah* returns to the house again, where it is accompanied by *shamadiyah*, *tahlil*, and *do'a* recited after the "sunset prayer" and food gifts. These food gifts are subsequently distributed among the participants. Particularly during the first seven nights, different food gifts mark the different parts of ritual reception of the deceased's *aruwah* as described in the summary above. The constitution of these different foodstuffs refers to the successive stages in the deceased's body's process of decay. For example sponge cake and *apam* cake signify that the dead body is in an advanced stage of decomposition. This indicates that the deceased's progeny partakes in a ritual communication of the former's substitute body: the deceased personal image (shown by the clothes and jewellery that he wore when still alive, for instance) as well as his body are represented in clothes and foodstuffs and distributed among his or her progeny. In other words, the aspects of the social person are transferred into gifts that are distributed to others. This fundamental idea of a person, whose constituent parts are distributed after his death among close kin and/or neighbours, is known in other Southeast Asian and Pacific societies as well (Coppet 1981; Platenkamp 1988a, 1988b, 1992, 1996, 2001, 2006; Barraud and Platenkamp 1990; Iteanu 1990; Manan 2015).

On the seventh night, the ritual series called *khanduri seunujoh* reaches its climax. The family release their feelings of sadness, loss, and frustration. The crying is no longer in regret of the death. Relatives, neighbours and close friends are served a large meal. Male reciters from various districts of South Acèh arrive to recite the Qur'ān for a full night for the deceased's benefit. The more famous the reciters are, the more monetary gifts are distributed by the deceased's family. These events constitute the critical point of transformation. The *aruwah* of the deceased releases its ties to the house, its worldly debts, and the people in turn release their pent-up emotions. The period between death and the seventh day corresponds to what Hertz called the "intermediary period", which is represented in some Indonesian societies by the exposed body of the dead. Hertz's insight is that, in these societies, the exposed, decomposing body is the material basis for the passage of the soul (Hertz 1960: 29); like the Gayo in Central Acèh (Bowen 1984: 27), the people of Blangporoh also see decomposition as a natural process and perform similar key rituals, but without exposing the dead body.

The mortuary ritual is not just performed for the bereaved family but most importantly to help release the *aruwah* of the deceased from its ties to the house, so that the remaining family can stop mourning. All the *khanduri* are to finalise the transformation and to accomplish a slow gradual unbinding of the *aruwah* to the house, the community, and the social world. Each time the gifts are properly distributed, the *aruwah* moves away from its house, domestic surroundings or village, and moves toward Allāh. In the meantime, the putrefaction process of the dead body continues. This indicates that there is a significant relation among the practice of commemorative *khanduri*, the reciting of *shamdiyah*, *tahlil*, and *do'a*, and the delivery of the *aruwah* to its own world, the transfer of gifts and the putrefaction process of the corpse before it finally disappears altogether into soil. The mortuary ritual therefore continues to have a lot to say about death and constitutes an intricate process of exchange, in which parts of the deceased person in spiritual, physical and objective modalities are fragmented and transferred amongst those social relationships that constitute the elementary network of which he was part of during his lifetime.

References

BOWEN, J.R. (1984) Death and the History of Islam in Highland Aceh. *Indonesia* 38: 21-38.

BOWEN, J.R. (1993) *Muslim through Discourse, Religion and Ritual in Gayo Society*. Princeton: New Jersey.

BARRAUD, C. and PLATENKAMP, J.D.M. (1990) Ritual and Comparison of the Societies. *Bijdragen tot de Taal, -Land-en Volkenkunde, Rituals and Socio-Cosmic Order in Eastern Indonesian Societies* Part II Maluku 146(1): 103-124.

COPPET, D. DE (1981) The Life-Giving Death. In HUMPHREYS S. C. and KINGS H. (eds.), *Mortality and Immortality: The Anthropology and Archaeology of Death*, pp. 175-204. New York: Academic Press.

FEDERSPIEL, H. M. (1995) *A Dictionary of Indonesian Islam.* Athens, Ohio: Ohio University Press.

HERTZ, R. (1960) *Death and the Right Hand.* New York: Free Press.

HUSDA, H. (1988) *Ketahanan Adat dan Budaya Nagan*, Banda Aceh: Ar-Raniry Press.

ITEANU, A. (1990) The Concept of the Person and the Ritual System. An Orakaiva View. *Man* n.s. (25)1: 399-418.

MUHAIMIN A.G. (2004) *The Islamic Tradition of Cirebon, Ibadat and Adat Among Javanese Muslims*, Jakarta: Religious Research, Development, and Training.

MANAN, A. (2015) *The Ritual Calendar of South Aceh, Indonesia*, Münster: MV-Verlag.

PLATENKAMP, J.D.M. (1988a) *Tobelo, Ideas and Values of a North Moluccan Society.* Leiden: Repro Psychologie.

PLATENKAMP, J.D.M. (1988b) Myths of Life and Image in Northern Halmahera. In CLAESSEN, H.J.M. and MOYER D.S. (eds.), *Time Past, Time Present, Time Future. Essays in Honour of P.E de Josselin de Jong*, pp. 148-167. Dordrecht and Providence: Foris.

PLATENKAMP, J.D.M. (1992) Transforming Tobelo Ritual. In COPPET D. de (ed.), *Understanding Rituals*, pp. 74-96. London: Routledge.

PLATENKAMP, J.D.M. (1996) The Healing Gift. In HOWELL, S. (ed.), *For the Sake of our Future; Sacrifice in Eastern Indonesia*, pp. 318-336. Leiden: Centre of Non-Western Studies.

PLATENKAMP, J.D.M. (2001) Temporality and Male-Female Distinctions in the Tobelo Vocabulary of Relationships. In ALÈS, C. and BARRAUD, C. (eds.), *Sexe relative ou Sexe Absolu? De la distinction de sexe dans les Sociétés*, pp. 241-266. Paris: Éditions de la Maison des Science de l'homme.

PLATENKAMP, J.D.M. (2006) Spirit Representations in Southeast Asia: A Comparative View. In LAUGRAND, F. and OOSTEN, J. (eds.), *The "Nature of Spirits": Human and Non-Human Beings in Aboriginal Cosmologies*, pp. 99-129. Quebec: Les Presses de l'Université de Laval.

Cosmological Cycles: Continuity in Inuit Society in North Eastern Canada

JARICH OOSTEN[†]

Introduction

When I was appointed to teach religious anthropology at the Institute of Cultural Anthropology and non-Western Sociology in Leiden 1970, that institute had still a great reputation for structural anthropology. I soon realised that in structural terms that position was quite weak. In fact, only P.E. de Josselin de Jong was keeping the tradition alive in his own impeccable style, while most members of staff were moving into other new directions such as development sociology and feminism. But for the survival of structural anthropology there was much potential in a group of young students associated with the famous room 607 where they had their inspiring meetings and debates. Jos Platenkamp was one of the leading protagonists of this group and immediately impressed me with his creativity, his intelligence and the rigour of his thinking. We always got on well and even though I had no specific expertise on Eastern Indonesia I was happy to become involved in the supervision of his research in that area.

In the early 1970s it became necessary to organise research in coherent clusters to acquire financing. In Leiden we developed a cluster called CASA (Cognitive Anthropology and Structural Anthropology). Despite this rather broad title we were mainly focusing on French structuralism, especially the magnificent work of Claude Lévi-Strauss. We would probably have continued to do so, had not Jos Platenkamp and Danielle Geirnaert started an intensive collaboration with Daniel de Coppet and Cécile Baurraud of the French research group ERASME that was very much inspired by the work of Louis Dumont. Notably the notions of idea/value and hierarchy stood central in this approach. I found their ideas interesting and stimulating, but at the same time I was uncomfortable with the rather strict way in which these notions were applied.

The assumption was that ideas and values were organised in a rather straightforward hierarchical framework. I thought this assumption is questionable as in my perceptions societies are not only well organised but also full of problems and contradictions. Participants tend to be aware of them, but they cannot solve them and may even get resentful at attempts by outsiders to fill in the gaps for them and turn an insurmountable problem into a neat puzzle that can easily be solved. In this paper I discuss some of these problems with respect to Inuit society. It is in many respects a result of the many fruitful and entertaining discussions I had with Jos and testifies to his influence on my work.

The Inuit of Nunavut: A Hunting People

In 1999 Nunavut, 'our land', became an independent territory. It was predominantly inhabited by Inuit (about 23000), the native population of the area, and about 2000 *qallunaat,* white people, most of them residing temporarily in the North. Today the Inuit live in permanent settlements that may vary from a few hundred to a few thousand people. Before the 1950s most Inuit still led a nomadic existence as hunters living where the game was available. In the 1950's and 1960's the Canadian administration encouraged Inuit to settle in permanent communities around the trading and mission posts. In these settlements schooling, health care, and other facilities could be provided and the Canadian administration considered the movement from the camps out on the land to these settlements a necessary transition to a more civilized life. The modern settlements brought together people from different hunting areas and even today people still feel connected to their original hunting grounds and hardly identify with the modern communities they inhabit now. Inuit feel strongly about the need to maintain their identity as a hunting people, but in practice hunting becomes more and more difficult as people no longer live close to the game. They often have to travel great distances to reach it. Modern hunting requires equipment such as snow scooters and motorboats that can hardly be financed out of the yields of subsistence hunting. Thus a hunter may need a salaried job in the community to be able to afford the necessary equipment. It has become hard to be a hunter without an additional income. Inuit feel that they can only remain Inuit and retain their culture by hunting. Through hunting not only relationship to the land and the animals, but also social relationships take shape.

Searching for Distinctions

In his famous paper on the seasonal variations of the Eskimo (originally published in 1904) Mauss (1979) examined the social morphology of Eskimo societies. In these societies the volume, density and distribution of the population varied in different seasons and areas. Along the Canadian Arctic coast people used to congregate in small winter camps, often comprising fifty or sixty people. In summer these camps usually broke up as individual families spread over the land to hunt on a family basis. The winter camps were usually constituted by a few related families. Mauss argued that no clear tribal organisation existed among the Inuit and that the settlement, not the tribe was "the true territorial unit" in this field of study.

> "The settlement has more than just a name and a territory; it also possesses a linguistic unity as well as a moral and religious one. Although these two categories may appear initially to be unrelated, we have purposely linked them because the linguistic unity to which we wish to call attention has a religious basis and is related to ideas about the dead and their reincarnation" (Mauss (1979: 28).

Mauss suggests that the tradition not to use the name of a deceased or even words related to that name until that name has been given to a newborn baby defines each settlement in terms of a limited number of names. He is well aware of the mobility of individual Inuit moving from one settlement to another, but assumes that this is always caused by vitally urgent necessities and that this variation does not really affect the principle that that each settlement constitutes a fixed and defined social unit.

The winter camps were the locations where intensive collaboration and social interaction between different families took place and here the most important rituals such as the Inuit winter feasts were celebrated. In that respect each winter camp effectively represented the whole society in its ritual relationships with the animals and spirits that inhabit the land, the sea and the sky.

Boas (1888: 425) already recognised that Inuit could not be easily divided into tribes or other distinct groupings, "In my opinion a great difference between these tribes never existed. Undoubtedly they were groups of families confined to a certain district and connected by a common life." The Inuit themselves connected people to the place or area where they were living. Thus the Nattilingmiut where people living at the 'place where there are seals' (derived from *nattiq*, seal), the Iglulingmiut 'at the place where there are igloos' (derived from *iglu*, snow house). In different winter seasons the location of the camps might vary considerably. Between different areas existed dialectical variation, but there was also considerable migration. Perceptions of other groupings varied greatly. Mathiassen (1928: 23) writes,

> "My old friend Takornaq, who accompanied me on my journey to Ponds Inlet, expressed it in this manner: that if she were to have a new husband, it was all the same to her whether he was an Aivilingmio, an Iglulingmio or a Tununermio; if need be, a Qaernermio or an Akudnermio might do; but a Netsilingmio; never!"

Although the Aivilingmiut feared the Nattilingmiut, this did not prevent the two groups from mixing, and today a community such as Naujaat has a mixed population of Nattilingmiut and Aivilingmiut. Although Nattilingmiut were feared by Iglulingmiut and Aivilingmiut, Nattilingmiut in turn feared Kitlinermiut.[1]

1 See Rasmussen (1931: 50), "I had not been there very long before they began to talk about their neighbours to the west, the Kitlinermiut, whom they all seemed to fear."

Hunters, Ancestors and Prey

Inuit inhabit and own the land. They used to survive by hunting the game animals of the land, the sea and the sky. The most important prey are the caribou and sea mammals such as seals, walrus and whales. Animals inhabit the land, the sea and the sky, but they do not own it. They are there to be hunted, and if they are not hunted their numbers will dwindle. The land and the sea are also inhabited by many non-human beings (*inungituut*). People have to know where they live and how to deal with them when they meet them. The land is covered by the graves of the ancestors. Their descendants and namesakes pay them respect and visit the graves to have tea and bring small gifts. The *tarniit* of the ancestors reside in a land of the deceased in the sky or on the bottom of the sea, but they may also roam out on the land.

The old Paallirmiut woman Kibkârjuk told Rasmussen (1930: 78),

> "Heaven is a great land. In that land there are many holes. These holes we call stars. In the land of heaven lives pan·a (the woman up there) or tap·aʒuma inua (the one that rules over, or owns, what is up there). There is a mighty spirit, and the aŋatkut hold that it is a woman. To her pass the souls of the dead. And sometimes, when many die, there are many people up there. When anything is spilt up there, it pours out through the stars and becomes rain or snow. The souls of the dead are re-born in the dwellings of pan·a, and brought down to earth again by the moon. When the moon is absent, and cannot be seen in the sky, it is because it is busy helping pan·a by bringing souls to earth. Some become human beings once more, others become animals, all manner of beasts. And so life goes on without end" (ihɔqaŋ·it·ɔq).

Thus the continuity of life is preserved. In this chapter we will focus on the role of three agencies in the preservation of the continuity of life: human beings, ancestors and game.

Hunting

Animals are thought to give themselves up willingly to be killed by the hunter. Simionie Akpalialuk from Pangnirtung stated,

> "We have to go out and get [food] [...]. I just can't go to the store and pick it up [...]. [Our] relationship to the land is very important because the land is alive, the animals and the sea itself; and you are interacting [with them] [...] Going to catch a whale, the feeling that you get is never what you get going down to the store. Going down to the store, you're on dead ground: all the buildings are dead. All items in there are dead, there's nothing alive [...] but here in the North, it's a lot more [...] You're not only just going there just for food, you're going there to interact with nature, to understand more of the animals. A whale gives of itself: it's an animal that you feel fully about, it's something that gives itself up to you and that's an important thing in our beliefs, that's still very strong" (Freeman 1998: 42).

If the animals were unwilling to let themselves be killed the ancestors might intercede. Michel Kanajuq from Qamanittuaq related how the deceased would interfere on behalf of their descendants:

> "Our ancestors thought that all the people who died went to the underworld instead of up above, and that they went to Nuliajuk the mother of the caribou. When the caribou did not want to leave, their faces would change into human-like faces, and the people there would whip them so that they would go to where people were living. An elder would make a small whip, and whip the land, and give it to Nuliajuk and say, 'Here is your whip. Now let us have some caribou.' That is what they used to do to get the caribou to go where they lived" (Laugrand and Oosten 2009: 79).

The killing of game is not considered as an act of violence of a human being towards an animal but as a meaningful act in which hunter and animals are connected as partners. The game provided meat for food, skins for clothing, and blubber for fuel. Bones, sinews bladders were all used to make implements. If the game did not show itself to the hunters, people would starve, and finally people might die. Whether animals would allow themselves to be killed, depended to a large extent on the behaviour of the hunter, his wife and the community as a whole. Hunters would only be successful if people behaved respectfully towards the animals and observed the correct ritual injunctions.

A successful hunter should share his catch with his hunting partners, his family, his housemates and the community where he lived. Nobody should be left hungry after a catch. This obligation to give was, and still is, all important. There is no obligation to give something back. But each hunter has to give, regardless of the fact whether he has received something in the past. Complex schemes of sharing existed in various communities. In 1956 Van de Velde published a study on the sharing of seals caught in *mauliqpuq* hunting[2] among the Nattilingmiut. The hunters addressed each other with the name of the particular part of meat they were entitled to when the meat was shared. A hunter, A, who received a shoulder from another hunter, B, if he caught a seal, would call him his *aksatkoliga*, 'my shoulder'. Hunter B would also receive a shoulder if A was successful so the address *aksatkoliga*, was reciprocal. Van de Velde provides fourteen reciprocal terms referring to the different parts of a seal: *niakroga*, 'my head', *tunnerdjuga*, 'my kidney', *sennerara*, 'my side', etc. The partnerships between the hunters were organised in terms of the parts of the body of the seal and the body of the seal thus encompassed the system of social relationships organising *mauliqpuq* hunting. The example shows clearly how the social relationships were shaped through the body of the seal that had made itself an object of sharing by allowing itself to be killed by the hunters. The notion that through hunting animals sociality is created is still very much alive.

---

2 Hunting at the breathing holes of the seals.

In 1998 four Inuit students presented a paper at the IXth Inuit conference in Copenhagen entitled *Natsiq, the seal; an integral part of our culture*. The students state in their paper:

> "Much has been said, and some have gone on to say, that because we do not hunt the seal in the traditional way, we are no longer true Inuit and do not solely rely on the seal for our subsistence. It is true that we could probably live in today's world without the seal, as long as we would have money to go to the local Northern store where we could buy many things, except for the seal and the caribou. The seal, however, provides us with more than just food and clothes. It provides us with our identity. It is through sharing and having a seal communion that we regain our strength, physically and mentally" (Aaju et al. 2002: 1).

Animals and human beings were thought to share a common origin and in many respects a common nature. But they were also fundamentally different. Animals are prey and human beings are hunters. Hunters should not become prey to the game.

Shared Origins

Animals and human beings may often share a common origin, but the relations that take shape in the myths of origin vary considerably. Some animals have always shared the land with human beings. According to the Nattilik woman Nalungiaq,

> "In the very first times there was no light on earth. Everything was in darkness, the lands could not be seen, the animals could not be seen. And still, both people and animals lived on the earth, but there was no difference between them. They lived promiscuously. A person could become an animal, and an animal could become a human being. There were wolves, bears, and foxes, but as soon as they turned into humans they were all the same. They may have had different habits, but all spoke the same tongue, lived in the same kind of house, and spoke and hunted in the same way" (Rasmussen (1931: 208).

The animals that Nalungiaq specified were all land animals and predators. In the darkness there was no difference between human beings and animals in terms of human appearance, language, and habits. Their relationship was marked by promiscuity.

Animals as well as human beings were thought to have a *tarniq* that survived them after death. The word *tarniq* is derived from the root *taaq-*, dark and is best translated as 'shade', or 'miniature image', but missionaries as well as anthropologists often translate it as 'soul'. It was usually located in a vital organ such as the kidneys or the liver. The *tarniq* was invisible to ordinary human beings but it could be seen by *angakkuit*, shamans. Knud Rasmussen (1929: 58-59) relates,

> "The soul, tarniŋa or inu·sia, is that which gives to all living things their particular appearance. In the case of human beings it is really a tiny human being, in the case of the caribou a tiny caribou, and so on with all animals; an image, but very much smaller than the creature itself.
> The inu·sia [meaning 'appearance as a human being'] is situated in a bubble of air in the groin; from it proceed appearance, thoughts, strength and life, it is that which makes the man a man, the caribou a caribou, the walrus a walrus, the dog a dog, etc."

Life depends on this miniature image inside the *pullaq* (bubble of air). When a person dies, the *tarniq* survives. According to Felix Pisuk from Kangiq&iniq, "When I die my *tarniq* will no longer be a bubble. It will become the way you see me" (Kolb and Law 2001: 1).

The *tarniq* differentiated the appearance of a being. The *tarniq* of a human being was a miniature human being, and that of a caribou a miniature caribou. The health and vitality of a being depended on its *tarniq*. It might leave the body temporarily, but if it did not return, the separation between body and *tarniq* would become fatal. Differences between the *tarniq* of men and animals could only be observed when there is 'light', preferably shamanic vision *qaumaniq* (derived from *qau*, light). When light was created the differentiation of men, raven, wolves, bears and foxes took shape. They no longer transformed into each other or lived promiscuously.

Some animals originated from human beings. A well-known myth relates how the sea mammals came into being. Thus the Iglulingmiut relate that the sea woman Uiningumasuittuq, 'the woman who does not want to marry a husband', was invited by a strange hunter to board his kayak. He turned out to be a stormy petrel in human appearance who took her to a far country. They lived together for some time. Later her father came to fetch her and persuaded his daughter to come back with him. When they travelled home in their *umiaq*, boat, the petrel pursued them, creating a storm with its wings. The father did not want to return his daughter to the bird and pushed her out of the boat. When she clung to the boat with her hands, her father cut off her fingers, joint after joint, and they turned into seals, walrus and whales as they fell in the water. Then the woman sank to the bottom of the sea. She became the sea woman, the owner of the sea as well as the sea game. The myth exists in many variants specifying different animals originating from the fingers: seals, square flippers (bearded seals) and beluga (a more recent Iglulik version recorded by Alexina Kublu), whales, seals and bearded seals (in a South Baffin version provided by Boas 1888: 584), seals, bearded seals, walrus and whales (according to another South Baffin version recorded by Peck, Laugrand, Oosten, and Trudel 2006: 363), or fjord seals, bearded seals and walrus (in a Nattilik version recorded by Rasmussen 1931: 381-382). Thus seals, bearded seals, and whales were usually considered as transformations of the fingers of the sea woman. Rasmussen (1931: 226) relates,

> "Whenever people have been indifferent towards her by not observing taboo, she hides all the animals; the seals she shuts up in her iŋ·aut: a drip-basin that she has under her lamp. As long as they are inside it, there are no animals to hunt in the sea, and mankind has to starve; the shamans then have to summon their helping spirits and conjure her to be kind again."

The *angakkuq* would have to descend to the house of the sea woman on the bottom of the sea. The lamps would be extinguished and he would descend to her dark dwelling place on the bottom of the sea to calm her down and appease her. Once he had managed to do that and the Inuit had confessed all their transgressions and offences, she would be prepared to set the seals free again.

Thus the land predators are defined in terms of a metaphorical relationship with human beings as fellow hunters whereas the sea mammals are defined in terms of a metonymical relationship as prey to the hunters.

Other animals originate from transformed human beings. Ivaluardjuk (Rasmussen 1929: 164) related that once a group of children were caught in a ravine. The parents could hear the children weeping and singing,

> "The children kept on crying, but as they did so, they were suddenly changed into guillemots, which came flying out through crevices in the rocks. And that is how guillemots were first made."

The story of the origin of the ptarmigan as told by Ivaluardjuk (Rasmussen 1929: 164-165) relates that an old woman frightened her grandchild so much by a story that it changed into snow bunting.

> "[...] and now her grandmother sat there in despair at not having been able to catch her. And she sat there alone on the sleeping place and kept on saying: 'Oh, my dear little grandchild, oh, my dear little grandchild.' And she sat there weeping, and kept on wiping her eyes. At last her eyes were all red and bloodshot. Then she took her sewing bag and fastened it round her neck, and put the needles into her kamiks, and then suddenly she fell to cackling and became a ptarmigan. Then she spread her wings and flew away. And from her come all the ptarmigan."

Stories also relate that people might change into animals after abuse or to avenge themselves on those who had mistreated them. Various stories exist relating how human beings, especially women, transformed into bears to avenge themselves.[3] In those stories the transformation appears to be a consequence of abuse or neglect and implies a transformation of a human being into an animal from which no return is possible.

3 See Boas 1901: 251-252 'The Woman who transformed herself into a Bear'; 252- 254 'The Woman who became a Bear and killed her Enemy'; 307-308 'Origin of the Black Bear''; see also Rasmussen 1929: 267-259.

Inuk and Inua

Animals can adopt a human appearance, but this does not transform them into human beings. They can do so because they have an *inua*, 'its person, inhabitant or owner'. The word *inua* is derived from the word Inuk, person, (pl. Inuit, people). Inuk and Inuit are by no means universal terms that can be applied to all people. Indians (*Allait*) and white people (*Qallunaat*) are not Inuit. They do not speak *Inuktitut,* 'like an Inuk', and they do not hunt like Inuit and most importantly, they are not inhabitants of the land. They do not belong to it. The notion of Inuit involves a notion of ownership that is ambiguous; people own the land and belong to it at the same time.

Whereas Inuit are owners, animals are owned. The *inua* of an animal is not necessarily specified. The *inua* of a dog or a kayak is the person who owns that dog or kayak. Places and animals often had an *inua. Inua* is a relational term. Thus, one may speak of *tariup inua,* 'the person of the sea', but there is no point in speaking of an *inua* or *inuat* (pl.) as a general concept. The concept can only be used when the being or object it refers to is specified. The word *inua* was also used for particular beings associated with the moon, the sky and the sea. The *inua* of the moon originally was a human being. After an incestuous relationship with his sister the two siblings were transformed into sun and moon. The *inua* of the sky was originally a giant baby who was mistreated and went up into the sky and was associated with the weather, especially the wind. The *inua* of the sea was the sea woman, usually considered as the mistress of the sea mammals. Inuit mythology also often refers to a mother of the caribou who is sometimes identified with the owner of the sea mammals.

The *inua* of an animal or a place could take the appearance of a human being, but such an *inua* was not really a human person. Stories relate how human beings visited places, where the animals live in human appearance. Although ordinary people might be fooled, *angakkuit* would perceive that such an *inua* was not a real human person. Rasmussen (1929: 271 ff.) relates several stories about people who visited bears and wolves in human appearance. The stories about the great traveller Kiviuq provide many narratives relating how Kiviuq married an animal in human appearance (e.g. a goose, a fox). These animals would usually not retain their human appearance but finally adopt their animal shape again. To what extent the *inua* of an animal, or the *inua* of the sea 'owns' animals is an issue of debate among Inuit themselves. Ollie Ittinuaq, an elder from Rankin Inlet, observed in a workshop:

> "I don't know if the animals have an owner. I have heard that animals belong to everyone. I have heard you can hunt them. I have never heard of anyone other than Nuliajuk that had any power over them. Mothers have power over their young because they [the young] cannot do things on their own. They follow their mother everywhere. My mother's grandfather was one of many people who saw Nuliajuk. [...] I cannot say who has power over animals. I have heard they were

> created by God. I have only heard about this recently" (Oosten and Laugrand 2002: 114).

Tivi Etok (Weetaluktuk and Bryant 2008: 163) from Nunavik stated,

> "There is a saying that goes like this: 'An animal roaming free on Earth does not have an owner. Therefore he who catches it becomes the owner.' That is an old saying. For instance, suppose I have a wife and I catch an animal. When I bring it to my wife's kitchen, then and only then does the animal belong to a person. Animals wandering the Earth do not belong to anybody. Only when someone catches the animal does that person get authority over it. It's up to him if he wants to host a feast. That was the law; the way it used to be."

Elders agree that animals have to be respected. If the rules of respect are not observed either the *inua* of the animal, the *inua* of the sea, the sky or the moon will retaliate. George Kappianaq (Iglulik Oral History database, IE-330) related,

> "It is said that the river and the lake each have an *inua* [it's person] who have control over the fish in that lake, and who tend to get upset very easily who will stop the fish from running. If people have verbally contended over the fish, because they are humans, the lake's *inua* can stop the fish for a period; then later on the fish can return."

Today the concepts of *inua* and owners of the game such as the sea woman and the caribou woman are no longer used. In the course of the 20th century, the Inuit of Northeast Canada adopted Christianity and integrated their traditional beliefs in the framework of the new religion. Animals to be hunted were no longer provided by non-human beings living out on the land and in the sea, but by God. The various prohibitions rules (*pittailiniit, tirigususiit*) required by these non-human beings no longer had to be observed, because God had created the animals for the benefit of human beings. Respect for animals remained a central issue in Inuit hunting. Inuit are convinced that if you do not respect animals or abuse them, animals are aware of this and will retaliate.

Respecting the Animal

According to Rasmussen (1929: 84), "The sea spirit Takanaluk demands taboo for all sea animals because they were made from her fingers." When she withheld the game because of all the transgressions, an *angakkuq* had to descend to her house at the bottom of the sea and he had to appease her and if necessary wrestle with her so that she would set the animals free again.

The animals that were hunted and killed were sentient beings and like human beings they are endowed with a *tarniq*. This was why it was important for people to respect the game. The old Aivilingmiutaq Ivaluardjuk stated:

> "The greatest peril of life lies in the fact that human food consists entirely of souls. All the creatures that we have to kill and eat, all those that we have to strike down

> and destroy to make clothes for ourselves, have souls, like we have, souls that do not perish with the body, and which must therefore be propitiated lest they should revenge themselves on us for taking away their bodies" (Rasmussen 1929: 57).

Traditionally the *tarniit* of animals were supposed to incarnate again whereas the *tarniq* of a human being would go to a land of the dead. It was expected that the shade of an animal that had been killed would reincarnate again. If it had been treated well and had been respected by the hunter and his wife, it would allow itself to be caught again by the same hunter who had shown it respect.

The mortality of the *tarniq* was a topic of discussion in Inuit society. Oosutapik, one of Peck's informants in Cumberland Sound, told Peck that many *tuurngait* (helping spirits of the shamans) were able to kill the souls of the animals, so that they could be easily killed by the hunters (Laugrand, Oosten, and Trudel 2006: 432, 436). However, another of Peck's main informants, Qoojessie, emphatically denied that this was possible, explaining that souls were immortal and could not be killed. (Laugrand, Oosten, and Trudel 2006: 383-384).

The willingness of the game to be killed could be explained in reference to its wish to change its body. Rasmussen (1929: 58) relates,

> "Animals have in reality no objection to being killed by human beings, as long as the rules of life are observed by the latter. It may even happen, and not infrequently, that an animal will approach a human being, actually desiring to be killed by that particular person. An animal may perhaps be tired of being what it is; and since its soul cannot change its envelope until the body has been killed, it is natural that animals should sometimes wish to die. The great danger in killing animals commonly hunted lies in the fact that there is hardly a single human being who has kept the rules of life and lived throughout in accordance with the laws laid down by the wisdom of his forefathers. Therefore it is said that the greatest danger lies in the fact that unclean and often guilty human beings have to depend entirely on the souls of other beings for food."

Many Inuit believe that animals offered themselves to the hunter in order to be killed. If animals were hunted they would multiply; if they were not hunted, they would feel neglected and disappear as George Kappianaq recalled,

> "Some animals can make themselves available. I suppose they want to be caught, so they make themselves available, or if they want to be turned to another species, so some animals make themselves available" (Iglulik Oral History Database, IE-454).

Peter Alogut, from Coral Harbour and Simeonie Akpik from Lake Harbour, stated,

> "The mussels, especially, multiply more when they are being harvested. Also I wish they could stop saying that the whales are decreasing in numbers. The Inuit people still believe that the animals will always show up even if you don't see them in a particular area for a long time. Inuit know about that, but the Renewable

resource people keep telling us that the animals are decreasing in numbers. They can even tell you how many animals there are, but nobody knows how many animals are really left. In our traditional knowledge, when you use the animals according to their purpose, they will always prosper and reproduce themselves as you need them" (McDonald, Arragutainaq and Novalinga 1997: 60).

The missionaries taught the Inuit that only human beings have souls and this point of view was then adopted by Inuit although there was always some debate. Thus Piugaattuq stated that game animals do not have souls, and added, "From what I have heard from Ittusarjjuat, I think that wolves might have souls, I am not certain" (Iglulik Oral History Database, IE-453). Thus the killing of game is a way of respecting the animal. It is also a dangerous operation and therefore people have to observe rules of respect pertaining to the animals.

Names

A child only became a complete person after it received an *atiq*, a name, usually from a deceased relative. The word *atiq* means 'name' as well as 'namesake' and therefore refers to a relationship, not to an essence. After a child was born, it received a name in order to become a human being. A child could be killed before, but not after it had received a name (Turquetil 1968: 43).

Some Nattilingmiut thought that people who had died wanted to go on living on earth and would choose the newborn person in whom they wanted to live again (Rasmussen 1931: 220). During the naming process, the people would try to determine which deceased person wanted to be reborn. When the child was not satisfied with its name and continued crying, a new name had to be given.

According to Boas (1901, 32), the Inuit at Cumberland Sound assumed that the soul of a deceased person entered the body of its namesake and remained there for about four months:

> "It is believed that its presence strengthens the child's soul, which is very light, and apt to escape from the body. After leaving the body of the infant, the soul of the departed stays nearby, that it may re-enter the infant in case of need."

The Iglulingmiut believed that the deceased gave constant support to their living namesakes: "We obtain aid from our namesakes when we strive with all our strength against any peril," and "Our namesakes and the namesakes of our namesakes are near us all" (Rasmussen 1929, 148). The Iglulik *angakkuit* said that on their spirit flights, they could see behind each human being a procession of spirits aiding and guiding as long as the rules of life were observed, but when the rules were not observed and a man was tempted to act in a way that the dead disapproved of, then all the invisible guardians turned against him as enemies and he was lost beyond hope (Rasmussen 1929: 59).

People inherited their strength and skills from their namesakes. The namesake relationship enabled people to share in the qualities of their ancestors and thus to become strong and successful as human beings. When people became ill, they were often given a new name to support them (Boas 1907: 239). It was advantageous to have more than one name. Among the Nattilingmiut, it was common to have five to ten names. The old woman Manilaq told Rasmussen (1931, 221),

> "I am named Manelaq because when that name was mentioned I came out of my mother's womb. Taqalekitaq was a name I once got while I was ill, and it mended me [...] I was held up solely by names. It is because of names that we breathe, and it is also because of them that we can walk on our legs. Through all these names I have grown old. I have withstood the attacks of shamans and all the dangers that would otherwise have uprooted me from the dwelling places of man."

The name plays a central part in maintaining the continuity of society, connecting the Inuit to their deceased namesakes. It was especially the name that made a human being into a social person and distinguished him or her from animals and other beings that did not have names defining their social identity. The continuity of a community was maintained through the transfer of the names of the deceased to their descendants. While the deceased dwelled in the lands of the dead, their individual skills, abilities, and appearance were passed on by their names. The name did not refer to an essence or intrinsic quality of the individual being named, but rather to the deceased namesake. It was a relational identity. The namesake relationship also implied that the social relationships of a deceased person could be transferred to his or her namesake.

The deceased who wished to be reborn might inform a pregnant woman or her relatives about their wishes. Fanny Arngnasungaaq from Qamanittuaq explained,

> "If that person has been dreamt of more than once [...] and if you know the person well [...] the person in your dreams is letting you know that she wants you to be her [parents]" (Bennett and Rowley 2004: 4).

Kublu from Iglulik (Kublu and Oosten 1999: 69-70) related her own experience how the deceased may claim a child as an *atiq*:

> "A few days after my daughter was born I dreamt that Andre [Nattiq, Attaarjuat, Utuqqaq] Aqatsiaq came to visit me and my new-born baby in the hospital. Just then, the hospital registrar came along to register my daughter's birth. I filled out everything but kept the space for the father blank. Andre told the registrar to write his name in the blank. I told him that I did not want his name on the paper, that I did not want him with me, that I did not want him to be named in my daughter. He then said: "If you just use my church name Andre and my government name Aqatsiaq, then your child doesn't have to use any of my names." I again said to him that I did not want him with my daughter at all, that she was not going to be named after him. At that point I was woken up by my daughter crying. A few days after, I got out of the hospital. Andre's father Aqaattiaq held my daughter in his arms and said: "Angutiq atigijumajarauna" [I want her to be my *angutiqati*,

paternal cousin]. His son had been named Nattiq after his paternal uncle Qutittalik/Arraq's son."

Even though Alexina did not give Andre's name to her daughter, forms of address expressed the connection between her daughter and Andre.

> "My daughter does not carry Andre's name Utuqqaq, but she did call his mother *nuliarainnuk* (wonderful wife) because he had been named after her first husband Attaarjuat also" (Kublu and Oosten 1999: 70).

The meaning of the dream could not be ignored when Andre's father addressed the child as his *angutigati*. It is better for the child if a parent complies with the wishes of the deceased and the elders.

The introduction of Christian names did not really affect the old system, as Inuit continued to identify with their ancestors through names and integrated the old traditions into the new naming system. According to the Oblate missionary Turquetil, Inuit were familiar with the idea that the spirits of the deceased in heaven wished to help their descendants, and when everyone had received a Christian name, they felt happy: "I asked someone whether he liked his name. Oh yes, he said, Joseph was good and his name will live in me and he will help me" (quoted in Scheffer 1950, 97).There was some opposition to the old system, however, especially among Anglican missionaries. Samuel Arnakallak from Mittimatalik (1993 interview by F. Laugrand) recalled,

> "I remember that one of the first missionaries here [Mittimatalik] didn't want that we have our Inuit names anymore because those names were from the dead (I mean to be named after a dead person). But when he left, we got our names back again."

And Trott (1988, 71) observes with respect to the community of Ikpiarjuk,

> "There has been intense pressure on the part of missionaries to eliminate the naming practices as it requires "pagan" beliefs to continue. The naming system as a system of relating to other people has remained in place."

Each name carries a history as well as a network of relationships. Through the names passed down, society reproduces itself and connects to its past. A complex system of terms of address ensures that the relationships embedded in the names are acknowledged. Kublu related how she called her grandmother, Iqallijuq *panik* (daughter) because she was named after Kublu, Iqallijuq's father:

> "My mother's mother Iqallijuq, calls me *ataatakuluk* [fatherkuluk]. She is my *panik*. Her two step-sisters Uriuqtuq and Nattiq, Kublu's actual daughters, both called me *ataata*. Kublu's oldest son Maktaaq from his second marriage to Ataguttiaq calls me *ataata*, and I call him *irniq*. His wife Kupaaq is my *ukuaq* [daughter-in-law]. She does not have the same name as my father. I have been able to find out that Kublu's mother's name had been Quliit, because I called one of "my" [i.e. my grandfather Kublu's] children *anaanakuluk* [motherkuluk]. I don't remember my mother's father Ukumaaluk, but his *atiqs* call me

ningaukuttuk [brother(cousin)-in-law], because of Kublu's marriage to Ataguttiaq" (Kublu and Oosten 1999: 68).

Today young Inuit have to rely on elders to explain to them the meaning of names and the terms of address they imply. A reconstruction of the past reveals a person's relationship to those whose names she or he was given. Still, the importance of ancestors has by no means been properly recognised. Inuit emphasise that they owe their survival to their ancestors. The ancestors not only live on in the lands of the dead but also out on the land where they are buried and in the children who are named after them.

Whereas the continuity of animals was ensured by a reincarnation of nameless *tarniit*, the continuity of human beings was organised by a succession of *atiit*. To what extent there was a connection between the *tarniq* and the *atiq* is unclear. Sometimes it is suggested, but it is never strongly marked. Boas (1901, 234) tells the story of a hunter who recognised the soul of his deceased mother in a fox and therefore could not kill it. In the famous story of Arnaqtaaqtuq told by Celestine Iqijjuk from Kivalliq, a rejected fetus reincarnated time and again in different beings, such as dogs, seals, and caribou, until it was finally reborn as a human being.[4] The Arnaqtaaqtuq story starts with a miscarriage suggesting what may happen when rules are transgressed, but no suggestion is made that this is a normal pattern. Human beings are connected to the ancestors through their names, whereas animals lack such a connection and live in a continuous circle of reincarnation. History is part of human society not of the world of anonymous animals whose *tarniit* continually reincarnate.

The Dynamics of Inuit Cosmology

Inuit perceive hunting as essential to the continuity of society. By hunting game people connect in the right way to the land and its animals. Hunting allows them to create society by sharing the meat yielded by the hunt. Animals and human beings may share many features. Thus animals can appear as human beings and like human beings they have a *tarniq*. The *tarniq* as well as the *inua* were visually marked: the *tarniq* was a miniature image of a being and could only be seen by shamans; the *inua* was the human appearance of a being and only shamans could perceive its true nature.

Animals may have their own rules or their own shamans and so they are not devoid of culture, but they do not have names that connect them to their ancestors. Dogs may have names, but these names do not connect them to their ancestors but to their owners, who select the names. In the case of human beings, the names are ultimately selected by the ancestors who wish to be renamed. Once their wish is fulfilled they will support their living namesakes.

4 See Blaisel and Arnakak (1993) for a detailed analysis of this myth.

Whereas animals die and are reborn as nameless beings, human beings are always connected to their ancestors.

When the *tarniq* is not connected to a name, the distinction between animals and human beings is at stake. In an account by the *angakkuq* Ava of the descent of the shaman to the sea woman a connection between seals and the deceased comes to the fore. Rasmussen (1929: 125) relates,

> "Meanwhile, the members of the household pass the time by singing spirit songs in chorus, and here it may happen that the clothes which the shaman has discarded come alive and fly about round the house, above the heads of the singers, who are sitting with closed eyes. And one may hear deep sighs and the breathing of persons long since dead; these are the souls of the shaman's namesakes, who have come to help. But as soon as one calls them by name, the sighs cease, and all is silent in the house until another dead person begins to sigh. In the darkened house one hears only sighing and groaning from the dead who lived many generations earlier. This sighing and puffing sounds as if the spirits were down under water, in the sea, as marine animals, and in between all the noises one hears the blowing and splashing of creatures coming up to breathe."

The anonymous dead appear as sea mammals in the séance and are released by the participants in the séance naming them. Thus the boundary between human beings and animals is preserved by the knowledge and use of the names of the deceased. Ava's account suggests that the *tarniit* of the deceased may transform into sea mammals. By naming them the relationship between the *tarniq* and the *atiq* appears to be restored.

The hunting ideology of Inuit society implies that animals become prey to the hunters and are reborn as animals again returning to the same hunter time and again. Thus the animal is caught in a hunting cycle in which it participates in Inuit culture as its meat is shared, shaping the social relationships in Inuit society and its skin is transformed in the beautiful and warm furs that allow the Inuit to survive in the cold.

The fate of human beings is not the same. They are not supposed to be killed by animals. By respecting the ancestors they represent through their names and respecting the animals they hunt they become true human beings and owners of the land and its animals. After death their *tarniit* should travel to the land of the dead and their names should be passed on to their descendants. The community should observe the rules of respect relating to the deceased. Otherwise a deceased person may not reach the lands of the dead but instead turn into an evil spirit roaming the land and intent on avenging itself on the culprits responsible for its fate. Iglulik traditions relate that people may also join the *ijirait,* a non-human people closely connected to the caribou.The power of the ancestors is acknowledged in the naming traditions of the Inuit. But the ancestors may not only be represented in the names, but also in the anonymous *tarniit*, transforming into game animals. Thus the ancestors may encompass the two poles in Inuit ideology, hunters and prey. Yet their role as ancestors supporting their namesakes is primarily valued whereas the idea that they may transform into game animal is only obliquely referred to and mainly a topic of tales and shamanic rituals. Thus the association of the ancestors with the *tarniit* is subordinated to their association with the names that are passed on to their descendants. The ancestors are everywhere. They may reside in the lands of the dead, their *tarniit* may roam the land when they did not reach that destination and they may even live on in the animals. But what matters most is that they live on in their namesakes and descendants who maintain the flow of life by killing game animals so people will survive and society will be preserved.

References

AAJU, P. et al. (2002) The Seal: An Integral Part of Our Culture. *Études Inuit Studies* 26(1): 167-174.

BENNETT, J. and ROWLEY, S. (eds.) (2004) *Uqalurait: An Oral History of Nunavut.* Montreal and Kingston: McGill University Press.

BLAISEL, X. and ARNAKAK, J. (1993) Trajet rituel: Du harponnage à la naissance dans le mythe d'Arnaqtaaqtuq. *Études Inuit Studies* 17: 15-46.

BOAS, F. (1888) *The Central Eskimo.* Sixth Annual Report of the Bureau of American Ethnology. Washington. (Reprint edition in 1964, Lincoln: University of Nebraska Press).

BOAS, F. (1901) The Eskimo of Baffin Land and Hudson Bay. From Notes Collected by Capt. George Comer, Capt. James S. Mutch, and Rev. E.J. Peck. *Bulletin of the American Museum of Natural History* 15(1): 1-370.

BOAS, F. (1907) Second Report on the Eskimo of Baffin Land and Hudson Bay. From Notes Collected by Capt. George Comer, Capt. James S. Mutch, and Rev. E.J. Peck. *Bulletin of the American Museum of Natural History* 15(2): 371-570.

FREEMAN, M.R. et al. (1998) *Inuit, Whaling and Sustainability.* Walnut Creek: Altamira Press.

IGLULIK ORAL HISTORY DATABASE. IE-interviews. Consulted in Iglulik, Nunavut.

KOLB, S. and LAW, S. (eds.) (2001) *Dreams and Dream Interpretation.* Inuit Perspectives on the 20th Century Series (Vol. 4). Iqaluit: Nunavut Arctic College.

KUBLU, A. and OOSTEN, J. (1999) Changing Perspectives of Name and Identity among the Inuit of Northeast Canada. In OOSTEN, J. and REMIE, C. (eds.), *Arctic Identities: Continuity and Change in Inuit and Saami Societies*, pp. 57-85. Leiden: Centre of Non-Western Studies.

LAUGRAND, F. and OOSTEN, J. (eds.) (2009) *The Ethnographic Recordings of Inuit Oral Traditions by Father Guy Mary-Rousselière.* Iqaluit: Nunavut Arctic College.

LAUGRAND, F., OOSTEN, J. and TRUDEL, F. (2006) *Apostle to the Inuit: The Journals and Ethnographic Notes of Edmund James Peck. The Baffin Years: 1894-1905.* Toronto: University of Toronto Press.

MATHIASSEN, T. (1928) *Material Culture of the Iglulik Eskimos.* Report of the Fifth Thule Expedition, Vol. 6. Copenhagen: Gyldendalske Boghandel.

MAUSS, M., in collaboration with BEUCHAT, H. (1979) *Seasonal Variations of the Eskimo.* London: Routledge and Kegan Paul.

MCDONALD, M., ARRAGUTAINAQ, L. and NOVALINGA, Z. (eds.) (1997) *Voices from the Bay. Traditional Ecological Knowledge of Inuit and Cree in the Hudson Bay Bioregion.* Ottawa and Ontario: Canadian Arctic Resources Committee and Environmental Committee of the Municipality of Sanikiluaq.

OOSTEN, J. and LAUGRAND, F. (eds.) (1999a) *Introduction. Interviewing Inuit Elders series* (Vol. 1). Iqaluit: Nunavut Arctic College.

OOSTEN, J. and LAUGRAND, F. (2002) *Inuit Qaujimajatuqangit: Shamanism and Reintegrating Wrongdoers.* Inuit Perspectives on the 20th Century Series (Vol. 2). Iqaluit: Nunavut Arctic College.

RASMUSSEN, K. (1929) *Intellectual Culture of the Iglulik Eskimos. Report of the Fifth Thule Expedition 1921-24, Vol. 7(1).* Copenhagen: Gyldendalske Boghandel.

RASMUSSEN, K. (1930) *Observations on the Intellectual Culture of the Caribou Eskimos: Iglulik and Caribou Eskimo Texts. Report of the Fifth Thule Expedition 1921–24, Vol. 12(2-3).* Copenhagen: Gyldendalske Boghandel.

RASMUSSEN, K. (1931) *The Netsilik Eskimos: Social Life and Spiritual Culture. Report of the Fifth Thule Expedition 1921-24, Vol. 8(1-2).* Copenhagen: Gyldendalske Boghandel.

SCHEFFER, L. (1950) La valeur religieuse des religions païennes. In *L'Union missionnaire du Clergé*, pp. 91-98. Québec.

TROTT, C.G. (1988) *Structure and Pragmatics: Social Relations among the Tununirrusirmiut.* PhD thesis, University of Toronto, Toronto.

TURQUETIL, A. (1968) Religious Rituals and Beliefs. In Valentine, V.F. and VALLEE, F.G. (eds.), *Eskimo of the Canadian Arctic*, pp. 43-48. Toronto: McClelland and Stewart.

VAN DE VELDE, F. (1956) Les règles du partage des phoques pris par la chasse aux aglus. *Anthropologica* 3: 5-15.

WEETALUKTUK, J. and BRYANT, R. (2008) *Le monde de Tivi Etok. La vie et l'art d'un ainé Inuit.* Quebec City: Multimondes, Institut Culturel Avataq.

Challenging the Stranger: The Dynamics of Constructing Social Identity in Hmong Origin Myths

CHRISTIAN POSTERT

'Strangers' are an undeniable fact of Hmong life in historical times and present-day Laos. The original area of Hmong settlement was in the southern provinces of China until the midth of the last century. At this time, the suppression of a wide-spread Hmong rebellion by Imperial Chinese troops forced thousands of them to flee southwards and to take refuge in the northern mountainous areas of later Laos (Lemoine 1972a: 15). Coming from a multi-ethnic Chinese setting and immigrating into another, but quite differently composed multi-ethnic setting in Indochina, the Hmong encountered and still encounter a variety of higly diverse ethnic groups. This gives rise to a variety of questions concerning the relations of 'own' and 'strangers': What Hmong categories exist to cope with this tremendous diachronic and synchronic variety of 'strangers'? How are relations to different categories of 'strangers' valorised, legitimised and challenged in myth? In what way are conceptions of mythical Hmong 'selves' and 'strangers' interrelated? How are myths modified in view of the present situation, leading to a reinterpretation of the past and serving as models for future actions?

Hmong 'Strangers'

All relations of Hmong to non-Hmong ‚strangers' are conceptualised as relations to *mab-suav*. This term might be translated as "ceux qui ne sont pas Hmong" (Bertrais 1979: 397) or "people other than the Hmong" (Tapp 1989a: 130). In this connotation, it seems to be most closely related to our conception of the ‚stranger' and is flexible enough to include both the notorious Chinese ‚stranger' that is deeply involved in framing Hmong history and fate for at least the last 2000 years and the ‚strangers' encountered only recently since the Hmong migration into Indochina 150 years ago, that is the Lao, Mon-Khmer or Europeans.

Mab-suav is a composite term that can be further dissected. *Mab* means „all ethnic groups who are non-Miao [non-Hmong] and who are not in a position of authority" (Lyman 1974: 178).[1] In the context of China, *mab* referred

1 The connotation of not being in authority appears to be confirmed by Johnson's explanation of the term *tub mab tub qhev* as 'to designate a king's slaves or a rich man's retinue of servants' or 'indentured servants' (Johnson 1985: XX).

to the Yi, a Tibeto-Burman group including the Lisu, Lahu, Akha and Naxi (Tapp 2001: 12). In Laos, "strangers" fitting into this category are the Khmu *máng khuamû*[2] and the Phi Tong Lïang *máng qù* resp. in general the *mab daum*, an encompassing category for all Mon-Khmer groups (Lyman 1974).

In general the term *suav* refers to the Chinese, but might be used with the wider meaning of „étranger; tous ceux qui ne sont pas Hmong" (Lyman 1974: 308; Bertrais 1979: 397; Tapp 1989a: 130; Johnson 1985: XXI). Like in the expression *sua fângtshî* „the Chinese living in China who are in a position of power", this term seems to attribute a position of higher authority (Lyman 1974: 308). Thus, Lolo (Norsu) Landlords have been called *suav*, Europeans might be called ‚white *suav*' (*suav dawb*), probably referring to their dominant position in colonial French Indochina (Tapp 1986: 94; Tapp 2001: 12). Even the Lao dominating the nation state in today's Laos, who are usually referred to as *nplog*, might be called *suav maib*[3] (Bertrais 1979: 397).

During my fieldwork[4], the term *suav* was almost exclusively reserved for denoting ethnic Chinese. However, a first and cautious comparative analysis of data from different authors suggests that Hmong rather tend to use the concepts of *mab* and *suav* as open and flexible social categories instead of as fixed and descriptive taxonomies. Thus, the categories of *suav* and *mab* might subsume diverse 'strangers' who themselves would not consider themselves to belong to one category of people, for instance as in the case of the Chinese, Europeans and Lao.

Apparently, there is a tendency to hierachise relations to ‚strangers' according to authority: in the first case, the ‚stranger' related to the Hmong as *mab* appears to be granted a lower status and hierachical subordination in terms of authority. In the second case, the ‚stranger' related to Hmong as *suav*

2 Different authors in different times use different transcription systems for different Hmong dialects and categories of strangers. As these heterogenous terms for a variety of categories of Hmong strangers are not easily interchangeable and/or determinable, I stick to the original terms and transcriptions in the framework of this essay.

3 In the case of the Lao, the data is more ambivalent as the term *suav maib* seems to be only part of children's language (Bertrais 1979: 397). Lyman who has done research in Thailand cites the statement of an informant as follows: 'The Miao left China and meeting new peoples, called them *máng*. This term included the Lao, Yon (Lanna), Khamuk, T'in, etc., but the peoples the Miao had already known in China (e.g. the Yao) were not called *máng*.' (Lyman 1974: 178). However, this statement seems to be in contradiction to the data gathered by other scholars and to myths in which Lao *nplog* and Chinese *suav* seem to be characterised by rather similar attributes. Thus, Johnson decided in the translation of a myth to translate the term *suav* both as 'Chinese' and 'Lao' (Johnson 1985: 289). For further discussion, I refer to the analysis of various myths in the framework of this essay.

4 My own data presented in this essay have been collected in 14 months of fieldwork from 2000 to 2001 in a Hmong highland village in Luang Prabang Province, Laos. This data collection and subsequent analysis were supported by a grant from the German Research Council (DFG) in the framework of the interdisciplinary Research Group "Cultural Diversity and the Construction of Polity in Southeast Asia: Continuity, Discontinuity, and Transformation" (FOR 362) at Münster University, Germany.

is attributed a higher status in terms of power. Equality in relations to the ‚stranger' does not seem to be a possibility: ranking is the necessary attribute of Hmong relating to 'strangers'.

Mythical Processes of Hierachisation between ‚Self' and ‚Stranger'

One focus of the broad research activities of Josephus Platenkamp are proto-typical social and cosmological relationships that need to be established for a person in order to acquire completeness (Platenkamp 1988: 23-31). In this context, he presented an influential comparative analysis of three myths about the incomplete person, recorded in insular Southeast Asia (Platenkamp 2004, 2005). My paper is inspired by this approach in analyzing how Hmong myths of incomplete persons articulate the embeddedness of social in cosmological relations and basic principles of social differentiation. Particularly, it is in myth that the ranking between *mab*, *suav* and Hmong is explained, legitimised and challenged. The first myth presented in part 2.1 has been recorded by Mottin in Thailand. It describes the coming into being of various ethnic groups and a hierarchisation of these with the Hmong in a position of authority. Another myth presented in part 2.2 has been recorded by Tapp in Thailand. It describes the reasons why eventually the Hmong have lost their superiority and are now subordinated to the *suav* in terms of power.

Hmong Origin Myth of ‚Self' and ‚Strangers'

A very frequent and popular motif occurs in diverse versions of an origin myth: all different ethnic groups (in some versions all Hmong clans *xeem*) originate from the union of a brother with his sister, the only survivors of a great flood. Versions of this myth have been recorded in Laos (Lemoine 1972a: 184-185; Lemoine 1972b: II 100-101, II 106-109, II 112-115), in Thailand (Mottin 1980: 33-35, 39-41, 57-59) and among Hmong refugees from Laos (Johnson 1985: 115-117). Another version translated into French by Mottin (1980: 57-59) is summarised in the following:

> "A brother and his sister are the only survivors of a great flood. When the brother proposes that they should live as husband and wife now, the sister reminds him of the incest taboo. Therefore, they separate in order to find potential marriage partners from another clan elsewhere. After a long and unsuccessful journey, they happen to meet each other in a plain. In the absence of alternatives, they decide to marry this time. After a year, the woman gives birth to a gourd that is growing continuously for three years. While the woman is working in the fields, the man hears furious cries and sounds of a battle from within the gourd. To see what happens inside, he takes a red hot pike and pierces the gourd. A Khmu leaves the gourd, all blackened by the soot at the edge of the hole. As there are still voices from within the gourd proclaiming "I want to leave first, I want to leave first!",

the man takes a hammer and crushes another part of the gourd, thereby releasing a Hmong. As the noise from within still does not cease, he finally takes an axe and smashes the gourd in a thousand pieces thereby freeing a Lao who is the final one. This is why the Khmu are the eldest and still have a black skin from the soot of the hole, the Hmong the middle ones and the Lao the youngest" (see Mottin 1980: 57-59 for the longer French version of this myth).

What happens in this myth? For analytical reasons, two parts can be distinguished:

Procreation

A sister and a brother are the only ones to survive a big flood. As all potential marriage partners have perished, they are compelled to marry in violation of a very strong incest taboo. After a year, the sister gives birth not to a child, but to a big gourd that continues to grow after birth (see also Lemoine 1972b: II 108, 114). In other myths, the product of incest is described as a white gourd without arms and legs, but with a mouth to eat (Lemoine 1972a: 184), as a child or gourd without head, face, arms and legs (Mottin 1980: 35, 41), or as an egg-shaped smooth stone without arms or legs (Johnson 1985: 117).

Apparently, this first part of the myth is playing with the logical extremes of the incest taboo. On the one hand, a radical obeyance of the rules of exogamy would lead to the end of humankind; marriage thus seems to be legitimate in this situation. And in fact, brother and sister are able to procreate. On the other hand, their offspring is not a child, only a gourd with a mouth to eat that manifests itself especially in its ability to grow. Thus, the myth points to the consequences of a transgression of the rules of exogamy. The offspring is incomplete and formless. Without a face or human features, it lacks social differentiation and social identity.

In the first part of the myth, one basic aspect of every person is introduced: to have a growing body with regenerative capacities like eating and digesting. An excursion into Hmong cosmology might serve to suggest a semantic field evolving around fertility, growing, eating and digesting which are all closely associated with the flowing of an essential cosmological constituent called *pa* that can best be translated as 'breath' or 'vapour'. The origin of this 'breath/vapour' is usually attributed to the 'breath/vapour of the earth' *pa av* that can be seen as emanating from the moist earth when the sun is shining. In the cultivation cycle, this 'breath/vapour of the earth' is absorbed by the rice plant and accumulated especially in the grains. When serving a meal, the 'breath/vapour of the rice' (*pa mov*) is visibly evaporating and is regarded to be the quintessential nourishing quality of the rice. Thus, cooked rice that has cooled down is believed to be derived of its nourishing faculty so as to leave one hungry even after abundant meals, in the long run probably leading to starvation despite big amounts of cold rice being eaten. Sacrifices to ancestors follow the same logic: hot and evapourating rice is put on the table as an offering to the ancestors who are supposed to come along to incorporate the gift by sniffing the nourishing *pa* of the rice (*hnia pa*). Once this is accomplished and the rice cooled down, the remaining cold rice could be eaten by other members of the houshold, but is rather considered to be useless and given to dogs.

If man eats hot rice, the 'breath/vapour of the rice' (*pa mov*) can be transformed into the 'breath/vapour of the blood' (*pa ntshav*) that similarly can be seen. When somebody cuts himself or herself on a cold day it is evapourating from the blood that drops on the ground. The presence of an abundant amount of 'breath/vapour of the blood' in the body becomes manifest in its ability to procreate, move and to grow. The opposite, sickness and death, is attributed to a severe loss of 'breath/vapour of the blood'. The process of growing older corresponds to a steady 'drying up' *qhuav* of a man's 'breath/vapour of the blood' until death.

In the origin myth, having *pa* is attributed to having a relation to a regenerative couple of parents. Indeed, in the Hmong view of conception, sperm *kua phev* is valued primarily for its fertile *pa*: if sperm has cooled down only once, it looses its procreative capacity as its *pa* got lost: 'sperm in man's body is hot, has *pa*. If it falls on the ground it is cold in a minute, does not have *pa* anymore, cannot make a child'. The same applies to menstrual blood that is valorised for its 'breath/vapour of the blood' (*pa ntshav*): if both menstrual blood and sperm get fused in the uterus, the 'house of the child' (*lub tsev me nyuam*), they form a growing 'fistful of foam' (*ib thooj npuas*). At this stage of development, the fetus still lacks a face, arms and legs. The 'fistful of foam' is just a growing amorphous mass fed by the mother's 'breath/vapour of the blood'. It cannot be regarded as a human being as it still lacks social identity: 'when sperm and blood are living together for a month it is not yet complete, the child is only a fistful of foam'. Thus, abortion at this stage of pregnancy was conceived to be unproblematic by my informants.

It is only in the first three months of pregnancy that the 'fistful of foam' gains human features, thus introducing another set of relations deemed to be indispensable to exist as a human being; these will be discussed presently.

In the context of *pa*, no social differences can be attributed to diverse types of 'strangers'. Thus, as my informants stressed, all different ethnic groups are regarded to have the same kind of *pa* indistinguishable from another. This is conceived to be the common origin of mankind, exemplified in this myth in the existence of one big growing and digesting gourd giving birth to all different ethnic groups on earth. As the myth states in the first part, ethnic groups do not differ in their aspect of regenerative life: no sociality is introduced on this level; different social persons or groups cannot be distinguished in terms of regenerative life. At the same time, the need to complement this aspect of the person by other aspects to acquire a full social identity is stressed. The gourd is not yet complete, lacks sociality.

Social Differentiation

The solution to this dilemma is presented in the second part of the myth, i.e., by means of the introduction of social differentiation. The socialising aspect of the person is attributed to relations to the father/ancestry. It leads to a progressive social differentiation according to relative age and visible differences (colour of skin) and involves violent competition between brothers. Nevertheless, whereas diverse versions of this origin myth do not differ very much in their first part that is concerned with procreative fertility, two different strains of versions can be discerned in the second part of the myths, even though they are playing with very similar motives. The origin myth given above is representative of the first strain of versions and will be analysed exemplarily in the following (see comparable versions in Lemoine 1972b: II 106-109, II 112-115).

Whereas the sister leaves the scene and goes to work on the field, the brother stays at home. In the second part of the myth, he is no longer called ‚brother' *nus* (identitfying him in relation to his sister *muam*), but ‚man' *niag txiv* focussing on his being a male ancestor to the offspring.[5] What seems to be introduced in this version is a focus on patrilineal ancestry, thus introducing a set of relations that are deemed indispensable for a social person to exist.

The 'man' hears the sounds of a battle from within the gourd. As he listens to the noise he is able to distinguish diverse voices disputing who will be the first to come out of the gourd. His curiosity raised, he takes a red-hot pike and pierces a hole into the fruit. The first one who subsequently leaves the gourd through the opening is a Khmu *mab khuj hmub* (Mottin 1980: 58); in another version, it is a *mab daum* representing a broader category of people that encompasses diverse Mon-Khmer groups conceived to represent the autochthonous population of Laos (Lemoine 1972b: 108, 115). In all these versions, their skin gets blackened by the soot at the edge of the hole. The second one to leave the gourd is usually a Hmong; the third one is either a *suav* (Chinese) or a *nplog* (Lao) (Lemoine 1972b: 108). Another version reverses this order, the Chinese coming out of before the Hmong, the last one leaving the gourd being a French (Lemoine 1972b: 115). Finally, a recent version recorded by myself in 2001 sets the order of birth as follows: first *mab daum*, then Hmong, then Lao, Thai, Chinese, American, French, Japanese, ‚and a lot of other countries', pointing to the fact that this myth is assumed to present a universal order encompassing all ‚strangers' met by Hmong in diverse historical circumstances.[6]

5 In the end of the myth, he is even called 'patrilineal grandfather' *yawg*, nevertheless translated as 'man' by Mottin (Mottin 1980: 58-59).

6 In a second strain of versions of this myth, the mythical gourd as the product of incest does not give birth to a set of brothers, but is slit up into slices and distributed onto various spots like plants or geological formations of the countryside (Lemoine 1972a: 184-185; Mottin

The succession of the brothers leaving the gourd corresponds to a ranking according to relative age: generally, the Khmu who has left the gourd first is considered to be the eldest brother, the Hmong the intermediate and the Chinese, Lao and American the younger brothers. Thus, the relative distinction in terms of age of same sex siblings introduces an asymmetry in the relations of Hmong and diverse ‚strangers': Khmu > Hmong > Lao, Thai, Chinese, American, French, Japanese.

The woman gives birth to a growing gourd that lacks social attributes. Social differentiation comes into being when regenerative life becomes complemented by a second constituent. It is only after a second act of violent birth-giving by the man piercing or crushing the gourd with a pike, a hammer and an axe, that social differences are introduced in the myth, in some versions in the form of ethnic differences, in some versions in the form of different Hmong clans. According to these myths, social differentiation into different ethnic groups and into different Hmong clans apparently seem to share basic principles of mythical sociogenesis.

Another interesting version of the myth recorded by Tapp in Thailand is worth citing. It deals primarily with the coming into being of different Hmong clans. Again, the underlying problem, that is social differentiation out of a unifying principle of regenerative fertility, appears to be similar. Thus, its analysis might serve here to examine in detail by what processes social identity is constructed:

> "There was one woman pregnant with child, and it was this which gave rise to all the Hmong clans; before that there were none. That woman gave birth to a baby son within the placenta, and the placenta, cut with a knife, produced twelve children, and they exclaimed,'That's lucky! (*hmoov*)'. It was only since then that we were called Hmong. And it was those twelve children which formed the twelve Hmong clans, because all the children born after that time took their father's names, forming the clans. The truth is that although all the Hmong were born from that one mother, people don't think about it, and so we are unable to love each other. When the Hmong remember this, we will be able to love each other" (Tapp 1989a: 75).

In this version of the myth, the gourd is substituted by a baby son in a placenta. The common origin of all Hmong in the context of regenerative fertility is

---

1980: 33-35; 39-41; Johnson 1985: 115-117). In contrast to the first set of versions explaining the origin of different ethnic groups, the second strain of versions seems to be preoccupied with social differentiation into different Hmong clans. Thus, after a period of rest, the pieces of the gourd turn into the ancestors of different Hmong clans who subsequently adopt clan names that form rhyming couplets with the names of the places they were emerging from (Lemoine 1972a: 184-185; Johnson 1985: 115-117). However, in two versions, the additional emergence of *suav* and *mab* out of the cut pieces of the gourd is reported without ranking them hierarchically in terms of respective age as exemplified in the first strain of versions (Mottin 1980: 33-35; 39-41).

associated with love, peace and non-competition among these as the concluding statements clearly indicate.

The way sociality is introduced in this myth is highly revealing: the cutting of the placenta is accompanied by the emergence of 12 children who are exclaiming '*hmoov*!' (Tapp 1989a: 75). Sociality seems to be closely associated with a very important concept in Hmong cosmology, that of *hmoov*. This concept is difficult to translate as it covers a broad semantic range. Usually, it is translated as 'fortune, luck, fate' (Heimbach 1979: 67), 'la chance, le mérite, le destin' (Bertrais 1979: 77) or as 'to be abundant, plentiful (of wealth or happiness); luck, fortune (esp. good fortune)' (Lyman 1974: 183). Furthermore, it is connotated with ''aggressivité' en terme de la morale de productivité et de mercantilisme de notre société moderne' (Lemoine 1972a: 101); to 'make *hmoov*' *ua hmoov* thus implies 'to be fortunate, be prosperous; have a superabundance of worldly goods' (Lyman 1974: 183). It is especially associated with patrilineal relations as in 'le père laisse ses mérites [*tseg koob hmoov*] au fils'.[7] Thus, the concept of *hmoov* covers a wide semantic field reaching from 'competitive spirit' over 'luck' to 'wealth'. *Hmoov* proves to be highly differentiating in social terms. As my Hmong informants stressed: all social entities differ from another in terms of their *hmoov*. Ethnic groups do not have the same *hmoov*, different clans *xeem* within an ethnic group do not have it either; even households in a lineage are distinguished in terms of their respective *hmoov*. There are many different ways of manipulating one's socio-cosmological relationships in order to increase one's *hmoov* that cannot be described in detail; one of these is to carefully preserve one's patrilineal ancestors' heirlooms, especially silver necklaces. However, the competetive quality of *hmoov* is evidenced by the fact that informants are rather reluctant to talk about any of these strategies as, as one of my informants put it, 'nobody wants to teach the ways of *hmoov* to other people'.

A brief look at Hmong conceptions of the constitution of the person further articulates the notions of a common, socially non-differentiated origin in terms of regenerative life and a subsequent social differentiation in terms of *hmoov*. As some of my data on the Hmong views of conception and pregnancy indicate, the complementing part of the 'fistful of foam' in the mother's womb seems to be the 'shadow/image *plig*'[8] (*plig duab*). Thus, whereas the coming into being of visual social identity seemed to be closely associated with the 'shadow/image *plig*', the process of growth was attributed only to the *pa* of maternal blood (Postert 2003).

The coming of a 'shadow/image *plig*' is usually attributed to the first or second month of pregnancy; its ritual equivalent is the 'calling of the *plig*' *hu*

7 *Tus txiv tseg koob hmoov rau tus tub* (Bertrais 1979: 77).

8 In the literature, *plig* is most often translated as 'soul' or 'spirit'. As I do not want to introduce connotations of a European-Christian notion of 'soul', I will stick to the term *plig* in the following.

plig three days after birth. This ritual cannot be described in detail here; for the present purposes, it suffices to note that it is in this ritual that the child is acknowledged to have *hmoov* (Postert 2003). In the same ritual, a name is given to the child that is especially important for boys. Usually, patrilineal ascendants choose the name of a deceased patrilineal ancestor who is supposed to be especially "famous" or „particularly lucky or respected". These qualities are conceived to be transmitted from ancestors to children through the name[9] *lub npe* (Bernatzik 1947: 53, 54; Symonds 1991: 170; Chindarsi 1976: 106). However, if there are respected strangers around, it is another option to delegate the choice of an auspicious name to them (Savina 1930: 225; Chindarsi 1976: 66). Tapp specifies that in this case, the stranger should definitely be „of some standing" (Tapp 1989b: 68). Even the name of the stranger himself might be chosen to make the child "as big and as mighty" as the stranger himself (Bernatzik 1947: 53). Lemoine explicitly summarises that name giving should only involve strangers or patrilineal ancestors who are supposed to have a lot of *hmoov* (Lemoine 1972b: II 103).

To summarise: First of all, the mythical emergence of a living, growing and digesting gourd/stone/placenta seems to be associated with relations of progeny to a couple of father and mother and, through these, to a cosmological constituent *pa* representing the all encompassing origin of mankind. Thus, in this context, relations between different social groups should be marked by mutual peace and love. Secondly, the mythical emergence of social differentiation out of an amorphous gourd/stone/placenta seems to be closely associated with patrilineal ancestry, the violent piercing or cutting of an amorphous living mass into social and visible differences in a context focussing on *hmoov*, that is competitive spirit, luck and wealth.

The relation of the Hmong as elder brother to the Lao/Chinese as younger brother is emphasised in myths further developping relations between 'own' and 'stranger'. Thus, the myths usually stress the authority of the elder brother over the younger in terms of rulership. Accordingly, in mythical past, the Hmong were the ones to dispose of the office of the mythical ruler *huab tais* far in the north in a mythical China; the Chinese/Lao were the ones to obey. Attributes of this rule were the possession of land as well as books written in the mythical Hmong script preserving wisdom from ancient mythical times; both are associated with the perpetuity of a mythical social order (Tapp 1989a: 75, 122, 126, 131, 159).

Re-Ranking in Terms of Mythical Brother Rivalry

The theme of competitive relations between brothers in the context of their *hmoov* is of vital importance in a myth describing further mythical processes

9 Mottin notes that *lub npe* might as well be translated as „l'image de la personne" (Mottin 1982: 15).

of re-ranking following the original act of creation of different social categories. The relationship of the Hmong to their elder brother, the *mab daum*, is neglected in a further implication of the myth. According to my informants, the Khmu *mab daum* do not know how to manipulate their relations to their ancestors in order to raise their respective *hmoov*. Thus, they sold all their ancestors' heirlooms to the Lao, marginalising themselves in terms of *hmoov* (Postert 2003). As a matter of fact, many of the itinerant workers in Laos are Khmu who are in shortage of land or addicted to drugs and have to seek employment as paid workers in wealthier villages, especially among the Hmong. Although in general, relations to Khmu are marked by mutual respect, there are widespread prejudices among the Hmong that Khmu tend to be 'lazy' and 'work less than the Hmong' (Johnson 1985: 21). In all geomantic myths, the reincarnation of the *huab tais* is expected to be a topic for competition only among the elder brother Hmong and the younger brother Chinese or Lao, silently bypassing the eldest brother *mab daum*.

The mythical rivalry that follows upon the creation of different ethnic groups should be viewed against the background of the Hmong geomantic system, in which the location of the graves of ancestors determines the destinies of their respective descendants.

In Hmong cosmology, a higher place on a mountain should ideally always be in the north. Thus, there is only one Hmong word for both ‚north' and ‚uphill' *saa ped*, that is translated by various authors as „côté haut" (Lemoine 1972a: 132), „upland" (Chindarsi 1976: 19) or „summit of the sky" *hau-ndû* (Lyman 1974: 384). The equivalent term for ‚south' and ‚downhill' is *saab nraad*, translated as „vers le bas", „downland" or „bottom of the sky" *qáng ndû*. If one is looking for a place with a lot of *hmoov* one is usually referred to the north and higher areas, whereas regenerative fertility is usually attributed to the south and lower areas. Space obviously serves here as a system of coordinates that orders topographical elements in hierarchical relationships to another according to basic idea-values of Hmong cosmology.

Space basically determines human fate: in the Hmong geomantic system, there are ‚dragon veins' *loojmem* presumed to run through the country (sometimes alternatively called ‚roots of the mountains' *cab roob*; Tapp 1989a: 166). These ‚veins' have no obvious morphological substrate and can be determined only by examining the relational arrangement of the surrounding mountain peaks, slopes and valleys. All these elements are rather permanent aspects of the geomantic system that nevertheless allow for a dynamic circulation: in the ‚dragon veins', the breath *pa* of a mythical dragon is supposed to flow down from north to south after sunset and to reverse its direction with sunrise (Tapp 1989a: 153).

The importance of the geomantic system becomes apparent in conceptions of death and reincarnation. If a descent group buries a dead on a very good ‚dragon vein' *loojmem* far in the north on a spot with a lot of *hmoov*, it is assumed that in the process of the decaying of the corpse different aspects of the dead person become detached from the grave and attach themselves to the flow of the ,breath of the dragon', recirculate, and reincarnate to constitute new persons (Chindarsi 1976: 89; Tapp 1989a: 152). The *hmoov* of such a new person is regarded to be a function of the *hmoov* that could be attributed to the respective burial site. One thing is particularly important: the recirculation of the different aspects of the person is only possible if the dead and his belongings in the grave have completely decayed. Thus, it is imperative for every Hmong to put only perishable objects of the dead person into the grave.

To retain the office of the ruler *huab tais*, the relevant social group has to look for an auspicious site in the landscape. Therefore, in the present and in the mythical past, searching for appropriate burial sites is thus a rather secret and competitive matter as a variety of Hmong myths depict (Mottin 1980: 107; Tapp 1989a: 131-135, 156-157, 159-160).

Inversion of Mythical Hierarchy

The next myth that will be discussed here has been recorded by Tapp in Thailand (Tapp 1989a: 159-160). In this myth, the brother hierarchy is reversed: the Chinese are described to be the elder brother, the Hmong as the younger one. This inversion of ranking suggests that the age precedence of the Hmong should not be regarded as being part of their identity *sui generis*, but as an expression of the fact that in mythical times they were the ones to rule their brothers. In the present time, however, the Hmong have to get accostumed to the fact that they no longer are in a position of authority, but the Chinese resp. Lao are. How can such a 'cognitive dissonance' be dealt with? On the one hand, it can be either maintained by continuing to stress the primacy of the Hmong brother over the Chinese/Lao, thus keeping up the tension between an ideal mythical past and present-day circumstances; this is the way apparently most often chosen. On the other hand, the cognitive dissonance would disappear if the mythical past could be modelled according to present-day circumstances. The acknowledgement of the present dominance of the Chinese/Lao might then lead to project it on the mythical ideal, thus rewriting a mythical past in accordance with a presumed primacy of the Chinese as elder brothers. Tapp gives a myth exemplifying such an inversion occuring within the myth itself. The myth starts by saying that 'In the earliest times the Chinese and the Miao were one family. The Miao was the older, the more powerful, and the more respected brother, and the Chinese was the younger.' After describing processes of separation, the myth ends with 'The Chinese descendants have grown more and more powerful, so that the Miao are now the younger and weaker brothers and the Chinese are the older and stronger brothers.' (Tapp 1989a: 161). Thus, stressing Hmong's mythical status of being the elder brother seems to deligitimise the present-day's dominance of the younger brother, the Lao/Chinese. Stressing the Hmong's mythical status as younger brothers seems to confirm the legitimacy of present-day's dominance of the elder brother, the Lao/Chinese. However, whereas the interpretation of the order of hierarchy between brothers in past and present may vary, in both cases the underlying model, as exemplified in a variety of myths, appears to be essentially the same.

> „Once two brothers went to look for good *loojmem* for their own graves. They both found the same good place, and each of them told their sons,'When I am a hundred and twenty years old, you must bury me here.' No matter whether they are Hmong or Chinese, they cannot be buried in the same place. But they were buried very close together. Because the older brother was richer than the younger, gold was placed under his neck, and stone under the neck of the younger brother, since after three years have passed, he whose neck has first broken will first go to be a king in heaven. And when the father is king in heaven, the son will be king on earth.
> The elders teach us to plant flowers on graves, and the flowers on the younger brother's grave blossomed before the flowers on the grave of the older brother. When the older brother's sons saw this, and saw that the younger brother's neck was breaking, they took away the stone and replaced it with the gold. When three years later the younger brother's sons came to visit their father's grave, they found the grave of their uncle already open, for he had gone to be a king in heaven. But they realised what bad men their cousins were – and since that day Hmong and Chinese have never got on together. The Hmong moved away from the Chinese, and did not wish to speak the same language any more. The Chinese wanted the Hmong to speak Chinese, but the Hmong did not want to and departed" (Tapp 1989a: 159-160).

The two brothers, Hmong and Chinese, are buried at the same spot identified as a ‚good place', in geomantic terms, a place with a lot of *hmoov*. It is obvious from the beginning that this will lead to an extraordinary competition because the dead will have to fight for the same geomantic point of reference, the ‚ownership of land' (Chindarsi 1976: 91). It is only one of them who will be able to take profit of that place – the rivalry could hardly be put in stronger terms.

Before a recirculation and reincarnation of the various aspects of the dead take place, all objects that have taken part in constituting the dead's sociality and that were put into his grave have to decay completely. If this is about to happen, life can be generated again in his own descendants: as an omen of this ongoing process, the flowers start to blossom on the grave of the Hmong brother. The Chinese who had put a piece of gold into the grave of their father were ignoring the fact that imperishable objects made by man prevent any process of rotting and recirculation. Realising their ignorance, the Chinese who are about to lose cheat the Hmong: they exchange the imperishable piece of gold as a social product of man against the stone that is likewise imperishable, but not man-made and belongs to the extra-social realm of the cosmos, not disturbing the process of decay of the corpse (another version of that myth mentions a bowl of iron and a bowl of wood being exchanged secretly; Mottin 1980: 107). The Hmong descendants come too late to prevent the cheating and have to realise that it will be the Chinese who are going to be ‚king in heaven' *huab tais ntuj*. They lose all insignia proving their legitimate claim to the office of the mythical ruler *huab tais*: the Hmong have to leave mythical North, ‚the summit of the sky', and escape south to ‚the bottom of the sky', thus reproducing the inversion of hierarchy topographically. This alienation from mythical origin is accompanied by the loss of the Hmong books and script embodying original Hmong wisdom: through Chinese perfidy, all books of the escaping Hmong fall into a river (Mottin 1980: 163). The Hmong script is burnt by the Chinese or has to be eaten by the starving Hmong on their long flight to the south (Tapp 1989a: 75, 122). Alienated from a mythical origin associated with rulership and a superabundance of goods like silver, there are only two options that remain for the Hmong in their present marginality: first, to engage in exchange with the now-dominant ethnic group[10], or second, to reinstall their mythical claim on authority. Both options have been chosen by different Hmong groups in different times.

The competitive aspect inherent in relationships between same sex siblings in the context of their *hmoov* thus leads to the inversion of the mythical ideal and to the illegitimate dominance of the younger brother that lasts up to the present day. It is contrary to the ideal of the authority of the elder brother over the younger one. This ideal is normative for every male Hmong in everyday relationships between same sex siblings (Lemoine 1972a: 171). Thus, a certain tension exists between everyday kinship relations and a mythical younger

10 One possibility of relating to a dominant stranger in exchange is of course marriage, as many examples from insular Southeast Asia convincingly demonstrate (see Platenkamp 1996: 217-218 for an overview). Although marriage between Hmong and *suav* is rarely described in myth, some Hmong clans *xeem* are attributing their origin to uxorilocal marriages between Hmong women and Chinese immigrants. Thus, the clan Lauj traces its origin back to a Chinese, who was not able to have children and subsequently married and reproduced with a Hmong woman; such a marriage is also referred to when explaining the coming into being of the clan Xem 50 years ago (Tapp 1989a: 169). Actually, many Chinese soldiers who had formerly been fighting for the Kuonmintang in the Chinese civil war subsequently escaped into Northern Indochina where they engaged in horse caravans trading with the mountain villages. In the 1950's, many of them settled down in larger Hmong villages and were succeeded in mobile trade by local Lao (Johnson 1985: 155-156, 311). However, Tapp reports similar accounts of clan genealogies among the Hmong in China (Tapp 2001: 203, 212). A reversion of this model is expressed in another myth: in times of Hmong dominance, it was the Chinese king who gave his daughter in marriage to the victorious Hmong (Tapp 1989a: 169-170).
The flexibility of genealogies comes into view in another case: the locally dominant clan Wang stresses its descendance from a Hmong man married to a Chinese. However, informants from another competing Hmong clan oppose to this view and claim that Wang's founding ancestor himself was Chinese and his wife Hmong, in this case apparently in the attempt to delegitimise Wang's claims on dominance (Tapp 2001: 328-330).

brother who cheated his elder one to defeat and subordinate him. The moral impetus to restitute the mythical ideal and to reinstall the Hmong as elder brother in a position of authority over the younger brother is obvious. This seems to account for a latent Hmong disposition for messianic rebellion as exemplified in diverse attempts to reinstall mythical dominance of the Hmong as elder brothers against diverse 'strangers' that were always in a position of authority, thus fitting into the category of *suav* as depicted in the beginning of this essay (see Culas 2005 for a detailed analysis of Hmong messianism).

Some versions of the myths in question reinterpret the outcome of the mythical brother rivalry in different terms. In one version, Hmong and Chinese after a fierce competition agreed to divide the land. Whereas the Hmong demarcated their areas with twigs and leaves, the Chinese chose stone walls to demarcate their land. When the wind swept the boundaries of the Hmong away, they lost all claims on land and had to accept the dominance of the Chinese (Tapp 1989: 165). Again, the category of the Khmu *mab daum* does not appear in the rivalry of brothers.

The only myth I know that explicitly acknowledges all brothers was recorded by myself in 2001 in Laos: here, it is *Saub*, one of the highest deities of Hmong cosmology who gives birth to a gourd. After piercing it, a Khmu appears first, then a Hmong, finally a Lao. *Saub* prepares a mound of earth and covers it with cloth, a heap of iron and a packet of paper. Then he calls the brothers to choose. The Lao and Hmong give precedence to their elder brother, the Khmu who takes the cloth – only to realise that it barely suffices to make loin-cloths for himself and his kin. The Hmong chooses the iron – and is obliged to go and work on the fields. The Lao finally takes the paper – and thus gets the script, the insignia of rulership. This course of events is still very much in line with other versions legitimising the dominance of the youngest brother. However, the myth continues. When the Hmong realises that the script fell into the hands of his younger brother, he returns to *Saub* to ask for script as well. Thus, *Saub* gives him a packet of paper. As the Hmong realises that working the land and studying the script at the same time is more than he can manage, he decides to eat the paper. And the myth concludes: 'now whatever the Hmong are doing they do it better [than the other groups mentioned] because they have already the script in their stomach'. Whereas the Lao have a script only in their possession (as exterior property that can be taken away from them), the Hmong incorporated their script as an inalienable part of their self.[11] For this reason, as the myth concludes, Hmong are assumed to be more

11 However, Tapp gives another myth recorded in Thailand with a very similar course of events whose implications might be interpreted differently. In this myth, the Hmong who have to cross a river are afraid that their books might get wet. In contrast to a merged social category of the '*Mab Suav* (Chinese and others)' who cross the river carrying their books on the head, the Hmong eat theirs out of hunger (Tapp 1989a: 122). The feeling of thereby loosing more than gaining seems to be expressed in the concluding statement: 'That is the

capable than any other ethnic group to acquire a high education and to be appointed to high offices in any polity, thereby gradually replacing individuals of other ethnic origin in positions of authority.[12] During my fieldwork, many of my informants conceived higher competitive qualities to be specific for Hmong compared to other ethnic groups. Thus, we might be dealing here with a mythical legitimisation of this view: eating the script would then have to be interpreted as incarnating these insignia of dominance into the very essence of Hmong identity. Relations to strangers thus seem to imply neglect as in the case of the *mab* or competition as in the case of the *suav*. According to this myth, challenging the stranger would then be an unavoidable disposition regarded to be inherent in Hmong identity. From this point of view, the notorious Hmong disposition to millenarian rebellion might promise success in the long run as it can be interpreted as an endeavour that is in accordance to the just and eternal order of a mythical past.

Conclusion

I have been discussing different contexts in which Hmong relations to strangers assume a different form and different attributes.

Firstly, in the context of 'breath/vapour' *pa*, social differences between strangers are subordinated to a unifying principle of regenerative life as exemplified in the mythical gourd. Thus, competitive relations between strangers should ideally be subordinated to the common origin they all share.

Secondly, in the context of *hmoov*, different social groups start to differentiate from another exemplified in the transformation of the growing gourd into the successive appearance of brothers ranked according to relative age. In this context, Hmong tend to distinguish between two broader social categories of strangers: *mab*, that is groups categorised as 'elder brothers' in mythical past, and *suav*, groups categorised as 'younger brothers' in mythical past. Ranking according to relative age implies the dominance of the elder brother over the younger. Neglecting the category of *mab*, diverse versions confirm the myth-

reason why now we can only be clever inside, in our hearts and only remember in our hearts, not in books. But before that, we had books of our own.' (Tapp 1989a: 122).

12 This view is reinforced by assumptions that Lao people do not know how to make effective use of the geomantic system. According to my informants, ordinary Lao people do not even know about the very existence of ‚dragon veins'. Lao Buddhist monks are different in this respect: on the one hand, Hmong informants state that monks know how to read and how to manipulate the geomantic system, for instance when building their temples on strong dragon veins. But on the other hand, they are not able to actually take profit from this knowledge as they do not bury their dead, but burn them. Thus, they keep the corpse from a process of decaying in earth that is perceived to be a precondition for an effective recirculation of the different aspects of the person. Expectations were uttered that the better knowledge of the Hmong in the manipulation of the geomantic system will in future invert the relations of dominance between them and the Lao.

ical ideal that Hmong as elder brothers should be entitled to power and dominance, their younger brothers *suav* to subordination. However, the *suav* manage to reverse this ranking through trickery in a fierce geomantic competition.

Apparently, Hmong conceptions of the 'own' and the 'stranger' *suav* are deeply interdependant. Tapp stresses that Hmong identity is to a large extent defined by constructing contrast and competition between the Hmong and *suav* (translated by Tapp as 'Han/other') (Tapp 2001: 12-13, 420-21, 428-29). This category of the 'stranger' is not a phenomenon constructed once and for all, but something that continually feeds back onto the reading of the past. Whether the 'stranger' in the present (be it the Chinese, be it the Lao) is perceived to be vulnerable or too powerful might very well find expression in a flexible reinterpretation of the mythical past: in the first case perhaps stressing the priority of the Hmong as mythical elder brothers, in the second inverting the mythical ranking and assuming the precedence of the 'stranger' *suav* in terms of brotherhood. Thus, when discussing myths, we are indeed dealing with extremely dynamic relationships between past and present. These are strategically modified all the time in view of the present situation, steering back towards a reinterpretation of the past. Mythical past is tightly connected to future action as it becomes a model in diverse messianic rebellions when the mythical dominance of the Hmong as elder brothers is sought to be reestablished. Thus, the Hmong challenging of the stranger has to be seen as a very dynamic process connecting past, present and future in a synchronic and meaningful whole whose legitimacy is always at disposition and constantly renegotiated according to present-day circumstances.

References

BERNATZIK, H.A. (1947) *Akha und Meau. Probleme der Angewandten Völkerkunde in Hinterindien*. Innsbruck: Kommissionsverlag.

CHINDARSI, N. (1976) *The Religion of the Hmong Njua*. Bangkok: The Siam Society.

CULAS, C. (2005) *Le Messianisme hmong XIXe et XXe siècle. La dynamique religieuse comme instrument politique*. Paris: Éditions du CNRS.

JOHNSON, C. (1985) *Myths, Legends and Folktales from the Hmong of Laos. With Explanatory Notes on Hmong Culture, Customs and Belief*. St. Paul: Macalester College.

LEMOINE, J. (1972a) *Un village Hmong Vert du Haut Laos. Milieu technique et organisation sociale*. Paris: Éditions du CNRS.

LEMOINE, J. (1972b) L'initiation du mort chez les Hmong. I. Le chemin II. Les thèmes III. Les thèmes. *L'Homme* 12(1): 105-134; 12(2): 85-125; 12(3): 84-110.

LYMAN, T.A. (1974) *Dictionary of Mong Njua, a Miao (Meo) Language of Southeast Asia*. The Hague: Mouton.

MOTTIN, J. (1980) *Contes et légendes Hmong Blanc*. Bangkok: Don Bosco.

MOTTIN, J. (1982) *Allons faire le tour du ciel et de la terre. Le chamanisme des Hmong vu dans les textes*. Bangkok: White Lotus.

PLATENKAMP, J.D.M. (1988) *Tobelo. Ideas and Values of a North Moluccan Society*. Leiden: Repro Psychologie.

PLATENKAMP, J.D.M. (1996) North Halmahera: Non-Austronesian Languages, Austronesian Cultures? In BREMEN, J. van, GODINA, V. and PLATENKAMP, J.D.M. (eds.), *Horizons of Understanding: An Anthology of Theoretical Anthropology in Europe*, 206-223. Leiden: Research School CNWS.

PLATENKAMP, J.D.M. (2004) From Partial Persons to Completed Societies. *Zeitschrift für Ethnologie* 129: 1-28.

PLATENKAMP, J.D.M. (2005) Des personnes incomplètes aux sociétés accomplies. *L'Homme* 174: 125-160.

POSTERT, C. (2003) *Rituelles Handeln verstehen: Soziale Morphologie und ritueller Zyklus der Person bei den Hmong in Laos und Thailand.* Unpublished Ph.D. dissertation, Department of Social Anthropology, Münster University.

SAVINA, F.M. (1930) *Histoire des Miao.* Hong Kong: Missions Étrangères de Paris.

SYMONDS, P.V. (1991) *Cosmology and the Cycle of Life. Hmong Views of Birth, Death and Gender in a Mountain Village in Northern Thailand.* PhD-Thesis, Brown University.

TAPP, N. (1986) Geomancy as an Aspect of Upland-Lowland Relationships. In: HENDRICKS, G.L., DOWNING B.T. and DEINARD, A.S. (eds.), *The Hmong in Transition*, pp. 87-96. New York: Center for Migration Studies.

TAPP, N. (1989a) *Sovereignty and Rebellion. The White Hmong of Northern Thailand.* Oxford: Oxford University Press.

TAPP, N. (1989b) Hmong Religion. *Asian Folklore Studies* 48: 59-94.

TAPP, N. (2001) *The Hmong of China: Context, Agency and the Imaginary*. Leiden: Brill.

“Making the Spirit”: The Potential of Knowledge from Elsewhere

ALMUT SCHNEIDER

One of the themes that are of central concern to Jos Platenkamp and that he explores in several of his writings is the status of, and the value ascribed to, persons, objects and ideas of foreign origin and identity. His interest was particularly in ‘the stranger’ as somebody – or something – conceived and valued as part of culturally specific discourses and actions. Be it in his inquiries into the ‘differential origins’ of Eastern Indonesian societies (Barraud and Platenkamp 1990), into ‘incorporative rituals’ of the Lao society (Platenkamp 2010a) or in a comparative perspective in Germany (Platenkamp 2004, 2014), he pursued the question of how objects, ideas and beings of ‘strange’ origin are integrated into a society, how they contribute to its foundation and reproduction. This interest, among a number of others, Jos has been sharing for many years with his Parisian colleagues and allies from the ERASME research group with whom he worked intensively in the mid 1980’s and has stayed in an intellectual exchange ever since. In the 2000s I received my post graduate training in Paris and worked for several years with the same research group. This shared experience – even if separated in time, is, I feel, a real privilege for me. The following pages shall be a small reminder of this.

Objects, persons and ideas that come from elsewhere are a creative focus in almost all social contexts of the Gawigl-speakers of the Western Highlands Province of Papua New Guinea.[1] Sweet potato gardens, for example, need external contributions in the form of vines that come from other, often far away locations, and it is women who are responsible for this movement. Another example are groups of brothers, living together in ‘men’s houses’, depending on wives to come to them from other, distant places. This is crucial in order to remain distinct from other ‘men’s houses’ and to ensure reproduction over time. And: Pigs and money – they are never local, but are always coming from somewhere else in an almost perpetual movement in the course of ceremonial exchanges. Obviously pigs and money are also always going somewhere else, as are out-marrying sisters and planting material from one’s own sweet potato mound, brought to a brother’s wife’s garden.

Besides gardens, pigs, money, or “men’s houses”, it is the ritual of “making the spirit” (*kuro tekem*) that is another example for an outside element as a

1 I am grateful to the DAAD (Bonn) for funding my fieldwork of 20 months, to the National Research Institute (Papua New Guinea), and to the office of the Administrator of the Western Highlands Province, for their support during my stay in the country.

necessary impetus to social regeneration.. This ritual was performed by the Gawigl to reinstall general fertility and the reproductive capacities of a whole local group, to ensure a renewed increase in crops, pigs and people.[2] What came from the outside in this example, are not, as in production or exchange, objects or people but spirits and ritual knowledge.

Houses and Names

Until some forty years ago local Gawigl groups performed the ritual "making the spirit" roughly once every generation. As for other changes, people tend to rationalise its suspension with the implantation of Christianity.[3] My account is a reconstruction fed from several sources: conversations with numerous people at my fieldsite Marapugl, from other places in the Tambul district and the neighbouring Nebilyer valley, two missionary ethnographies of the 1940s and 1960s that give detailed descriptions on comparable events in the neighbouring Hagen area (Vicedom 1943-48, Strauss 1990 [1962]), and anthropological accounts of regional variations of the ritual in the Hagen (Strathern 1970, 1991a, 1991b) and Enga (Wiessner and Tumu 1998) area.

The setting is a familiar one for the Highland region: no villages, but houses and hamlets scattered on the mountain slopes or in the valley bottoms, and in between sweet potato gardens, needed in large quantities to support men and the semi-domesticated pigs, further up the mountains, the forest, area of vast garden clearings and home to marsupials, birds and spirits. The hamlet is the smallest social unit, three to four houses, often grouped around a ceremonial ground; at a distance of ten to twenty minutes by foot, often crossing small ridges and streams, going along sweet potato and coffee gardens or through pockets of secondary forest, one reaches the next hamlet. Several of these form a residential group called "men's house", *yie ulge*, i.e., put very shortly, groups of classificatory brothers. Several men's houses in turn constitute a bigger unit, the "one name" *imbi teglu*.[4] At the periphery of this locality

2 For a proper description and analysis of this ritual see the last chapter in Schneider 2016. For this paper I owe very much to Jos' generous, patient and perceptive input, some of which he gave in a Parisian café.

3 One often hears statements such as: "Today we no longer take care of making the spirit because now we go to church." It remains to be seen whether this is a permanent or a temporary replacement of rituals.

4 Earlier writings translate the expression "men's house" with 'lineage' and "one name" with 'clan'. I prefer to retain more of the vernacular expression, though for better readability, in the abbreviated form House and Name. In Tambul, large Names would have around 1200 people, small ones around 500. Houses can be as small as 30 members but can also include 200 people.

of a Name is the forest, cut by a few paths leading to the neighbouring Name groups.[5]

Ceremonial exchange feasts are the primary occasions for Houses to manifest themselves, and to become visible to other Name-groups. The scale of the events serves as a measure of their capacity to assemble and generate wealth. In Tambul, such feasts are linked to life-cycle exchanges and signify an almost constant movement of pigs and (modern) money either between individuals or between Houses. I sketch these exchange feasts briefly, as they will be instrumental for understanding the ritual.

Marriage exchanges, consisting of living pigs and large sums of money link brothers of an in-marrying woman with her husband and his brothers and mark the beginning of a relationship between Houses of two different Names that lasts for two generations. Several years after the marriage, a significant prestation of living pigs and a large amount of money[6] follows. It is called *magadl,* and is given to the mother's brother of the couple's children for their growth and wellbeing. The relationship between the two Houses that had begun with the marriage, terminates with yet another of these *magadl* prestations. At the death of the couple's son, his House now offers a *magadl* to the descendants of the mother's brother, thereby contributing to transforming the deceased into an ancestor (a "planted spirit", *kuro imbo*). The main task in organising these different prestations, for the children of all the brothers of a House and for the dead, is the lengthy process of assembling pigs and money, within the House and (above all) outside of it, appealing to relatives and affines for help. Apart from work in the gardens, assembling marriage prestations or *magadl* is the major activity of Gawigl men, either as a main donor or as a supporter.

The larger scale Name groups are less visible than the Houses with their regular public appearance at the occasion of the ceremonial exchange feasts (*magadl*) they give. And if Houses are groups of 'brothers' that share common ancestors, Names are groups of 'brothers' that share common residence and the rule of exogamy. Occasions for Names to appear as such are infrequent and irregular: warfare (or rather feuds), compensation payments (following

5 People of one Name share residence on a specific stretch of land; they do not share 'common descent'; the smaller "men's houses" *do*, it would appear, to a certain extent and are composed of classificatory brothers (a more elaborate discussion of this phenomenon can be found in Schneider 2016). Immigration from House members of other Names is quite frequent, and becoming a member of a men's House is usually a matter of one generation only.

6 Giving numbers is difficult as they depend on the particular wealth of a House at a given time and on their capacity to mobilise resources outside the House. But it is safe to say that the fathers of the children and their brothers (i.e. the House) need several months to assemble the prestation and this always includes activating their kinship networks, spreading to other Houses.

war), voting for the local political member or – in the past – organising the ritual of 'making the spirit' would be such occasions.

The Stranger-Spirit in the Making

When had there been such occasions? When appeals to the ancestors were wearing off, thus proving less and less effective, it was by "making the spirit", and thus, going beyond the 'planted spirits', that the Gawigl attended to the recurrent atrophy that was afflicting the land and its inhabitants: persistent bad harvests due to draught, frost or flooding, insects attacking tubers, flying foxes eating away bananas and pandanus, people and pigs quarrelling, becoming thin, falling ill and dying early. Manifestations of the spirit itself – either in dreams or in the form of an unusual stone, turning up at unlikely places – encouraged the decision of people of a Name to execute the big ritual. This decision had far reaching consequences: The life of both men and women of the Name was for a period of eighteen to twenty-four profoundly changed, in a way magnified and virutally scaled up.

We must understand the name of the ritual, "making the spirit" in the sense of 'making something new, that was not there before'. Each time the ritual was accomplished, it 'made' a new spirit, yet unknown, not local (*imbo*). If ritual is usually understood as something that has to do with repetition (Iteanu 2004, 2015; Bell 2009), there is definitely an element of non-repetition here: each spirit that was 'made' in such a ritual, was only made once by each Name. It is the ritual itself that located and brought a new named spirit, previously unknown, to this Name. From the perspective of Gawigl, the ritual was never done in an identical way, each time it was replaced by a new version, executed for an unknown spirit, whose characteristics people only got to know during the performance of the ritual itself. The spirit as well as the knowledge of its performance came from the outside, from a neighboring Name that had already completed the ritual, and has obtained the knowledge in turn from another Name. Its origin was unknown and foreign. This was indeed an ambulatory, circulating (Strathern 1991) ritual, which was transmitted from Name to Name. If ever there was an element of repetition in this ritual, it was the fact of accomplishing it every generation. Its main features were known, but details considered crucial for "making the (new) spirit", were ignored and only learned during the performance of the ritual itself. A ritual specialist (*agla yie*) from a Name who had already performed the ritual transmitted the necessary knowledge.

To overcome local atrophy one required an external, non-local contribution of ritual knowledge. This distinguished this particular ritual from the life-cycle rituals mentioned earlier, which are considered to be thoroughly local. Also different was the scale of "making the spirit", in particular in relation to duration and scope. A whole Name was involved in the execution of the ritual which lasted approximately one to two years.[7] This long period had three quite different phases: The first consisted of the closure of a Name (or the virtual stopping of all of its outside relations), which began with the arrival of the ritual specialist, directing some ritual preparations (like establishing an enclosed ceremonial ground 'for the spirit', burying the stones representing the spirit) and the first seclusion of the men. This phase lasted approximately for two months. It was followed by a very long phase, that I call 'the time of growth', extending over twelve to eighteen months. Finally came the climax of the ritual, the shortest, but the most intense phase: the second, three day long seclusion of the men and the final feast, when huge amounts of pig meat were distributed to hundreds of visitors.

The particular knowledge of the foreign ritual specialist concerned the two phases of seclusion, and the preparations that preceded them. During this time the ritual specialist came to stay with the men of One Name and instructed them on each single step. The participants were left in the dark about the reasons behind the sequence of unusual actions they performed according to these instructions (like going into the forest in single file, eating synchronously, staying awake all night, repeating phrases of an unintelligible language in unison, etc.). Only one or two local experts (*agla yie*) were put completely into the picture and became apprentices of the external expert. Since the ritual would never be repeated in their own Name, they acquired the knowledge to transmit it to another One Name group. This was spatial, not temporal transmission.

Putting the Circulation on Hold – and a Feast

It is due to the way in which I gathered the information on the ritual that I am not focussing on the details of what happened during the seclusion of the men.[8] What can be retained from these phases is the non-repetitive, singular character it lends to the ritual as well as the necessity of movement from an exterior, taking the form of a spirit and of ritual knowledge that is unfamiliar and of which the ultimate origin is unknown. Not focussing, however, on what people said had been the heart of 'making the spirit' (i.e. the phases of seclusion, when the men were 'with the spirit') made a more external view possible. I was thus/then able to look on the ritual as an extended period of time and as a sort of up-scaling of an entire Name.

This became particularly visible during the lengthy 'time of growth'. The objective of this phase was the maximal augmentation – in number and in size

7 See Schneider (2007) for my argument for considering the ritual period to be much more inclusive, and therefore longer, than Strathern, who puts his focus on the concluding phase of the ritual.

8 The accounts of participants were by and large inconclusive and the ritual experts I met did not want to tell me all the particulars.

– of the pigs looked after by the entirety of the Name. The daily activities of both men and women increased considerably: in view of the incoming pigs, new sweet potato gardens were established and pig houses built. The most relevant and arduous task, however, consisted of finding the animals themselves and keeping them for many months despite the circulation they were usually subject to. Unlike ordinary times, during this period the men of a Name needed to coordinate their exchange activities outside the Name, in particular with affines, maternal kin and other exchange partnerships they had built up over the last years. This alone was rather unusual – such a cooperation usually took place in the much smaller social units of Houses, but even rarer was the redirection of the flow of pigs that characterised this stage of the ritual. The challenge was to organise, or rather channel, all the exchange relationships in a way as to having pigs come into the Name without leaving it again (see figure 1).

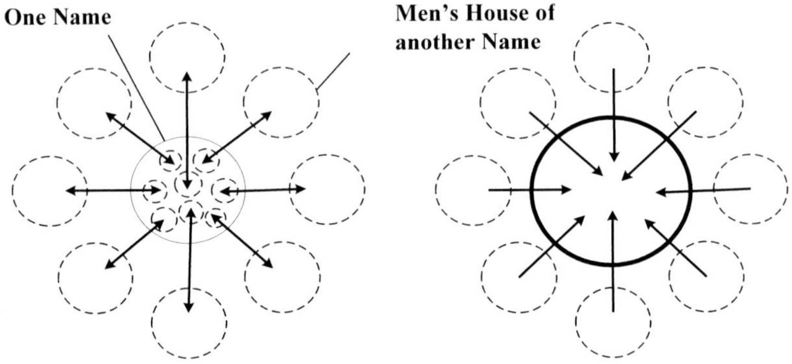

Fig.1 Exchange relations in ordinary time (left), and during the ritual "making the spirit" (right)

The most important strategy men used was to temporarily reorganise the life-cycle exchanges mentioned above. One type of exchange, that in which pigs were received, needed to be promoted and accelerated. The other type, that in which pigs were given away, had to be delayed and put off till after the ritual was completed. In fact the main work men had to accomplish during this 'time of growth' was negotiating. All those affines from whom they expected to receive prestations, that is to say affines through their sisters were persuaded and pressed to come up with the pigs they were due in connection with a *magadl*.[9] On the other hand, those prestations the men of the Name were due to give to affines on their wives' side, wife's brothers or children's mothers

9 In ordinary times the final handing over of pigs was (and is) notoriously delayed for weeks or sometimes months.

brothers, were put on hold. Thus, two complementary skills of persuasion were needed: inciting some people (affines on the sister's side) to give a *magadl* prestation earlier than they had planned, and making others (affines on the wife's side) wait much longer than they had anticipated to get theirs.

Apart from the kinship relations, the individual exchange partnerships of the performing Name were rechannelled according to the same principle; those owing pigs were persuaded into giving them as soon as possible, those to whom one owed oneself animals were put off. In fact, what happened during the long ritual phase of the 'time of growth' was a very mundane form of putting the circulation of pigs on hold – at least with regard to the Name performing the ritual.

The effect of temporarily 'receiving while not giving' was that the Name gradually 'filled up' with pigs. If it was mostly men who did the work of persuading their exchange partners into giving pigs, it was mostly women who had the responsibility of looking after and feeding the growing pig herds. Even in the context of preparing *magadl* prestations, smaller events, limited to a House, and shorter in preparation, the gradual accumulation of pigs put a strain particularly on women.

After the long months of accumulation and when the Name decided that they had enough animals for the final feast, the ritual specialist arrived once again to lead the men through the final phase of the ritual. They would spend some days in reclusion to 'be with the spirit' (*kuro pea modlmo*) and the local ritual specialists would learn the actions and incantations to coordinate this interaction. Then came the complex preparations for a very large scale feast for which all the animals would be slaughtered, cooked, cut into pieces and distributed to the guests. Cooking stones and firewood was needed in enormous quantities, a platform surrounding the ceremonial ground of the Name had to be build, women had to harvest vegetables to feed the hundreds of guests, men had to assemble their body decoration and had to organise the butchering of the pigs, a task they could not execute themselves since they would be with the spirit just before the distribution of the meat.

On the day of the final feast, a large crowd of people gathered on the ceremonial ground of the Name.[10] Contrary to other feasts, participation was not restricted to certain groups of people. Everybody who knew about the feast – and given the lengthy preparation period of the 'time of growth' it was very widely known in the region – could attend and expect a share of meat. This climax of the spirit ritual was a feast of profusion and self-manifestation of a Name. A visible example – the huge quantity of meat serving as an index – of

10 One might recogn ise a regional variant of the 'Female Spirit Cult' on which Andrew Strathern published a number of articles (1970, 1979, 1991a, 1991b, Stewart and Strathern 1999), and, in what concerns scale and profusion of the event one is reminded of other ethnographic descriptions in the wider region (Rappaport 1968; Brown 1978; Wiessner and Tumu 1998; Wiessner 1999).

their ability to "make a new spirit" and to regenerate the reproductive capacity of their land and their people.

• • •

I think Jos Platenkamp knows how much my understanding of the Gawigl 'making the spirit' owes to his own work on ritual, starting with Barraud and Platenkamp 1990 till Platenkamp 2010a and b. His careful untacking and analytical reassembling of rituals taught me to see the larger picture in the Gawigl spirit ritual. Coming back from the field, I had found it particularly puzzling, since the ceremonial exchanges, which I had actually constantly seen among the Gawigl were so dominant in the data I brought back. But the 'stranger-spirit', necessary to the general regeneration of people, animals, plants and the land itself also had a significant impact on those ceremonial exchanges. And they have to do with the paradox (so it seems to the anthropologist) inherent in this situation: A Name decided to perform a spirit ritual because its general situation was bad and kept on getting worse. But the performance of the ritual itself, notably the time of growth, required an enduring expenditure of energy of the Name's members. It demanded a particular kind of strength of the men to persuade the exchange partners, of the women to take care of (more) gardens and (more) pigs, and, even, of the land to cope with the increase in gardens. And another paradox: on the day *after* the final feast, that just publicised the successful regeneration of the Name so dramatically, it found itself with empty hands, without any pigs and hardly any vegetable left to eat. In fact, the strain put on kinship relationships and the network of exchange partners, by accelerating some prestations and delaying others, had the effect of strengthening and re-defining these relations. Otherwise they would hardly have met the challenge that arose after the final feast when there were no pigs to give immediately, but still, everybody considered the Name and its Houses as particularly strong future givers of *magadl* prestations.

References

BARRAUD, C. and PLATENKAMP, J.D.M. (1990) Rituals and the Comparison of Societies. In BARRAUD, C. and PLATENKAMP, J.D.M. (eds.), *Rituals and Socio-Cosmic Order in Eastern Indonesian Societies, Part II: Maluku*, Special Anthropological Issue, *Bijdragen tot de Taal-, Land- en Volkenkunde* 146(1): 103-123.

BELL, C.M. (2009) *Ritual: Perspectives and Dimensions.* Oxford, New York: Oxford University Press.

BROWN, P. (1978) *Highland Peoples of New Guinea.* Cambridge: Cambridge University Press.

ITEANU, A. (2004) Partial Discontinuity: The Mark of Ritual. *Social Analysis* 48(2): 98-115.

ITEANU, A. (2015) Recycling Values: Perspectives from Melanesia. *HAU: Journal of Ethnographic Theory* 5(1): 137-150.

PLATENKAMP, J.D.M. (2004) Über die Gesellschaftliche Relevanz der Ethnologie. In BERTELS, U. et al. (eds.), *Aus der Ferne in die Nähe. Neue Wege der Ethnologie in die Öffentlichkeit*, pp. 21-32. Münster/Berlin/New York: Waxmann Verlag.

PLATENKAMP, J.D.M. (2010a) Becoming a Lao Person. Rituals of Birth and Socialisation in Luang Prabang, Laos. In BERGER, P. et al. (eds.), *The Anthropology of Values*, pp. 180-200. Delhi: Manohar Publishers.

PLATENKAMP, J.D.M. (2010b) Political Change and Ritual Tenacity: The New Year's Ritual of Luang Prabang, Laos. In ITEANU, A. (ed.), *La cohérence en anthropologie sociale*, pp. 193-234. Paris: Editions de la Maison des sciences de l'homme.

PLATENKAMP, J.D.M. (2014) Strangers, the State and the Self in Germany: A Comparative View. *Austrian Academy of Sciences Working Papers in Social Anthropology* 27: 1-9.

RAPPAPORT, R. (1968) *Pigs for the Ancestors: Ritual in the Ecology of a New Guinea People*. New Haven and London : Yale University Press.

SCHNEIDER, A. (2007) Aspects de l'espace-temps dans un rituel des Gawigl (Papouasie-Nouvelle-Guinée), Online-Publication on the web site of the *Réseau Asie* (3e Congrès): http://www.firsturl.de/sGhe7yL.

SCHNEIDER, A. (2016) *La vie qui vient d'ailleurs : Mouvement, échanges et rituel dans les Hautes-Terres de la Papouasie-Nouvelle-Guinée*. Berlin: Lit.

STEWART, P. and STRATHERN, A. (1999) Female Spirit Cults as a Window on Gender Relations in the Highlands of Papua New Guinea. *Journal of the Royal Anthropological Institue* (n.s.) 5: 345-360.

STRATHERN, A. (1970) The Female and Male Spirit Cults in Mount Hagen. *Man (n.s.)* 5: 571-585.

STRATHERN, A. (1979) Men's House, Women's House: The Efficacy of Opposition, Reversal, and Pairing in the Melpa Amb Kor Cult. *Journal of the Polynesian Society* 88: 37-51.

STRATHERN, A. (1991a) Circulating Cults in Highland New Guinea: Pointers for Research. *Australian Journal of Anthropology* 2: 98-107.

STRATHERN, A. (1991b) Fertility and Salvation: The Conflict between Spirit Cult and Christian Sect in Mount Hagen. *Journal of Ritual Studies* 5: 51-64.

STRAUSS, H. (1990 [1962]) *The Mi-Culture of the Mount Hagen People; transl. G. Stürzenhofecker and A. Strathern*, University of Pittsburgh: Ethnology Monographs Series, No. 13.

VICEDOM, G. (1943-48) *Die Mbowamb. Die Kultur der Hagenberg-Stämme im östlichen Zentral-Neuguinea (Vol. 2)*. Hamburg : de Gruyter und Co.

WIESSNER, P. and TUMU, A. (1998) *Historical Vines: Enga Networks of Exchange, Ritual and Warfare in Papua New Guinea*. Bathurst: Crawford House Publishing.

WIESSNER, P. (1999) A Collage of Cults. *Canberra Anthropology* 22(1): 33-65.

Infusing Life and Enlisting Spirits: Some Indonesian Shipbuilding Rituals

ROBERT WESSING

Introduction

In this contribution I look at the rituals conducted during the building of ships in Indonesia, focusing on the area around Sulawesi and on the Madurese, in both Madura and their diaspora in East Java, the latter including some recent (2014) data from Puger. The literature on Indonesian ship building can be divided into two broad categories: 1) the question to what degree ships are models for the organisation of society at some level, or *vice versa* (e.g. Manguin 1986; Domenig 2014: 472-475), and 2) detailed, almost step-by-step accounts of the building process itself, describing the rituals but neglecting to consider the question to whom or what the rituals and associated offerings are addressed, and how the ship is thought to benefit from them. Ships are said to be alive and have personality, without a mention of how this came about during what is often described as their birth (cf. Horridge 1985: 79-80). Here I address this latter matter, considerations of space forcing me to ignore the former.

The literature generally calls these activities rituals (*upacara*), without further differentiation. *Upacara*, however, covers both rituals pertaining to 'religious' matters, and civil ceremonies: the word is used to speak of the act that 'marries' male and female components of the ship, its animation, and its transfer from the builder to the buyer, actions that address different concerns and audiences.

In an article on construction sacrifice in Indonesia, Wessing and Jordaan (1997) distinguished between appeasement and animation rituals. Appeasement, in which a spirit is manoeuvred into ceding some of its property, involves a degree of exchange: "if we give you this, you'll give us that". Animation rituals, to the contrary, do not involve exchanges, but enlist the aid of a spirit, sometimes by first creating it. Once enlisted this spirit strengthens or gives life to the structure it is infused into.

Another problem concerns the idea of 'making an offering', which can cover diverse activities. Correlated with the idea of a ritual exchange, the item offered can be seen as a sacrifice (*korban*) for which a *quid pro quo* is expected: the gift obliges the god or spirit to reciprocate (cf. Kristensen 1961: 97). However, this is not necessarily so, and my informants denied that God or ancestors could be forced into such a position: people owe them worship or veneration and respect, but they are not obliged to reciprocate. "God", Lehman (1996: 47) writes, "can be solicited but no more. God can, indeed, not be coerced". Spirits may be another matter, however, since people regularly try to make deals with them.

In a ritual context we often hear the expression *sesaji*, which is glossed with ritual offering. Yet the root word *saji* only means to offer something, either to a guest or a spirit, e.g an act of hospitality for which no *quid pro quo* is expected. Thus, starting with the idea that the Indonesian *menyajikan* or Javanese *nyajekane* express hospitality, a *sesajen* becomes something offered to an invited guest, whose very *presence* happens to aid in the achievement of a goal. The offerings to spirits or ancestors, whether ritual flowers, incense, or prayers are not sacrifices: they are offerings that invite (*panggil*) the entity to be present, and show it hospitality. It is a making present of the invitee in the sense of Kristensen (1961: 10). People are sure (*yakin*) in their hearts that having been invited, they are certain to come, perhaps demonstrating a faith not justified by Lehman's conclusion, though inviting a spirit may not be the same as coercing it. In any case the invitation indicates that the spirits' presence is considered an important part of the carefully enumerated (see below) ritual assemblage. How they are called or invoked is a matter of local custom and sensibility: some use prayer, others flowers, etc.

The details of the ritual actions performed can be found in the literature (e.g. Rato 1991; Pelly 1977). I shall only mention the relevant features of each, and attempt to place it in a framework suggested by the above discussion.

Felling Trees

Felling the tree used for the keel involves considerable ritual action (Pelly 1977: 92; Horridge 1985). Its spirit-owner or -guardian must be appeased so that it will consent to the tree being removed (cf. Wessing and Jordaan 1997: 104-105) and not cause mischief during the building (cf. Wessing 1978: 103-104). The shipwright, an expert in dealing with the sometimes dangerous spirit powers, fells the tree for the keel, uttering invocations and sacrificing a chicken to the spirit to obtain its permission (Arief and Abbas 2001: 47). After trimming it, the keel-log is hauled to the shipyard, whose tutelary spirits have been given 'appropriate offerings' (Horridge 1986: 5).

The interaction with the spirit of the tree as well as possibly with the tutelary spirits of the shipyard is one of appeasement. The ritual is carried out, and the sacrifice is made in the hope that the spirit will be satisfied and not interfere. This sacrifice is not a *sesaji* since neither of these sprits is invited to be present. Indeed, it is hoped that the tree's spirit will find new lodgings, and that those of the shipyard will remain quiet.

As Horridge (1986: 8, 12) points out, the old rituals have been declining. In Puger (East Java) shipwrights now import their lumber from Kalimantan, and assume that appeasement was carried out there. Appeasing spirits not being permitted by Islam, in Puger appeals are made to various Muslim saints, a ritual that also takes care of the tutelary spirits of the shipyard. An informant, denying the existence of appeasement, said:

> "When building a boat, one must ask permission from Nabi Ilyas, who owns the earth. Next one appeals to Nabi Sulaiman, who owns the fish. Then to Nabi Nuh, who owns the boats, to Nabi Khidir, the owner of the sea, and last to Nabi Ibrahim, the owner of the Ka'bah in Mecca. The shipwright asks permission from these saints."

Thus, rather than relying on appeasing spirits, the blessing of Islamic saints is invoked. This appeal takes place during a *sandhingan*, a ritual at which Javanese informants request the presence of spiritual entities, here the Muslim saints. A communal meal is usually part of the ritual. A *sandhingan* is not an appeasement, since no nature spirits are addressed or involved. It is also not an animation ritual, because nothing is infused at this stage. Rather, it creates a context of safety or wellbeing for the building activity. This situation comes about through the assembly of locally specified requirements, which here include the presence of the saints. These saints do not in themselves suffice: both Rato (1991: 19-21) and my informants gave detailed lists of what was required during the performance of the rituals, thereby narrating the solution to the problem at hand (Wessing 1978: 163-166). It is, therefore, the conjunction of their presence with that of other guests and carefully listed and enumerated items and conditions that specifies and brings about the desired result (cf. Kristensen 1961: 9-10; Beatty 1999: 31, 37).

Laying the Keel

Not all non-Islamic rituals have disappeared, however, especially those concerned with the life of the ship, which is thought to be influenced by powers ritually infused into it, especially during the laying of the keel and the 'birth' of the new vessel. Among the Bugis and in Puger the creation of the keel assembly, which consists of a base to which bow and stern parts are attached, is presented in terms of a marriage (Mad. *makabin lenggi*), with an emphasis on the sexual, fertility aspects of the joining (Rato 1991: 16; cf. Pelly 1977: 96). The ship's keel and hull are often considered female, while the stern and the controlling rudder assembly are male, guided by the male captain; both female and male must be present for the relationship to come to fruition (De Jonge 2002:14).

In Sulawesi the shipwright begins the 'procreation' of the ship by bringing together the keel's component parts using mortise and tenon connections. Prayers are said and incense is burned, as he cuts a mortise (a 'vagina') into the bow-end of the symbolically female keel base (Pelly 1977: 94) (see figure 1). He next makes the end of the bow-piece into a projecting tenon (a 'penis'), to fit into the mortise. The chips and shavings are saved, and will later serve to protect the vessel from danger.

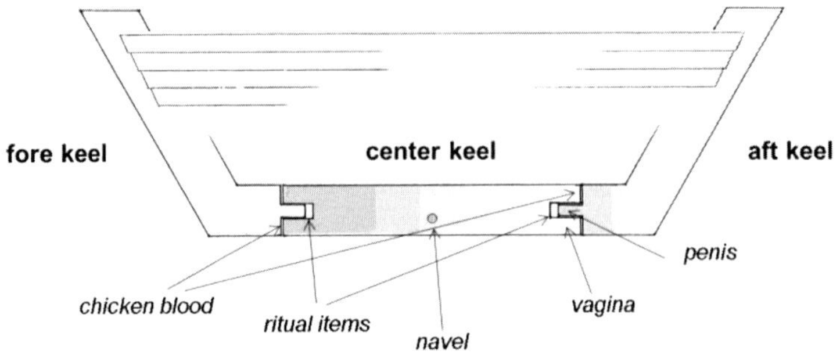

Fig. 1: 'Procreation' of the ship

Though some in Puger now deny it, before the mortise and tenon are joined, a small white cloth bundle is or was inserted into the mortise. It contains a bit of gold, a bit of iron for strength, a bit of copper, and some rice, coconut, and a leaf from the tree from which the keel base was made (Horridge 1986: 5).[1] When this bundle is in place, the tenon of the bow piece is forced in, after which the blood of a chicken is applied on the joint (cf. Pelly 1977: 94). A small ritual meal (*selametan*) of rice cakes concludes the activity.

The ('sexual') union of the keel is the moment of conception of the new vessel's spirit (*nyawa*) (Southon 1995: 100),[2] and the creation of the new spirit's home (Horridge 1986:10). Beside the insertions, the Bajau offer pepper-plant leaves, betel nut and lime, coins, and cigarettes at each of the joints (Stacy 2007: 51). Although the inserted items are now said to bring luck and avert misfortune (Rato 1991: 17; Arief and Abbas 2001: 50), it is likely that they are what brings the ship to life, and later give it an 'indwelling spirit' (Horridge 1986: 3).

God and Islam are also involved in these proceedings (Rato 1991: 25-7).[3] The shipwright may invite an Islamic scholar or a *haji* to recite prayers for the ritual's successful completion (Irawan 1982: 72). The Bajau indeed consider the ship to have been created by God, and the prophets are thought to reside

1 In Pesisir (East Java) *témpé* (fermented soybean cake) is included either among the items inserted in the mortise or in the ritual meal that follows, symbolic of femaleness or *yoni* (Rato 1991: 25). The pronunciation of *témpé* is close to that of *tempik* (female genitals), making this a word-play on the female nature of the keel.

2 Since this is when the ship comes to life, this spirit is equivalent to a person's *nyawa* (life spirit), not to be confused with its *semangat* or *ruh* (personality), which is imbued later (Wessing 2006: 72-73).

3 In Sulawesi a shipwright attempted to catch angels (Horridge 1985: 79), perhaps to use as life-giving sprits.

in the joints, and are appealed to in prayers during the offerings (Stacy 2007: 51-52): the ship is literally prayed together (cf. Errington 1996: 183-184). The joining of the keel is concluded with a small meal, the shipwright's share of which is offered to the spirits of the shipyard (Horridge 1985: 79), perhaps to assure their quietude.

The joining of the keel involves no expectations of *quid pro quo* or even relations with local or tree spirits, these having been dealt with beforehand. It is therefore not an appeasement ritual. Rather, the 'copulation' of the keel is surrounded with protective formulae, during which the required elements are brought into conjunction, resulting in the 'conception' of a new body, the ship, into which an animating spirit can later be imbued, either through an animal or human sacrifice, or through the invocation of God, the prophets, and the angels. Similar to the enlivenment of Buddhist stupas,[4] these invocations and prayers suffice to imbue powers and life into the structure: they animate the incipient vessel. It is not clear if the sacrificial animal is told what is expected of it, though elsewhere this is certainly the case (e.g. Schefold 2012: 193, 305). The ritual meal at the conclusion of the laying of the keel should perhaps be seen as a *sesajen*, welcoming the life-giving, animating spirit into its new home, one that will need to be watched over and strengthened (Horridge 1986: 3).

Barzanji and Songkubala

The keel having been laid, the ship now exists in essence, and only remains to be given a hull and superstructure: these are technical tasks, rather than ritual ones.[5] Once these tasks are completed, on the night before the ship is launched, a flurry of ritual activities takes place aboard it. The first of these, in Sulawesi preparatory to giving the new ship a 'navel', is a dual effort. One part is a recitation by community leaders and the building crew of the excellent qualities of the Prophet Muhammad (*barzanji*) (Pelly 1977: 98). This recitation evokes the presence of the spirits of Muslim saints and even that of the Prophet himself. *Barzanji* is part of a prayer for God's blessing (Ad Diba' 2012: 15-21), usually read at rites of passage. The man leading the recitation invites the men to stand and say the prayer for the Prophet (*salawat nabi*), blessing him and inviting him to be present and welcoming him (Al Barzanjie n.d.: 105-

4 Snodgrass (1985: 354) writes that in "bringing to life" Sinhalese Buddhist stupas, relics of the Buddha deposited in the stupa "works as if the treasure of the [...] [Buddha's] Dharma were deposited there". These deposits "enliven the stupa because they are imbued with the qualities [...] of the Dharma".

5 The planks with which the hull is built up are heated with fire to make the wood pliable and more easily bendable. In Puger a secret ritual (*upacara ageni*) is reportedly conducted near the fire, about which Rato (1991: 16) was unable to obtain further information. I was unable to verify this.

111). The men greet each other, and it is hoped that the Prophet will come or at least send his blessing. According to an informant, he is thought to be there, accompanying the person doing the greeting. This is followed by much praise for the Prophet. A *sesaji* of sweetmeats is placed aboard the boat to welcome the Prophet and the saints.[6]

Simultaneous with this, a traditional pre-Islamic Bugis ritual takes place, called *songkubala* (opposing danger; Pelly 1977: 98). While the men recite the songs of praise, a ritual specialist recites prayers or *mantra* while stirring a container filled with water and 'some leaves' or 'various magical ingredients'. When the men conclude the *barzanji*, the ritual specialist sprinkles water from his pot around the ship and over the vessel's captain and his crew, each of whom also drinks some of it (Arief and Abbas 2001: 50; Pelly 1977: 98). Afterwards all those present, hopefully including the Prophet and the saints, share a meal.

Neither of the reports on Madurese ship building (Irawan 1982; Rato 1991) mentions either *barzanji* or *songkubala*. However, an informant spoke of a *sandhingan*, which also is an invitation to spirit entities and Muslim saints to come. Irawan (1982: 74) only writes that prior to the ship's launching a *ruwatan* takes place, an exorcism to remove danger, thus creating the condition of safety, similar to the *songkubala*.

The descriptions of these two events do not mention the involvement of spirits, malicious or benign. The *barzanji* (or *sandhingan*) invites the Prophet, with the hope that this will benefit the proceedings. The participation of the notables and the saints, and the shared meal, demonstrate harmony, both within the community, and between it and Islam generally. The ritual meal that follows is a *sajian* (hospitality offering), and does not create an obligation, which informants said is impossible, though they have faith that the effects will be positive. The *songkubala* or *ruwatan* also makes no reference to spirit-beings. In combination with the invoked saints it creates a protective fence (*pagar*) that shields the ship and the participants from danger (cf. Wessing 2010).

Birthing the Ship

Within this ambience of safety, one of the most important transitions in the building process now occurs: the drilling of a hole through the ship's keel, allowing the vessel to be 'born' under the guidance of the shipwright, who in some places is said to act as both mother and midwife.[7] This is characterised as either cutting a new-born's umbilical cord, or as giving the ship a navel

6 Just like the Buddha and his teachings (Snodgrass 1985: 354), having the Prophet attend the event would be equivalent to the presence of Islam as such.

7 http://uun-halimah.blogspot.com/2007/12/perahu-pinisi.html (Last accessed 19 April 2014).

(*pusar*), which it must have at its birth, the launching (Horridge 1985: 79; Arief and Abbas 2001: 51).

The shipwright marks a spot at or near the centre of the keel. Some trays of offerings (*sesaji*) are placed around it. Then, having said a prayer, and using a special chisel or drill (Horridge 1985: 79; Stacy 2007: 52-53), he cuts a hole through the keel. The implication is that this hole goes straight through: the chips and shavings, "even those that fall under the boat" (Horridge 1985: 79), are collected and will later serve to protect the vessel during danger, much like the umbilical cord of a child is used when it is sick (Wessing 2006: 68-70). Others claim that the hole may run diagonally (Horridge 1985: 79), the angle predicting the life of the ship (Pelly 1977: 93, 99-100).

Southon (1995: 102-103), while calling it a navel (*lambe puse*), likens the hole to a vagina, and the drill-bit to a penis, their intercourse resulting in the birth of the ship. Both he and Stacy (2007: 53) claim that this is when the ship is 'given life', though both note that the navel is the point of attachment of the vessel's spirit (*semangat*), which is not the same thing as its life (*nyawa*). The infusion of *semangat* comes about through a blood sacrifice: the comb of a chicken is cut slightly, and its blood is allowed to drip onto the marked spot before it is drilled (Stacy 2007: 52-53) or through the hole after it is made (Pelly 1977: 100).

Southon (1995: 105) writes that in Buton the chickens are sacrificed after they have eaten from rice scattered along the keel. The sacrifice may also be a goat or a bull,[8] or, in the past, perhaps a human being. Horridge (1985: 80) cites an "early Spanish account [that] describes how in the Philippines a human prisoner was crushed under the advancing [boat] to increase the strength of its spirit", and Southon (1995: 104) writes about a boat's being launched over seven pregnant young women in "an attempt at the transference of human lives to inanimate things" and to assure good cargoes. These reasons are related since a ship without spirit will not to attract cargoes (Southon 1995: 102). Possibly in the past a human sacrifice was thought necessary for large, important ships, though this has long since been left behind, or been replaced with animals (Wessing and Jordaan 1997).

Both Southon (1995: 104) and Horridge (1985: 80) state that these sacrifices aimed to transfer "human lives to inanimate things" or to "increase the strength of its spirit". Note that this human sacrifice together with that of the chicken at the navel, the goat and/or buffalo at the launching, and perhaps parts of the ritual meal, replicates the set pointed out by Wessing and Jordaan (1997: 118-120). As was noted there, the elimination of human sacrifice due to religious change either left the actors with an incomplete set, or a changed set with a new principal offering.

8 http://id.wikipedia.org/wiki/Pinisi (Last accessed 19 April 2014).

With the infusion of its spirit, the vessel becomes a social entity and is sometimes said to promenade, strut, show off, and laugh (Manguin 1986: 196): it has *semangat* (personality). In Sulawesi at this point the 'supernatural residents' of the ship are fed (Pelly 1977: 100-101). This involves the slaughter and chopping up of a newly hatched chick, which is mixed with a chopped up banana flower (symbolically a heart). Dabs of this mixture are placed at the navel, the keel joints, and the mast, while the rest is scattered about the ship. This may be a feeding of the ship's *nyawa*, acquired at the laying of the keel, and its spirit, located at the navel. Whether this strengthens them is unclear.

In Puger the idea of an actual navel seems to be unknown. Informants in 2014 denied that a ship needed a spirit. Yet conceptually the navel is there (cf. Rato 1991: 44): shown a diagram of a fishing boat, an informant pointed at a spot near the centre of the keel as the location where the offerings should be placed, "forward and aft of the *pusar*" (navel), with yellow rice strewn along the keel, similar to the practice in Buton and reminiscent of the final ritual mentioned by Pelly. Yet the informant denied the sacrifice of a chicken. The yellow rice, often an ingredient in rice cones at rituals, combined with the *sandhingan* probably replaced the bloody sacrifice, in keeping with the emphasis on Islam. Rice is symbolically female, appropriate to the vessel's female keel, along which it is strewn. It also infuses and sustains a person's spirit (Wessing 2006: 64-67), casting doubt on the assertion that the ship does not need a spirit: the rice could simply be a substitute for previous sacrifices. An informant added that rice cones and *sesaji* are nearly the same: rice cones are considered Islamic, and can thus be seen as Islamic *sesajen*, inviting the presence of the ancestors and saints.

With the creation of the navel, the ship has become a place in the sense of Smith (1992) with the newly created animating spirit as its focus. No 'deal' is made with this spirit, and it is not appeased. In addition, God and the saints are invoked through a *sandhingan*, in which they join the builders and the owner in eating the rice cones at a ritual meal (Rato 1991: 18). The ship is now finished and is transferred to the new owner at a small ritual meal at which a religious leader says prayers. No spirits are involved in this event, which is primarily a formal and harmonious transfer of the vessel between the parties involved (Irawan 1982: 72).

Launching the Ship

Whether or not a spirit was thought to have been infused at the navel, one Puger informant still considered it necessary for the ship to be infused with life. Before it is launched, he explained, a banana plant, complete with bananas and a banana-heart, is tied to the mast, along with two ritual coconuts and a stalk of black sugarcane. These depict "a man's testicles and penis. The banana plant, symbolically female, gives the ship life". Later he said that the banana plant, coconuts, and sugarcane are only placed on the ship at its actual launching (cf. Rato 1991: 18), when it is married to its owner (Rato 1991: 25) or to the sea (Horridge 1985: 80): these ingredients are indeed symbolic elements at Javanese and Madurese weddings.

At the launching, the new owner slaughters a water buffalo to feed the helpers, its blood being "spread on the front section of the ship's keel" (Pelly 1977: 101), which may be a form of the sacrifice that infused a spirit in the ship, or an attempt to strengthen the ship's life-force (cf. Kreemer 1956: 183). In Puger a prayer is then said after which guests board the ship for its trial voyage. Back in port, the proceedings are concluded with a ritual meal during which the *tumpeng* used at the creation of the navel are eaten, and a communal prayer for blessing is said (Rato 1991: 18).

Conclusion

We see then that rather than forming a uniform pattern of a *quid pro quo* exchange, the rituals discussed here are solutions to problems, in which the shape of each is determined by both the nature of the question and the social context. The emphasis on Islam may have shifted the emphasis from sacrifice toward hospitality, except where this was impossible. This shift was facilitated by the nature of the rituals themselves: even in a non-Islamic context the infusion of life and a spirit were ideally based on invitation rather than coercion.

The most general context is indeed set by Islam, as can be seen in the appeal to various saints, both initially and during the *barzanji* recitation. The saints' blessings and a condition of safety having been obtained, the focus shifts to more specific problems such as appeasing the spirit owner of the trees to be felled. This is done in a business-like manner by giving them a small sacrifice for which a *quid pro quo*, e.g. permission, is expected. The possibility of the spirits' refusal does not seem to arise, which may say something about the superior position of human beings vis-à-vis spirits, especially from an Islamic point of view.

At the shipyard tutelary spirits are said to need offerings in order to prevent their mischief. Relations with them differ from those with the tree-spirits. All that is obtained from them is an absence of mischief, and they are *invited* to attend a ritual the aim of which is achieving the condition of well-being (Rato 1991: 15, 18). Following Kristensen (1961: 10), this stated goal, which includes specifying the entities one is to have peace with, makes the spirits' presence one factor among others that together define the condition of *selamet*: the spirits are active agents in their own pacification.

The laying of the keel and the creation of the navel differ from the appeasement ritual in that they recruit a spirit rather than pacifying one. In both operations the ship is animated. It is first given life through the copulation of the parts of the keel, and, when its 'body' is formed, it is given an animating spirit.

In both, the purpose of the ritual is stated in the presence of specified kinds and amounts of ingredients and guests (cf. Rato 1991: 17-21; Beatty 1999: 31-32). Together these define the goal to be achieved (Kristensen 1961: 181, 338). These rituals, then, bring about a desired end through the conjunction of specified components: they are performative in the sense of Austin (1975), and create the situation through appropriate actions.

The degree to which these rituals are still performed depends on local Islamic sensibilities. Combinations are possible, and in any case, the effect is much the same whether the ship is enlivened by a shanghaied spirit or by invoked saints. In the end, many would probably agree with an informant in East Java that "God, ancestors, and guardian spirits are probably all names for the same thing, and what you call them depends on your needs".

References

AD DIBA', A. (2012) *Diba'. Makna Jawa Pegon & Terjemahan Indonesia*. Surabaya, Al-Miftah.

Al Barzanjie, Ja'far (n.d.) *Terjemah Al Barzanjie.* Jakarta: Pustaka Amani.

ARIEF, M. and ABBAS (2001) *Pinisi: Perahu Khas Sulawesi Selatan*. Makassar: Proyek Pembinaan Peninggalan Sejarah Purbakala dan Permuseuman Sulawesi Selatan.

AUSTIN, J.L. (1975) How To Do Things With Words. In URMSON, J.O and SBISA, M. (eds.), *How To Do Things With Words*, pp. 1-6. Cambridge: Harvard University Press (2nd ed.).

BEATTY, A. (1999) *Varieties of Javanese Religion. An Anthropological Account*. Cambridge: Cambridge University Press.

DE JONGE, N. (2002*) The Religious Art of the Southeast Moluccas. Masterpieces from the collection of the National Museum of Ethnology and their Cultural Context*. Leiden: Digital Publications of the National Museum of Ethnology.

DOMENIG, G. (2014) *Religion and Architecture in Premodern Indonesia*. Studies in Spatial Anthropology. Leiden: E.J. Brill.

ERRINGTON, J. (1996) Lifting up the Prayer-Beam: a Javanese House-Building Ceremony. In HEADLEY, S. (ed.), *Towards an Anthropology of Prayer. Javanese Ethnolinguistic Studies*, 171-193. Aix-en-Provence: Publications de l'Université de Provence.

HORRIDGE, A. (1985) *The Prahu: Traditional Sailing Boat of Indonesia.* Singapore: Oxford University Press.

HORRIDGE, A. (1986) A Summary of Indonesian Canoe and Prahu Ceremonies. *Indonesia Circle* 39: 3-17.

IRAWAN, B. (1982) Perahu Tradisional Madura, Benda Koleksi Museum Sumenep. In MADURA V., *Kumpulan Makalah Lokakarya Penelitian Sosial Budaya Madura 1981*. N.p., Proyek Peningkatan Sarana Pendidikan Tinggi, Departemen Pendidikan dan Kebudayaan R.I.

KREEMER, J. (1956) *The Karbouw. Zijn betekenis voor de volken van de Indonesische archipel*. 's-Gravenhage/Bandung: Van Hoeve.

KRISTENSEN, W.B. (1961) *Symbool en Werkelijkheid: Godsdiensthistorische Studiën*. Zeist: W. de Haan/Arnhem, Van Loghum Slaterus.

LEHMAN, F.K. (1996) Can God be Coerced? Structural Correlates of Merit and Blessing in Some Southeast Asian Religions. In KAMMERER, C.A. and TANNENBAUM, N. (eds.), *Merit and Blessing in Mainland Southeast Asia in Comparative Perspective*, 20-51. New Haven: Yale University Southeast Asia Studies.

MANGUIN, P.-Y. (1986) Shipshape Societies: Boat Symbolism and Political Systems in Insular Southeast Asia. In MARR, D.G. and MILNER, A.C. (eds.), *Southeast Asia in the 9th to 14th Centuries*, 187-213. Singapore: Institute of Southeast Asian Studies/Canberra, Research School of Pacific Studies, Australian National University.

PELLY, U. (1977) Symbolic Aspects of the Bugis Ship and Shipbuilding. *Journal of the Steward Anthropological Society* 8(2): 87-106.

RATO, D. (1991) Prahu Madura: Ritual Pembuatan dan Simbolisme Hiasannya. Studi Kasus di *Puger dan Besuki*. Jember: Bidang Kajian Madura, Universitas Jember.

SCHEFOLD, R. (2012) *Wees Goed voor je Ziel. Mijn Jaren bij de Sakuddei*. Amsterdam: Nieuw Amsterdam.

SMITH, J.Z. (1992) *To Take Place. Toward Theory in Ritual*. Chicago: University of Chicago Press.

SNODGRASS, A. (1985) *The Symbolism of the Stupa*. Ithaca: Southeast Asia Program, Cornell University.

SOUTHON, M. (1995) *The Navel of the Perahu: Meaning and Values in the Maritime Trading Economy of a Butonese Villlage*. Canberra, Department of Anthropology, Research School of Pacific and Asian Studies, Australian National University.

STACY, N. (2007) Boats to Burn: Bajo Fishing Activity in the Australian Fishing Zone. Canberra, ANU E Press.

WESSING, R. (1978) *Cosmology and Social Behavior in a West Javanese Settlement*. Athens: Ohio University Center for International Studies, Southeast Asia Program.

WESSING, R. (2006) *A Community of Spirits: People, Ancestors, and Nature Spirits in Java*. *Crossroads* 18(1): 11-111.

WESSING, R. (2010) Porous Boundaries: Addressing Calamities in East Java, Indonesia. *Bijdragen tot de Taal-, Land- en Volkenkunde* 166(1): 49-82.

WESSING, R. and JORDAAN, R.E. (1997) Death at the Building Site: Construction Sacrifice in Southeast Asia. *History of Religions* 37(2): 101-121.

Historical Explorations

From Buffer State to Market Place: Migration, Ethnicity and Development in the Lao Border town of Müang Sing

VOLKER GRABOWSKY

Jos Platenkamp belongs to the very few scholars of Southeast Asian cultures who did intensive social-anthropological field work in both insular and mainland Southeast Asia. His increasing interest in the cosmological system of the lowland Lao of Luang Prabang since the late 1990s coincided with my own move from Vientiane to Münster in autumn 1999 when, as a trained historian, I embarked on an intellectually stimulating cooperation with Jos Platenkamp sharing his interest in the society and culture of the Lao people. It was the time when I had fallen in love with a Tai minority group in the far north of Laos, close to the Laos-China border.

Located in the extreme northwestern corner of the Lao People's Democratic Republic, the district of Müang Sing has become one of the main tourist destinations in landlocked Laos. Hardly accessible to foreigners for a long time, due to wars, revolutions and social dislocations that haunted the region especially during the Cold War period, this borderland has gradually recovered since the late 1980s. It belongs to an increasingly dynamic region known as the "Economic Quadrangle", linking Laos, China, Burma and Thailand (see Walker 1999). My first encounter with Müang Sing was in 1992 when I discovered a strange map in the Thai National Archives depicting the layout of a fortified walled town called "Chiang Khaeng", the construction of which had just been completed in 1885 (Grabowsky 1999: 288-289). When I moved to Laos in the summer of 1996 to teach at the newly founded National University of Laos in Vientiane, I soon got the chance to visit that place which is more commonly known as "Müang Sing". The remains of the old earthen wall were still visible, especially at its northern and eastern corners. At the time of my first visit in November 1996 the town of Müang Sing was still an agglomeration of four villages without electricity, except for the compound of the German Society for Technical Cooperation which had been running an "Integrated Food Security Programme" for the highland populations in the district of Müang Sing at that time. Some eight years later, in March 2005, the town had become more urban with more than a dozen hotels and guest houses, a few banks and a number of shops, sprawling along the main street leading to the Lao-China border ten kilometres further north. Almost all houses now had electricity supplied from Müang Phong in neighbouring China, and satellite dishes could be seen everywhere. Another eight years later, in February 2013, the process of transformation from a cluster of villages to a small urbanised/urbanising centre for businesses, tourism and services had made significant progress and seemed almost completed at the time of my most recent field (February 2016), with numerous new shops, production sites, restaurants, hotels and even several entertainment venues built in the town.

As recent sociological and anthropological research could demonstrate, the dynamics of occupation and appropriation of new social and economic places by displaced migrant populations has led to the emergence of new towns in many border districts of northern Laos.[1] Müang Sing, I would argue, is definitely one of these towns. In my paper I will explain from a historical perspective how processes of migration and inter-ethnic relations had a formative influence on Müang Sing and the surrounding countryside and how they had an impact on social and economic development. My analysis is largely based on local chronicles and historical accounts reflecting the indigenous point of view, or at least the perspective of those members of the local elite who were able to express their ideas about the past, present and future of their particular *müang* – to use the traditional Tai term for polity. Apart from the various extant versions of the Chiang Khaeng Chronicles (Grabowsky and Renoo 2008; Lafont 1998), I heavily rely on a history of Müang Sing written from 2002 until 2004 by Mai Thamdi Thammavong, during his tenure as acting district chief of Müang Sing. Mai Thamdi's still unpublished manuscript contains many insights into the political developments in Müang Sing during the Indochina wars (1945-1975) and after the victory of the Communist forces. Though the author remains largely loyal to Party and State, he is by no means uncritical of certain measures undertaken by the Pathet Lao movement. Other sources on which I rely are official statistics and field data taken by myself and other scholars – such as Paul Cohen and the late Houmphan Rattanavong – over the last two decades. I will first provide some basic information on the ethnic makeup of Müang Sing district, then discuss how the migrations and inter-ethnic relations influenced political developments in Müang Sing since the late nineteenth century. Finally, the dramatic socio-economic changes of the last twenty years will be analysed against the background of demographic dynamics.

1 See, for example, the research by the French scholar Vantina Bouté who is currently studying the social dynamics of the development of medium-sized border towns in northern Laos, giving special emphasis on the emergence of new actors and social groups who are taking advantage of opportunities pertaining to the socio-ecnonomic development of these localities. See the website of Seatide (Integration in Southeast Asia: Trajectories of Inclusion, Dynamics of Exclusion): http://www.seatide.eu/?content=researcher-staff.

Ethnic Diversity in Müang Sing

Not much is known about the demographic history of Müang Sing and its ethnic make-up prior to the nineteenth century. The fertile rice-growing plain of Müang Sing is watered by the Nam Sing and Nam Dai rivers and covers an area up to 25 km in length and 19 km in width. Müang Sing was once a dependency of Chiang Khaeng. This small principality which extended over both banks of the Mekong River was founded probably in the late fourteenth or early fifteenth century. Its first capital was built in the vicinity of present-day Ban Chiang (Siang) Khaeng, a Tai Lü village situated on the left (Lao) bank of the river. Four stupas, now in ruins and overgrown by jungle, are witnesses of Chiang Khaeng's former splendour and prestige in the sixteenth and seventeenth centuries when Chiang Khaeng was the centre of a Burmese military zone in the upper Mekong valley. The Chiang Khaeng Chronicle (Wat Ta Pao version) mentions Müang Sing as one of 20 village clusters belonging to the core area of the Chiang Khaeng polity during that period (Grabowsky and Renoo 2008: 16, 111-112). Local tradition has it that for most of the seventeenth and eighteenth centuries Müang Sing became abandoned by its inhabitants, due to continuous raids from rival polities in the south.

The same oral tradition has it that due to the initiative of Nang Khemma, a woman of aristocratic background from Chiang Khaeng, the valley of the Sing River was repopulated in 1792. Müang Sing prospered again, but less than half a century later the bulk of its population had disappeared. Most people were deported to and resettled in the Tai Yuan principality of Nan. Large-scale resettlement campaigns by the ruler of Nan against Müang Sing and other smaller *müang* in the region were carried out in 1805/06, 1812/13 and, again, in 1838. For the next forty years the plain of Müang Sing laid waste and abandoned. Siamese archival documents indicate that after a futile attempt in 1866/67, Kòng Tai, the ruler of Chiang Khaeng at that time, started preparations for the refounding of Müang Sing in 1877/78.[2] From far and near people were moving into the plain of Müang Sing. The bulk of the immigrants were either Tai Lü from Müang La and Müang Phong in Sipsòng Panna[3] or a mixture of Tai Lü, Tai Nüa, and Tai Khün people from Chiang Tung. Among these three Tai speaking groups the Tai Khün were the smallest. Most of them settled in Chiang Kok, Müang Kang and Müang Lòng (nowadays Müang Lòng district) in the valley of the Ma river which connected the plain of Müang Sing to the small commercial port of Chiang Lap, an isolated outpost of Chiang Tung on the right bank of the Mekong River.

2 Lacroze (1994: 6) believes that a famine which afflicted Siang Khaeng [here: Müang Yu] in 1877 caused the population movement to Müang Sing.

3 As for the relationship between Sipsòng Panna and Chiang Khaeng, see Liew-Herres, Grabowsky and Wichasin 2012; Yanyong and Ratanaporn 2001.

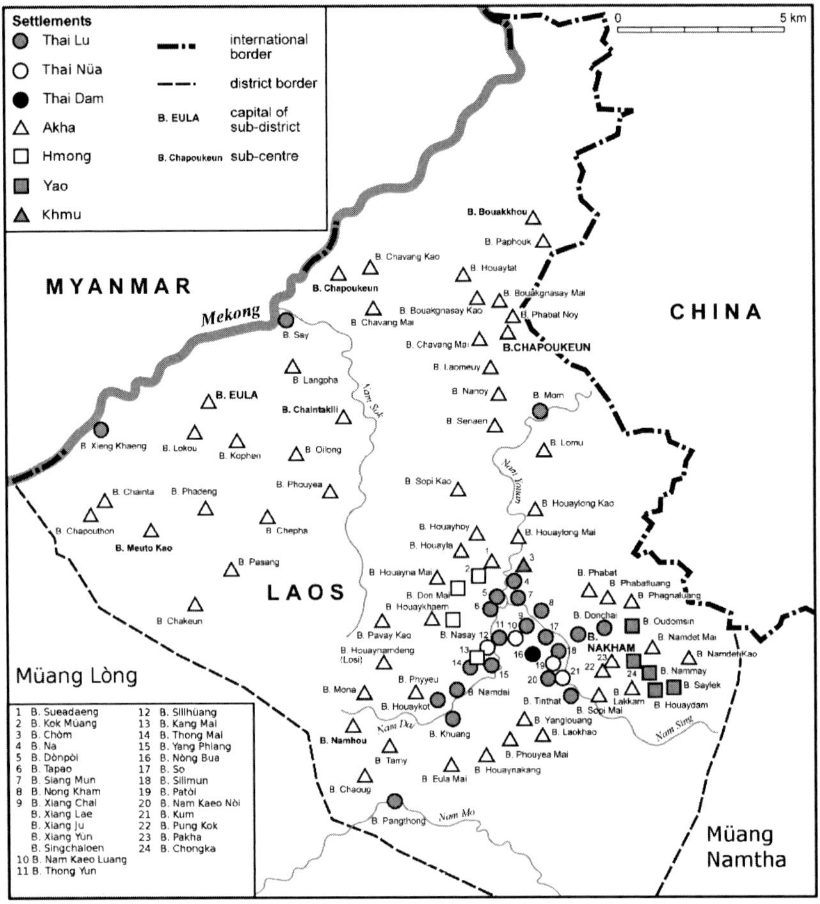

Fig. 1: Ethnic Groups in Müang Sing District (c. 2013)
© Volker Grabowsky

Tai Nüa. Among the first settlers who arrived in the plain of Müang Sing were a group of people called Tai Nüa or "Northern Tai". Some Tai Nüa probably arrived as early as 1878 (Saimöng 1981: 270). According to local tradition, the Tai Nüa of Müang Sing originate from Chiang Ku (Jinggu) and Müang Lò in the Szemao (Simao) district, which is situated to the north of Sipsòng Panna in southern Yunnan (Wityada 2003: 25). They had fled this district not long before 1878 when Chinese troops were scouring the borders in search of remnants of the Taiping Rebellion (1851-1864).[4] The ruler or *chao fa* (lit., "Lord of Heaven") of Chiang Tung, who granted the Tai Nüa refugees shelter in his territory, regarded these brave and industrious people as ideal settlers for the

4 "Mission du Haut Mékong …", CAOM, GGI, No. 20162. The Taiping rebellion was led by a Hakka named Hong Xiuquan (1814–1864) who founded the "Heavenly Kingdom of Great Peace", called *Taiping Tianguo*.

deserted plain of Müang Sing (Ganesco 1904: 5). Their main village was Ban Nam Kaeo Luang, situated on the outskirts of Müang Sing town. It was in this relatively densely populated village that the British troops, who temporally occupied Müang Sing in 1895/96, established their camp.

Comprising roughly 550-600 persons in 1896/97, at the beginning of French rule, and almost 1,000 persons in the early 1920s, the Tai Nüa formed a strong element among the lowland population. In spite of the pressure exerted to assimilate them into the Tai Lü mainstream, the Tai Nüa of Müang Sing have preserved their language and culture until the present day. Most of them are bilingual (Tai Nüa and Tai Lü), and with pride they preserve old mulberry paper manuscripts, which are written in a script completely different from the Dhamma script of the Tai Lü but almost identical to the script used in Dehong (Yunnan). The Tai Lü variant of the Dhamma script is also used for inscribing Buddha images. Of the thirty-two ethnic Tai villages that are in the plain of Müang Sing, five villages are still inhabited by Tai Nüa.[5] According to the latest census, Tai Nüa make up one-twelfth of the total and one-fifth of the Tai population in Müang Sing.

Tai Lü. They inhabited almost all river valleys in "Western Chiang Khaeng" and also formed the bulk of the settlers in the plain of Müang Sing. As one British visitor observed, in the early 1890s "a great part" of the plain was "not yet cleared of jungle".[6] In other words, Müang Sing still possessed a large potential for further agricultural and demographic expansion. Only a few years later, Pierre Lefèvre-Pontalis felt overwhelmed by "rice fields show[ing] themselves everywhere" when he entered "the splendid plain of Muong Sing" (2000: 296). The Tai Lü settlers who, along with their Tai Nüa brethren, were gradually transforming the landscape of Müang Sing, belonged to three different groups:

a.) The first group arrived at Müang Sing together with Chao Fa Sali Nò Kham when he transferred his capital from Müang Yu. They came from areas in "Western Chiang Khaeng".

b.) The second group were descendants of former inhabitants of Müang Sing who had been deported to Nan during the first half of the nineteenth century. Until 1891, only about one-fourth of the 2,000 captives had returned and settled down in Müang Sing. Plans to repatriate one-third of the remaining 1,500 persons in 1893 were rejected by the Siamese Ministry of the Interior (Grabowsky 1999: 248).

5 Apart from Ban Nam Kaeo Luang, these are the villages of Ban Silihüang, Ban Kum, Ban Thong Mai, and Ban Pang Kham (Ban Pa Tòi). The first four villages did already exist by the turn of the twentieth century with almost half of the population living in Ban Nam Kaeo Luang, with 260 inhabitants one of the largest villages in the plain of Müang Sing at that time. See COAM, GGI No. 20162 (section: Territoires de Muong Sing).

6 Scott 1891: 18 [British Library, IOOC].

c.) The third and by far largest group consisted of Tai Lü who immigrated from adjacent districts of Sipsòng Panna, notably from Müang Phong and Müang La (Warry 1895: 13). Oral tradition says that some migrants came from places even farther away; migrants from Müang Ham founded the village of Ban Yang Piang, and Ban Hua Khua was founded by people from Müang Kai. Voluntary migration from these places continued during the French period. Political instability and dynastic feuds were the main reasons for people to move from Sipsòng Panna to Müang Sing. In 1905 alone, a single group of 300-400 people from Müang Phong asked the French authorities for permission to settle down in Müang Sing.[7]

In the course of the twentieth century, a large number of Tai Lü, along with members of other ethnic groups, fled to Huai Sai and to the area of Chiang Khòng in northern Thailand during the Lao civil war in the 1960s and 1970s (see next section), while the political disturbances in China during the "Great Leap Forward" (1958) and the "Cultural Revolution" (1966) triggered a wave of Tai Lü refugees from Sipsòng Panna to Müang Sing – who either took the direct route via Müang La and Müang Phong or the route via Müang Yòng and Chiang Lap in Burma. These more recent migrations changed the composition of the Tai speaking population in the plain of Müang Sing significantly.

Non-Tai groups. As I have argued in a previous article (Grabowsky 2003), "Kha" highlanders[8] and "Tai" lowlanders formed a symbiotic relationship based on both ritual and economic exchanges. In Chiang Khaeng as well as in other Tai polities of the region, the Kha possessed ritual functions as lords of the territorial guardian spirits and held a strong economic position. They controlled precious forest products, such as elephant tusks (ivory), rhinoceros horn, incense, honey, opium and cardamom, which were sent as tribute to royal courts or were exchanged for cloth, knives, pottery and other consumer goods. They also controlled natural resources such as iron and copper mines. The Chiang Khaeng Chronicle[9] reports two armed conflicts between the Kha and the ruling Tai elite – one in the early seventeenth and one in the late eighteenth century – providing rare evidence from indigenous sources of the Kha's political and economic strength.[10] Indigenous sources rarely mention specific ethnic groups among the highland population and prefer generic terms such as "Kha" or "Lua". An undated Tai Lü manuscript from the plain of Müang Sing

7 "Rapport de l'administrateur Sérizier sur le territoire de Muong Sing – les états Shan britanniques et les Sip-Song Pan-na chinois", CAOM, GGI, No. 22299.

8 'Kha', literally 'slave, servant', is a generic term used by the politically dominant Tai to denote a variety of ethnic groups living in the highlands, disregarding their specific languages and cultural traditions.

9 The various versions of the Chiang Khaeng Chronicle are discussed in detail in Grabowsky and Renoo 2008: 59-64.

10 Chiang Khaeng Chronicle-Wat Ta Pao version, ff. 9/3-9, 11/14-12/4, cf. Grabowsky and Renoo 2008: 99-100, 103-104; cf. Grabowsky 2003: 114-115.

mentions, in the context of manpower mobilisation, the following hill peoples: Kò Dòi (Akha), Yao, Maeo (Hmong), Lantaen (Lantèn), Kui, and Musoe (Lahu).[11] Among these groups, the Akha were by far the most numerous, followed by the Yao and, lagging far behind, the others. An early French visitor made the interesting observation that the Akha, who even surpassed the Tai Lü in numbers, were considered as the clients of the highest Tai Lü authorities, whereas the Yao, along with the Hmong, were attached to the chief of the Tai Nüa, the second largest Tai group (Raquez 1902: 296). Thus, in the hierarchical multi-ethnic polity of Chiang Khaeng, the various hill peoples sought protection from different leaders of the Tai ruling class.

Among all highlanders, the Akha had the best reputation as hard-working men. Whenever laborious and unpleasant work had to be done in the plains, the Tai authorities recruited the Akha to do it for them. It is reported that the moat surrounding Müang Sing "was entirely dug by them, and dug partly in the unhealthy season, when the mortality was very great" (Warry 1895: 25). Cohen points out that even nowadays the Tai of Müang Sing still "depend almost entirely on the Akha for the heavy task of clearing land for new rice fields and for the digging of ditches" and the Akha are also employed for the transplanting and harvesting of rice. The Akha thus supplement the traditional labour recruitment of the Tai based on cooperative exchanges among family members and neighbours (Cohen 2000: 193).

> "While the Akha are numerically superior in population [...], the Tai lowlanders have dominated them politically and economically since the founding of the principality of Muang Sing in the 1880s" (Cohen and Lyttleton 2002: 3).

The Akha were called Kò or Kha Kò by their Tai neighbours. By the late nineteenth century they were widely spread over most of the hills in the Upper Mekong valley as far south as Hòng Tai Lük (near Chiang Saen?). They lived almost entirely on the mountains separating the plain of Müang Sing from the Mekong, whereas the Yao population dominated the hills close to the border to Müang Phong (Sipsòng Panna). In the mountainous areas of the Müang Lòng district, Akha villages were mixed with settlements of Lantèn, Lahu, and other groups. In the market of Müang Sing, situated outside the town, two hundred metres from the southern gate, Chinese caravan traders and numerous merchants from diverse ethnic groups came together every five days to exchange goods (McCarthy 1900: 160).

Though the Tai Lü formed the majority of the population in the lowlands, other Tai groups, such as the Tai Nüa and Tai Khün, also played an important role in the society and economy of Chiang Khaeng. Half of the population

---

11 The manuscript is entitled *Paweni hit kòng chao fa sali nò kham moeng sing*. [The customs and rules of Chao Fa Sali Nò Kham of Müang Sing. The English title is used in the following footnotes.] Presently it is in private hands in Ban Na Kham, a village situated close to the town of Müang Sing. The mention of the hill peoples is on f. 23/4.

were highlanders and they substantially contributed to the local economy as producers of cotton, salt, and iron. Moreover, they provided manpower and tribute and paid taxes that were needed to maintain the *chao fa*'s state apparatus. The various tribes living in the highlands, mostly known under the generic name "Kha", were organised in separate administrative units called *buak* and *tò*, the function of which, however, is not entirely clear and still needs to be studied in depth. The supreme administrative council, called *nüa sanam*, supervised the central government in the capital, the surrounding villages in the large and the fertile plain of Müang Sing. Furthermore, it coordinated the administration of the more distant *müang*. Among these *müang*, those situated in the valley of the Ma River, i.e., Müang Nang, Müang Kang, and Müang Lòng (in today's Müang Lòng district), were strategically most important as they linked up the post-1884 core area of Müang Sing with the outlying *müang* on the west bank of the Mekong River. This is reflected in contemporary Siamese archival sources usually listing the *müang* of Sing, Nang, Kang, and Lòng as one inseparable unit: Sing-Nang-Kang-Lòng.[12]

Siamese officials who visited Müang Sing in 1891 estimated the total number of Tai speaking households in the plain of Müang Sing and other lowland areas at "more than 600". They were outnumbered by the more than 1,000 households of highland groups. These figures include the present-day district of Müang Lòng which was, compared to Müang Sing, much less populated. More reliable and precise figures come from population statistics provided by district authorities and the German Society for Technical Cooperation (Gesellschaft für Technische Zusammenarbeit – GTZ) in the mid-1990s. The ethnic composition of Müang Sing district at that time was as follows:

12 See, for example, Thai National Archives, R.5 M.58/194; R.5 M.58/195, and R.5 M.58/196.

ethnic group	*1* *number of villages*	*2* *number of inhabitants (absolute)*	*3* *number of households (absolute)*	*4* *inhabitants / households*	*5* *number of inhabitants (in %)*	*6* *inhabitants (in %) GTZ- data*
Akha	72	10,756	1,976	5.44	46.29	50
Tai Lü	26	7,555	1,558	4.85	32.51	28
Tai Nüa	5	1,729	326	5.30	7.44	8
Yao	5	755	288	2.62	3.25	4
Tai Dam	1	242	88	2.75	1.04	1
Lao Sung (Hmong)	2	2,091	344	6.08	9.00	8
Lao Thoeng (Khmu)	1	109	23	4.74	0.47	0
Total	112	23,237	4,603	5.05	100.00	99

Table 1: Müang Sing – Population and Ethnic Groups (c. 1995)[13]

It is interesting to note that in Müang Sing district all Tai speaking ethnic groups combined are far outnumbered by the Akha, the largest highland group who make up almost half of the total population. All Tai villages are situated in the plain of Müang Sing, except for three isolated Tai Lü villages which are situated along the east bank of the Mekong River: Ban Sai, Ban Bò, and Ban Siang Khaeng. The last settlement was the old capital of the principality of Siang Khaeng that once extended to territories on both sides of the river. It is difficult to reach via land. The fatiguing march from Müang Sing to the old capital lasts not less than two full days, passing through various Akha villages.

Müang Sing: Historical Developments

Chiang Khaeng was a small and sparsely populated multi-ethnic polity. Ruling over less than 20,000 subjects of whom only one half were Tai Lü or Tai Nüa, the *chao fa* of Chiang Khaeng presided over a small and manageable state

13 Apart from the last column, all figures were kindly provided by the district administration of Müang Sing. Columns 4 and 5 were calculated by the author. The data is from the GTZ sponsored "Müang Sing Integrated Food Security Programme." They are based on data provided by the district office (from 1995). It is unclear why columns 5 and 6 differ. Bold letters indicate groups belonging to the Tai-Lao language family. Cohen (2000: 180) mentions only 68 Akha, but 3 Hmong villages.

apparatus. Born in 1846,[14] Chao Fa Sali Nò Kham was a strong and charismatic personality. Apart from the mythical founding prince of Chiang Khaeng, Chao Fa Dek Nòi (Grabowsky and Renoo 2008: 21-32), the refounder of Müang Sing is considered the most important ruler in the history of Chiang Khaeng. In indigenous as well as Siamese and Western sources he is portrayed as a just ruler who was close to the people and their needs. To a certain extent, this picture might be a reflection of nostalgia. However, epigraphic evidence suggests that the *chao fa* was not only concerned about the political, social, and economic issues but also with the spiritual well-being of his subjects. He sponsored the construction of monasteries and built close relations with the local Sangha, the order of monks and novices. Several inscriptions on Buddha statues, found in the region of Müang Sing and documented by the author of this article, mention the *chao fa* as the main initiator of the donations. Sometimes the ruler himself carved a Buddha statue, such as is stated in an inscription (LNT, No. 121, Grabowsky collection) documenting the donation of a wooden Buddha image at Wat Siang Chai or Wat Luang, the main monastery of Müang Sing town, just next to the royal palace:

1. In the *rawai set* year, CS 1248, in the first month, on the full-moon day, on a Wednesday, the Tai [say] a *dap pao* day,[15]
2. the royal faithfuls, Phaña Luang Caiwong, who has the highest rank,
3. and all his children and grandchildren came [to order] the casting of this Buddha image
4. in order to protect the religion until its end after five thousand years. [The image] is donated
5. to visit his own elder wife, I Khan Kaeo, in heaven. *Sudinnaṃ*
6. *vatta no dānaṃ nibbāna paccayo hontu no nicaṃ dhuvaṃ dhuvaṃ dhuvaṃ.*[16]

During the two decades of his rule (1881–1901), Chao Fa Sali Nò Kham oversaw the transfer of the capital from Müang Yu to Müang Sing (1885/86), the recognition of Siamese suzerainty (1889), the partition of his realm by British and French imperialist powers (1896), and the integration of the eastern portion into French Indochina. For three years, between 1893 and 1895, the two main European colonial powers attempted to create a buffer state along the eastern and western banks of the Upper Mekong in order to separate the frontiers of their Southeast Asian domains by a few dozen miles. The core of this

14 Chiang Khaeng Chronicle-Wat Ta Pao version, f° 70/12, see Grabowsky and Renoo 2008: 170.

15 1248 Karttika 15 = Thursday, 11 November 1886. The nearest *dap pao* day was 1248 Karttika 5 = Monday, 1 November 1886.

16 "Well donated is our gift. May it become a condition for [the attainment] of nirbbāna. [May it serve] us continuously and for evermore."

"buffer state" or "état tampon" would have been the principality of Chiang Khaeng. As British and French designs on such a buffer state were incompatible, the secret negotiations eventually failed. Subsequently, the principality of Chiang Khaeng was divided along the course of the Mekong. Within a few years the principality of Chiang Khaeng, which had managed not only to survive but even to prosper in the flexible framework of traditional patterns of tributary relations, ceased to exist when the "geo-body" of the colonial and post-colonial states had to be defined along clearly marked borderlines. The western, more populous part of Chiang Khaeng (8,000 km^2) fell under British suzerainty while the eastern part, including the capital Müang Sing, came under French control (6,700 km^2).[17]

At the beginning of Chao Ong Kham's rule in 1901, the French restructured the inflated local administration along functional lines. Departments of Agriculture, Finance, Justice, Military Affairs, and Public Works were formed. Furthermore, the number of officials (*tao khun*) heading these Departments was reduced from seventy to only twenty-five, including the eight members of the *hò sanam*. Along with these measures fiscal and judicial reforms were introduced to narrow the differences with the rest of French controlled Laos.[18] Thus French policy contributed to an increase in tensions in Müang Sing, even more so since resentment against French rule was already simmering before the implementation of the most far-reaching reforms. In November 1900 a group of Tai Lü from Müang Phong (Sipsòng Panna), altogether sixty to seventy men who 'were taking weapons under their bags', arrived at the market of Müang Sing and tried to stir up the local population against the hated "Kala" (Europeans). The secret plot was discovered and the French mastered the situation. Müang Sing remained an autonomous principality under French protection. In the end Mòm Ong Kham tried to seek independence from France with the support of Chinese armed groups who, in 1914, had invaded and made insecure large parts of northern Laos. On 6 April 1916, when the French toppled the *chao fa* who was accused of being "guilty of common law offenses and political crimes"[19], the last ruler of Müang Sing, who had fled to Sipsong Panna, and his heirs lost all their privileges.

As a consequence, France abolished the special status of Müang Sing as a protectorate and integrated the region into the province of Haut-Mékong with Luang Namtha as it centre. In 1941, this province was integrated into the

17 Grabowsky and Renoo 2008: 2.

18 For details about the administrative and financial restructuring, see Chiang Khaeng Chronicle, Vat Ta Pao version, ff° 72/1-73/9, cf. Grabowsky and Renoo 2008: 171–73; Chiang Khaeng Chronicle, Izikowitz version, ff° 69/4-70/10, cf. Lafont 1998, 250–53. See also: 'The chronicle about how …', f° 7/13-17, cf. Grabowsky 1999: 270.

19 The arrêté signed by the Résident Supérieur du Laos notes, '[le] Chao-Fa Ong Kham est déclus de tous les titres, droits et privilèges qu'il tenait du Gouvernement Français. Cette déchéance s'étend à ses héritiers'. Quoted from *Bulletin Administratif du Laos*, 1916, p. 178.

Kingdom of Luang Prabang. The southern and western parts of Müang Sing, with a number of Tai Lü settlements stretching out into the Ma River valley, became a separate district called Müang Lòng. French troops were permanently stationed in Müang Sing. Near the southern corner of the town the French erected a military fort, destroying the wall in that sector. Moreover, the colonial administration built an asphalt road from Müang Sing to the Chinese border. Parts of the road were running very close to or directly on the southeastern front of the town wall. On the outer side of the "China Road" the French built a market and behind that market a new brick wall. The "urban" population shifted gradually towards this road, the town's economic lifeline. Although the market and the brick wall behind were completely destroyed during the fighting between Chinese Kuomintang troops and Franco-Lao forces in 1945/46, the vital importance of the "China Road" did not decline. Since 1954, the newly built market of Müang Sing has again been attracting traders from far and near. During the first decades under French rule Müang Sing expanded. In addition to the first monastery, Wat Luang Ratchathan, also called Wat Chiang Cai, three other monasteries were founded within the town wall. They were all placed in symmetric rotation to Wat Luang (in relation to the centre of the town).

After the dismantling of the French colonial forces in Indochina by the Japanese in the spring of 1945, the French-installed administration in Müang Sing did not collapse immediately. However, the power vacuum following the capitulation of Japan in August 1945 was filled by the Kuomintang regime which sought to expand Chinese influence throughout northern Laos and even annex the whole region of Müang Sing. Mai Thamdi's history of Müang Sing contains a long chapter providing interesting details about how the Kuomintang occupying forces, comprising some 50 well-armed soldiers who were stationed outside the town at Ban Hua Khua, were chased away in a people's uprising under the leadership of Mai Yawong (1893-1977), a local nobleman who held the position of district chief of Müang Sing at the time of Chinese occupation. Mai Thamdi, a former Parthet Lao cadre, argues in his history of Müang Sing that this victory was only possible through the cooperation of all ethnic groups and social classes: "Wherever there is solidarity, victory will be achieved" (Mai Thamdi 2002/04: 62). Between 1946 and 1953 a French military post was consolidated in Müang Sing and a Lao military company was installed. This company was led by four French officers and corporals, 2–3 ethnic Lao officers, one Khmu officer and soldiers from various ethnic groups, including Tai Dam (from Hua Phan province), Tai Lü, Khmu, Akha, and Chinese from Yunnan (Chazee 1999: 4). From 1953 until 1960, the French administrative official Pierre-Bernard Lafont, who later became an influential scholar of Lao Studies in France, lived in the town of Müang Sing. He provided one of the rare first-hand accounts about the town and the surrounding countryside during the era of the Royal Lao Government (RLG). In 1960, the cluster of villages which formed the town of Müang Sing comprised 204 households with 980 inhabitants. They were almost all Tai Lü, apart from 5–6 Tai Khün monks and businessmen and 5–7 Chinese caravan traders. At that time the plain of Müang Sing was inhabited by roughly 1,500 Tai Nüa and 4,500 Tai Lü, whereas the vast majority of the highland people were Akha, except for a few Yao and Lanten villages and one "Kui" (presumably Khmu) village (Lafont 1973).

Among the half dozen Tai Khün living in the town was an affluent merchant from Chiang Tung called U Mao who at the age of fifteen had come to Müang Sing together with his mother, a widow, and his younger brother. Over the years U Mao, born in 1894, became one of the richest traders in Müang Sing whose influence was not only based on his business network but to a much larger extent relied on his cultural capital as he made numerous substantial donations to monasteries over a period of fifty years, from 1932 until his death on 19 October 1982. In recognition of his meritorious deeds, King Sisavong Vong bestowed upon U Mao the honorary title "Phaya Sekong" (Houmphan et al. 2006: 117). Unlike many supporters of the royalist region, U Mao alias Phaya Sekong did not leave Müang Sing after the Pathet Lao seized the town in 1962 nor did he flee the country after the founding of the Lao P.D.R. in December 1975. On the contrary, he continued his religious donations (sponsoring the copying of manuscripts and donating them to monasteries, financially supporting the construction or restoration of stupas and Buddhist temple buildings, etc.), as well as his philanthropic projects (such as the building of schools and roads) even in the economically difficult years of the late 1970s.[20] A diary written on a mulberry paper manuscript in concertina format writes about Phaya Sekong's life, starting in Tham-Lao script (on page 74):

> "Phò Thao Mao passed away in the *tao set* year, [CS] 1344, on the third waxing day of the first month[21] at five o'clock [in the afternoon] when his heart stopped to beat. His original name was U Mao, his new name was Phaya Sekong. His original home was Chiang Tung; together with his brother he came to Müang Sing. U Mao came to Müang Sing as a novice at the age of fifteen. [His younger brother] U Yi was still a child of twelve when he arrived. U Mao and U Yi came [to Müang Sing] in the year CS 1266 (AD 1904/5)."

and continuing in modern Lao script (on page 75):

> "The house and the land of Phaya Sekong were sold by Bun Wichai to the state [owned] Lan Xang Bank. Lan Xang Bank handed it over to the Cultural [Office]."

20 This manuscript, starting with an entry on 15 October 1932, is now kept by Nan Cai Saeng (Ban Nam Kaeo Luang), a retired government official who had been director of the district museum until 2009.

21 1344 Karttika 3 = Tuesday, 19 October 1982.

In the early 1960s, Müang Sing turned into a battlefield when the RLG came under attack by pro-communist Pathet Lao troops who eventually conquered the town on 5 May 1962. During this attack the monastery of Chiang Yün, close to the French fortress, was destroyed. The forces loyal to the regime in Vientiane had to withdraw to Huai Sai on the Mekong opposite the Thai port of Chiang Khòng. The Ban Wiang Nüa version of the Chiang Khaeng Chronicle written by Nòi Insongkariyawong in the late 1980s provides the best summary of the immediate consequences of the Pathet Lao victory:

> "Müang Sing was liberated by the Lao Patriotic Front[22] on 3 June 1962. At that time the following officials, military officers, and police officers were serving in Müang Sing: 1. Administration: Mr. Chanpheng Maniwong was the chief of the district of Müang Sing.[23] 2. Military: Captain Lagui with 25 soldiers. 3. Police: Warrant Officer Inthoeng Lakkhamkaeo with fewer than 10 policemen. In addition there was Cheng Seng with fewer than 10 remnants of the Chao Kocek (Chiang Kaichek) troops. All three groups fought against the Lao Patriotic Front unsuccessfully; they fled to Ban Khwang and from there they withdrew until they reached Müang Huai Sai. Joining their forces they reorganised the administration in the territory of Müang Huai Sai up to Ton Phoeng at Müang Müng [near the confluence of] the Nam Yu river [and the Mekong River]."[24]

According to a "History of Müang Sing" composed in August 2002 by Mai Thamdi, a local historian from Müang Sing who became district chief in the 1990s and retired in 2004, the Pathet Lao leaders in Müang Sing committed "ultra-leftist mistakes" provoking the flight of thousands of Tai Lü and Tai Nüa villagers to RLG-controlled areas along the Mekong valley and to Thailand (Mai Thamdi 2002/04: 53). Though hundreds of Tai Lü families from Sipsòng Panna – mostly from the districts of Müang Phong, Müang La, and Müang Long – fled to Müang Sing after the Great Leap Forward (1958) and, once again, at the height of the Cultural Revolution (1966/67), Müang Sing suffered from a net loss of inhabitants during the Lao civil war which ended in 1973.[25] Moreover, the rude and arrogant behaviour of the local Pathet Lao leadership, heavily influence by Maoist ideology, also drove the Yao villages near the border to China into open rebellion and directly into the arms of the RLG who armed the Yao with American weapons.

---

22 In Lao: *Naeo Lao Haksat*, the mass front organisation led by the Communist Lao People's Revolutionary Party.

23 The Lao term *chao müang* is now used to designate the chief of a district during French colonial rule.

24 Chiang Khaeng Chronicle, Ban Wiang Nüa version, ff° 106/10-17, cf. Grabowsky and Renoo 2008: 207-208.

25 Cohen (2000: 190) arrives at a similar conclusion stating that the loss of population which resulted from the mass exodus of Tai lowlanders to royalist strongholds in Bokaeo province were "only partly compensated by the influx of Lue from Sip Song Panna as a result of the collectivisation campaign in China."

> "The chief of the [Pathet Lao] administration at that time, Mai On, a native from Müang La [in Sipsòng Panna], exercised power through the gun barrel menacing the people of Müang Sing. Un Nok Kho was the head of the police. [...] The police chief and Mai On from Müang La were the people who applied the ideology of the Chinese Cultural Revolution in Müang Sing. People were arrested and executed without their guilt being proven. This imposed much hardship on the inhabitants of Müang Sing who became very fearful and fled their ancestral homes seeking refuge behind the front line in areas controlled by the Royal Government forces in Bo Kaeo province."[26]

The history of the anti-Communist revolt in the mid-1960s, as recorded for the first time by Mai Thamdi, calls into our memory an almost forgotten sideshow of the Lao civil war. The Yao rebellion of Müang Sing is hardly mentioned in any studies of modern Lao history. Jean Deuve who has written the most detailed accounts of the war in Laos mentions briefly that the Americans and their Thai allies set up a base on Thai territory in Chiang Khong, just opposite of Huai Sai, to recruit and train Hmong, Yao (Mien), and Khmu soldiers in secret guerilla units against the pro-Communist forces (Deuve 1984: 242). Thus we have to rely on Mai Thamdi's account of the history of Müang Sing to learn more about the background of the Yao rebellion in Müang Sing. In 1963/64, the inhabitants of two Yao villages, situated in close proximity to the border with China, were molested by a battalion of Pathet Lao soldiers under the command of Kham Lek, an ethnic Akha. The conservative Yao villagers got angry as they were forced to pay "financial contributions" to the revolutionary forces. As a result many of them joined the anti-communist secret units set up by the CIA. They were joined by members of other ethnic minorities living in the province of Luang Namtha, including Hmong, Khmu, Tai Lü, Tai Nüa, and even some Akha. Mai Thamdi frankly admits that "most people of Müang Sing did not yet understand the political line of the Lao Patriotic Front and adopted a neutral position. Many of them still appreciated the old way of thinking and admired the Americans as heros. They did not realise the vicious scheme of the Americans to rule our nation like a colony" (Mai Thamdi 2002/04: 50). It seems that the Pathet Lao forces soon became isolated and even were encircled in the plain of Müang Sing for quite some time, as the multi-ethnic "Yao rebels" managed to gain control of most of the hills surrounding Müang Sing in early 1965. Fighting continued through most of the year, and the Pathet Lao forces managed to survive and gain back the military initiative through the support of North Vietnamese support troops (*ai-nòng wiatnam*). Since c. 1968 the whole district of Müang Sing was again firmly in the hands of the Pathet Lao with only several isolated Tai Lü villages on the left bank of the Mekong River, including Ban Siang Khaeng which was

26 Mai Thamdi 2002/04: 53.

reconquered by the Royal Lao Army in c. 1964 and since then remained under the control of the Vientiane government.[27]

After the armistice agreement of Paris (January 1973) the situation in Müang Sing improved with many refugees returning from Huai Sai and other areas controlled by the Royal Lao Government. After the founding of the Lao People's Democratic Republic (Lao PDR) in December 1975, the political and military situation remained "peaceful" (*mi khuam sangop*), as Mai Thamdi emphasises. However, due to the worsening relations between Laos and China in the aftermath of the Vietnamese invasion of Cambodia, tensions with China increased over the 1980s. In 1986, the anti-Vientiane insurgence trained and equipped by Chinese Special Forces in Yunnan province, made frequent incursions into the district of Müang Sing. Finally, "in October 1988, the local militia, coordinating its actions with Battalion 32 [of the Lao People's Army], were able to capture 78 Lao exile soldiers, killed 10 ten soldiers, while 200 others surrendered" (Mai Thamdi 2002/04; 57). Since then the security situation improved again.

In contrast to the practice of the RLG and the Pathet Lao in the periods 1953–1962 and 1962–1975, respectively, the government of the Lao PDR mostly appointed local cadres rather than outsiders as district chiefs, taking into consideration the ethnic make-up of the area. The list of members of the district committees set up in the years 1982–2002 indicates that all major ethnic groups in Müang Sing were represented, although the district chiefs were either Tai Lü or Tai Nüa, while their deputies were often Akha. However, due to internal problems within the district leadership, the nature of which is not explained by Mai Thamdi, the district committee elected in 2004 was directed by a cadre from Luang Namtha district, who was a member of the Tai Yuan ethnic group, and two deputies, one of them being originally from Phong Saly (probably a Phu Nòi) and the other a Tai Dam from Luang Namtha (Mai Thamdi 2002/04: 67).

Social and Economic Changes since the 1990s

The economic and social development in Müang Sing stagnated until the late 1980s. The implementation of the GTZ (now: GIZ)-supported "Müang Sing Integrated Food Security Programme" in c. 1993 contributed to a sustained economic boom which was also supported by ecotourism (see Neudorfer 2006). During my first visit to Müang Sing in November 1996, already half a dozen guesthouses frequented by Western backpackers had already opened. Until May 1998, several more guesthouses were added, and a local branch of the "Banque pour le commerce exterieur Lao" had opened next to the morning

27 The main stupa (*that luang*) of the monastery (Vat) of Siang Khaeng was renovated in BE 2509 (AD 1966) by using cement transported from the Lao capital Vientiane. See inscription of That Luang Siang Khaeng.

market opposite the district office. However, the town as well as the surrounding villages did not yet have electricity in those years. The GTZ office compound was the only electrified place in the town due to the organisation's own generator.

Already in 1993 the district office of Müang Sing had come forward with an ambitious development plan. This plan closely follows the original quadratic layout and keeps the delineation of the street system intact. The plan envisages a much higher population density than is the case at present. It depends obviously on the hope that the urban population will increase steadily, at least by 2.6% per annum which was the average for the province of Luang Nam Tha in the late 1980s, but probably even much higher.[28] The total urban area of 83.3 ha, which includes roughly 20 ha outside the old town wall, would be used as follows:

exploitation	*area (in ha)*	*area (in %)*
administration	12.0	14.4
business (incl. housing)	6.4	7.7
trade (market)	1.9	2.3
living quarters	48.0	57.6
park	15.0	18.0
Total	83.3	100.0

Table 2: Müang Sing Development Plan
Source: Adapted from District Office (Müang Sing) 1993

In an earlier evaluation I argued that should this plan ever be realised in full, the population of Müang Sing would increase considerably and the number of houses would almost be doubled (Grabowsky and Kaspar-Sickermann 1997: 137-138). Furthermore, the population would become more evenly distributed with much of the waste land, especially in the quarters of Siang Yün and Siang Lae, being covered with new private or public buildings. In addition, two large park areas with extensive flower gardens and restaurants were planned, one in the northern corner (Siang In and Siang Lae) and the other outside the eastern corner near the district hospital. This bold vision underlines that, in the early 1990s, the district and provincial authorities anticipated Müang Sing to become both a major attraction for tourists and a regional economic centre. In the mid- and late 1990s, however, it seemed still far too optimistic to see

28 Statistics for Müang Sing vary. During the 1988-1992 period the average increase of population in the town was less than 1.5% per annum. It seems that both town and district of Müang Sing were net emigration areas during the two decades following the founding of the Lao PDR (December 1975). This factor retarded population growth that was below the provincial average. One has to wait and see whether the assumption of the district administration that this trend could be reversed in the future is realistic. As regards to population figures of Müang Sing see District Office (Müang Sing) 1993.

Müang Sing as the future hub of an "economic quadrangle" linking trade routes between Burma, China, Laos and Thailand. Furthermore, I surmised that a combination of trade with nearby China and "alternative tourism" – exploiting the ethnic diversity and cultural richness of the district – might appear attractive and promising. In this respect, the re-discovery of Müang Sing as an historic township, connecting Laos with her neighbours China and Burma, could become a meaningful unique selling point for the border district.

When I returned to Müang Sing in March 2005 after a break of six years, the changes I realised surpassed even my most optimistic expectations. Now almost all houses in the town and in adjacent villages had electricity delivered from nearby Müang Phong and Müang La in the southeast of Sipsòng Panna, China. Most houses already had satellite TV receivers on their roofs. The road infrastructure in the valley of Müang Sing had improved considerably, and the three major roads connecting Müang Sing with Luang Namtha, the Chinese border at Müang Mang, and Chiang Kok (via Müang Lòng), respectively, were now paved with asphalt. Whereas it took me almost eight hours by car from Müang Sing to Siang Kok in early 1998, at the end of the dry season, seven years later the same trip of the 75 km long road lasted one hour and a half.[29]

As of 2005, the physical presence of China was not yet really visible in Müang Sing. Eight years later, by mid-February 2013, Müang Sing has changed considerably. One of the four quarters (*chiang/siang/ceng*) of the area within the old city wall, Siang Lae, had a large bus station built next to the new bustling Morning Market. The old morning market, situated along the main road leading to the Laos-China border, had burned down before 2005. Instead of rebuilding it, it was decided to move the market further inside the town. Near the old, now abandoned morning market a smaller "evening market" has opened where each evening some food is sold by local vendors. A significant number of traders, though not yet the majority, are not local Tai Lü but Chinese (ethnic Chinese as well as Sipsòng Panna Tai Lü) from the other side of the border.[30] The seat of the District Administration which was previously a shabby one-storey building just opposite the old market at the main road to the Lao-China border, has been transferred to a representative and spacious three-storey building which, however, looks a bit oversized for the seat of a district. Next to this new main administrative building, situated directly

29 According to Houmphan (2008: 263), the construction of the Müang Sing-Siang Kok road, known as Road 17B, started in 1996 and was finished in 2000.

30 Walker (1999: 73) states: "A second north-western border crossing has been established between China and Laos. Though this is officially only a local border crossing, it services some long-distance trade that heads down the old caravan route to Xiangkok and onward to Burma and Thailand. The old Lü settlement of Mengla is still the predominant Chinese-Lao trading town, though it has been radically transformed by central Chinese investment and immigration. A border trading village, with neatly laid out rows of concrete shop-houses, has also been established at Mohan, just a few kilometres north of Boten."

in the heart of the town where in the old days of Chao Fa Sali Nò Kham the ruler's palace (*hò kham*) stood, the Lao Development Bank has its quarters; and on the opposite side of the road the Agricultual Bank is based in another large building. Finally, a petrol station opposite to Vat Siang In and close to the Bus Station bears testimony to the fast economic developments in Müang Sing.

Taking an hour of walking along the main streets and the smaller sidewalks within the interior of the town (*wiang*) of Müang Sing, I was stunned to see how many Chinese have settled since 2005, as before the percentage of Chinese had still been very low. As my visit took place in the week following the Chinese New Year, the decoration in front of private houses betrayed the China affiliation of the owners or inhabitants, as the local Tai Lü, unlike their Tai Nüa neighbours, do not celebrate Chinese New Year that long. Especially in Siang Yün and Siang Cai, but also in Siang In, the Chinese presence seemed to be quite significant, amounting to at least one-tenth of the whole population. Most of the more than two dozen shops, including several shops selling motorbikes, were apparently run by Chinese. Most surprising was a supermarket called the "Chinese Laos Supermarket" or shortly "Big C China". This relatively large supermarket, situated in Siang Lae quite near the Morning Market, has been run already since 2009 (interview by Ms. Zhou Hanli with the owner, 23 February 2016). It is frequented by numerous Chinese living and working in Müang Sing. In 2015, a second, smaller Chinese-run supermarket had opened nearby around the corner. Hardly any of the vendors in the two supermarkets, except one local woman, could speak Lao or Tai Lü reasonably well. Large tracts of land along the road from the China-Laos border near Udomsin to Müang Sing and from there to Müang Lòng and Siang Kok have been rented in long-term contracts to Chinese companies planting either sugar cane or bananas there. The labour force for the plantations are partly local highlanders, notably Akha but also Khmu, and partly Chinese contract workers (some of them living near the plantations without proper residence permits). Large trucks transporting sugar cane and bananas leave Müang Sing every ten minutes in the direction of the nearby border station.

Movement related to economic exchanges between Müang Sing and Sipsòng Panna has increased dramatically during the last decade. Trade between the districts of Müang La and Müang Phong in southern Sipsong Panna and the districts of Müang Sing and Müang Lòng has become quite impressive. When I arrived at Ban Seo in February 2013, a small Tai village of roughly sixty households on the outskirts of Müang Phong, to take a rest at the paternal house of Ai Nò La, a young Tai Lü colleague from Chiang Rung, a pick-up truck had just stopped in front of the house. The car was carrying dozens of rice bags each of which contained fifty kilograms of sticky rice. I was surprised to learn that the rice traders came from Ban Dòn Chai, the last Tai Lue village in Müang Sing district before reaching the Laos-China border only 5 km further to the east. The two traders, a couple, said that they came to Müang Phong twice a week to sell rice. In Müang Sing they would buy one bag of sticky rice for 200 yuan and then sell it in China for 240 yuan. Thus their profit margin would be 40 yuan for each rice bag. As their car could carry up to forty bags their total profit would be up to 1,600 yuan (200 euros) for each trip. Müang Phong has turned into a rice-deficit area as former rice land has been converted into banana groves, because bananas are now much more profitable than rice.

Most of the Chinese presence in Müang Sing district is concentrated in the town of Müang Sing whose population has more than doubled during the last two decades, whereas the population of the whole district has increased only by 60 percent (from 23,200 to 37,100 inhabitants) in the same period. The nearly five-fold increase of population within twenty years in the town quarter of Siang Yün is most remarkable. Here we find the largest concentration of modern stone houses, many of which are inhabited by Chinese migrants. The influx of migrants from China is a major issue of concern among the local people in Müang Sing who fear that ethnic Chinese might at one time in the future form the majority of the district's urban population. The economic boom, along with the expansion of the tourist industry, seems to have silenced this public resentment, at least for the time being. According to recent statistics (2012), the four town quarters of Müang Sing comprise a population of 2,800 inhabitants to which the newly founded settlement of Singchaloen (539 inhabitants) – directly bordering Siang Chai and Siang Yün – and the rapidly expanding village of Hua Khua (735 inhabitants) should be added. Thus we arrive at a total population of more than 4,000 people in Müang Sing, town and suburbs, of which an estimated 20 percent are migrants from China. This population is rapidly rising. Most recent data, obtained through interviews with Chinese residents, indicate that by early 2016 around 1,000-1,200 Chinese have been living with a legal residence permit in Müang Sing district, mainly in the urban centre. Interestingly, most Chinese migrants do not originate from Yunnan, as one may expect, but from the provinces of Hunan and Szechuan further north. In general they are young and middle-aged adults; some of them have their very young children staying with them but all school-age children would be educated in China. Though a large portion of the townpeople still work full- or part-time in the agricultural sector, trade and commerce, as well as tourism, are now dominating. The town has become more urban than ever before, or, as one Lao scholar has aptly remarked, the town of Müang Sing, which "was only a large village" in the past, has now "become a small town and an important center for business, tourism and services" (Houmphan 2008: 263).

Town quarter	Inhabitants in 1993		Inhabitants in 2013	
	Absolute	in percent	Absolute	in percent
Siang Chai	546	44,6	749	26.7
Siang Yün	303	24,7	1,400	49.9
Siang Lae	220	18,0	321	11.4
Siang In	156	12,7	338	12.0
Total	1,225	100,0	2,808	100.0

Table 3: Population of Müang Sing town
(Source: District Office (Müang Sing) 1993, in: Grabowsky and Sickermann 1997: 133).[31]

In the rural areas of Müang Sing the increase of population over the last two decades has been remarkable as well. A comparison of the census data of 2013 with the incomplete population figures from the early 1990s shows some differences between the demographic development in the plain of Müang Sing and in the mountainous areas. In the latter region we can observe a growing concentration of the hill people in a decreasing number of villages, as smaller settlements have been abandoned in favour of larger villages situated closer to the main roads. There are also a few cases where isolated small Tai settlements were abandoned. One example is the isolated Tai Lü village of Ban Bò, situated on the left bank of the Mekong north of Ban Siang Khaeng. In the early 2000s, the 39 villagers (1997) were resettled to Müang Mòm, another Tai Lü outpost close to the Laos-China border, but larger (2013: 430 inhabitants) and better connected by road to the district centre.

Another phenomenon which has become visible in Müang Sing, as elsewhere in the country, is the emergence of multi-ethnic villages. In the plain of Müang Sing such villages have expanded considerably since the mid-1990s. This trend can be observed in other parts of the country, as well. Eva Sevenig, studying the multi-ethnic (Khmu-Samtao) village of Ban Hatnaleng in Nalae district, Luang Namtha province, observes: "In general, multi-ethnic villages have been on the increase in Laos, and especially in the North, since the 1990s" (Sevenig 2015: 4). This does not mean that before the 1990s all villages in Müang Sing were strictly mono-ethnic. Almost all these villages also included members of other ethnic groups, mostly men and women who had married local people and where usually called 'sons-in-laws' (*khoei*) and 'daughters-in-law' (*saphai*). Whereas there was only a single mixed village at the time of my first visit in 1996, namely Ban Nòng Bua (Tai Dam with a Phu Nòi minority of 30–40 percent),[32] the plain of Müang Sing has experienced immigration from the northeastern Lao provinces of Phong Saly and Hua Phan.

31 Data pertaining to population figures in 2013 are obtained from the provincial administration of Luang Namtha. I thank Joseba Estevez (Münster) for providing me with these data.

32 Interview with Mai Thamdi, former acting district chief (*chao müang*) of Müang Sing, 23 February 2014.

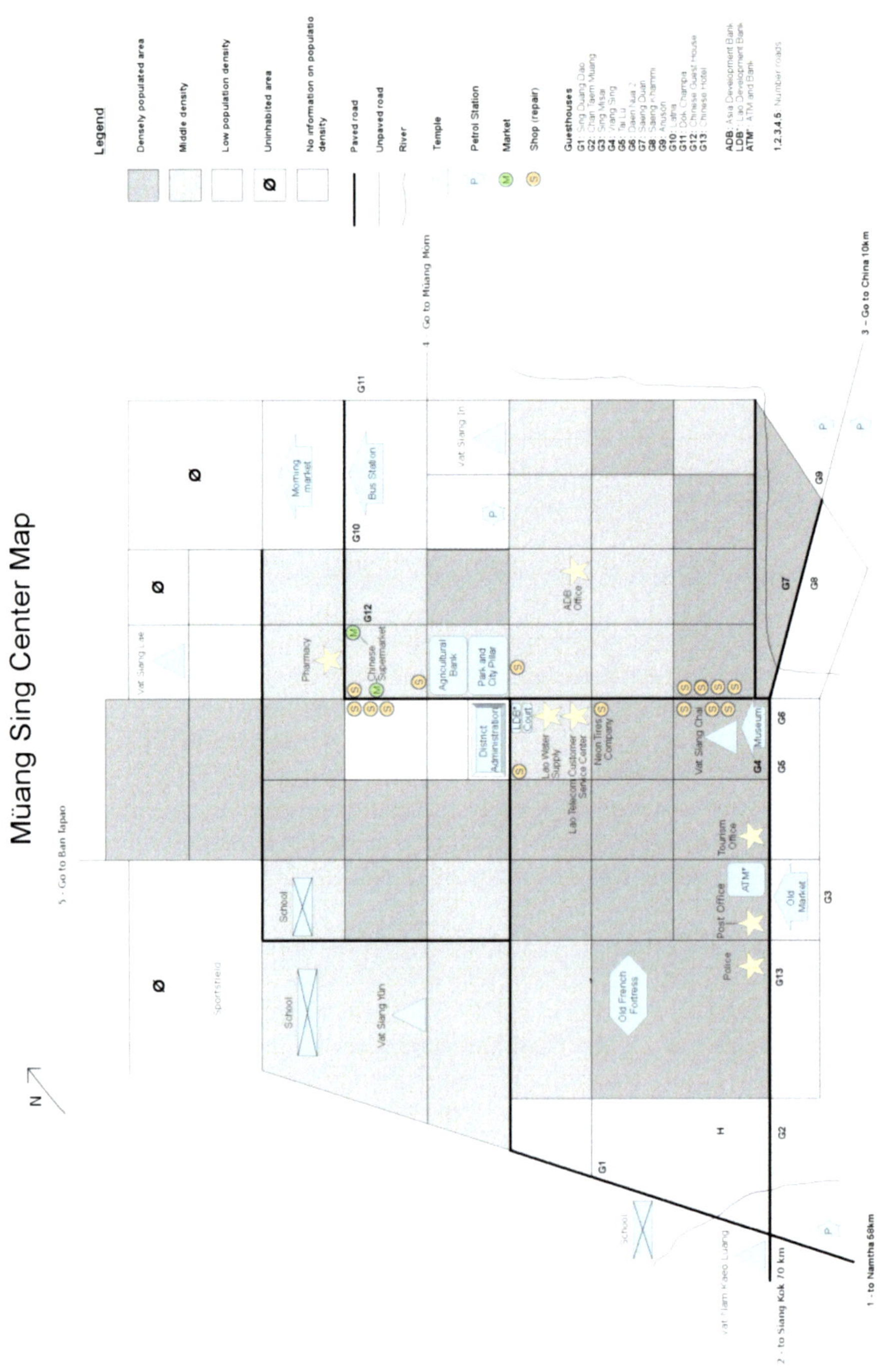

Fig. 2: Map of the Town of Müang Sing (2016)
© Volker Grabowsky

After the establishment of a military garrison near this village, numerous soldiers from all over the northern region of Laos and their families have settled in Ban Nòng Bua. In the late 1990s it was still a village of less than five hundred inhabitants, but as of 2013 Ban Nòng Bua counted already 335 households with slightly more than 1,600 people, the largest single settlement in the district.

In some cases completely new villages were founded to accomodate the growing immigration, notably from Phong Saly and Hua Phan provinces. Examples are Ban Singchaloen (next to Siang Yün and Siang Chai outside the old city wall) and Ban Sing Phatthana (on the route to Müang Lòng and Siang Kok). Many inhabitants of Ban Sing Phatthana (2013: 359 people) are the families of soldiers stationed at Ban Nòng Bua. Not all the new villages, settled by immigrants northeastern Laos, are multi-ethnic. For example, Ban Huai Kot, situated at the edges of the plain of Müang Sing on the road going to Ban Siang Khaeng, was founded in 1993. In early 2005, this village counted 38 households (2013: 45 households). All 229 villagers (2013: 224 inhabitants) were 'Tai U', i.e., Tai Lü from Müang U Nüa in Phong Saly province who migrated to Müang Sing because of lack of rice-farming land available in their old home.[33]

Ethnically mixed is now also the village of Ban Hua Khua (Tai Lü: Ban Ho Kho), literally 'Bridgehead Village', an old Tai Lü settlement whose founding even predates the founding of Müang Sing town. During the last two decades, Ban Hua Khua has expanded considerably. Before 1962, it was a small village of less than ten houses, as the former district chief, Mai Thamdi, recalls.[34] In the mid-1990s Ban Hua Khua was still small but since then the number of households increased to 146 by 2013, the village now counting 735 inhabitants. According to local informants, the village had "more than 100 households".[35] Asked about the ethnic composition of the settlement, I was told that more than half of the population were Phu Nòi, a Sino-Tibetan ethnic group who mainly live in Phong Saly province. The vast majority of the Phu Nòi in Müang Sing who arrived here during the Lao civil war are Buddhists. Besides, there were up to twenty Tai Daeng and about the same number of Tai Dam households. The members of these two groups practice ancestor worship (*napthü sasana phi*), whereas the slightly more than ten Tai Lü families are devout Buddhists. A few Hmong, 'Lao Thoeng', and Tai Nüa families also lived in the multi-ethnic village. Due to the heavy presence of non-Buddhist groups in Ban Hua Khua, the old village monastery, whose architecture looks

33 Personal field notes; Statistics of the provincial government of Luang Namtha (2013).

34 Interview with Mai Thamdi, former acting district chief (*chao müang*) of Müang Sing, 23 February 2014.

35 Personal field notes; Statistics of the provincial government of Luang Namtha (2013), kindly provided by Joseba Estevez.

impressive from the distance, seemed to have been neglected when I visited the place in 2013 and 2014. Vat Ban Hua Khua had only one monk, a fifty-year-old man from the Phu Nòi ethnic group who had been ordained, first as novice then as monk, for already thirty-five years; there were five novices, mostly Phu Nòi boys, under his supervision.

Conclusion

Müang Sing district in the northwestern Lao province of Luang Namtha is a good case study to demonstrate how the political upheavals and socio-economic changes of the last 120 years affected the life of the people at the periphery of one of poorest nation-states in Southeast Asia. In pre-colonial times Müang Sing was the seat of the small multi-ethnic buffer state of Chiang Khaeng, maintaining close political relations to other Tai polities in current Laos, Burma (Myanmar), China, and Thailand. The rivalry of European colonial powers – France and Britain – along with the enduring weakness of regional powers – Siam and China – resulted in the break-up of Chiang Khaeng and the integration of the principality's eastern part, which included the plain of Müang Sing, into French Indochina. Migration has shaped the ethnic and social diversity of Müang Sing from the beginning. Whereas the ruling elite came from the Tai Lü and Tai Nüa ethnic groups who lived as rice farmers in the lowlands, the highlands were dominated by the Akha. The war in Laos from 1958 to 1973 uprooted large segments of the population in Müang Sing district. The losses caused by the flight of numerous Tai Lü and Tai Nüa villagers to Huai Sai and to third countries, following the seizure of Müang Sing by Pathet Lao forces in June 1962, were hardly compensated by the influx of Tai Lü refugees from Yunnan during the years of the Chinese Cultural Revolution. In the age of globalisation the borders, especially with China, have became more and more porous, and the once poor and underdeveloped district of Müang Sing an increasingly prosperous border area. Trade, business, and tourism have given rise to an economic boom which has been accompanied by immigration from other parts of northern Laos and from neighbouring China. The emergence of multi-ethnic villages in the plain of Müang Sing and the transformation of the district's administrative centre into an urbanised space is another result of this ongoing process.

Acknowledgement

I am indebted to Mai Thamdi Thammavong, the former acting district chief of Müang Sing as well as to Nan Chai Saeng, the former curator of the local museum. Furthermore, I would like to thank Renoo Wichasin, Apiradee Techasiriwan, Zhou Hanli, Khamvone Chayavong, Walter Kaspar-Sickermann, Paul Cohen, Wulf Raubold, Bounleuth Sengsoulin, and Joseba Estevez for their support during various stages of my long-term research (1996-2016). Finally, I would like to express my gratitude the German Research Foundation (DFG) for funding my fieldwork and archival research in its early early stage and to the European Commission's 7th framework programme for supporting it int its final stage (in the framework of the project Integration in Southeast Asia: Trajectories of Inclusion, Dynamics of Exclusion – SEATIDE).

References

I. Archival sources

British Library, India Office & Oriental Collections (IOOC), London:

SCOTT, J. G. (1891) Memorandum by Mr. J. G. Scott, Officiating Superintendent, Shan States, on a Visit to the Kyaington-Chiengmai Boundary to Meung Hsing and to Chieng Hung. In *Scott Papers* [Mss Eur F 278/73].

Centre des Archives d'Outre-Mer (CAOM), Aix-en-Provence, Gouvernement Général de l'Indochine (GGI), Cabinet du Gouvernement Général :

No. 20162: "Mission du Haut Mékong. Rapport du lieutnenant de vaisseau Jacquemart commandant la canonnière 'le Lagrandière' au Gouverneur Général de l'Indochine, sur le Muong Sing et les pays environnants, 4 juin 1899."

No. 22299: "Rapport de l'administrateur Sérizier sur le territoire de Muong Sing – les états Shan britanniques et les Sip-Song Pan-na chinois."

Grabowsky Collection :

Pawat phün müang sing khwaeng luang namtha ປະຫວັດພື້ນເມືອງສິງ ແຂວງຫລວງນໍ້າທາ *(The History of Müang Sing, Luang Namtha* province). Lao manuscript by Mai Thamdi, 19 August 2002.

Public Record Office (PRO), London:

WARRY, Mr. (1895) [Report sent] to the Chief Secretary to the Chief Commissioner, Burmah. (Moulmein, 10 May 1895) In *Further Correspondence respecting the Affairs of Siam*, Part VII: July to December 1895 [PRO, FO 422/43].

II. Published Sources and Secondary Literature

BULLETIN ADMINISTRATIF DU LAOS. Année 1916. Hanoi: Imprimerie d'Extrême-Orient.

CHAZEE, L. (1999) *The Peoples of Laos: Rural and Ethnic Diversities*. Bangkok: White Lotus.

COHEN, P.T. (2000) Resettlement, Opium and Labour Dependence: Akha-Tai Relations in Northern Laos. *Development and Change* 31: 179-200.

COHEN, P. T. and LYTTLETON, C. (2002) Opium-Reduction Programmes, Discourses of Addiction and Gender in Northwest Laos. *SOJOURN* 17(1): 1-23.

DEUVE, J. (1984) *Le Royaume du Laos 1949-1965* (Histoire événementielle de l'indépendance à la guerre américaine). Paris: École française d'Extrême-Orient.

GRABOWSKY, V. (1999) Introduction to the History of Müang Sing (Laos): The Fate of a Lü Principality. *Bulletin de l'École française d'Extrême-Orient* 86: 233-291.

GRABOWSKY, V. (2003) Chiang Khaeng 1893–1896: A Lue Principality in the Upper Mekong Valley at the Centre of Franco-British Rivalry. In GOSCHA, C.E. and IVARSSON, S. (eds.), *Contesting Visions of the Lao Past, pp.* 37-70. Copenhagen: NIAS Press.

GRABOWSKY, V. (2008) The Tai Polities in the Upper Mekong and Their Tributary Relationships with China and Burma. *Aséanie* 21: 1-53.

GRABOWSKY, V. and KASPAR-SICKERMANN, W. (1997) The Town of Müang Sing: Past and Present. *Lanxang Heritage Journal* 2(4): 125-157.

GRABOWSKY, V. and WICHASIN R. (2008) *Chronicles of Chiang Khaeng: A Tai Lü Principality of the Upper Mekong*. Honolulu: Center for Southeast Asian Studies, University of Hawai'i.

HOUMPHAN R. (2008) The Changing Livelihoods of the Akha in Muang Sing and Muang Long in Laos. In McCASKILL, D. et al (eds.), *Living in a Globalized World: Ethnic Minorities in the Greater Mekong Subregion,* pp. 259-276. Chiang Mai: Mekong Press.

HOUMPHAN, R. et al. (ຫູມພັນ ຣັດຕະນະວົງ ແລະຄະນະ) (2006*) Pawat that luang siang tüng lae pawat müang sing ປະຫວັດທາດຊຽງຈືງ ແລະປະຫວັດເມືອງສິງ (The History of the Chiang Tüng Stupa and the History of Müang Sing).* Tai Lü Community in the United States.

LACROZE, L. (1994) *Monographie de deux provinces septentrionales du Laos: Phong Saly et Houa Khong*. Les cahiers de Péninsule no. 1. Metz.

LAFONT, P.-B. (1973) La plaine de Muong Sing. *Cahiers d'Outre-Mer* 101 (1): 39-53.

LAFONT, P.-B. (1998) *Le Royaume de jyṅ khèṅ: Chronique d'un royaume tay lœ2 du haut Mékong (XVe - XXe siècles)*. Paris: l'Harmattan.

LEFÈVRE-PONTALIS, P. (1902) *Voyages dans le Haut Laos et sur les frontières de Chine et de Birmanie (Mission Pavie Indochine 1879-1895, tome V)*. Paris: Ernest Leroux.

LEFÈVRE-PONTALIS, P. (2000) *Travels in Upper Laos and on the Borders of Yunnan and Burma (The Pavie Mission Indochina Papers, Vol. 5)*. Bangkok: White Lotus (English translation of Lefèvre-Pontalis 1902).

LIEW-HERRES, F. M., GRABOWSKY, V. and WICHASIN, R. (2012) *Chronicle of Sipsòng Panna: History and Society of a Tai Lü Kingdom, Twelfth to Twentieth Century*. Chiang Mai: Mekong Press.

MCCARTHY, J. 1900. *Surveying and Exploring in Siam with Descriptions of Lao Dependencies and of Battles against the Chinese Haws*. London: John Murray (reprint Bangkok: White Lotus, 1994).

NEUDORFER, C. (2006) Globale Diskurse, lokale Prozesse: Problemfelder im nachhaltigen Tourismus. In REUTER, J., NEUDORFER, C. and ANTWEILER, C. (eds.), *Strand Bar Internet: Neue Orte der Globalisierung*, pp. 61-83. Münster: Lit.

RAQUEZ, A. (1902) *Pages laotiennes. Tome 2*. Hanoi: Schneider.

SAIMÖNG MANGRAI, S. (1981) *The Pāḍaeng Chronicle and the Jengtung State Chronicle Translated.* Michigan Papers on South and Southeast Asia, No. 19. Ann Arbor: The University of Michigan, Center for South and Southeast Asian Studies.

SEVENIG, E. (2015) Social Mobility in Baan Had-Naleng: When the Valuation of Communality allows for a Demarkation in a Multi-Ethnic Village in Northwest Laos. *Dorisea Working Paper*. [Online] Dorisea Database (16): 1-18. http://www.dorisea.de/de/dorisea-working-papers. [Accessed: 26th January 2016].

WALKER, A. (1999) *The Legend of the Golden Boat: Regulation, Trade and Traders in the Borderlands of Laos, Thailand, China and Burma*. Honolulu: University of Hawai'i Press.

WITYADA, T. (วิทยา ทองแดง) (2003) Thai nüa müang sing ไทเหนือเมืองสิง (The Tai Nüa of Müang Sing). *Müang Boran Journal* วารสารเมืองโบราณ 29 (3): 22-32.

YANYONG C. and SETHAKUL R. (ยรรยง จิระนคร และรัตนาพร เศรษฐกุล) (2001) *Prawattisat sipsòng panna* ประวัติศาสตร์สิบสองปันนา *(History of Sipsòng Panna)*. Bangkok: Amarin.

The British Interregnum in Java (1811-1816): An Experiment in Supranational Fraternal Governance

Roy Jordaan

Introduction: The Conquest of Java and the Establishment of British Rule

Napoleon Bonaparte's annexation of the Netherlands to the French empire and the extension of French and British military operations to the Indian Ocean and Asian waters were the direct reason for the British invasion of Java in August of 1811. Contrary to the orders of the Court of Directors of the East India Company, Lord Minto, Governor-General of British India, did not stop at the destruction of Dutch fortifications and the disarmament of the Dutch-French military, but decided on the permanent occupation of Java by British forces, as well as the incorporation of the island into the Asian part of the British empire.

Lord Minto entrusted Thomas Stamford Raffles with the actual administration of Java and its dependencies. The latter's administrative experiences, however, did not extend beyond the function of Secretary to the Governor of Penang, a recent British settlement on a small island off the Malay coast that the East India Company rented from the Sultan of Kedah. Decisive for Lord Minto's appointment of Raffles as Lieutenant-Governor of Java was his representative's outstanding performance as *avant-courier* or 'forerunner' of the British invasion. Raffles' intelligence-gathering work and subversive plans against the Dutch from his base in Malacca had been covered up under the wordy designation of "Agent of the Governor-General with the Malay States".

With the appointment of the relatively inexperienced Raffles, Lord Minto clearly took a great risk. It is understandable therefore that prior to returning to Calcutta he thoroughly instructed Raffles and wanted to give him a leg-up. One of his first decisions in this connection was the formation of a small mixed British-Dutch administrative body to support Raffles' rule, namely the British Council in Java or 'Java Council' for short. Raffles also received substantial support from the Dutch in other administrative fields and in some personal matters.

The aim of this paper is to explain how the remarkable examples of British-Dutch co-operation and mutual assistance to be described below actually came to pass, as well as the motives and principles on which they were founded. Given that all four members of the Java Council were Freemasons, one is justified in suspecting that membership in the Order was regarded as an important qualification. The composition of another British-Dutch administrative body, the Mackenzie Land Tenure Commission, was also dominated by Masons.

Although it has long been known that Raffles became a Mason after his arrival in Java, different opinions about the timing of his entrance into the Craft have obscured the significance of Freemasonry in the formation of supranational forms of governance and co-operation. Closer examination of Raffles' admission, and his subsequent elevation to the degree of Master Mason, will not only shed light on this matter and on his special bond with prominent Dutch brethren, particularly Nicolaus Engelhard, but also correct some flaws in the current biographies of Raffles.

Raffles' Entry in Freemasonry

Very little is known about Raffles' admission to the Order of Freemasons, although this is not surprising for a society traditionally associated with secrecy and mysteriousness. However, Raffles' initiation has also been left unexplored owing to vague and contradictory statements about its dating in the literature published by Dutch East Indies lodges (e.g. *Gedenkboek Vrijmetselarij* 1917; De Visser Smits 1931; Stevens 1994). It appears that we are only certain about the year of Raffles' advancement to the rank of Master Mason in Surabaya in July of 1813. His admission as Entered Apprentice and passing to the degree of Fellow Craft are often assumed to have taken place simultaneously a few months earlier, even though this would have been somewhat unusual, but not exceptional, for lodges endorsing the Saint John's system. Far more important is that the significance of Freemasonry membership for both the rapid rapprochement of, and the cooperation between the British and the Dutch, was overlooked because of the obscurity of the dating. This explains the mistaken opinion that there is not much to be said about Freemasonry in the Dutch East Indies during the British interlude (Stevens 1994: 85).

Anthropologist Nigel Barley – former Keeper at the British Museum and organiser of a Raffles exhibition – appears to be best informed about Raffles' membership in the secondary English literature. He writes:

> "Before he left [Java], Lord Minto did more than lay down the general political lines Raffles was to follow, he engaged in a very paternal action that showed his close personal relationship to Raffles. He sponsored him as a Mason. Freemasonry was rife in both the English and Dutch trading companies. [...] The list of Masons is inevitably incomplete, and the role of Freemasonry in empire is a huge book that remains to be written, but we do know that *all* Raffles"immediate associates were members. He eschewed both the Batavia lodges at that time and joined a small lodge called "*Virtutis et Artis Amici*" on the Pondok Gedeh coffee plantations near Bogor. A fellow member was his Dutch ally, Muntinghe, and his future arch-enemy General Gillespie. Raffles never mentions his Masonic activities, but accounts of Freemasonry in India and London dot the pages of his *Java Government Gazette*. From the records, we know that he took it seriously enough to rise

> to the third degree of Freemasonry during an overnight stay in Surabaya in 1813. Shortly before leaving Java he "received Perfection in the Rose Croix Chapter [of] '*La Vertueuse*' in Batavia, in the company of his secretary and his banker" (Barley 1991: 85-86).

This rather brief summary leaves a number of interesting questions unanswered. For instance, why did Lord Minto 'sponsor' Raffles only after their arrival in Java, and why didn't Raffles join the Craft during his earlier five-year stay in Penang? It stands to reason that the timing depended on the successful completion of the invasion, justifying the trust that Lord Minto had put in Raffles. Nevertheless, the initiation took place on Javanese soil and not in an ambulant lodge on a British ship, such as the flagship HMS *Modeste* on which Lord Minto, Raffles and other leading figures in the military operation had sailed to Java.

Before explaining when the initiation took place, we need to know more about the two existing lodges in Batavia, and why the lodge on a coffee plantation outside the capital ultimately was chosen for Raffles' initiation. It is not readily apparent which sources led Barley to contend that Raffles would have eschewed the Batavia lodges *La Vertueuse* and *La Fidèle Sincérité*. Indeed, it is precisely his cordial relations with Dutch Masons that is highlighted in Dutch sources. This is particularly true of his relationship with *La Vertueuse*, a lodge long dominated by members of the *Ancien Régime*, i.e. from the closing days of the VOC, such as Nicolaus Engelhard, Willem Jacob Cranssen and Willem Wardenaar. These old timers he seems to have preferred over those from the immediately preceding French-Dutch period of Marshall H.W. Daendels and General J.W. Janssens. None other than Engelhard, Past Master of the *La Vertueuse* lodge, was the founder of *Virtutis et Artis Amici,* the lodge where Raffles was initiated.

After being sacked by Daendels in 1808, and with his former position as Governor of Java's Northeast Coast having been formally abolished, Engelhard withdrew from public and semi-public functions, and moved to his Pondok Gedeh estate near Buitenzorg (Bogor), south of Batavia. It is presumed that the temporary *Virtutis et Artis Amici* lodge that he founded on his estate was primarily intended for the use of Dutch landowners in the region.

It seems likely that Engelhard, whose keen political insight often prompted timely responses, had contacted Lord Minto immediately after the fall of Batavia. If his administrative credentials would not have sufficed for this purpose, Engelhard certainly would not have hesitated to strengthen his request for a private appointment by resorting to secret Freemasonry symbols and codes, as we know he did in other critical circumstances. Once they had met, it would not have been difficult for the brethren Minto and Engelhard to conceive of ways in which Dutch Masons in Java could be of assistance to the British rule over Java, at least during the period in which the Netherlands had not yet been liberated from French-Napoleonic rule, and the future of the occupied Dutch territories abroad had not yet been determined through international peace treaties.

Former Enemies Become Collaborating Brethren

But what would have been the reason for the Dutch in Java to collaborate with the British invaders? If we rely on the early Raffles' biographers, the answer to this question seems obvious. Reginald Coupland, for instance, makes it appear that it simply was a matter of Dutch nationals coming around to the altered political situation. After the Dutch army had been defeated and disarmed, the Dutch must have realised that only the British could protect them against the revolt and massacres that the native population was presumed to have in store for them. Additionally, there was Lord Minto's Proclamation of 11 September 1811 that little would be changed in the circumstances of Dutch citizens in Java, who were led to believe they would enjoy exactly the same rights as British subjects. In view of the above, "it would be plainly impolitic not to make friends with the [British] newcomers, and well-nigh suicidal to plot and scheme against them. Social relations, therefore, were quickly established" (Coupland 1934: 40). Furthermore, there were the so-called Kew letters of the exiled Prince William of Orange, in which he called upon the governors of Dutch territories abroad to surrender their domains to the King of England (Wurtzburg 1954: 57).

Although these arguments cut ice, they do not satisfactorily explain why close collaboration between British leaders and the Dutch could be achieved so quickly. Raffles' biographers Charles Wurtzburg (1954) and Astrid Spierings (2013) pointed out that the British themselves had recognised the need for Dutch support, prompted by the enormous administrative challenges they were facing. After all, the British occupying force was both relatively small and homogeneous: less than ten thousand men had been involved in the invasion of Java, and more than half of this expeditionary force would be pulled out after the completion of its conquest. Facing the remaining British personnel were a far great number and variety of native peoples, dispersed over a very wide and barely known area (a few years earlier the British had already taken over the administration of other Dutch possessions in the area such as Malacca, Padang, and the Moluccas). As Wurtzburg (1954: 175) succinctly writes, "there were far too few English to go round". The British, therefore, could use every support in the form of administrative experience, ethnographic knowledge, and local contacts from Dutch government officials, provided that they were not suspected of pro-French leanings and were willing "to take the oath of allegiance to his Britannic Majesty and of fidelity and obedience to the Government of Java". Many of the Dutch, senior government officials in particular, were attracted to the British proposals as this would allow them to continue working and secure their livelihoods and assets. It was, in other words, of mutual advantage to establish relationships involving collaboration and exchange.

But how to separate the wheat from the chaff; that is to say, those who were favourably disposed towards the British rather than pro-French, and who were the most capable and talented Dutch administrators amongst the supposedly indolent rank and file? Engelhard was the man who could help the British in this selection. Until a few years earlier, he had been the second highest Dutch government official, next to the Governor-General, having already enjoyed a long career involving the performance of various administrative functions (De Haan 1910: 77-87). Moreover, he was related to two former VOC Governor-Generals through both descent and marriage, and he knew most other senior government officials personally, or was otherwise acquainted with their virtues and vices. Additionally, the former Commissioner of Native Affairs and Governor of Java's Northeast Coast was well-informed about internal Javanese affairs, particularly those involving the royal courts of Mataram in Central Java.

But there was more. Engelhard also was a Mason and a longstanding member of the *La Vertueuse* lodge. To a greater extent than the other Batavian lodge, *La Fidèle Sincérité*, this lodge had long been dominated by prominent government officials and citizens. Engelhard, as said, was Past Master of *La Vertueuse* and in 1798 had been appointed Deputy Grand Master of the Netherlands East Indies by the Grand Lodge in Holland. What has thus far been overlooked is that his leading position in the Brotherhood is what enabled him to continue to exert his influence because Freemasonry would be an important (yet undisclosed) selection criterion in collaborative British-Dutch projects.

To clarify the role of Freemasonry during the British interlude, I have to recall Barley's statement about the 'huge' book about the relationship between Empire and Freemasonry that 'still had to be written'. This book has in the meantime been published by J.L. Harland-Jacobs (2007). Her study makes clear that Masons played an important role in the establishment of the British Empire, so much so that Harland-Jacobs called them the 'Builders of Empire'. That the relationship between Masons and lodges of different nations could be friendly and mutually supportive, even when those countries were at war, is also relevant. The reciprocal visits of British and Dutch Masons in Bengal in the 1770s and in South Africa in the 1790s are prime examples of this. Apparently, ideas and values of Freemasonry, such as the doctrine of the universal brotherhood of man, mutual assistance, and conviviality, enabled the Masons involved to supersede, at least temporarily, the national and political dividing lines, and forge transnational bonds.

Raffles is almost entirely missing from Harland-Jacobs' study. His name is mentioned once and only briefly in connection with the founding of Singapore by order of the new Governor-General of British India, Lord Moira, who also was a Mason. Lord Moira provides another example of a Masonic 'Builder of Empire'.

Let us return to Java. For the senior Dutch government officials who were Masons, the prospect of collaboration with the British must have been enticing, particularly after 'the fearful days of Daendels' when their activities were severely curtailed and a number of prominent Masons, including Engelhard, were arrested and imprisoned on charges of British sympathies and plans of conspiracy (De Geus and De Visser Smits 1931: 158). To lay the foundation of mutual trust between the actors in the Java Council, Raffles, on the British side, and H.W. Muntinghe, on the Dutch side, were simultaneously initiated into the first two degrees of Entered Apprentice and Fellow Craft at the lodge at Pondok Gedeh, in the presence of Lord Minto and Engelhard. Considering the date of Minto's departure from Java, the initiation must have taken place before 19 October 1811. Among the other participants in the secret masonic ceremonies, apart from Gillespie (already a Mason), were the earlier mentioned Cranssen (former Governor of Ambon, who had succeeded Engelhard as Worshipful Master of *La Vertueuse*), Willem Wardenaar (former member of the Council of the Indies and member of *La Vertueuse*), and H.A. Parve (former member of the Council of the Indies and member of *La Vertueuse*). Cranssen was the second Dutchman elected to the Java Council, presumably also nominated by Lord Minto on Engelhard's advice.

According to Engelhard, he had himself been approached by Lord Minto to join the Council, but had turned down the offer for 'political reasons' without stating what these were. Considering, however, that Engelhard had once been praised for repelling a British assault near Batavia, it would have been odd for him to now declare loyalty to the British cause. In the literature it is said that he kept 'aloof' and that he had neither pro-French nor pro-British leanings – being foremost a Dutch nationalist and royalist. It is also possible that he kept aloof simply because of his poor command of the English language (see De Haan 1910: 84). Be that as it may, his relationship with Raffles remained friendly, while Engelhard would not have been Engelhard if he had not tried to continue to exert influence, even if from backstage. Thus, he acted as the intermediary in the division of General Janssens' property at Buitenzorg, and in the sale of the remaining furniture and effects to the British. Several years later he would perform a similar service for Raffles, following the latter's departure for England.

To demonstrate his trust in the Dutch, Raffles provided Lord Minto with a description of the appeal of the oath of allegiance to the inhabitants of Batavia, and the merry conclusion of the oath taking by Muntinghe, Cranssen and other high-ranking Dutch government officials on 21 October 1811. His report says that

> "As soon as it was known that the oaths might be taken, the public offices were crowded from morning till midnight with the inhabitants. [...] The late members

of the Council [i.e., the disbanded Council of the Indies] came forward in a body; and after taking the oaths before me, I am sorry to add, got most jovially tipsy at my house in the company with the new Councillors" (Coupland 1934: 40).

The reconstruction of the chronology of events is confirmed by what Raffles later disclosed about the formation of the Java Council, namely that "Muntinghe and Cranssen were both selected by the Earl of Minto to be members of the said Council". To continue with the remark: "Of the wisdom and benevolence which determined the late Earl of Minto to place two members of the Dutch nation at the Board of the British Council in Java, it is unnecessary to speak." (Raffles 1817: xii).

Apparently, the mutual relationship became so harmonious that some Masons put forward the suggestion of having an Englishman (presumably Raffles) elected as the Worshipful Master of *La Vertueuse* lodge. As this was going too far for Engelhard, he decided to advance his own candidacy and was elected unanimously.

But Engelhard did play a mediating role in Raffles' advancement to the degree of Master Mason, as attested by a letter he had sent to the lodge *De Vriendschap* (Friendship) in Surabaya, where the ceremony took place on 5 July 1813. Muntinghe was simultaneously elevated to Master Mason. Notice worthy is the presence of the name of Th. O. Travers in a register of Masons in the Dutch East Indies in the period between 1760 and 1860 (see De Visser Smits 1931: 298-309). Following his name is the year 1813, which suggests that Raffles' Secretary had either been initiated or elevated on the same occasion in *De Vriendschap* lodge. The same may hold for several other associates of Raffles whose names are also recorded in the register, all in 1813. Is it possible that they went through a collective initiation or elevation ceremony, similar to what happened at the occasion of Raffles's Perfection in the Rose Croix Chapter in 1816?

In passing, it may be remarked that the Masonic ceremonies in Surabaya took place in a new lodge building and that the construction of the building "had originally been supervised by Mackenzie, Colonel in the British Army". This was almost certainly Colonel Colin Mackenzie, who was in Surabaya during this period. Since he had been assigned other tasks for Lord Minto and Raffles, and was a busy man, it is no surprise to learn that "the actual architect was Johan Willem Bartholomeus Wardenaar, Captain in the British Engineering Corps (he was a Dutchman, but this is the year 1812, during the British Interregnum). Wardenaar [son of the earlier mentioned Willem Wardenaar] is the well-known draughtsman of the plates in Raffles' book [*The History of Java*]' (De Geus and De Visser Smits 1931: 172). A memoir of *De Vriendschap* lodge states that 'among those entertained were men such as Mackenzie and [G.P.] Baker who had distinguished themselves during the English Interregnum in the field of science" (*Gedenkschrift* 1934: 27).

Other Fields of Supranational Fraternal Co-Operation

Dutch and British Masons also collaborated in other activities. For reasons of space I cannot do more here than briefly refer to a few examples and mention the names of the Masons who were directly involved. Firstly, the task of the Mackenzie Land Tenure Commission was to gather information about agricultural and economic conditions in Java as a preparatory step for land reforms and taxation. Aside from the Chairman, Mackenzie (a Mason, see above), the Commission comprised the Dutch members F.J. Rothenbühler (former Commander of Java's East Hook; masonic affiliation unknown), Johannes Knops (Mason, member of the Semarang lodge *La Constante et Fidèle*) and P.H. van Lawick van Pabst (Mason, *La Vertueuse*). The input of the Dutch members was considerable, according to the research by John Bastin (1954), one of Raffles' leading British biographers with a sound knowledge of Dutch historical sources.

Deserving separate mention are the scientific activities of Raffles and Mackenzie. Soon after the fighting in Java had ended, they were able to devote part of their time to the pursuit of their wide-ranging scientific interests. Raffles and Muntinghe were elected respectively as the president and vice-president of the Batavian Society of Arts and Sciences, an institute founded and dominated by Masons (Van der Veur 1976). In their respective investigations of archaeology and art history, Raffles and Mackenzie received much assistance from Dutch fellow Masons, both directly and indirectly. For instance, in both their collections is found a manuscript by Van Boeckholtz about the history of Java, along with an English translation and additional notes by Mason J. Ekenholm. Van Boeckholtz' essay may have inspired Raffles' *History of Java*, next to the well-known study of William Marsden on the history of Sumatra. Further, both Raffles and Mackenzie owned copies of Javanese manuscripts dealing with traditional historiography (*babad*) originating from the private collections of Engelhard and Van Middelkoop, who also arranged for the translation of some of these manuscripts (Weatherbee 1978). Van Middelkoop was the Worshipful Master of lodge *De Vriendschap* in which Raffles was raised to the degree of Master Mason. Then there is the large number of cartographic maps, plans and drawings of Hindu-Buddhist temples by Dutch draughtsmen and fellow Masons such as Van Boeckholtz, H.C. Cornelius, P. Beetjes, and J.W.B. Wardenaar. The greater part of this extensive material is now kept in the British Library and the British Museum.

The final example of British-Dutch fraternal collaboration is the founding of the Java Auxiliary Bible Society, with Raffles assuming the position of President and Engelhard that of Vice-President. Apart from three Reverends, the members of the Society's Committee comprised of eight prominent Dutchmen, all of whom were Masons (*Java Government Gazette* 11 June 1814).

Personal Bonds Between Leading Dutch and British Freemasons

The contributions of Dutch Masons were not confined to administrative activities and scientific investigations; on some occasions their support assumed a more personal character. Take, for instance, the involvement of some of the leading Dutch and British Masons, including Engelhard and Raffles, in the purchase of plots of governmental land by private individuals. Meant to counterbalance the financial deficits of the Treasury, the participation of Engelhard and some well-informed members of the Government and of the Mackenzie Commission as buyers were bound to raise suspicions among other bidders (Bastin 1954: 85). These controversial transactions would later be brought to the attention of the East India Company by Gillespie, following his transfer to Calcutta. Although Raffles would formally be acquitted from the corruption charges levelled against him, the affair would remain a stain on his reputation and a hurdle for his biographers. "Perhaps the most likely explanation for the land scandal", according to Tim Hannigan (2012: 258), "was that Raffles had been taken for a ride by a cabal of dodgy Dutchmen, steeped in the old ways of Batavian society and in collusion with the murky McQuoid". A weak point in this explanation is the presumed naiveté of Raffles as an innocent investor. Engelhard, for one, alleged that he had acted in line with Raffles' fancies at the time to stay on in Java if the British interlude would happen to come to an end (Sollewijn Gelpke 1889: 42). Engelhard was willing to take over Raffles' share at the original price when the latter, having heard of Gillespie's charges against him, wanted to get rid of his share as quickly as possible. Hannigan might have phrased his explanation somewhat differently had he known that the 'cabal' in this case comprised several Dutch and British brethren.

A more meritorious example of Dutch support was the hospitality extended to Raffles after the death of his wife Olivia, and the reception of the delayed message of the death of his patron Lord Minto, which had brought Raffles to the edge of a mental breakdown. The Dutch Masons Cranssen, Engelhard and Pieter van Heemstede Cappelhoff then provided accommodation for Raffles at their estate in the cooler climate of the Priangan highlands (Wurtzburg 1954: 373).

As for the merits of Muntinghe and Cranssen as members of the Java Council, it is worth quoting Raffles from his Foreword to *The History of Java*:

> "At this critical juncture these two gentlemen declared for England and the ancient order of things; and to the influence of their decision and conduct is to be ascribed, not only the cordiality and good understanding which soon prevailed between the English and Dutch, but in great measure also that general tranquillity of the country, without which the retransfer of it to the rule of its former masters might have been practical" (Raffles 1817: xii).

Please take note of the peculiar expression "[to declare for] the ancient order of things", which I take for a covert allusion to their adherence to the precepts of Freemasonry.

The Ending of the British-Dutch Freemasonry Governance Experiment

So far as can be ascertained, the collaboration between British and Dutch went relatively smooth, especially during the first two years. Regarding the functioning of the Java Council, however, an unworkable situation gradually arose as a result of mounting tensions and disagreements between Raffles and Gillespie. Their frequent clashes are for a significant part to be attributed to diverging factional interests of a civilian administrator and a military commander, as well as to an incompatibility of personalities, but perhaps also to their respective nationalities and social backgrounds, namely English middle class against Irish-Scottish gentry (Wurtzburg 1954: 186; Hannigan 2012: 250-252). The Dutch members of the Java Council and fellow Masons proved utterly helpless in alleviating this situation.

As for the collaboration of Masons at the lower levels of administration, such as the Mackenzie Commission, communication problems stemming from a poor command of English on the part of Dutch members seem to have resulted in various delays and mutual frustrations (Bastin 1954: 125-126). This fact and the calling back of Mackenzie to India influenced Raffles' decision to stop the land tenure research and dissolve the Commission.

Nationalistic sentiments and considerations arose over time. It became increasingly evident that Raffles was anxious for Java to remain British, even after it became known that the island and its dependencies would be restored to the Dutch under the Treaty of London in 1814. A telling example is Raffles' secret order to Captain G. P. Baker to chart the southern coast of Central Java and to look for suitable anchorages – an assignment that can only be explained by their possible future use in British military operations against the Dutch on this relatively unprotected side in their line of defence (Hannigan 2012: 325). Rumours of other suspicious activities led Engelhard to contact the Dutch government with a plea for Raffles' early recall. This is what happened: in December 1815, Raffles was succeeded as Lieutenant-Governor by John Fendall, who assumed diplomatic responsibility for the formal restoration of Java and its dependencies to the Dutch. On 25 March 1816, Raffles departed for England, never again to set foot on his beloved island Java.

References

BARLEY, N. (1991) The Duke of Puddle Dock; Travels in the Footsteps of Stamford Raffles. Harmondsworth: Viking/Penguin Group.

BASTIN, J. (1954) Raffles Ideas on the Land Rent System in Java and the Mackenzie Land Tenure Commission. 's-Gravenhage: Nijhoff.

COUPLAND, R. (1934) Raffles 1781-1816. London: Milford.

Gedenkboek Vrijmetselarij (1917) Gedenkboek van de vrijmetselarij in Nederlandsch Oost-Indië 1767-1917. Soerabaia: Van Dorp.

Gedenkschrift (1934) Loge "De Vriendschap" No. 35 N.C. in het oosten van Soerabaja. Gedenkschrift uitgegeven bij gelegenheid van het 125-jarig bestaan, 28 november 1934. Malang: Nimef.

GEUS, A. de and VISSER SMITS, D. de (1931) Beknopte geschiedenis der Vrijmetselarij in Nederlandsch Oost-Indië. In VISSER SMITS, D. de (ed.). Vrijmetselarij; Geschiedenis, maatschappelijke beteekenis en doel, 153-179.

HAAN, F. de (1910) Priangan. Batavia: Bataviaasch Genootschap van Kunsten en Wetenschappen.

HANNIGAN, T. (2012) Raffles and the British Invasion of Java. Singapore: Monsoon Books.

HARLAND-JACOBS, J.L. (2007) Builders of Empire; Freemasons and British Imperialism, 1717-1927. Chapel Hill: The University of North Carolina Press.

RAFFLES, TH. S. (1817) The History of Java. London: Black, Parbury, and Allen.

SOLLEWIJN GELPKE, J.H.F. (1889) De landerijen onder het Engelsche Tusschenbestuur verkocht en het verbod van heerendiensten aldaar. Ceribon: Bisschop.

SPIERINGS, A. (2013) Winnaar en verliezer: de reputaties van Thomas Stamford Raffles en Nicolaus Engelhard naast elkaar gezet. Unpublished MA thesis, Leiden University.

STEVENS, T. (1994) Vrijmetselarij en samenleving in Nederlands-Indië en Indonesia, 1762-1961. Hilversum: Verloren.

VEUR, P.W. van der (1976) Freemasonry in Indonesia from Radermacher to Soekanto, 1762-1961. Athens: Ohio University Center for International Studies.

VISSER SMITS, D. de (ed.) (1931) Vrijmetselarij; Geschiedenis, maatschappelijke beteekenis en doel. Soerabaia: Van Dorp.

WEATHERBEE, D.E. (1978) Raffles' Sources for traditional Javanese Historiography and the Mackenzie Collections. Indonesia 26: 63-95.

WURTZBURG, C.E. (1954) Raffles of the Eastern Isles. Singapore: Oxford University Press.

The Heron and the Snail: A Visual Pun in a Javanese Manuscript of the *Jawharat al-Tawḥīd*

EDWIN P. WIERINGA

This contribution[1] to the *Festschrift* in honour of Jos Platenkamp may be perceived as an example of work done by one of those "isolated scholars sitting in reading rooms and staring at very old books, wondering how they were made and what to make of them" (Machan 2003: 1005). The aphorism of a Tobelo elder, which functions as the motto of Jos's PhD thesis, also holds true for what I will try to achieve in the following pages, viz. "Everything is meaningful, it is just that we, people on earth, can often only guess" (Platenkamp 1988). The book at which I have been staring for some time now is in fact not so very old. It is a Javanese handwritten copy made in Surakarta in 1919, containing an extensive Javanese commentary upon the *Jawharat al-Tawḥīd* ("Jewel of Divine Oneness"), which is a well-known Islamic theological work on the Unity of God, originally written by Ibrāhīm al-Laqānī (d. 1041/1631).[2] This most influential Ash'arite creedal work is studied throughout the entire Islamic world.[3]

This also includes insular Southeast Asia. At the end of the nineteenth century, Snouck Hurgronje (2007: 307) already observed that "a Malay Commentary on the *Jauharat at-tauhîd* (by Ibrâhîm al-Laqânî) after a manuscript written in Sambas" was printed in Mecca. According to Mohd. Nor bin Ngah

1 First drafts of this paper were presented previously on several occasions, among others at the 14th Manassa Congress in Yogyakarta, 11-13 September 2012 ("*Kitab Johar Tokid* salinan *abdi carik* Ki Sastragupita: Penjawaan Kitab *Jauharat at-Tauḥīd* karya Ibrāhīm al-Laqānī (w. 1041H/1631M)") and at the Association franco-indonésienne Pasar Malam in Paris, 15 January 2013 ("Lire une enluminure: Le Héron et l'escargot dans un manuscrit islamique javanais"). I am grateful to the respective audiences for interesting discussions. The final version of this article was written during my stay as Special Visiting Professor at Tokyo University of Foreign Studies (TUFS) in the winter semester of 2015-2016. I would like to thank Assoc. Prof. Yumi Sugahara for inviting me to Japan and Prof. Koji Miyazaki (TUFS) for acting as my academic host. My special thanks go to Mrs. Katherine Maye-Saidi for proofreading the article.

2 Duly listed in GAL II: 412, i.e. Brockelmann (1949: 412).

3 Cf. El-Rouayheb (2015: 276) who reckons it among the most widely studied works in Islam. According to Loimeier (2009: 188-189), this was the basic text in Zanzibar in the late nineteenth century for advanced students in theology. It is still on the curriculum of Islamic university education in contemporary Daghestan, see Bobrovnikov, Navruzov, and Shikaliev (2010: 155). An English translation is available at http://marifah.net/articles/JawharatalTawhid.pdf.

(1983: 1), the Malay translation of the *Jawharat al-Tawḥīd* belongs to "the most popular and widely used *Kitab Jawi*," i.e. Islamic theological books in Malay script. Several translations and commentaries in local languages (Malay, Javanese, Sundanese, and Madurese), which are still available in print, testify to its continuing popularity until the present day among Muslims in insular Southeast Asia.[4] As Van Bruinessen (1990: 252) observes, the *Jawharat al-Tawḥīd* is still memorised by students (*santri*s) who also study its commentaries:

> "One of these is Ibrāhīm al-Bājūrī's *Tuḥfat al-murīd*. An anonymous Malay scholar and two Javanese *ulama*, Saleh Darat of Semarang and Ahmad Subki Masyhadi of Pekalongan, wrote extensive commentaries in their regional languages, which are commonly known by the same title of *Jauharatut Tauhid*. Saleh Darat's Javanese commentary, especially, is interesting in that it reflects contemporary Javanese views and concerns."

Although it would be very tempting for this card-carrying philologist to dwell on the contents of the 1919 manuscript which consists of no less than 589 pages thus constituting a most extensive commentary raising all kinds of questions on its place in the textual genealogy of the *Jawharat al-Tawḥīd* in Javanese, I will concentrate my attention on its colophon for it is the accompanying animal "vignette" with its intriguing image of a (quasi-)mirrored heron and a snail which most set me wondering.[5] Assuming that this marginal art is not mere decoration, I herewith present my guesswork as to its possible meaning. The present article is very much a pioneering essay, for as Annabel Teh Gallop (2005: 160) points out,

> "Islamic manuscript art from Southeast Asia is still barely known, either as a subset of Islamic art or as a distinctive Southeast Asian art form. This is largely because, in the Malay world, the study of manuscripts has traditionally been the domain of philologists rather than art historians, but it also reflects the uniquely private nature of manuscript art."

On the latter aspect, Gallop (2005: 160) astutely observes that

> "manuscript art is, literally, a closed book. Hidden within the covers of manuscript volumes, illumination was only ever encountered and savoured by a very small elite – the readers, owners or patrons of the book."

In the case of the 1919 Javanese manuscript, only a few eyes may have been able to view it. Its patron and first owner was the Solonese courtier Raden Ngabehi Kartahartana. The second owner appears to have been R.H.S.

4 Cf. Hermansyah and Zulkhairi (2014). See also the catalogue of Kawashima (2015: 122-125, 127-128) and the handlist of Heer (2012: 11, 27, 64).

5 In the beginning of the manuscript the original author of the text is called "Seh Ibrahim Alkani [*sic*]." Furthermore, several theological works are mentioned, *inter alia* "kitab Bajuri anggitanipun Seh Ibrahim" (page 4) which must be Ibrāhīm al-Bājūrī's *Tuḥfat al-murīd.*

Wongsodinomo, a well-known *batik* entrepreneur in Solo (see his stamp on the title page, illus. 1). After his death, the manuscript ended up for sale at the local flea market where my late friend Jan Hendrik ("Henk") Nederhand (1941-2012) bought it on my behalf in 1994 for a handsome amount of rupiahs. The title page informs us that the text is called *Kitab Johar Tokid* which was copied by Ki Sastragupita on the order of Raden Ngabehi Kartahartana (illus. 1).[6] The copying was begun on 20 May 1919 and finished on 4 September 1919.[7] As the copyist explains in the introduction, he "carried out the command" (*anglampahi ḍawuhipun*) of his master, Raden Ngabehi Kartahartana, who wished to possess a Javanese copy in Javanese script of this text. There already existed a version written in *pegon* (Javano-Arabic script), but none in Javanese script were for sale.[8] Ki Sastragupita was the copyist (*jurucitra* or scribe) who accomplished this task for Raden Ngabehi Kartahartana.

The latter was a high-ranking aristocrat at the Solonese court who was also known as Raden Ngabehi Tirtapraja, and an author in his own right.[9] Together with Raden Ngabehi Djajasupana he had written a biography of the grand-vizier of Solo, *Laksitarja* (1885) and the *Sĕrat anyariyosakĕn lampahanipun Baktiyar* which was based upon a Malay prototype, viz. *Hikayat Bakhtiar* (1888).[10] He also adapted an episode of the *Thousand and One Nights*, viz. *Pĕṭikan cariyos sewu: Alidin Abu Samad* (1892).[11] Furthermore, he wrote non-fictional works: *Kawruh Kewan* on animal science (1885; 1894) and *Gĕgĕlangan ilmu alam* on natural science (1891). He also seems to have been behind the rewriting of a version of the historiographical work *Babad Bangun Tapa* (1917?) on the exiled Solonese ruler Pakubuwana VI (d. 1849).[12]

6 The Javanese text of the title page reads: *Punika Kitab Johar Tokid/kagunganipun/Raden Ngabehi Kartahartana/sri nugraha pangkat 5/ing Surakarta Adiningrat/katanḍan abdinipun jurucitra/Ki Sastragupita/Tĕmbung kita kanggenipun wontĕn ing kitab Johar Tokid atĕgĕs kowe utawi sampeyan.*

7 The date of the beginning is mentioned on page 1, viz. 19 Saban 1849 Anno Javanico (i.e. 20 May 1919 Common Era) with the chronogram *trus* [9] *sucining* [4] *brahmana* [8] *tunggal* [1]. The date of its finishing is mentioned on the final page (589), viz. 8 Besar 1849 Anno Javanico (i.e. 4 September 1919 Common Era).

8 The Javanese text on page 1 reads: *Purwanipun kula Ki Sastragupita anĕḍaki bab Johar Tokid, sĕrat pegon kasĕrat Jawi, anglampahi ḍawuhipun bandara Raden Ngabehi Kartahartana, kĕpengin kagungan ingkang sĕratan Jawi, baḍe tumbas botĕn wontĕn ingkang sade ingkang sampun sĕratan Jawi...*

9 His works are listed on the website http://www.worldcat.org/identities/viaf-290439190/ (last accessed: 26th November 2015).

10 The catalogue of Pigeaud (1970: 261) describes the *Laksitarja* as a text on Adipati Sasranagara, the grand-vizier of Surakarta. Ricklefs (2007: 144) mentions it as a text on "the life of *Patih* Sasranĕgara and the bestowing of the star of the Lion of the Netherlands on him."

11 Information from http://www.worldcat.org/title/pethikan-carios-sewu-lalampahanipun-alidin-abul-samad/oclc/66272913 (last accessed: 26th November 2015).

12 For further details on R.Ng. Kartahartana, see Wieringa (1994: 44-46, 166-169).

The main text of the *Jawharat al-Tawḥīd* ends on page 580, followed by a kind of *captatio benevolentiae* on pages 580-581, in which the copyist turns to God (*Gusti Allah*), asking for forgiveness for all the mistakes which he has made (illus. 2).[13] Eight lines in Arabic script finally close the text with a conventional prayer in Arabic. Rather remarkably, the copyist who displays a trained hand in writing Javanese script, was apparently not very used to writing Arabic characters which make a very insecure impression. The final three lines contain the year of the Islamic calendar, viz. 1337 (equivalent to 1918-1919 of the Common Era) in Arabic and Latin ciphers, and the copyist's name Sastragupita in Javano-Arabic script, likewise written in a rather inexperienced way.

But what is the meaning of the image of the mirrored heron and snail on the final page (illus. 2)? It has crossed my mind that we might be dealing here with a *candrasĕngkala mĕmĕt,* i.e. a chronogram in the form of a picture in which the numerical values of its constituent elements express the year of the book's compilation. If the four animals represent four digits, the only possible year would be 1919 Common Era. The heron normally belongs to the category of chronogram words for "one" as it is a bird (*manuk*, see Bratakesawa 1928: 126), although it is a winged creature (*ĕlar* or "wing" denotes "two" in the standard lists, see Bratakesawa 1928: 153; Noorduyn 1993: 302 and Teeuw 1998: 376).[14] The snail must then have the numerical value of "nine," but it is not included in the standard lists. Another possibility, viz. such a chronogram sentence as *cangak mangan keyong iku* ("the heron eats that snail") is also rather problematic. At least two digits can be filled in: as convention prescribes, the sentence has to be read backwards, so the chronogram word for the millennium is *iku* ("that") which of course has the numerical value of "one," whereas the chronogram word for the decade is *mangan* ("to eat") which normally has the numerical value of "three" (see Darmosoetopo 1976: 124).[15] This leaves us with 1-x-3-x which might point to the Islamic year 1337,

13 The Javanese text reads: *Ḍuh, ḍuh Gusti Allah, manawi wontĕn lĕpating panyĕrat kawula, saha kecering tĕtĕmbungan ingkang cewet dereng kawula sĕrat tuwin kiranging aksara utawi sanḍangan pangkon miwah pĕjahan sarehning kawula wau bangsa langip botĕn langkung kawula namung nyuwun gĕnging pangapura, mugi sampun andadosakĕn ca- [page 581] cading badan kawula, makatĕn malih para nupiksa ugi paring pangaksama kados panyuwun kula inginggil wau. Amin, amin, amin, ya Rabul Ngalamin.*

14 The words in chronograms have a fixed numerical value. For example, words associated with "eyes," "eyebrows," "ears," "arms," and "hands" mean "two." The philologist Noorduyn (1993: 304) explains that the reason why words for "fire" have the numerical value "three" goes back to the old Indian idea that there are three kinds of sacrificial fires (*tryagni*). However, the popular Javanese reasoning is that fire has three colours (yellow, red, and black) and three characteristics, viz. burning, smoking, and providing heat, see manuscript Cod. Or. 2177 (2) in Leiden University Library. For a good English-language introduction to Javanese chronograms, see Ricklefs (1978: 239 ff.). The most comprehensive explication is still Bratakesawa (first published in 1928).

15 For a discussion of the chronogram word *iku* as "one," see Noorduyn (1993: 303, 311).

but the only bird with the numerical value of seven seems to be the goose (*banyak* or *angsa*, see Bratakesawa 1928: 89; Pigeaud 1938: XII).

Turning to symbolism, it is of little use, I think, to rely upon general reference works, e.g. Ferber (2007) or Butzer and Jacob (2008), in order to find a possible answer. The internet with its myriad possibilities falls in the same category. For example, on a website discussing the symbolism of the heron, we find that this animal is "symbolic in many cultures," after which a number of examples are enumerated: in ancient Egypt the heron is "honored as the creator of light"; in China it "represents strength, purity, patience and a long life"; among native American tribes the heron is perceived as "a symbol of wisdom."[16] Such lists can go on and on, of course, but rarely if ever do we find information in such general compendia on Indonesian examples. The few Indonesian handbooks, however, are not very helpful either. For example, Van der Hoop (1949) includes many animals, but not the heron or snail. However, Van der Hoop (1949: 226) draws attention to the shell which "in ornamentation occurs mostly in the Hindu period, and is usually given wings as the symbol of Vishnu." As this shell very much looks like a snail, Sunaryo (2009: 118) has copied the images and interpretation of Van der Hoop, but suggesting that we are dealing here with a snail.

An Indonesian colleague of Gadjah Mada University in Yogyakarta once told me that only advanced students at the post-master level were allowed to write theses on manuscript art, because the temptation of *hineininterpretieren* was particularly high in this field. Browsing the bookshelves, thumbing tomes, or simply surfing the web, it is not too hard to find out that the marginal image of the snail is ubiquitous in medieval Western manuscripts where it is, among others, linked to female genitalia (Camille 1992: 35). Add to this that the word "bird" is both in Malay/Indonesian (*burung*) and in Javanese (*manuk*) a colloquial expression for "penis" (or as Stevens and Schmidgall-Tellings (2004) rather pedantically insist, "an infant's penis") and you will know what kind of analysis to expect.

Contrary to the positive symbolic value which is attached to the heron in many cultures, this bird has a bad reputation in Javanese literature, in which it is emblematic of the hypocrite: its immaculate appearance is only outwards, hiding its treacherous character.[17] This image can already be found in Old Javanese literature (e.g. *Tantri Kāmandaka* where the heron beguiles fish into trusting him), but is also regularly evoked in New Javanese (Islamic) literature (see Drewes 1966: 342 n. 19 with references). The heron is often compared to

---

16 See http://www.symbolic-meanings.com/2007/10/08/symbolism-of-the-heron/ (last accessed: 26th November 2015).

17 About its positive symbolic value (grace, calmness, intelligence, tactfulness, etc.), see e.g. the website http://www.buzzle.com/articles/what-does-a-heron-symbolize.html (last accessed: 26th November 2015).

the *wiku ĕnḍog* or egg-saint, i.e. white outside but completely differently coloured inside (Drewes 1966: 342 n. 19). Once again, the danger of *hineininterpretieren* looms large. Guided by courses of critical theory (post-colonial studies and the like), a student might easily interpret the image of the hypocrite together with the image of the proverbially slow snail as an indirect criticism directed at the high-ranking nobleman who commissioned the manuscript. Should a veritable Muslim not have been able to read the original text in Arabic, not needing a Javanese intermediary? Is this, then, our student might ask, perhaps a "weapon of the weak" – to use a powerful "euphoric couplet" in Indonesian studies?[18]

Of course, such an interpretation would be much too bold, completely oblivious to the fact that "the contemplation of the work of art on the page remained an intimate, one-to-one activity" (Gallop 2005: 160-161). The image was only intended to be seen by its patron and first owner and hence it is hardly conceivable that a servant could in this way poke fun at his master in the margins. It seems far more likely that the image somehow had to do with the scribe himself. The use of two rather negative symbols would fit the colophon expressing the copyist's modesty, which once more would lower his position, being fully in accordance with Javanese etiquette.

The place to look for additional information is the field of proverbs and fables, but it took me quite some time to find useable clues. Dictionaries of proverbs proved to be of no help, whereas the collections of fables mainly confirmed the image of the treacherous heron. For example, in the fable of the crow and the heron it is explained how the heron once turned the tables on the crow so that the formerly beautiful crow became black and ugly whereas the heron took its place as a good-looking white bird (Puteh and Said 2003: 113, 276-277). The fable of the heron and the crab (Puteh and Said 2003: 201) is a retelling of the well-known *Pancatantra* tale, which is also known from the above-mentioned Old Javanese *Tantri Kāmandaka*, in which the wicked heron eats almost all the fish in the pond until the crab finds out what has been going on. There is little on the snail which may be of relevance for our purposes, apart perhaps from an East Javanese folktale about the "golden snail" (*keyong emas*), in which Princess Candra Kirana is put under a spell and transformed into a golden snail. She will only resume her human form again when she finally meets her destined husband Raden Inu Kĕrtapati (Komandoko 2013: 173-175). This could be interpreted, if one so wishes, as an illustration of the idea that appearances are deceiving, just as in the case of the heron, but here without the negative connotation.

18 This phrase was first coined by Scott (1985). See Cribb (2005) for its well-worn use in Indonesian studies.

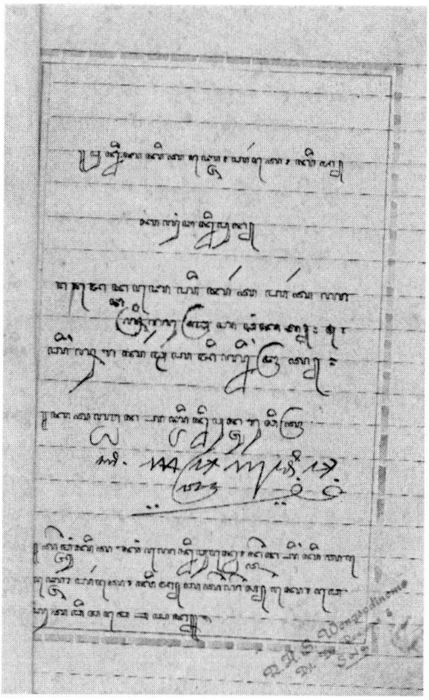

Fig. 1: Title page with the signature of the copyist Sastragupita (in Javanese script (line 7) and underneath (line 8) in Javano-Arabic script (gh-f-y-t or ghupita). In addition, the stamp of its second owner, R.H.S. Wongsodinomo who lived in Djalan Dr. Radjiman 8 (in Solo).

Fig. 2: Colophon written in a triangular block with (quasi-)mirrored image of the heron and the snail

However, what is the idea behind the combined image of the heron and the snail? One could perhaps think of a race between unequal contestants, i.e. variants of the tortoise and the hare type,[19] but I fail to see a connection with the text; other folktales may, however, provide a possible solution. For example, in the collection of Othman Puteh and Arifin Said, which is said to be an anthology of Malaysian folktales but in fact contains examples that can be found in the entire region of insular Southeast Asia and beyond, there is an aetiological tale about the reason why the heron is so meagre. Once upon a time, the arrogant heron tried to trick king Sulaiman (Solomon) so that he could eat seven times a day, an act for which he was punished following which he had great difficulty finding food, and therefore becoming very meagre (Puteh and Said 2003: 284-285).

A variant of this folktale of the arrogant heron has been made famous in European literature through the fables of Jean de la Fontaine (1621-1695), in which he tells the story of the choosy heron who declines to eat until he finds the perfect fish, eventually at the end of the day, however, ending up with just a small snail. La Fontaine provides the following moral: *Gardez-vous de rien dédaigner* ("Beware of disdaining anything").[20] A similar message, or so it seems to me, might have been intended by Ki Sastragupita. In words he had already expressed his excuses about possible mistakes and the accompanying image of the heron and the snail may be seen in this same perspective of humility: please accept this humble offering. As visitors to Javanese households know, the common invitation to eat together is expressed (in polite Javanese) as *(mangga) ḍahar sawontĕnipun* or (in Indonesian) as (*mari* or *marilah,* i.e. "let's) *makan seadanya* which Stevens and Schmidgall-Tellings (2004: 6) translate as "to eat what is before one, take potluck." This is the common self-denigrating way to minimise the quality of the offering which is usual throughout Indonesia, known as *berbasa-basi* or observing the conventional pleasantries (cf. DuFon 2004: 65, 68-70). This would also apply to a most elaborate meal, well known to the Dutch as *rijsttafel* ("rice table").

Lack of space does not permit further study of "the world of exchange" in Javanese literature. Could Ki Sastragupita have known La Fontaine's fable or some version of it? Due to the flood of Dutch schoolbooks which were translated into Malay and Javanese from the middle of the nineteenth century this is not unlikely. However, little is known about the literary horizon of readers and writers at the early twentieth-century Solonese court. Thompson's well-

19 For versions from all over the world, see e.g. http://www.pitt.edu/~dash/type0275.html (last accessed: 26th November 2015).

20 The story can also be found on the internet as "children story" and its lesson is summarized as follows: "Do not be too hard to suit or you may have to be content with the worst or with nothing at all." See http://www.childrenstory.info/children-stories/fables/the-heron/ (last accessed: 26th November 2015). For a recent edition of La Fontaine's fable of the heron with beautiful illustrations, see Biberon (2015: 106-109).

known motif-index of folk-literature intriguingly notes a reference to India under U125.1. "Heron (crane) loses interest in heaven if there are no snails," but I am unaware of variants in (Old) Javanese literature.[21]

Broadly speaking, there are two approaches to marginal art: firstly, relegating it to the "menial position of 'pure decoration'," and secondly to consider it as "full of 'meaning'" (Camille 1992: 31). When Ki Sastragupita added his image to the text, he and God knew what it entailed. Now, only God knows, and we, people on earth, can only guess. Or as Indonesian Muslims conventionally like to conclude: *Wallahu alam bissawab* ("only God knows what the real truth is").

References

BIBERON, M. (2015) Fables. La Fontaine illustré par ... Grandville, Boutet de Monvel, Lorioux, Doré, Rabier, La Nézière, Fraipont, Bouchot, Oudry, Vimar, Gélibert, Sabran, Klein, etc.. Rennes: Editions Ouest France.

BOBROVNIKOV, V., NAVRUZOV, A. and SHIKALIEV, S. (2010) Islamic Education in Soviet and Post-Soviet Daghestan. In KEMPER, M., MOTIKA, R. and REICHMUTH, S. (eds.), *Islamic Education in the Soviet Union and its Successor States*, pp. 107-167. London and New York: Routledge.

BRATAKESAWA, R. (1928) *Katranganing tjandrasengkala.* Batavia: Volkslectuur. http://www.sastra.org/bahasa-dan-budaya/38-kamus-dan-leksikon/313-condrasangkala-bratakesawa-1928-610 [Accessed: 26th November 2015].

BROCKELMANN, C. (1949) *Geschichte der arabischen Literatur.* (Band II). Leiden: Brill.

BUTZER, G. and JACOB, J. (2008) *Metzler Lexikon literarischer Symbole.* Stuttgart: Metzler.

CAMILLE, MICHAEL (1992) *Image on the edge: The margins of medieval art.* London: Reaktion Books.

CRIBB, R. (2005) Circles of Esteem, Standard Works, and Euphoric Couplets. *Critical Asian Studies* 37(2): 289-304.

DARMOSOETOPO, R. (1976) *Laporan penelitian: Peninggalan-peninggalan kebudayaan di lereng barat Gunung Lawu.* Yogyakarta: Universitas Gadjah Mada.

DREWES, G.W.J. (1966) The Struggle between Javanism and Islam as illustrated by the Sĕrat Dĕrmagandul. *Bijdragen tot de Taal-, Land- en Volkenkunde* 112(3): 309-365.

DUFON, M.A. (2004) *Mari belajar sopan santun Bahasa Indonesia manual.* https://scholarspace.manoa.hawaii.edu/bitstream/handle/10125/10578/LT34m.pdf?sequence=2. [Accessed: 26th November 2015].

EL-ROUAYHEB, K. (2015) *Islamic Intellectual History in the Seventeenth Century. Scholarly Currents in the Ottoman Empire and the Maghreb.* Cambridge: Cambridge University Press.

FERBER, M. (2007). *A Dictionary of Literary Symbols. 2nd Ed.* Cambridge: Cambridge University Press.

21 Thompson's motif-index is also available on the internet, see https://www.ualberta.ca/~urban/Projects/English/Content/u.htm (last accessed: 26th November 2015).

GALLOP, A.T. (2005) Islamic Manuscript Art of Southeast Asia. In BENNETT, J. (ed.), *Crescent Moon: Islamic Art and Civilisation in Southeast Asia*, 158-189. Adelaide: Art Gallery of Southeast Asia.

HEER, N. (2012). *A Concise Handlist of Jawi Authors and Their Work.* Seattle, Washington: n.n. http://faculty.washington.edu/heer/handlist23.pdf. [Accessed: 26th November 2015].

HERMANSYAH and ZULKHAIRI (2014) *Transformasi syair Jauharat at-Tauhid di Nusantara.* Denpasar: Pustaka Larasan.

KAWASHIMA, M. et al. (2015) *A Provisional Catalogue of Southeast Asian kitabs of Sophia University.* 2nd Ed. Tokyo: Sophia University. http://dept.sophia.ac.jp/is/iac/en/library_info/collection/kitab2nd.pdf. [Accessed: 26th November 2015].

KOMANDOKO, G. (2013) *Koleksi terbaik 100 plus dongeng rakyat Nusantara.* Yogyakarta: Cabe Rawit.

LOIMEIER, R. (2009) *Between Social Skills and Marketable Skills: The Politics of Islamic Education in 20th Century Zanzibar.* Leiden: Brill.

MACHAN, T.W. (2003) Review of Andrew Taylor, Textual Situations. *Speculum* 78:1005.

MOHD. NOR BIN NGAH (1983) *Kitab Jawi: Islamic Thought of the Malay Muslim Scholars.* Singapore: Institute of Southeast Asian Studies.

NOORDUYN, J. (1993) Some Remarks on Javanese Chronogram Words: A Case of Localization. *Bijdragen tot de Taal-, Land- en Volkenkunde* 149(2): 298-317.

PIGEAUD, Th. (1938) *Javaans-Nederlands handwoordenboek*. Groningen: Noordhoff.

PIGEAUD, T. (1970) *Literature of Java.* (Volume III). The Hague: Nijhoff.

PLATENKAMP, J.D.M. (1988) *Tobelo: Ideas and Values of a North Moluccan Society.* Leiden: Repro Psychologie.

PUTEH, O. and SAID, A. (2003) *Himpunan 366 cerita rakyat Malaysia*. Kuala Lumpur: Utusan Publications.

RICKLEFS, M. (1978) *Modern Javanese Historical Tradition: A Study of an Original Kartasura Chronicle and Related Materials.* London: SOAS.

RICKLEFS, M. (2007) *Polarising Javanese Society. Islamic and Other Visions (c. 1830-1930).* Singapore: NUS Press.

SCOTT, J.C. (1985) *Weapons of the Weak: Everyday Forms of Peasant Resistance.* New Haven: Yale University Press.

SNOUCK HURGRONJE, C. (2007) *Mekka in the Latter Part of the 19th Century. Daily Life, Customs, and Learning. The Moslims of the East-Indian Archipelago.* Leiden: Brill.

STEVENS, A.M. and SCHMIDGALL-TELLINGS, E. (2004) *A Comprehensive Indonesian-English Dictionary.* Athens, Ohio: Ohio University Press.

SUNARYO, A. (2009) *Ornamen Nusantara. Kajian khusus tentang ornamen Indonesia.* Semarang: Dahara Prize.

TEEUW, A. (1998) An Old Javanese Poem on Chronogram Words. In JANSE, M. (ed.), *Productivity and Creativity: Studies in General and Descriptive Linguistics in Honor of E.M. Uhlenbeck*, 369-392. Berlin and New York: Mouton de Gruyter.

VAN BRUINESSEN, M. (1990) Kitab kuning: Books in Arabic Script used in the pesantren Milieu. *Bijdragen tot de Taal-, Land- en Volkenkunde* 146(2/3): 226-269.

VAN DER HOOP, A.N.J. Th. à Th. (1949) *Indonesische siermotieven. Ragam-ragam perhiasan Indonesia. Indonesian ornamental design*. Bandoeng: Nix.

WIERINGA, E.P. (1994) *Babad Bangun Tapa. De ballingschap van Pakubuwana VI op Ambon 1830-1849*. Leiden: Universiteit Leiden.

Applied/Developmental Anthropology

The Mediation of Intercultural Competence as a Form of Applied Anthropology

URSULA BERTELS

"Only ethnology can teach us to regard ourselves through a foreigner's eyes." This was the final sentence of Jos D. M. Platenkamp's contribution to „Über die gesellschaftliche Relevanz der Ethnologie" ["On the social relevance of ethnology"] at the conference „Aus der Ferne in die Nähe – Neue Wege der Ethnologie in die Öffentlichkeit" [From a far to here – new ways of bringing ethnology into the public"] in February 2002 (Platenkamp 2004: 31).

Referring to examples from the Moluccas in Eastern Indonesia and of the Lao Lum in Laos Platenkamp explained that among these people „das Fremde eine qualitative Vervollständigung der Gesellschaft als Ganzes darstellt" (*ibid.*: 28) and suggested that ethnology should devote to the question as to the position that 'foreignness' occupies within the value system of German society (*ibid.*: 31). The conference was organised by Ethnologie in Schule und Erwachsenenbildung (ESE) e.V. (Anthropology in School and Adult Education), which cooperates with the Institute of Ethnology of the University of Münster. Platenkamp is both a member and the academic advisor of this association.

Regarding oneself through a foreigner's eyes – as Platenkamp suggested – is an important element of intercultural competence. Thus the quotation given at the beginning underlines the role of anthropologists in the mediation of intercultural competence.

Intercultural Competence – An Ethnological Approach

In times of globalisation, intercultural competence has become a key qualification for living and working in multicultural societies. Imparting intercultural competence to pupils, students and adults by means of intercultural education has become increasingly important in order to prepare them to live and work together with people from different cultural backgrounds.

Politicians have recognised the importance of intercultural competence for a multicultural society. Thus, already in 1996 the Minister of Culture of the Federal Republic of Germany recommended to include intercultural competence into school curricula (Sekretariat der Ständigen Konferenz der Kultusminister der Länder 1996).

In contrast to pedagogy, social pedagogy, and psychology German ethnology was relatively late to take part in the discussion about intercultural competence. However, this does not mean that ethnology is unable to make relevant contributions. In contrast, as the science of cultural foreignness or „als Wissenschaft des Umgang zwischen Eigenem und Fremden" (Antweiler 1999: 19) especially ethnology can add new aspects to the discussion. Important aspects of intercultural competence have been dealt with in ethnology for a long time.

Thus the contribution of ethnology already starts with the fundamental reflection about the term culture which is constantly referred to within intercultural learning and which is partly disputed. Ethnology focuses both on the information about other societies and the way in which other societies are looking at us and conceptualises these elements as a natural basis for a successful intercultural encounter. Moreover, ethnology can analyse concepts of societies that are dealing with the relationship between natives and foreigners and thus allows for the study of attitudes towards foreigners (Platenkamp 2004).

Within this context the different approaches of ethnology to understand other cultures are also important and a central aspect of intercultural competence. Especially the method called 'participant observation' is regarded as a condition to understand members of another culture (Jost 1998: 212). The self-reflection of the researcher in the intercultural context is also part of intercultural competence and requires a distance to both one's own culture and the examined one. Additionally, the awareness of one's own position has to been taken into account (Fischer 1992: 92).

Due to these conditions ethnology emerges as a suitable mediator between cultures. Its methods make it possible to understand the ideas behind people's thoughts and actions and render them intelligible to outsiders (Hauser-Schäublin 1997: 3-17).

Whereas ethnology has already provided relevant contributions to the topic of intercultural competence for decades, the term itself has only recently entered the academic discussion in the German speaking area. Especially Alois Moosmüller devotes himself explicitly to this topic (e.g. Moosmüller 2007).

During the last years the topic of intercultural competence has also gained prominence among ethnology students, particularly as they hope to find employment in this field.

From Theory to Practical Application

To do justice to the significance of anthropology in teaching intercultural competence, ESE was founded by anthropologists and pedagogues in Münster (Germany) in 1992. It has to be noticed that at this time Germany was facing several xenophobic attacks.[1] The association's aim is to convey anthropological knowledge about other cultures, but also to inform about the different

1 As an example one could refer to the incidents at the asylum application office in Rostock which is considered as one of the heaviest xenophobic attacks in modern German history.

ways in which other cultures look at us. The work of ESE consists of developing teaching materials that provide comprehensible ethnological information to pupils and adults. Thus, the team-members of the association conduct projects in schools, offer courses and seminars for teachers, and are involved in many other activities in order to promote and teach intercultural competence. Apart from the work in school and universities, ESE is also involved in producing publication on this subject. Anthropological material is incorporated critically and made understandable to the lay public; thereby increasing the latter's competence in the field of intercultural learning. This is achieved by engaging in scientific debates about theories of intercultural learning.

However, what is meant by referring to the term 'intercultural competence'? Before giving an answer it is necessary to deal with the following questions: What is the difference between multicultural and intercultural? And last but not least: how to define culture?

The Theoretical Background of the Work of ESE

Whoever deals with multicultural or intercultural aspects will soon notice that this field lacks universally applicable definitions. Terms like for example multicultural, intercultural and – as has become prominent during the last years – transcultural are interpreted differently. As Demorgon and Cordes (2006: 27) put it:

> „Multikulturell, transkulturell, leitkulturell, interkulturell – diese Begriffe drohen in ein sprachliches Imponiermaterial verwandelt zu werden, in welchem man sich um ihre Bedeutung und Tiefenschärfe nicht weiter kümmern muss."

According to Hinz-Rommel (1994: 36) the term *multicultural* describes a state whereas the term *intercultural* describes a programme and stresses its interactive aspects. The reason for criticising this definition is that the term *multicultural* – if it is defined like this – is based either on the condition of „die parallele Existenz weitgehend eigenständiger kultureller Lebenswelten in einer Gesellschaft" (Römhild n.d.: 2) or on the „Existenz klar unterschiedener, in sich homogener Kulturen innerhalb ein und derselben staatlichen Gemeinschaft" (Welsch 1995: 40). It is indisputable that homogeneous and segregated cultures of this kind do not exist anymore. However, from the point of view of numerous authors the advantage of this definition of the term *multicultural* is that it implies the acknowledgement of the culture of the foreigner (cf. Demorgon and Cordes 2006: 28). Objections against the definition of *intercultural* as put forward for example by Hinz-Rommel point in the same direction. According to Römhild interactions that are necessary for intercultural actions inevitably require „die Existenz geschlossener kulturelle Universen voraus, zwischen denen dann ein Austausch erst stattfinden kann" (Römhild n.d.: 2).

With reference to interculturalism Welsch even talks about a „Kugelverfassung der Kulturen" getting in contact with each other (Welsch 1995: 40). That is why he pleads for the term *transcultural*. In his opinion the plurality of possible identities and cross-border contours, on which the term is based, for him make possible „intensive Austauschprozesse zwischen den Kulturen, auch über die Kategorien Eigenheit und Fremdheit hinweg" (*ibid.*: 42). With this in mind he designs a concept of transculturalism, namely

> „ein anderes Bild vom Verhältnis der Kulturen. Nicht eines der Isolierung und des Konflikts, sondern eines der Verpflichtung, Durchmischung und Gemeinsamkeit. Es befördert nicht Separierung, sondern Verstehen und Interaktion" (*ibid.*: 44).

In contrast, for Demorgon and Cordes the acknowledgement of agreement and familiarity is more important when defining the term *transcultural*. „Nur in einer solchen um Allgemeinheit bemühten Transkultur können sich Menschen verschiedener Kulturen, Ethnien und Rassen [...] als Freie und als Gleiche anerkennen" (Demorgon and Cordes 2006: 32). This short presentation of the scholarly discussion on the three terms shows that their definition strongly depends on the way in which the term *culture* is conceptualised. Especially for ethnologists defining *culture* with all its facets and different contexts is very difficult.[2] Moosmüller (2004: 46) thus summarises: „Während die einen zupackend (oder naiv?) das Rätsel Kultur angehen, werden die anderen nicht müde, auf die Gefahren hinzuweisen, die auf den lauern, der das Rätsel lösen wird."

Knowing that a definition of the term 'culture' never can be comprehensive and applicable to all contexts, ESE works with the following definition:

- Culture is created by Man.
- It is constantly changing and does not follow rigid rules, nor is it confined by boundaries.
- People differ in their culture. They live and interpret it in their own terms.[3]

The working definition of ESE is based on a dynamic concept of culture which in the context of migration gains additional importance (James 1993: 258 quoted in Beck-Gernsheim 2007: 20). At the same time the ESE definition focuses on individual forms of culture which is experienced by everyone in the context they are situated in (see also Antweiler 2009: 56). The discussion of the terms *multicultural, intercultural*, and *transcultural* described above allows for the following classification:

2 Beer (2003) gives a good survey of the discussion of the term *culture* in ethnology. Interesting aspects of the ethnological view on the term *culture* as applied in transcultural training can be found in Moosmüller (2004) and Latorre (2004).

3 Further details can be found in Bertels at al. (2004).

A multicultural society is a society bringing together people from various cultural backgrounds. Each member of this society can simultaneously belong to several cultures. Between the cultures or individuals in a multicultural society there are both similarities and differences. The members of different cultures can take advantage of the similarities they share to create a transcultural space. However, in order to be able to cope with the differences it is necessary to communicate with each other interculturally. Interculturalism is a matter of 'negotiation'. Demorgon and Cordes (2006: 34) describe how difficult this 'negotiation' can be:

> „Interkulturelle Begegnung ist nicht nur Bemühen um Dialog, sondern zu ihr gehören unausweichlich auch Zusammenstoß (Schock), Kampf und Gewalt. Denn interkulturelle Verständigung erfolgt nicht nur durch die An-Erkenntnis des Anderen und Fremden (Multikulturalismus) und auch nicht nur durch das An-Erkennen des Gemeinsamen und Vertrauten (Transkulturalismus), sondern darüber hinaus auch durch die Bearbeitung der Zwischenräume und Zwischenperspektiven zwischen Eigenem und Anderem, Vertrautem und Fremden."

What is Intercultural Competence?

There is no clear definition of the term yet, but there are various interpretations which have been put forward by different disciplines, such as pedagogy, psychology, or communication studies (Bertels et al. 2004: 23-26). ESE has developed a definition of its own:

> "Intercultural competence is the capacity to cope directly or indirectly with other cultures. Intercultural competence is the ability to engage with members of different cultures, while achieving a high degree of understanding for the other. It is understood as a lifelong learning experience" (Bertels and Bußmann 2013: 33-34).

Some of the terms used in this definition can further be specified in the following way:

- Learning Experience: The acquisition of intercultural competence is seen as a lifelong learning experience, since cultural misunderstandings are inevitable.
- Directly or Indirectly: Directly means immediate contact and/or personal encounters with people from cultures, indirectly means dealing with 'knowledge' of these cultures, for example in the form of media reports or information given in textbooks.
- Understanding: Understanding is necessary in order to work and live with people from different cultures. A 'full' understanding of people from other cultures is not possible, since even the understanding of people from one's own culture can be difficult. Nevertheless,

through intercultural competence cultural differences can be bridged (Bartmann 2012: 30; Maletzke 1996: 34-36).

One of the basic points of intercultural competence is the confrontation with other cultures. This confrontation may result in a direct encounter. During seminars and workshops on intercultural learning it is also possible to put cultures to which the pupils or workshop participants have no connection in the center of the learning process. That is why the basic method used by ESE for teaching intercultural competence is entitled *Dritt-Kultur-Perspektive*, the "perspective of the third culture". The 'third-culture-perspective' serves as a theoretical approach and basic tool in the efforts of ESE.[4] This model suggests that one is dealing with cultures to which one does not have a personal relation. In a further step, this helps to learn how to deal with people from different cultural backgrounds in one´s home country. By looking at regions that are far away from the everyday world of the participants in Germany (e.g. Indonesia, Mexico or Pakistan), the group gets involved with different cultural points of view in a relatively unprejudiced way. If, however the cultural encounter reflects the culture of one of the participants who have a 'migration history'[5] the participant might feel 'exploited'. By using the 'third-culture-perspective' one can avoid this kind of situation. At the same time, participants with a migration history are given the opportunity to contribute to the conversation by integrating their own cultural insights if they wish to do so (Bertels and Bussmann 2014: 9-10).

The application of this method involves two basic steps:

1. The first step starts with the discussion of intercultural encounters using examples of geographically distant and culturally different countries, i.e. examples of encounters between people from different cultures with which participants have no relationship.
   For instance in a workshop one could imagine the encounter between a German and an Indonesian/Ghanaen. Participants learn that people from different cultures have different views. One can also approach different perspectives through role plays and communication techniques, such as trying to take over the view of the other. Thus, the participants learn, how to resolve intercultural misunderstandings. This knowledge later can be applied in the professional and everyday life.
2. The next step is the discussion between members of different cultures the participants of which have frequent contacts, for example,

4 This approach was developed with reference to Gudykunst, Hammer and Wisemann (1977).

5 The term "persons with a migration history" is used in Germany to refer to people (and/or immigrants and foreigners) who arrived after 1949 and where at least one parent was not born in Germany, and/or infants of immigrant families.

a conflict between a German and a Syrian person. To remedy this situation, participants should apply the experiences gained during the first phase of the exercise.

The method of the 'third-culture-perspective' provides participants with a tool to act from a neutral point of view and analyze themselves and others, their opinions and perceptions. These skills are considered as basic aspects of intercultural competence.[6]

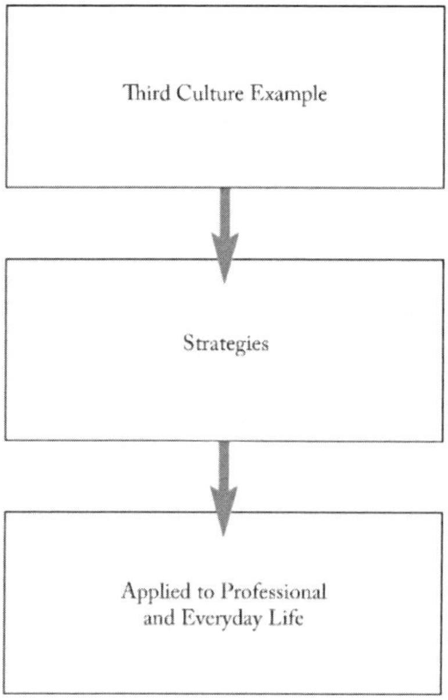

Fig. 1: Method of 'Third-Culture-Perspective'

By using the 'third-culture-perspective' the following learning objectives of intercultural competence should be achieved:

1. Collect information and develop interest

In order to achieve this learning objective, scientifically-based knowledge about foreign cultures is taught. Furthermore, people should be encouraged to acquire this kind of information themselves. In order to do so, intrinsic interest in other cultures is an important prerequisite. This interest can be fostered by giving authentic reports and descriptions to the persons.

6 In 2003 ESE won the „Preis für Innovationen" (innovation award) of the Deutsches Institut für Erwachsenenbildung (German Institute for Adult Education) for its concept of the 'third-culture-perspective'.

Methods to implement this learning objective can include various examples of intercultural encounters, reports of personal experiences, the use of role plays, images, and texts.

2. Initiate a change of perspective

As part of this learning objective persons should ask themselves the questions "How do others see us?" and "How do others see themselves?" and, finally, "How do we see others?" By using the 'third-culture-perspective' persons are first encouraged to look at their own culture which is not taken for granted anymore after other viewpoints have been taken into account. In addition, persons are encouraged to question their view of foreign cultures. The ultimate goal is that persons come to realise the relativity of each cultural point of view.

Methods to implement this learning objective may include the use of comic strips or personal reports, or quotations by members of other cultures.

3. Recognise and overcome ethnocentrism

Ethnocentrism is based on the assumption that the values and codes of one's own culture are universally valid. Ethnocentrism often implies that one's own culture is valued higher than others. To overcome ethnocentrism, persons are encouraged to deal with other cultures in order to realise that one's opinion is largely determined by one's own cultural upbringing and therefore has to be analysed critically.

Methods to implement the learning objective can be looking closely at prejudices and "quick judgments", and the analysis of discourses in travel guides, the press, or literature.

4. Reflect on situations of intercultural encounters

The main learning objective is the insightful encounter with members of other cultures. Situations of cultural encounters are analysed and reflected upon. Thus persons can develop strategies to enable almost conflict-free and successful communication.

Methods to implement this goal can be the analysis and reflection of the pupils' own experiences, or learning behavioural strategies, such as in role plays.

5. Promote attitudes and values

Attidudes and values (such as openness, tolerance, acceptance, or respect) are a necessary foundation to acquire intercultural competence. However, values and attitudes only change slowly, in the course of a long-term education (Schlöder 1988).

Although these five learning objectives are presented in a consecutive order, the latter cannot always be implemented in this way while mediating intercultural competence.

As the following learning unit for pupils will show, it is still frequently possible to integrate at least some of these goals into one class unit.

Applying the "Third-Culture-Perspective"*: The Example of Día de los muertos*

In the following some modules applied during the one-day school project *Día de los muertos* – the Mexican All Saints' Day – will be introduced in order to highlight the third-culture-perspective.[7] The day starts by confronting the pupils with a friendly smiling skull, *calavera Catrina*, who is considered to be the symbolic figure of the *Día de los muertos*. Then the pupils are asked to describe their impressions (educational objective 1: develop interest).

Next the pupils are requested to come forward with their associations regarding the presentation of skulls and skeletons in Germany. Terms like *poison*, *fear*, or *horror films* are frequently mentioned here. Pupils with a migration history are encouraged to name their associations which may be different. The pupils' associations are the compared with the description of *calavera Catrina* (educational objective 1: collect information).

In another module the pupils learn about the background and the performance of the Mexican All Saints' Day by means of a report by a Mexican girl (educational objective 2: convey information). Afterwards, the pupils are asked to describe how the dead of their own families are commemorated. Here the pupils with a migration history may again draw on different cultural models.

Having created a chocolate skull which is typical of the Mexican All Saints' Day, (see figure 2) the pupils are asked to engage in a role play by performing a dialogue between German tourists in Mexico and a Mexican selling chocolate skulls. In the context of this role play the pupils can make use of the information received earlier and take over different perspectives. The following dialog emerged:

> German male tourist:"Look at this, this is really bad taste!"
> German female tourist:"That's disgusting and they want to sell it!"
> German male tourist:"What on earth do they have in mind when they do this?"
> German female tourist:"Let's ask what it's all about. What do these skulls mean?"
> Mexican seller:"It's our tradition to sell theses skulls."
> German male tourist:"Why is it a tradition?"

7 For further information on the school day project, see the CD enclosed in the *Handbuch Interkulturelle Didaktik* (Bertels and Bußmann 2013).

> Mexican seller: “The skulls are a symbol of All Saints. That’s a family celebration for us. We believe that at this festivity the dead family members are also present. Isn’t there something like this in Germany,too?”
> German female tourist: “That’s not a custom in Germany. But thanks for the explanation. I’ll take two skulls.” (Bertels and Lütkes 2001: 458)

Having obtained an inside into the Mexican point of view the pupils are able to explain the meaning of the chocolate skulls which at the first side had a shocking effect (educational objective 2: practise the change of perspective and educational objective 4: reflect on intercultural situations).

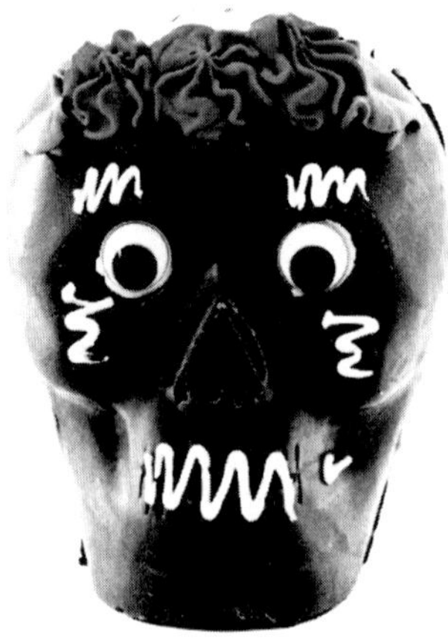

Fig. 2: Chocolate skull
Foto: Frank W. Koch, © ESE)

Conclusion

The example of the school day devoted to the theme *Día de los muertos* reveals that ethnological knowledge can help to achieve various goals in the transmission of intercultural competence:

- All pupils learn something new together.
- Important educational objectives are conveyed to the pupils (e.g. practice the change of perspective) strengthening their intercultural competence.
- Pupils with a migration history can voluntarily contribute with their own experiences.

In conclusion, ethnology has many possibilities to contribute to the maintenance of a multicultural society. “*Only Ethnology can teach us to regard ourselves through a foreigner’s eyes*” (Platenkamp 2004: 31) and thus take advantage of the encounter between the familiar and the foreign in a constructive way.

References

ANTWEILER, C. (2009) *Heimat Mensch. Was uns alle verbindet.* Hamburg: Murmann.

ANTWEILER, C. (1999) Ethnologie als gesellschaftlich relevante Humanwissenschaft. Systematisierung praxisorientierter Richtungen und eine Position. *Zeitschrift für Ethnologie* 123(2): 215-255.

BARTMANN. S. (2012) Nicht das Fremde ist so fremd. In BARTMANN, S. and IMMEL, O. (eds.), *Das Vertraute und das Fremde. Differenzerfahrung und Fremdverstehen im Interkulturalitätsdiskurs*, pp. 21-36. Bielefeld: Transcript.

BEER; B. (2003) Ethnos, Ethnie, Kultur. In FISCHER, H. and BEER, B. (eds.), *Ethnologie – Einführung und Überblick*, pp. 53-72. Berlin: Reimer.

BECK-GERNSHEIM, E. (2007) Wir und die Anderen. Frankfurt am Main: Suhrkamp.

BERTELS, U. and BUSSMANN, C. (2013) *Handbuch Interkulturelle Didaktik.* Münster u.a.: Waxmann.

BERTELS, U. (2011) Einleitung. In BERTELS, U. and HELLMANN DE MANRIQUE, I. (eds.), *Interkulturelle Streitschlichter – Interkulturelle Kompetenz als Schlüsselqualifikation für Jugendliche*, pp. 4-10. Münster u.a.: Waxmann.

BERTELS, U. et al. (2004) *Ethnologie in der Schule – Eine Studie zur Vermittlung Interkultureller Kompetenz.* Münster u.a.: Waxmann.

BERTELS, U. and LÜTKES, C. (2001) Perspektivenwechsel als Grundlage interkultureller Kompetenz – Erfahrungen des Projektes „Ethnologie in der Schule". *Internationale Schulbuchforschung* (Zeitschrift des Georg-Eckert-Instituts für internationale Schulbuchforschung) 4: 453-464.

DEMORGAN, J. and CORDES, H. (2006) Multikultur, Transkultur, Leitkultur. In NICKLAS, B., MÜLLER, B and CORDES, H. (eds.). *Interkulturell denken und handeln – Theoretische Grundlagen und gesellschaftliche Praxis*, pp. 27-36. Bonn: Campus.

FISCHER, H. (1992) Feldforschung In FISCHER, H. (ed.), *Feldforschungen – Erfahrungsberichte zur Einführung*, pp. 89-98. Berlin: Reimer.

GUDYKUNST, W., HAMMER, M. R. and WISEMANN, R. (1977) An Analysis of an Integrated Approach to Cross-Cultural Training. *International Journal of Intercultural Relations.* 1: 99-109.

HAUSER-SCHÄUBLIN, B. (1997) Blick zurück im Zorn: Ethnologie als Kulturkritik. *Zeitschrift für Ethnologie* 122: 3-17.

HINZ-ROMMEL, W. (1994) *Interkulturelle Kompetenz. Ein neues Anforderungsprofil für die soziale Arbeit.* Münster u.a.: Waxmann.

JOST, S (1998) Keine Angst vor Kultur – Eine Empfehlung. In PREISLER, G. et al (eds.), *Global denken – Lokal Handeln. Regensburger Beiträge zur Didaktik der Geographie*, pp. 207-217. Regensburg: Universität Regensburg.

LATORRE, P. (2004) Was heißt eigentlich „Kultur"? Anthropologische Fragen an die interkulturelle Kommunikationsforschung. In BOLTEN, J. (ed.), *Interkulturelles Handeln in der Wirtschaft. Positionen, Modelle, Perspektiven, Projekte*, pp. 45-67. Sternenfels: Wissenschaft und Praxis.

MALETZKE, G. (1996) *Interkulturelle Kommunikation.* Opladen: Westdeutscher Verlag.

MOOSMÜLLER, A. (2004) Das Kulturkonzept in der Interkulturellen Kommunikation aus ethnologischer Sicht. In LÜSEBRINK, H-J. (ed.), *Konzepte der interkulturellen Kommunikation.* Theorieansätze und Praxisbezüge in interdisziplinärer Perspektive, pp. 66-98. St. Ingbert: Röhrig Universitätsverlag.

MOOSMÜLLER, A. (ed.) (2007) *Interkulturelle Kommunikation – Konturen einer wissenschaftlichen Disziplin.* Münster u.a.: Waxmann.

PLATENKAMP, J. D. M. (2004) Über die gesellschaftliche Relevanz der Ethnologie. In: BERTELS, U. et al. (eds.). *Aus der Ferne in die Nähe – Neue Wege der Ethnologie in die Öffentlichkeit*, pp. 21-32. Münster u.a.: Waxmann.

RÖMHILD, R. (n.d.) Multi-, Trans- und Inter- ... Was ist eigentlich Kultur. Availale from: www.tu-darmstadt.de/wusgermany/stube/sminare/gericht/RoemhildReferat.htm.

SCHLÖDER, B. (1988) Soziale Vorstellungen als Bezugspunkte von Vorurteilen. In SCHÄFER, B. and PETERMANN, F. (eds.), *Vorurteile und Einstellungen*. Köln: Deutscher Institutsverlag.

SEKRETARIAT der Ständigen Konferenz der Kultusminister der Länder: Empfehlung „Interkulturelle Bildung und Erziehung in der Schule". Beschluss vom 25. November 1996. http://www.kmk.org/doc/beschlo/671-1_Interkulturelle%20Bildung.pdf [Accessed: 2nd December 2008].

WELSCH, W. (1995) Transkulturalität – Zur veränderten Verfasstheit heutiger Kulturen. *Zeitschrift für Kulturaustausch* 45(1): 39-44.

A Just Arena of Negotiating Identities During Social Change Processes. A Possible Pathway?

YULIA SUGANDI

> "Anthropologists exist to attest that the way we live, the values we believe in, are not the only ones possible, that other ways of life, other systems of values have allowed and continue to allow human communities to find happiness"
>
> (LÉVI-STRAUSS 2013: 37).

The World of Exchange, Processes of Social Change, and Justice

Processes of social change are arenas where different cosmologies collide. Prof. Jos Platenkamp's invitation to embrace the world of exchange, particularly the relational approach[1], has shaped my view about dignified processes of social change that are sensitive to vulnerable groups, minorities and indigenous peoples. Contributing to knowledge about the stakes of self-determination for such groups became a priority in my anthropological work. This view is rooted in my PhD work, which focused on the conceptual construction and valorisation of the collective social identity of the Hubula, the indigenous people living in the Palim valley of Papua. It explores how this identity is expressed in ritual actions, and in the production and exchange of cultural artifacts, and looks at the way in which the Indonesian State and the Roman Catholic Church have impacted upon and transformed it (Sugandi 2014: 12). As my PhD supervisor, Jos taught me the art of listening with empathy and respect in doing my ethnographic research. These skills are essential to enter the world of exchange in Palim valley, and allowed me to understand the Hubula's ontological foundation and resilience in the processes of social change. The following article that is based on my ethnographic research develops one of the basic questions that Jos and I have been continuously discussing, the advocation of equity in processes of social change.

Development: Social Identities and Culturalism

Development was initially associated with economic growth, modernisation, distributive justice, and socio-economic transformation (Mobugunje 1989:

1 The relational approach concerned with how exchanges reflect and shape the identity of the social actors involved, as well as the relationship those actors have with each other, with the thing exchanges, and even with the act of exchange itself (Carrier 2002: 219).

35-50), pursued through approaches dominated by economistic and technical thinking. However, more recent development paradigms have shifted towards aiming for socio-political transformation, and the former goals of development are increasingly presented as its means (Mosse 2005: 237). This recent trend affirms, borrowing Borgstroem's term (1976: 88), an 'anthropologisation of development discourse' that emphasises human instead of economic factors. The actor-oriented approach in the ethnography of development emphasises the following: (a) the way in which the meaning of the development process and interaction is viewed by various development actors as well as negotiated in practice by the latter (Mosse and Lewis 2006: 9), and (b) the action and pattern of interaction of social actors as the foundation of development (Olivier de Sardan 2005: 63). Such an emphasis would redirect the analysis towards what has been neglected in development research (Baaz 2005: 2), by looking into power relations and identities rather than focusing exclusively on economic dimensions. Through its socio-political transformation, development provides a medium that confronts various cosmologies and is also a negotiation arena that forms social identities[2].

One of the elements that shapes the imbalanced power relations among actors involved in this socio-political transformation is culturalism. As Bayart (2005: 33) pointed out:

> "in one sense culturalism is like Beaujolais Nouveau; the inebriation it produces is often malign and leaves begind a hangover. However, the adage *in vino veritas* cannot be applied to culturalism [... as] culturalism maintains that a 'culture' is composed of a stable, closed corpus of representations, beliefs, or symbols that is supposed to have an 'affinity' [...] with specific opinions, attitudes, or modes of behaviour."

The analysis of culturalism might be directed to the local community in question. For instance, Timmer (2012: 21-40) negates culturalism in Papua altogether by arguing that community groups in Papua share characteristics with similar groups elsewhere in the region or wider world. Timmer considers different community groups (i.e. Papuan elite, Free Papua Organisation, and migrants) as groups with their own cosmologies or as groups with whom one can work for development or reconciliation. This approach involves working to analyise forms of stability and dynamism, rather than taking ethno-classifications as models when trying to research or work with a local culture for a development program (*ibid*). Such a research method can elicit patterns of interaction and cultural diversity, as well as common characteristics and overlaps between the cosmologies of local groups.

2 All human identities are, by definition, social identities; identifying ourselves, or others, is a matter of meaning, and meaning always involves interaction: agreement and disagreement, convention and innovation, communication and negotiation (Jenkins 2008: 17).

However, the analysis might also be focused on the dynamic relations among actors (both among local communities themselves and with external agencies) in the process of social change. Culturalism is adopted in inventing traditions and giving rise to a discourse on groups that are either developed or not yet developed. During the colonial period, culturalism was maintained to confirm social identities as described by Bayart (2005: 42-46):

> "the British administrators of Indirect Rule forged a quasi-aristocratic lifestyle in order to overcome the isolation in which they found themselves to preserve their dignity as 'civilised' people, whereas white miners in South Africa and Rhodesia developed a corporatist working-class culture that asserted their distinctiveness with regard to black workers. In Asia, the joint invention of tradition by the coloniser and the colonised is inseparable from the key concept of the village community. In the Dutch Indies, for instance, the administration relied on the *pamong desa*, the village chiefs, at the same time that it refused to assimilate the natives and promoted a form of colonial nationalism."

In the post-colonial period, culturalism was also adopted by various modern development agencies. Galeski (1976: 152) affirmed that in the 20th century evolutionism was no longer considered to be fashionable among social scientists who preferred to use the term social change. Nevertheless, culturalism prone to westernisation is part of the cosmology of modern development bringing about the following phenomena as impacts: (a) Marginalisation of cultural factors; for instance community development in the post-WWII period was associated with the development theories that were prevalent at that time, i.e. modernisation theories which placed considerable stress upon cultural factors. These were built on western cultures which constituted ideal examples of what a modern society should be, dismantling all the non-western 'traditional' structures, and cultural change was prescribed as the essential prerequisite for self-sustaining economic growth (Mayo 2000: 89); (b) Pacification followed by possible violence. Bayart (2005: 38) argued that such regressive definitions of modernity make identity related strategies *potentially* totalitarian mainly for two reasons: *first*, because the culture imagined to be authentic is defined in opposition to neighbouring ones that are seen as radically different, and because this alleged alterity entails a principle of exclusion whose logical result is ethnic cleansing. In such a view intercultural exchange is then deemed to involve alienation, a loss of substance, or even pollution; and *second,* because the imagined culture assigns to those individuals who are supposed to belong to it a simplified identity – one is tempted to call it an identity kit – that they are expected to endorse, if necessary via coercion; and finally these definitions of modernity involve stereotypes and horizontal inequalities. In her research among European development workers in Tanzania, Baaz (2005) noted that Eurocentrism and paternalism in development practice has influenced development practitioners, which is reflected in the partnership discourse itself. Such attitudes are rooted in the idea of Europe's general superiority (Baaz 2005: 111).

Ethnodevelopment against Technical Approaches

The application of culturalism in modern development programs can be illustrated by focusing on the process of social change among the Hubula, an indigenous people of Palim Valley, Papua. They live in the province of Papua, the western half of the island of New Guinea which is situated at the eastern end of Indonesia. Although the area is 3.5 times larger than the island of Java, only 1% of the total population of Indonesia live there (census 2010). There are over 250 small-scale 'traditional' societies scattered across the island. Although Papua hosts Freeport Indonesia's Grasberg copper and a gold mine, and has the fourth highest GDP per capita in Indonesia – mostly derived from industries exploiting natural resources – Papua has the highest index of poverty in Indonesia (census March 2014).

The Palim Valley is about 45 kilometres long and 15 kilometres wide and stretches along the Palim River, the main river flowing between the hills in the mountain of Papua. Palim Valley is located in Jayawijaya Regency, the most densely populated area in Papua (census 2010). The Dutch government and missionaries were present in the Palim Valley in the 1950s, while the influx of Indonesian migrants began after the 1960s. The Hubula of Palim Valley have been labelled by various stereotypes. The Hubula are part of the mountain people of Papua or indigenous highlanders who are often said to be hardliners having 'stubbornly' resisted the Indonesian government. Many areas in these highlands are labelled as 'red zones', which means that the Indonesian security forces assume that OPM (*Organisasi Papua Merdeka*; Free Papua Organisation) has bases in that area. This stigmatisation of the mountain Papuans is enhanced by other stereotypes. They are said to be isolated, primitive, 'Stone Age people', cannibals, war-like, and unable to live peacefully and harmoniously.

This stigmatisation is based on external perceptions according to which the Hubula of Palim Valley are an inferior people in need of help, which in turn has led to their marginalisation and the imposition of technical means by development agencies, valuing modern technology over their knowledge and values in order to bring about social change. In this purely 'technicist' approach Hubula's cultural rights are neglected. The program of modernisation that had been introduced was considered a proper compensation for the Hubula's ancestral land that was seized for development. Development in Palim Valley was equated with economic change and growth, and modernisation was the solution to local problems, without any reference to traditional values (Sugandi 2014). The development 'industry' is driven by short-term, measurable outcomes to demonstrate effectiveness. The value for money leads to inevitable tensions between a results-driven approach to development – as promoted by many development agencies – and the rather more 'slow-burn' ethnodevelopment approach relying on research on indigenous knowledge (Briggs 2013: 239), even though ethnodevelopment might counterbalance the culturalism brought about by modern development. Ethnodevelopment offers an approach to deconstruct the power relations in the development processes and implies 'development' involves no more than looking at 'one's own culture' straight in the eye (Bayart 2005: 10). Not all communities are of the same type, and we therefore need a differential understanding of 'community' (Moris 1996: 225-226), or as Strathern argues, "what makes the similarities is the effort to 'see' connections; what makes the dissimilarities is the 'recognition' of 'difference'". In order to understand the cosmology of indigenous people in Melanesia, Strathern put relations in the centre of her analyses, arguing that relations are what make people 'see' anything at all. The pattern of relations in Melanesia and its differences with the 'West' is important to understand issues regarding the land (Strathern 1999: 260).[3] In order to understand Hubula cosmology in its difference with modern development ideology, I shall scrutinise in this article how the relations produced by Hubula warfare can be seen as constituting one of the sources of their social identity. This article also attempts to explore the possible pathway to negotiate these identities by displaying some cases related to ethnodevelopment policies in other contexts.

Warfare as the Ontological Basis of Social Structure

The cosmology and social structure of the Hubula, similar to other Melanesian societies, is characterised by several abstract metaphysical concepts which are at the foundation of their system of social relations. Recent studies on Southeast Asian and Melanesian cultures have pointed to the importance of understanding indigenous metaphysical thought as a precondition for the analysis of social and ritual action (Woodward 2000: 220). White (1991: 241) described how narrative constructions of the past are central in the Solomon Islands for the construction of social relations. In the context of the Hubula, the central concepts include a moiety system, myths, and traditional warfare, forming the basis of Hubula rituals. War itself is a cultural or mythical condition, one that plays upon and activates the deepest meanings of a society where persons transcend individual existence and become part of a wider endeavour, sometimes of a cosmic scale (Eller 2009).

Anthropological research on warfare was dominated for a long time by evolutionist ideas (Otterbein 2000: 801-802). In the 1960s, the subject of pre-industrial warfare began to draw attention among those anthropologists who

3 Strathern describes that from a Melanesian perspective the land is not a separate entity from the persons who live on it.

saw aggression as being rooted in phylogeny, a product of biological adaptations for innate aggressiveness or territorial behavior. In the 1970s, *Ecological Determinism* saw primitive warfare as a part of a society's functional adaptation to particular environmental conditions, an adaptive response to population pressures and to limited resources (Robarchek 1989: 905-907). This type of evolutionary thought is incompatible with an interpretive analysis of the origin of warfare. As Thorpe (2003: 159-160) stated,

> "the materialist theory would be far more convincing if there were a significant increase in conflict with the adoption of agriculture. We therefore need to turn to historical factors and considerations of the specific societies concerned. Mesolithic conflict need not have been over economic resources. Warfare in early prehistory may well have arisen from matters of personal honour."

Robarchek (1989: 915) proposed that understanding the trajectory of this dispute requires to put the brute material 'facts' into psychological, social, and cultural contexts, and to attempt to understand how the situation was conceptualised and defined by the people involved. The practice of warfare is highly 'embedded' in social-cultural systems and as such may permeate many other aspects of life, so that the political dynamics, economic consequences, and sociological organisation of raiding also should be taken into account (Whitehead 2000: 835).

Diamond (2012), drawing on Broekhuijse (1967), describes warfare in an evolutionary manner, identifying different evolutionary stages of Hubula warfare without taking into account its embeddedness in a specific cultural logic. In the past Hubula warfare has been portrayed as an act of violence that is a function of the Hubula character, and as being the foundation of their collective pride (cf. Heider 1970; Koentjaraningrat and Simorangkir 1993). However, this ignores that warfare is also an exchange of violence that at the same time forms the basis of Hubula social structure. The exchange of violence involved in ritualised warfare begins with the taking of the enemy's ancestral emblems collected from victims previously killed in battle (*apwarek*), and the affirmation of ancestral emblems in the form of limestone pieces that recall ancestors killed in past battles *(kaneke*). The idea here is that relations with the spirits of ancestors who were killed in the course of war must be strengthened ritually, whereas the spirits of the enemies killed must be pacified. These sacred objects obtained in warfare are cared for by organising rituals involving the sacrifice of pigs. Thus, Hubula warfare involving an exchange of violence and of the 'sacred objects' gained from warfare is a cosmological intra-ancestral exchange of identities. These ancestral identities, as they are manifested in the sacred objects, are at the core of Hubula social structure.

Accordingly, Hubula traditional warfare is a central practice through which a number of essential relationships are defined: Firstly, there is the intimate relationship between the Hubula and their ancestors, as manifested in sacred objects, which were stored in a special traditional compound called *kanekela*.

This relationship is the basis of the traditional Hubula belief system. Secondly, there is an inalienable relationship between the Hubula and their ancestral land. Access to ancestral land and its natural products is also grounded in the history of warfare. Thirdly, the history of warfare also impacts on the rules regulating marriage, the wedding ritual and the associated exchange of objects. Basically, all Hubula relations, whether cosmological, socio-cultural, ecological, or social, are rooted in the history of warfare. In addition, conflict and warfare are intertwined with Hubula notions of fertility, i.e. the condition of the ancestral land, the productivity of the gardens, and the well-being of the people and the pigs.

Being Hubula: Becoming Civilised Indonesians and Christians

Culturalism might entail what Eller (2009) descibed as ethnocide[4], which aims at the deculturation and re-culturation of indigenous people so as to make them culturally indistinguishable from the dominant society. Clarke (2001) argued that the annexation of West Papua by Indonesia in the 1960s and the massacre of Chinese and other minorities following the collapse of the Sukarno government in 1965 showed that ethnic minorities, and to a lesser extent indigenous peoples, have been the victims of *ethnocidal* policies as governments attempted to promote a uniform national culture and/or failed to protect minorities. These policies are part of a trend to treat ethnic minorities and indigenous people in Southeast Asia as 'backward' and 'uninterested in socioeconomic progress' from the beginning of the 1960s onwards. At the same time a policy of assimilation and integration including sedentarisation (minorities are compelled to live in fixed and permanent settlements and to engage in local cultivation) was accompanied by discrimination against animist religious practices, giving rise to erosion of traditional lifestyles and livelihoods, and exploitation of natural resources by migrants, corporations, governments, or military forces (Clarke 2001).

Since their encounter with foreigners, the Hubula have tried to contest the power exerted by the state and religious authorities in the Palim Valley. From the perspective of the government, cultural change was seen as key to economic development (Mayo 2000: 91). The Dutch colonial government in Papua applied a similar strategy as the British colonial power by putting more emphasis on local structures and cultural specificities rather than simply defining 'civilisation' as represented by the colonisers, as the ultimate *raison d'être*. In order to pacify Hubula warfare the Dutch government organised peace festivals in the Palim Valley that involved local community leaders, especially those who were part of tribal confederations. The Fransiscans intervened using techniques of enculturation, thereby respecting the local culture

4 Ethnocide means destroying the culture or institutions of a group rather than the people themselves. It often hides other 'educational' endeavours. (Eller 2009).

and allowing the Hubula to continue their traditional practices and keep their sacred objects (Lieshout 2008). This situation changed, however, with the arrival of the Indonesian government in the1960s.

Soon after, in 1971-1973, the Indonesian government implemented operation *Koteka* (penis sheath), where the Indonesian armed forces and the civil government were engaged in activities designed to 'civilise' the Papuans. It was basically a military campaign aimed at coercing highland Papuans to abandon aspects of their indigenous culture, attend school, modernise economically, and adopt a more mainstream Indonesian identity. In 1972 representatives of the Indonesian government in the Jayawijaya Regency instructed the Hubula to leave their traditional way of life and become adherents of a recognised religion, to be chosen from the five major religions which had been recognised by the Indonesian government: Islam, Hinduism, Buddhism, and the Protestant and Catholic Churches. The above mentioned asymmetric power relations in shaping social change in the Palim Valley are also characteristic for other indigenous peoples in the world; for example, the Melanesian community in the Solomon Islands was faced with the attitude of European missionaries who sought to destroy ancestral shrines, whereas according to the indigenous approach they would have rather been blessed and/or 'baptised' (White 1991: 246).

Between Globalisation and Cultural Development

Clarke (2001: 433-434) argued that governments, donors, and civil society actors alike still have a lot to do as they strive to redefine the nation-state and development projects to accommodate the region's ethnic minorities and indigenous peoples. Although a transition from ethnocide to ethnodevelopment in Southeast Asia has taken place since the late 1980s, ethnodevelopment remains confined to a limited number of initiatives in the context of a broader pattern of domination. The 'globalisation' of economic, political and social structures continues to homogenise societies in the region and to marginalise minority groups. Rosa (2013) designates the forced assimilation of the Tao (an indigenous community in Taiwan) as a cultural genocide manifesting itself in a structural discrimination and suppression through a policy of Sinicisation[5] by banning Tao animism, language, myths and tales about tribal ancestors, as well as destruction of traditional Tao houses. In the extreme case of the Burmese state using oppressive ethnocracy to build a nationalism based on deculturalisation and ethnic cleansing against Karen people, Fong (2008) proposes liberation ethnodevelopment. This variant of ethnodevelopment advo-

5 Sinicisation means the forced assimilation of indigenous people in order to build up a Sinocentric and homogeneous society (Rosa 2013).

cates a bottom-to-top trajectory and views self-determination as a development process aimed at establishing an institutionally autonomous political unit.

At its limits, arguments on the position of local tradition in modernisation are polarised into: on the one hand the cultural relativism that nurtures the myth of the 'Noble Savage' (Otterbein 2000: 801-802), and on the other hand the paradigm of 'detraditionalisation'[6] (Helaas 1996: 3) involving the *replacement* of closed and 'cold', societies with open and 'hot' ones. The Hubula and other indigenous people that struggle against detraditionalisation in the process of social change require ethnodevelopment that is able to provide the space for tradition and modernity simultaneously. Here, 'tradition' should not be understood as being necessarily static, backward and conservative because there is 'a modernity' to tradition and 'a tradition' of modernity (Luke 1996: 116).

This also calls for the recognition of 'cultural development'. Initiatives to promote 'cultural development' tended to be marginalised, although they could contribute significantly to the creation of confidence building and the opening up of new opportunities for economic and social development (Mayo 2000: 98), an example being the vital role of Khat production in Uganda which while forging social identities – independently from ethnicity and religion – also promotes the economy. Therefore, Khat production is an example of a successful indigenous development that is supported by the local government (Beckerleg 2010: 143, 182-184).

The capacity to exercise self-determination whether individually or collectively is systematically related to wealth, in terms of both material and other resources, and market position (Jenkins 2008: 203-204). The Yali, another indigenous society living in the highlands of Papua, linked Christianity with modernisation and the improvement of their lives, giving rise to a local cargo movement. Thus, they had high expectations of obtaining more pigs and rich harvests if they converted to Christianity. The Protestant mission, therefore, not only offered religion, but also improved living conditions motivating the Yali to burn their sacred traditional objects (the latter being similar to Hubula's sacred objects derived from warfare). Things worked out differently in the Palim Valley because at that time the Hubula valued their customs and traditions higher than externally introduced religions. Despite being cautious when dealing with strangers, the Hubula were a proud, self-confident people who were not easily attracted to outside cultural elements and values. Up to the present the Hubula uphold strong relations with their ancestors still informing their social identity.

6 As a working definition, detraditionalisation involves a shift of authority: from 'within' to 'without'. It entails the decline of the belief in a pre-given or natural orders of things. Individual subjects are themselves called upon to exercise authority in the face of the disorder and contingency which is thereby generated. (Helaas 1996: 2).

Nevertheless, the Palim Valley is considered by outsiders to be a 'backward' area that needs to progress in terms of modern technology and economics. Instead of focusing on indigenous logic and conducting a 'deep' study of cultural values and practices, modern media events, such as cultural festivals and a local museum that turns ancestral emblems into folklore, were viewed as a solution to the problem of maintaining Hubula identity. The ancestral objects derived from warfare were not considered to be appropriate by the government and the Church. Instead, the government and the Church promote a type of change leading to individualisation and commodification, thereby alienating the Hubula from their socio-cosmic relations.

This commodity logic which is in line with individualism and impersonal modern capitalism, and which according to Morris (1996: 223-224) contributes to the loss of community, can be overcome by striving for a new sense of collective purpose, shared values, and the common good. Andolina et al. (2005: 678-702) point out that the notion of *ayllu* (community) in Bolivia, as a form of social capital, has become widely shared in ethnodevelopment networks. This trend follows social capitalism in building on indigenous people's 'cultural knowledge' to generate marketable projects in ethno-tourism, ethnobiology, and agriculture. If part of what has made *ayllus* distinct is being rooted in non-capitalist modes of production, the insertion of *ayllus* into a social neoliberal paradigm may undermine cultural differences (*ibid*).

An inter-cultural dialogue is thus possible when two major elements mentioned by Briggs (2013: 238) are being accommodated: Firstly, mutual respect at the local level – provided that indigenous knowledge is no longer seen as 'non-knowledge', secondly, the idea that indigenous knowledge enriches and widens current scientific understandings of technology. Only by understanding the cosmology of communities, their social changes and patterns of relations – as has been demonstrated in this article – a new path to maintain cultural identity as self-determination in the development process can be achieved. As Englund and Leach (2000: 7) observe:

> "what matters to people is not culture (even when threatened with its extinction), and it is not development (even when it would enhance their security and well-being). What matters is that they are heard and respected for their contributions to a world in which each of them is made to feel – locally and globally – that it is theirs both to protect and to create anew."

References

ANDOLINA, R. et al. (2005) Development and Culture: Transnational Identity Making in Bolivia. *Political Geography* 24: 678-702.

BAAZ, M.E. (2005) *The Paternalism of Partnership – A Post Colonial Reading of Identity in Development Aid.* New York: Zedbooks.

BAYART, J-P. (2005) *The Illusion of Cultural Identity.* Chicago: The University of Chicago Press.

BECKERLEG, S. (2010) *Ethnic Identity and Development – Khat and Social change in Africa.* New York: Palgrave Macmillan.

BORGSTROEM, B.-E. (1976) Thoughts on the Relevance of Social Anthropology for Development Planning: The Case of Nepal. In PITT, D.C. (ed.), *Development from Below – Anthropologists and Development Situations*, pp. 83-95. The Hague: Mouton Publishers.

BRIGGS, J. (2013) Indigenous Knowledge: A False Dawn for Development Theory and Practice?. *Progress in Development Studies* 13(3): 231-243.

BROEKHUIJSE, J.T. (1967) *De Wiligiman Dani – Een Cultureel-Anthropologische Studie over Religie en Oorlogvoering in de Baliem-vallei.* Unpublished PhD thesis. Utrecht: Rijksuniversiteit.

CLARKE, G. (2001) From Ethnocide to Ethnodevelopment? Ethnic Minorities and Indigenous Peoples in Southeast Asia. *Third World Quarterly* 22(3): 413-436.

DIAMOND, J. (2012) *The World until Yesterday.* London: Penguin Books.

ELLER, J.D. (2009) *Cultural Anthropology – Global Forces, Local Lives.* New York: Taylor and Francis.

ENGLUND, H. and LEACH, J. (2000) Ethnography and the Meta-Narratives of Modernity. *Current Anthropology* 41(2): 225-248.

FONG, J. (2008) Revising the Ethnodevelopment Model: Addressing Karen Self-Determination within the Context of the Failed Ethnocratic State of Military-Ruled Burma. *Ethnic and Racial Studies* 31(2): 327-357.

GALESKI, B. (1976) Conflict and Change as Aspects of Development. In PITT, D.C. (ed.), *Development from below – Anthropologists and Development Situations*, pp. 151-165. The Hague: Mouton Publishers.

HEIDER, K.G. (1970) *The Dugum Dani – A Papuan Culture in the Highlands of West New Guinea.* Chicago: Aldine.

HELAAS, P. (1996) Introduction: Detraditionalization and Its Rivals. In HEELAS, P. et al. (eds.), *Detraditionalization – Critical Reviews on Authority and Identity*, pp. 1-20 Oxford: Blackwell publisher.

JENKINS, R. (2008) *Social Identity.* New York: Routledge.

KOENTJARANINGRAT and SIMORANGKIR, V. (eds.) (1993) *Masyarakat Terasing di Indonesia.* Jakarta: Departemen Sosial, Dewan Nasional Indonesia untuk Kesejahteraan Sosial dan PT Gramedia Pustaka.

LÉVI-STRAUSS, C. (2013) *Anthropology Confronts the Problems of the Modern World.* Cambridge: The Belknap Press of Harvard University Press.

LIESHOUT, F. (2008) *Sejarah Gereja Katolik dan Pergumulan Masyarakat Balim dengan Injil. Yubileum 50 tahun Gereja Katolik di Lembah Balim*. Wamena: Dekanat Gereja Katolik.

LUKE, T.W. (1996) Identity, Meaning and Globalization: Detraditionalization in Postmodern Space-Time Compression. In HEELAS, P. et al. (eds.), *Detraditionalization – Critical Reviews on Authority and Identity*, pp. 109-133. Oxford: Blackwell.

MAYO, M. (2000) *Cultures, Communities, Identities – Cultural Strategies for Participation and Empowerment.* Basingstoke: Palgrave.

MORRIS, P. (1996) Community Beyond Tradition. In HEELAS, P. et al. (eds.*), Detraditionalization – Critical Reviews on Authority and Identity*, pp. 223-249. Oxford: Blackwell.

MOSSE, D. (2005) *Cultivating Development- An Ethnography of Aid Policy and Policy Practice*. London: Pluto Press.

MOSSE, D. and LEWIS, D. (2006) Theoritical Approaches to Brokerage and Translation in Development. In MOSSE, D. and LEWIS, D. (eds.), *Development Brokers and Translator – The Ethnography of Aid and Agencies*, pp. 1-26. Bloomfield: Kumarian Press.

OLIVIER de SARDAN, J.P. (2005) *Anthropology and Development – Understanding Contemporary Social Change*. New York: Palgrave Macmillan.

OTTERBEIN, K.F. (2000) A History of Research on Warfare in Anthropology. *American Anthropologist* 101(4): 794-805.

ROBARCHEK, C.A. (1989) Primitive Warfare and the Ratomorphic Image of Mankind. *American Anthropologist* 91(4): 903-920.

ROSA, E. (2013) Re-Learning of Traditional Knowledge in Times of Modernity. *Campus-Wide Information Systems* 31(1): 14-23.

SUGANDI, Y. (2014) T*he Notion of Collective Dignity among Hubula of Palim Valley, Papua.*

Münster: Verlaghaus Monsenstein und Vannerdat OHG.

STRATHERN, M. (1999) *Property, Substance and Effect. Anthropological Essays on Persons and Things.* London: Athlone Press.

THORPE, I.J.N. (2003) Anthropology, Archaeology, and the Origin of Warfare. *World Archaeology* 35(1): 145-165.

TIMMER, J. (2012) Menentang Kulturalisme di Tanah Papua. In SURYAWAN, I.N. (ed.), *Jiwa yang patah*, pp. 21-40. Yogyakarta: Kepel Press.

WHITE, G.M. (1991) *Identity through History: Living Stories in a Solomon Islands.* Cambridge: Cambridge University Press.

WHITEHEAD, N.L. (2000) A History of Research on Warfare in Anthropology – Reply to Keith Otterbein. *American Anthropologist* 102(4): 834-837.

WOODWARD, M.R. (2000) Gifts for the Sky People: Animal Sacrifice, Head Hunting and Power among the Naga of Burma and Assam. In HARVEY, G. (ed.), *Indigenous Religions: A Companion*, pp. 219-229. London: Cassell.

Website

OFFICIAL STATISTIC NEWS (2014) http://www.bps.go.id/

When "Development" enters the Village

SJOERD ZANEN

The relevance and success of Development Cooperation are often questioned, also by anthropologists, and this was a big issue in Platenkamp's early days at Leiden University. Jos preceded me in this field (Platenkamp 1978) with his desk study on the Jonglei canal. I had the honour to follow-up that study in the field, together with Bert van den Hoek (van den Hoek *et al* 1978)[1]. Our structural analysis of Nilotic cultures (Zanen and v.d. Hoek) was published as an essay in honour of P.E. de Josselin de Jong (de Ridder and Karremans, 1987). I continued in Development Cooperation, altogether living 18 years in East and West Africa. My thesis in this paper is that relevance and success of Development Cooperation are dependent on the way you conceive 'development'. I will show how 'development' often takes an unexpected turn. The paper is written from a structural anthropological perspective, and represents the 'participant's view' (P.E. de Josselin de Jong 1977). The question is what anthropology may reveal in the field of Development Cooperation what remains hidden for other disciplines.

Introduction

Donor organisations still promote development policies and strategies of which they assume that these are desirable and will automatically be shared in local contexts abroad. Donor ideas are often projected on local situations, even by local partner Community Based Organisations of northern co-financing Civil Society Organisations: one assumes that what is good for 'us' will also be good for 'them'. In practice, this situation is hardly ever realistic, especially not with respect to those on whom development cooperation focuses: the poor, who are the 'target groups' in government or NGO poverty alleviation projects and programmes. We all know that donors generally rely on 'logical', linear development planning models. Practice appears always more unruly and more unpredictable than was assumed; there appear to arise contextual, cultural, social, organisational or mentality problems which block applicability of nice ideas from the outside[2].

Herewith, we enter the world of perceptions, thinking, knowledge, ideas, communication, in a multi-cultural setting. This world is pre-eminently the

1 See also Vermeulen, in this volume.
2 Often, the failure of development projects is blamed on the recipients of aid in varying degrees of 'victim blaming'.

object of cultural or social anthropology. Without trying to capture 'anthropology' in a definition, anthropology concerns the above mentioned fields of knowledge, ideas and intercultural communication. In the field of development cooperation anthropology is also very much concerned with assumptions, prejudices, professional blindness, and cultural, social and psychological ignorance. Its role is, in the first place, to study what people think, what happens and why this happens – from village level up to the (inter)national level. In the second place alternative solutions may be proposed based on such anthropological knowledge.

An effective policy of poverty alleviation should be based on an analysis of poverty and the poor. Presently, quite surprisingly, this is hardly ever the case. It is true that the official poverty alleviation *discourse* refers to the poor, and that poverty is statistically defined, but in reality, poverty is hardly understood nor studied by policy makers. In African countries[3] all kinds of 'reforms' are introduced, funded by international donor agencies: in Tanzania, in 2001, 14 important reform programmes were funded by donors for at least 60%; the remaining 40% national funding had to prove that 'ownership' of the reforms lies with the national government. Through donor funding (with its conditions) the policies have been adapted to the donor's ideas and wishes. The 'reformed' institutional environment is, on paper, fit to serve as a vehicle for all kinds of global processes, transactions and interventions. Thus an effort is made to adapt the context to the aid, instead of adapting the aid to the context. This adaptation is often seen as 'modernisation', or (nowadays) 'globalisation'. But in practice, the 'reformed' state often lacks anchorage in society. In these societies one observes more and more the development of mechanisms to *avoid* the 'reforms' (an informal economy, fraud, criminality, "traditionalism", "terrorism", emigration on a big scale), at least among the poor masses. Obviously, some assumptions about 'poverty' were wrong. It would therefore be unrealistic to expect from the mentioned policies that they would resolve poverty.

What happens when a poor countryside comes into contact with so-called development (intervention), a process having taken place, or still taking place everywhere in Africa since some 40 years? Below, we will discuss a number of perceptions and processes with regard to 'development', 'poverty', and development 'interventions'. The case presented concerns the countryside of Burkina Faso inhabited by Mossi farmers· The analysis will shed light on local ways of thinking, local cultural and social 'logics', attitudes *vis-à-vis* the outside world (foreigners, Whites, donors, NGOs, 'projects'), and on the internal villagers' absorption capacity of development interventions.

3 And in others, but I restrict myself to Africa in this paper.

Burkina Faso: Development, Wealth and Poverty

The concept of 'development' with its strong focus on the future and its connotations of change and progress, cannot be easily found back in Mossi culture. There are no clear-cut, simple translations of the term development in the local language, *moore.* One can only describe development as 'deliverance', 'a good life', 'divine benediction', having a good reputation, being 'respected', collective well-being. These words do not refer to a process of material accumulation or 'progress', but they refer to a morally desired state of affairs: to find oneself in a state of grace, of harmony in the relation with the community and God. It is in such a situation that 'all will be well' *(nazemse),* that the rain will fall, that healing will take place, that the harvest will be good and livestock will reproduce. This view refers to the balance between material and spiritual well-being. Somebody who, within the 'traditional' way of life, succeeds in life and collects a certain affluence, is somebody who testifies of respectable behaviour, who has done as it socially should, who lives in harmony with the community, the forces of nature, the ancestors and God.

Affluence and poverty are couched in the terms *ra kaagre,* a 'rich person', and *talaga,* a 'poor devil' (Zanen 1996). The following properties are attributed to a *ra kaagre:*

> "He is rich, impressive, well respected; he is loved by God; he has a big family, a lot of wives and sons, he has employees, big fields and full granaries; he has not too much time to chat with visitors, but if he talks, he talks well; he rescues lives; he doesn't give money, he doesn't lend out, but he gives grain; he receives girls in return since he already possesses all other things – so his sons marry young; he is one who has all, he doesn't lack anything – if he has what he has it is thanks to God; he is feared by chiefs, since he himself is called chief by the people".

On the contrary, a *talaga* is:

> "an unlucky fellow, a pathetic, pitiful man; everything goes against him; he possesses a hostile *kinkirga* (ghost); he has no wife or his wife/wives and children have died; he doesn't know how to behave; he has diseases in his crops and his house catches fire".

In all these descriptions, the material, the social and the metaphysical seem to be indissolubly related.

In the dichotomy mentioned above does not figure the 'nouveau riche', *ligdi soba* ('owner of money'), or *nim bedre* ('big man') who is materially 'rich', but who is not necessarily respected. The concept refers to the material level alone, and so with the unuttered thought that this form of affluence may appear to be volatile. A *ligdi soba* has not obtained his wealth from within the Mossi socio-economic system, it comes from 'outside'. The only way to legitimise these riches obtained from elsewhere is to socialise them: in the first place to distribute them to support networks, but also to show generosity to his paid employees, not by paying them high salaries, but socially, by helping them when they encounter crises in their families, or on the occasion of baptisms and *funerals.* In other words, he should behave like a traditional chief, both admired and feared, who follows the social code and who reigns with authority, and respects the traditions *(roogm miki,* litt. 'born-found'). If he does not act in this way, he will become the object of sorcery, practiced by order of his jealous supporters (kin, neighbours, friends).

When we refer to modern 'development' we are usually focussed on such 'nouveaux riches', who seem to have succeeded in pursuing individual goals and to increase individual wealth like real entrepreneurs. We will come back to the fate of these *nim beda,* 'big men'.

We and the Others; Inside and Outside; Customary Power and Foreign Power

In this context, a second conceptual opposition is of interest, namely the difference between *mo tedo,* 'things of the Mossi', and *nasara tedo,* 'things of foreigners/strangers'. The distinction is broadly applicable. *Nasara tedo* is connected with everything which is not originally Mossi, and which is therefore not intelligible and controllable for the Mossi system of thought; it is potentially threatening and should be approached carefully. It is not necessarily bad or to be disapproved of and it remains to be seen how it could be matched with *mo tedo.* For instance: for traditional cures, or magical ends (typically *mo tedo),* modern things (like sea-salt or maggi, or sea-fish or aluminium utensils, a torch, or money) can definitively *not* be used. Certain wells are *mo tedo,* as special sacrifices were brought for them: in these wells no rubber buckets (made from lorry tyres, typically *nasar tedo)* may be used. On the occasion of marriage transactions all money is replaced by cowries *(mo tedo),* the traditional means of transaction. The chieftaincy village-quarter where the house of the chief is situated (the 'village of power' *(na tenga),* being very *mo tedo*) should not also contain the 'village development association', being 'a thing of the Whites', *nasara tedo.*

What we observe is not pure conservatism or the fear for modern elements, but rather the fear for the loss of the functioning of traditional things, because of mixing or contamination with new, foreign elements. Two systems of meaning do not easily mix.

In politics, there are nowadays also two forms of power: *mo naam,* the traditional power reserved for 'the people of power', and *nasar naam,* the power of strangers, foreign power, which does not come from God and can be acquired by 'anybody'. With these foreign forms of power the traditional elite has an ambivalent relationship. On the one hand it is threatening, but on the other hand it is a reality which can no longer be denied, and with which a compromise would perhaps be welcome in order to transform the situation into an advantage. This is the reason why traditional chiefs nowadays venture more and more into national politics, and also why one nowadays sees that educated civil servants are elected as village chiefs since they know the 'functioning' of the administration and how this knowledge could be utilised in favour of the village. These are examples of how the village community, in a careful way, starts an interaction with elements from the outside, thereby accommodating two systems of meaning.

'Development': The Interaction between Inside and Outside

From this perspective, 'planned development intervention', the totality of ideas and measures introduced by the outside world, obtains a special significance: it concerns the interaction between what is their own *(mo tedo)* and what is foreign *(nasara tedo)*. Foreign elements ('innovations' in the form of techniques, behaviour and new forms of organisation), if proven advantageous, could be incorporated in the culture, if necessary in adapted form or with another significance. During this process, in social life, adaptations will find place. Certain innovations will be valued by one network of actors, others by other networks. The outcome of the 'process of development' is therefore dependent on the power struggle around this conflict of interpretation in the 'arena' of the rural areas.

In the context of the conscious introduction of innovations, the developers have to realise that the so-called 'target group' must first of all organise a socio-cultural 'framework of reception', a social framework in which the actors will become motivated to adapt their economical behaviour, and a cultural or ideological framework in which such new economical behaviour becomes meaningful, or is legitimised. The local assessment of foreign elements, the attribution of a meaning to these and their possible adoption, is determined by the influence it exerts on the social and cosmic orders and their interrelations. For that reason there takes place a systematic 'translation' of interventions with, as a stake, the maintenance or the redefinition of prevailing social and socio-cosmic relationships. Usually, the spoils of development go to the dominant elite, unless a real power struggle has resulted into a 'revolutionary' outcome.

Among the Mossi of Burkina Faso, the actors in this process of reinterpretation are players in a socio-hierarchical field constituted by powers from above (God), and from below (the ancestors), and influenced by powers from 'the other side' (the spirit world). The field is kept up by a rural gerontocracy, authoritarian old men – the defenders of 'tradition'. Sorcery is common in this field. This strongly limits the individual's room for manoeuvring. It will be evident that the traditional multi-dimensional reality in which the 'target groups' operate, contrasts strongly with the donor's single-dimensional and single-directional development reality. The interface of these realities, in the case of 'planned intervention', may lead to a misty form of communication and to unexpected outcomes of the 'development'-process.

The Day-Break of Development Cooperation on the Local Level

Of course, the need for 'development' is urgent. There have been recent crises everywhere in Africa, either climatical, political, economical or social crises. And apart from needs, there are also opportunities, directly or indirectly related to the above mentioned crises which have attracted donors. With so-called development cooperation, a new dawn and new hopes glimmer in the village. But how to start the process of harvesting some of the seemingly boundless resources of 'the Whites'? A communication has to be created with the donors and a relationship developed with them, but by whom, and how?

In the countryside of Burkina Faso, in the 1980-s, it was often the educated youth who took the lead in contacting and communicating with donor agencies. Among them there was, and is, a strong desire to escape from the strenuous social and economical environment in the village. They long for more liberty and more individual wealth, more autonomy and room for manoeuvring. However, showing such ambitions, they soon found themselves confronted with adversity, jealousy, and invisible forces invoked by sorcery with which it was tried to drag them back into the lap of traditional Mossi society.

It is obvious that youths alone can not engage in 'development' by leaving out the rest of the village: 'development', as it is considered by all donors, is a collective affair and should be undertaken by communities, or at least organised groups. Therefore, it seems most urgent to start with a (non-traditional) *groupement,* an organisation with a management structure and a minimum of resources, which is capable of maintaining contacts with 'partners' and of organising the 'participation' required by most donors. The first thing they start with is to collect small sums of money from their members. The founding of the organisation, some initial 'good deeds' for the community and the collection and control of some funds functionally serve to show to the outside world that they are prepared and ready for 'development'.

Many local village associations and NGOs have come into being in this way, with the main aim of functioning as a structure of reception for 'development' entering the village. With their associations as a base, they undertook efforts to engage the rest of the village in their 'projects'. However, such recently organised groups signify, *qualitate qua,* a threat to traditional as well as (in Burkina Faso in the 1980-s) revolutionary authorities who used to control all social action in the village. Through the actions of 'development', meant to benefit the whole village, the careful balance of loyalties of which Mossi villages socially consist, was seriously threatened. Some villages experienced a decennium of conflicts which opposed successively, alternately or at the same time, 1. the old and the young, 2. the traditional and the revolutionary chiefs, 3. the 'people of power' and the 'people of the earth', 4. the quarters of a village, 5. the villages among each other, 6. the catholics and the protestants, 7. the 'animists' and the muslims, 8. the state agencies and the farmers, 9. the farmer's associations *in statu nascendi* among each other and the NGO, and finally individuals fighting for the control of the associations.

What happens when such a village entity, such a 'basket of crabs', starts a relationship with international donor agencies? In the first place, in such interactions, a hierarchy is established. Both parties had initially expected to enter into a relationship between equals, who act from the same perspective, but the relationship is and remains, in fact, between the superior donor and the inferior beneficiaries, the latter being the receivers of gifts. In this mechanism counter-gifts are not expected. Thus, a communication between unequals emerges, whereby neither party knows exactly the strategy of the other. Both parties, because of different cultural backgrounds, different interests and different intentions, represent different systems of meaning: there is a semantic difference in their way of thinking. This opens the way for the *ruse*, or cunning[4]. In the ambiguous field of development cooperation, cunning leads to an unofficial communication which does not exclude treachery, theft of each other's objectives and in extreme cases, to corruption.

The Logic of Ruse and Cunning

In West Africa in general, and in Mossi traditional culture in particular, the ruse is mastered by people who 'own' cunning *(silim soaba),* usually applied as conflict management in the community characterised by complicated social interrelationships.

Such a trickster is a mediator who acts in situations of ambiguity. He may profit from his acts of influencing (potential) adversaries. He may be a crook who escapes with the booty, but he is not considered wicked or a thief[5]. Instead he is considered a shrewd person, and sometimes a hero. To know the way how to use tricks or mediation tactics in public debate in matters of community interest is considered a quality which leaders should dispose of. When a village association has managed to attract a donor to fund certain community projects, controversies usually disrupt in the village community as certain categories of people consider themselves neglected, discriminated or ignored, or are just jealous about the achievement of their presumed competitors. In such a case the success of 'projects' depends on the degree of intra-communal harmony. If relative peace can not be established, 'projects' (partially) fail, as is visible in many Mossi villages: the primary school which is only frequented

4 In Dutch: de list.

5 Two famous *trickster* culture heroes in West Africa are Leuk le Lièvre (L.Senghor and A. Sadji, 1953) and Wangrin (Amadou Hampate Ba, 1973).

by children from some but not from all village quarters, or the health clinic, the vaccination campaign, the pump, and the vegetable garden all becoming objects of conflict in the village.

However, villagers (for whom *wuum taaba,* community peace, is so important) do not wish to show their inner conflicts to others, least of all to donor agencies. Conflicts remain hidden and invisible for foreigners. Therefore, when the donor representatives visit the village they remark an impressive demonstration of solidarity, with a massive presence of all couches of the village. This show of solidarity is very attractive to donors. During the show, the donors also receive (symbolical) presents which tend to charm or to oblige them, in a general sense. It literally creates 'good will' on the side of the donor. Such behaviour is part of the cunning of the NGO leader *vis-à-vis* the whole village, and, on the other hand, of the whole village *vis-à-vis* the outside world (Laurent 1998).

In the communication between village associations and representatives of donor agencies, both with their different systems of meaning – hence, so to speak, the communication between the 'blind and the deaf' – hidden agendas tend to develop. Since the donor agencies officially make the rules of the game, the beneficiaries, in order to receive the maximum of the fruits of 'development' (which never suffice to satisfy all needs), tend to do as if they stick to the donor's rules, but meanwhile pursue quite different goals. As such the ruse, the trick, enters the field of development cooperation. Towards the outside world inner oppositions are dissolved, and this leads to a ruse which is collectively shared, namely treason of the donor's principles. It is not treason of the donors with the aim to break the pact with them, but rather the expression of adherence to their own *reality, language and logic* at the cost of the donor's.

In matters of unpersonal, 'anonymous' development cooperation, the recipients do not attach a moral value to funds as gifts. They have been made to understand that a return gift (in material form, in contrast to symbolical gifts) is not appreciated. One does not consider as return-gift a 'correct' management of these gifts, and neither does one feel obliged by such standards as 'good governance', 'democracy', 'transparency', or 'accountability'. Ultimately, these norms and values are 'things of the Whites', their *reality, language and logic*, something *nasara tedo,* something to do with the Whites' "mission civilisatrice", for which rural Mossi do not feel obliged to account. In most areas of rural Africa such is the attitude toward the public sphere or public goods and means in general. The mentioned development cooperation norms are considered foreign. They are outwardly respected, displayed as 'virtuous intentions'. They are expressions of the development-language and frequently invoked as *mantras,* devoid of meaning, of, what I would call, a *donorcracy.*

A *donorcracy* is a country having succumbed to foreign aid and being dependent on it to survive. In order to receive that aid, it has adopted all that the donors wish to see and to hear – *seemingly*. It has adopted these things *pro forma* (its form, but not its content), while in reality these things have not or hardly changed. Silently, it adheres to its own 'projects', to its own 'agendas', or to its own *reality, language and logic*, norms and values, which may totally contradict the projects officially adopted, or the norms and values to which lip-service is paid. It is striking to realise that much of what the donors assume to be reality in 'development countries', appears to be based on wishful thinking (others would say: 'built on quicksand'), or on wrong assumptions.

The donor staff will not easily be enlightened by their educated counterparts or partner organisations about what is really going on. In the first place, these counterparts might live the same wished realities, but on the other hand it is not their vocation, nor is it in their interest to discourage the enthusiasm of the donors. A *donorcracy* derives its favourable reputation, its exhibited identity, from a façade, from a pretended reality, not from what lies behind it. What lies behind is another reality with another logic which is lived by the inhabitants of that country, the 'masses'. Confronted with donors, however, these civilians too are ready to propagate that façade-Iogic. The more skilled these countries are in such pretended behaviour, the more they become 'darlings of the donors'. The latter believe, or wish to believe, that their strategies are bearing fruit (Kabou 1991).

The mechanism described above is part of a construction of 'ruses' originating from the interface of differing systems of meaning and knowledge, or systems of 'logics'. The ruse, when intentionally employed for individual profit, is equivalent with what we call corruption.

'Modernity': The Penetration of 'Village' into 'City'

The ideal of ambitious villagers is to become 'urban oriented' or 'global oriented', and to transcend the image of 'farmer' and 'villager' which is linked with everything that contrasts with 'modernity' and which obstructs individual accumulation and personal development. The only way to achieve this is, first, to attract outside help, and, second, to create a distance *vis-à-vis* relatives and other close acquaintances.

Identification with 'modernity' does not automatically lead to freedom in the process of personal development. Even if, through foreign help, a certain degree of affluence is reached, the certainty of the lineage is not easily replaced by the certainty of 'modernity'. The latter would imply that the State creates anonymous chances for individuals. The State (at least in the nineteen eighties and nineties), although pretending modernity, could not meet the expectations of being the carrier of 'modernity'[6]. Instead, with continuing liberalisation, the erstwhile state-functions were replaced by paid services, either officially, or through the mechanism of corruption by government employees, leading to so-called *state cannibalism*, i.e. the state which illegally exploits its citizens.

The 'privatisation of the State', during the mentioned period, experienced by the poor in the rural, the semi-urban and the urban areas, has prevented the building up of feelings of security needed by individuals to invest in socio-economic development. Instead, ambitions have dried up and frustrations have fed feelings of fear, suspicion and jealousy. This has brought people who did not find protection and security with the constitutional state (which seemed to be reserved for the rich and mighty), to look for protection within easy reach of anyone, their relatives, Big Men. Consequently, this has led to a farewell to 'modern' logic, and the return to *magic* thinking, to the *reality, language and logic* of the village and to the dependence on the symbolical system fed by gifts, debts and manipulation, from which one wanted to escape (cf Dietz and Zanen 2009). This return to the logic of the village, even if one resides in the city, is a typical characteristic of the contemporary 'modernity' of the poor. It is an example of a return to traditional certainties where 'development' has provided none.

Once in this phase, social-economic accumulation is realised by an 'economy of envy' whereby the success of the one is considered as a loss or a defeat of the other. Henceforward, between jealous fellowmen, there is a competition in accumulation of riches *through* a competition based on the accumulation of magical strength. Sorcery, fear of sorcery and occult protection against sorcery create vicious circles of fear, distrust, revenge, rumours and accusations. Sorcery leads to a social crisis whereby social relations are guided by silent, magical warfare with the risks of violence and aggression. A balance is sought based on the hope of one's own sorcery harming the other, and the fear for violence which the other may be able to generate. Economic risers, ideally assumed to be the driving forces of the economy, become hostages in networks of competitors and dependents (often kin) who do anything to make their successful relatives keep to their lineage obligations, and thus distribute their 'foreign' wealth. Very often, promising enterprises or careers have failed because of the tensions built up by the 'état sorcière'. So kinship and sorcery are related: sorcery is the night-side of kinship, of which the dayside is social security.

---

6 In many African countrysides, following structural adjustment policies, the State failed to create favourable conditions for individual development, to provide basic services (education, health, clean water, credits). It also failed to guarantee social peace, social security and employment.

On a higher level, Big Men (*nim beda*), having succeeded in individual accumulation of wealth with the help or approval of their kin, and keeping their support networks happy, tend to be the cause of such a 'state of magics' because of the jealousy they arouse among similar, competing networks. In this way the entrepreneurial climate in communities is poisoned. Those who survive will really become the 'nouveaux riches' of the country, very powerful in all respects.

There are two, usually related, ways out of the imprisonment for debt. The first one is to emigrate to regions out of reach of competitors and lineage networks. This escape is indeed frequently tempted by the youths of the Burkina Faso Mossi plateau, who used to flee from the village secretly to try their chances on the coastal plantations of Ivory Coast. A second escape is to convert to another religion, either the Muslim or the Christian religion, which liberates the individual from the wrath of ancestors and the malign forces of nature. During the last twenty years conversion to Protestant sects has become popular. These provide resistance, or a strong immunity to secret, uncontrollable, magical forces evoked by jealous rivals.

Conclusion

The arguments presented in the foregoing sections show the dynamics of actual cultural and socio-economic development processes in a local context, triggered by intervention from outside.

Donors, their idea being to stimulate development processes, usually have little idea of what their actions and intentions bring about in the village. Neither are they equipped to understand the underlying reasons for the lack of progress in projects, or the project's lack of 'good governance' or 'accountability'. They are inclined to think that the foregoing is caused by a lack of knowledge and skills of the recipients, hence the need for all kinds of training, meetings, workshops, etc. For local recipients all (much willed) training courses, seminars, excursions, conferences and other meetings are sources of income because of the allowances (fr. *indemnités, per diems),* generously dispensed by donor agencies. Donors do this out of self interest: the need to have successful projects is most urgent! The course members are chosen by the local development *committee,* or *groupement,* or *association* because of their right to profit this time from the *allowances,* the spoils of development. And this pattern repeats itself on national level in development bureaucracies.

Recently (with so-called sector-wide approaches) donors have tried to solve the issue of the above described local 'problems'. They have chosen to keep their distance from local contexts, and thus to reduce 'transaction costs'. They now prefer to give funds to large institutions or partner-organisations of which they assume that they dispose of 'good governance' and hold them responsible for positive project impact. They are to be the 'owners' of large (sectoral) programmes, and it is assumed that such ownership guarantees acceptance and adoption of innovative ideas on the level of the 'poor'. As we have seen, it doesn't work that way. The villager's intention is to keep up the ambiguous relationship with donors, so as to be able to profit as long as possible from the spoils of development.

The story behind the façade is what anthropology and anthropological 'searchers'[7] may unearth 'when development enters the village'. Anthropological knowledge and analysis shed light on what happens behind the façades of *donorcracies* whether they are villages or nations. Such knowledge may explain some of the dynamics of development cooperation, considered from the *reality, language and logic* of the lifeworld of the poor. It provides insight in the way social change is rooted in culture, and in the effects of development interventions – with other words: insight in the impact of 'development'.

References

BROCH-DUE, V. and ANDERSON, D.M. (1999) Poverty & the Pastoralist: Deconstructing Myths, Reconstructing Realities. In ANDERSON, D.M. and BROCH DUE, V. (eds.), *The Poor Are Not Us. Poverty & Pastoralism in Eastern Africa,* pp. 3-19. Athens: Ohio University Press.

DIETZ, T. AND ZANEN, S. (2009) Assessing Interventions and Change among Presumed Beneficiaries of 'Development': A Toppled Perspective on Impact Evaluation. In HOEBINK, P. (ed.), *The Netherlands Yearbook on International Cooperation,* pp. 145-163. Assen: Van Gorkum.

EASTERLY, W. (2006) *The White Man's Burden. Why the West's Efforts to Aid the Rest Have Done so much Ill and so Little Good.* New York: The Penguin Press.

HAMPATÉ BÂ, A. (1973) *L'étrange destin de Wangrin.* Paris: Editions 10/18.

HOEK, B. van den, ZANEN, S. and LEEK DENG, P. (1985 [1978]) *Social-Anthropological Aspects of the Jonglei Development Projects in South Sudan.* Leiden: Institute of Cultural Anthropology (ICA Publication 67).

JOSSELIN de JONG, P.E. de (1977) The Participants' View of Their Culture. In JOSSELIN de JONG, P.E. de (ed.), *Structural Anthropology in the Netherlands: A Reader*, pp. 231-252. The Hague: Martinus Nijhoff.

KABOU, A. (1991) *Et si l'Afrique refusait le développement?* Paris: L'Harmattan.

LAURENT, P-J. (1998) *Une association de développement en pays mossi. Le don comme ruse.* Paris: Karthala.

PLATENKAMP, J.D.M. (1978) *The Jonglei Canal: Its Impact on an Integrated System in the Southern Sudan.* Leiden: Institute of Social and Cultural Studies (ICA Publication No. 26).

RIDDER, R. de and Karremans, J.A.J. (eds.) (1987) *The Leiden Tradition in Structural Anthropology: Essays in Honour of P.E. de Josselin de Jong.* Leiden: E.J.Brill.

SENGHOR, L. and SADJI, A. (1953) *La belle histoire de Leuk le Lièvre.* Paris: EDICEF.

7 'Searchers', in contrast to 'Planners', cf Easterly (2006).

ZANEN S. and HOEK, A.W. van den (1987) Dinka Dualism and the Nilotic Hierarchy of Values. In RIDDER, R. de and KARREMANS, J.A.J. (eds), *The Leiden Tradition in Structural Anthropology,* pp. 170-196. Leiden: E.J.Brill.

ZANEN, S. (1996) *Ontwikkeling: Idee en Handeling. De confrontatie van boeren- en donorstrategieën in de praktijk van ontwikkelingssamenwerking (Nederland-Burkina Faso).* Leiden: Research School CNWS.

Contributors

Wassef Al Sekhaneh is lecturer at Yarmouk University and Al Balqa University in Physics. Before, he has worked as assistant at the Duisburg-Essen University in Physics and Chemistry Departments and at the Max Planck Institute for solid state research in Stuttgart. He obtained his PhD in Social Anthropology at the Westfälische Wilhelms-Universität in Münster. He has conducted research on Bedouins in Jordan.

Gabriele Alex is Professor at the Department of Social and Cultural Anthropology, Eberhard Karls Universität Tübingen. She obtained her Master degree in Medical Anthropology and her PhD in Social Anthropology from Brunel University, West London. Before joining the University of Tübingen, she was teaching at the University of Münster, was Assistant Professor of Cultural Anthropology while also serving as Director of the Master's Program „Health and Society in South Asia" at the South Asia Institute, University of Heidelberg. After this she joined the Max Planck Institute for the Study of Religious and Ethnic Diversity as a research fellow with a focus on Medical Diversity. Dr. Alex's research interests include Medical Anthropology, Ethnicity, Healing and Identity, the anthropology of Youth and Childhood, and Visual Anthropology.

Cécile Barraud is a retired Senior Researcher at the National Center for Scientific Research (CNRS, France). She has conducted extensive fieldwork in the Southeast Moluccas, Eastern Indonesia, since 1971. Her research focused initially on understanding "houses" as social groups, then on rituals, exchanges and the comparison of ideologies. She specialised on kinship, gender and sex distinction. Among her major publications are "De la distinction de sexe dans les sociétés: une présentation ", in C. Alès et C. Barraud (éds. 2001), Sexe relatif ou sexe absolu? De la distinction de sexe dans les sociétés, Editions de la Maison des sciences de l'homme: 23-99, 2001, and "De la résistance des mots. Propriété, possession, autorité dans des sociétés de l'Indo-Pacifique", in A. Iteanu (ed. 2010), La cohérence des sociétés. Mélanges en hommage à Daniel de Coppet, Paris, Éditions de la Maison des Sciences de l'Homme : 83-146.

Helene Basu is Professor of Social Anthropology at Westfäelische-Wilhelms-Universität in Muenster. She holds positions of Principal Investigator at the Cluster of Excellence 'Religion and Politics in Pre-Modern and Modern Cultures' and in the Priority Program 'Cultures of decision-making' at the University of Muenster. Her publications include *Journeys and Dwellings – In-*

dian Ocean Themes in South Asia (2008 (ed.), *Von Barden und Königen. Ethnologische Studien zur Göttin und zum Gedächtnis* (2004), *Embodying Charisma* (co-edited with Pnina Werbner, 1998). She is co-editor of the *Encyclopedia of Hinduism* (Brill) and she has produced ethnographic films of which one deals with practices of healing madness through counter-sorcery (Spirits of Envy, 2012).

Ursula Bertels studied ethnology in Münster and Freiburg. The regional and thematic focus of her ethnological research is Mexico. Since 1986 she regularly carries out research there, and since 2003 she has been endowed with short-term lectureships. In 1992 she wrote her dissertation about "The Pole Dance in Mexico – historic development and current manifestation." Since 1994 she is working as lecturer on the subject of intercultural competence and migration. Since 2002 she heads as Chairwoman the "Association of Ethnology in School and Adult Education" (ESE e.V.). As part of her work at ESE she is responsible for third party funded projects on the subject of intercultural competence. She further organises and supports since 2006 the German-Mexican student exchange program for intercultural pedagogy on behalf of the Institute for Ethnology.

Joseba Estevez, is a PhD. candidate in Social Anthropology at the Institute of Ethnology of the University of Münster, Germany. He has conducted long-term social anthropological fieldwork on the role of the Lanten (Yao Mun) ritual experts in Luang Namtha Province, Laos from 2010 onwards. His research interests include ritual studies, exchange systems, forms of social organisation, social transformation, animism, and Daoism. He is presently managing the digitalisation of the Lanten textual heritage in Northern Laos in collaboration with the University of Hong Kong and the National Library of Laos financially supported by the British Library's Endangered Archives Programme (EAP 791).

Volker Grabowsky is Professor of Thai Language and Culture (Thai Studies) at the Asia-Africa-Institute of the University of Hamburg since October 2009. He is also head of the institute's Southeast Asia Department. Previously he was Professor of Southeast Asian History at the Institute of Ethnology, Westfälische Wilhelms-Universität Münster. From September 1996 until August 1999 he worked as a visiting lecturer of the German Academic Exchange Service (DAAD) at the Department of Lao Language and Literature, National University of Laos (NUOL) in Vientiane. He has specialised in the history and culture of the Tai peoples in mainland Southeast Asia and Southwest China. Recently he has embarked on long-term research projects pertaining to the manuscript cultures and Buddhist literary traditions of the Tai Lü and Lao. Together with Hans Georg Berger he edited the volume "The Lao Sangha and

Modernity: Research at the Buddhist Archives at Luang Prabang, 2005–2015" (New York/Luang Prabang: Anantha Publishing, 2016).

Roland Hardenberg is Professor of Social and Cultural Anthropology at the Institute of Asian and Oriental Studies of the University of Tübingen, Germany. He has conducted ethnographic fieldwork on divine kingship in the temple town of Puri (India, from 1995-1996), on sacrifices and marriages among shifting cultivators in the highlands of Odisha (India, from 2001-2003) and on funerals and graveyards of formerly nomadic people in the Tian Shan mountain range (Kyrgyzstan, from 2007-2008). His research interests include ritual studies, kinship, exchange systems and forms of social organisation. He is presently vice-speaker of the Collaborative Research Center (CRC 1070) on "ResourceCultures" and leader of a research project dealing with religious resources in South and Central Asia. Latest Monograph: The Renewal of Jagannatha's Body. Ritual and Society in Coastal Orissa. 428 pp. New Delhi: Manak Publications. 2011.

Andreas Hartmann is Professor of Volkskunde/European Ethnology at the Westfälische Wilhelms-Universität in Muenster.

Stephen C. Headley earned a B.A. degree (1965) in Oriental Studies (Chinese and Sanskrit) and an M.A. degree in Buddhist Studies (1969) from Columbia University. He continued his studies in Paris with a diploma in Sanskrit philology at the Ecole Pratique des Hautes Etudes (1972) and a doctorate in social anthropology under Georges Condominas at the Sorbonne in 1979. He also studied theology at Saint Vladimir's Orthodox Theological Seminary (Crestwood, New York, 1966-1969) and at the St Sergius Institute of Orthodox Theology in Paris (1969-1973). He worked at the French National Center for Scientific Research (CNRS) between 1981-2008, and between 1998 and 2008 he was working with the research team (ERASME) founded by Louis Dumont. He is presently retired from the CNRS and teaches ascetical theology at the Russian Orthodox Study Center in Paris. Between 1973 and 2005 Headley did extensive fieldwork in central Java. Between 2006- 2010 he taught in Moscow and undertook field work on parish life. He has published on Javanese Cosmology and Islam, and other topics related to the anthropology of religion.

André Iteanu is a social anthropologist. He is directeur de recherche and a member of the institute CASE at the French National Center for Research and professor (DE) at the Ecole Pratique des Hautes Etudes (Paris). He has carried out research among the Orokaiva of Papua New Guinea and among the marginalised youth in the northern suburb of Paris. His recent publications include "Comparison made radical: Dumont's anthropology of value today (*Hau*

2015, 5(1)) and *La cohérence des sociétés* (André Iteanu, 2010 ed. Éditions de la MSH, 480 pages).

Kirsten Jäger studied Social Anthropology, Political Science and German Philology at the Westfälische Wilhelms-Universität in Münster, Germany where she also received her PhD in Social Anthropology in 2015. The topic of her dissertation is "The Sultanate of Jailolo. The Revitalisation of 'Traditional' Polities in Indonesia". Her research focuses on the political anthropology of Southeast Asia, particularly Indonesia. She is affiliated with the Institute for Social Anthropology in Münster where she teaches undergraduates and graduates courses.

Roy Jordaan (1947) studied Non-Western Sociology at Leiden University. After his Master's he spent two years on the island of Madura, gathering material for his Ph.D thesis on indigenous Madurese medicine (awarded 1985). Then, for several years he was a researcher at the former Royal Institute of Linguistics and Anthropology (KITLV) in Leiden, and a lecturer at the Department of Social Anthropology of the University of Amsterdam. Subsequently he and his family lived in Jakarta as expatriates. While in Indonesia he became an independent researcher. Attracted to the Hindu-Buddhist temples of Prambanan in Central Java, he immersed himself in research in the field of 'classical' Javanese archaeology and art history. His present writings about Raffles and the British Interregnum are an offshoot of the latter research.

Sabine Klocke-Daffa is Senior Researcher in Social and Cultural Anthropology at the University of Tübingen, Germany. She has conducted intensive fieldwork in Namibia. Her research interests include cultural networks of sharing and exchange, gender, basic income, social security and formal insurances, and practical/applied anthropology. She is currently building up BA and MA courses in Applied Anthropology at Tübingen and is speaker of the special research unit on "Networks and Resource Complexes" within the Collaborative Research Center on "ResourceCultures" (SFB 1070) financed by the German Research Foundation. She has published on the social anthropology of Namibia (since 2001), basic income (2012), ethics in development anthropology (2013), public anthropology (2014), and networks and social resources (2015).

E. Douglas Lewis – a native of Texas and educated at Rice and Brown Universities (PhD, Australian National University) – taught anthropology in Australia for many years. His main subjects were kinship, alliance and social organisation, the anthropology of religion, and the evolution and neurobiology of consciousness, language and mind. Lewis began his ethnographic research on the peoples of Sikka on the island of Flores in eastern Indonesia in 1977, where he worked principally in the Tana 'Ai region of Sikka until 1994. Since 1994 he has carried out historical and ethnological research on Sikka and has lived in the town of Maumere when on Flores. His books include *People of the Source* (1988), *Kamus Sara Sikka – Bahasa Indonesia* ([A Sara Sikka – Bahasa Indonesia Dictionary] with M. Mandalangi Pareira, 1998), *Hikayat Kerajaan Sikka* ([*The Chronicles of the Rajadom of Sikka*] with Oscar Pareira Mandalangi, 2008), and *The Stranger-Kings of Sikka* (2010).

Dr. Abdul Manan, MSc., MA is lecturer of social anthropology at Adab and Humanity Faculty of The State Islamic University (UIN) Ar-Raniry, Banda Aceh, Indonesia. He received his BA in English language education at UIN Ar-Raniry in 1997, MSc in educational and system design at Twente University, Enschede-Holland in 2001, MA in Islamic Studies at Leiden University, Leiden-Holland in 2003, and Dr. Phil. in ethnology at Münster University, Münster, Germany, in 2010. Among his books are *Ritual Kalender Aneuk Jamee di Aceh Selatan* (Vol I) Ar-Raniry Press, 2012. *Rabbani Wahid: Bentuk Seni Islam di Aceh* Balai Pelestarian Nilai Budaya Aceh-Sumut, 2012. *Ritual Kalender Aneuk Jamee di Aceh Selatan* (Vol II) Ar-Raniry Press, 2013. *Syari'at Islam di Aceh (Realitas dan Respon Masyarakat)* Ar-Raniry Press, 2014 and *The Ritual Calendar of South Aceh, Indonesia*, WWU Muenster, Reihe X Band 22, MV-Verlag, 2015.

Denis Monnerie is professeur d'ethnologie at the Institut d'ethnologie and laboratoire DynamE of the Université de Strasbourg. He has done research in historical anthropology on Mono-Alu in the northwestern Solomon Islands and long term fieldwork in Arama in the northern part of Kanaky New Caledonia. His research interests are among others Oceania, the imbrications of local and translocal relations, and theories of exchange. He is a former member of the ERASME research team, where he first met Jos Platenkamp.

Anke Niehof was trained as an anthropologist and a demographer. In 1985, she obtained her doctorate at Leiden University on the basis of a thesis entitled "Women and Fertility in Madura, Indonesia." Since 1993 and until her retirement in April 2013, she held the Chair of Sociology of Consumers and Households at Wageningen University, the Netherlands. She has published extensively on household livelihood and food security, women's reproductive health, care, and impacts of HIV and AIDS in Africa. Between 2002 and 2012 she led the AWLAE Project, in which twenty women scholars from eleven African countries did their PhD studies at Wageningen University on topics relating to the gendered impacts of HIV and AIDS on rural livelihoods and food systems in sub-Saharan Africa. Gender and agency are important perspectives in her work.

Dirk J. Nijland (1938) was from 1971 till 1999 chief of the Section "Visual Etnography", at the Department of Cultural and Social Studies, University of Leiden. His duty was research and education in the possibilities of audiovisual media for Cultural Anthropology. His interest is in nonverbal communication and ritual. Apart from "*Tobelo Marriage*" he also made the ethnographic film *"Sacrifice of Serpents: The Festival of Indrayani, Kathmandu 1992/1994* (1997) with Bert van den Hoek and Bal Gopal Shrestha, and 3 ethnographic multimedia DVD's *Ashes of Life, The Annual Rituals of Laboya, Sumba 1996* (2007, IWF/TIB) with Daniel Geirnaert and Erik de Maaker.

Jarich Oosten studied History of Religion at the Universities of Amsterdam and Groningen, receiving his Ph.D in 1976. Between 1970 and 1997 he was Associate Professor at the Department of Anthropology at Leiden University (Netherlands), and from 1997 to 2008 Full Professor. From 1997 to 2004 he was the Director of the Research School for Asian, African and Amerindian studies (CNWS). Since 1994 he has concentrated his research on a series of projects on recording Inuit culture and history in workshops with Inuit elders. In 2008 he retired from the Department of Social and Cultural Studies continuing his research on Inuit culture and history in cooperation with his friend and colleague Frederic Laugrand of Laval University. Among his major publications are *The War of the Gods. The social code in Indo-European mythology*. London: Routledge & Kegan Paul (1985) and *Hunters,Predators and Prey. Inuit perceptions of animals* (co-author: Frederic Laugrand). New York: Berghahn (2015).

Georg Pfeffer completed his Ethnology studies at Freiburg University with a PhD dissertation on Punjabi Sweepers (1970). In 1976 his post-doctoral thesis (Habilitation) on the elite among Brahmins in the Indian province of Odisha was accepted in Heidelberg, where he became associate professor at the South Asia Institut in 1979 to move on to the Free University of Berlin as a professor from 1985 to 2008. Earlier he had founded the anthropology department of Quaid-i-Azam University, Islamabad in 1975. Between 1980 and 2002 he almost annually visited the tribal areas of western Odisha for two or more months to organise a major research project between 1999 and 2005 in that region. Much of his academic work has been devoted to kinship and to secondary funeral rites. In 2016 his book on *Verwandtschaft als Verfassung; unbürokratische Muster öffentlicher Ordnung* („Kinship as Constitution; non-bureaucratic patterns of public order") came out.

Christian Postert, M.D. and Ph.D., has a degree both in Social Anthropology and in Child and Adolescent Psychiatry and Psychotherapy. Since 2013, he is Professor of Child and Adolescent Psychiatry and Psychotherapy in the Study

Program Occupational Therapy, Department of Applied Health Sciences, University of Applied Sciences, Bochum, Germany. He has conducted anthropological fieldwork from 2000 to 2001 in a Hmong highland village in Laos supported by a grant from the German Research Council (DFG). After studying Social Anthropology, Medicine and a subsequent specialisation in Child and Adolescent Psychiatry and Psychotherapy at Münster University Hospital, he was Senior Physician and Research Director in the Department of Child and Adolescent Psychiatry and Psychotherapy, University Hospital Münster (2011-2012).

Laila Prager is Professor of Social Anthropology at the University of Hamburg. Previously, she worked as a researcher and senior lecturer at the University of Münster and Leipzig. She conducted field research among the Arab speaking Alawi in Turkey and Germany and among Bedouin societies in Syria and Jordan. Her research interests include migration, medical anthropology, anthropology of religion, and kinship and gender. Since 2009 she has undertaken field work in Syria and the UAE on heritage (museums, heritage parks, festivals, etc.), auto-orientalism, and the reinvigoration of tribalism. Her publications include *Die 'Gemeinschaft des Hauses'* (2010), *Bedouinity on Stage* (Nomadic Peoples 2014, 18) and *Displaying Origins* (Horizons in Humanities and Social Sciences 2015, 1).

Michael Prager studied anthropology at Leiden University and at the University of Heidelberg where he obtained his PhD in 1996. From 1996 onwards, he has taught as senior lecturer in anthropology at the Institute of Ethnology of the University of Muenster and in Leipzig (Germany). He has conducted field research in the Bima regency of Sumbawa island (Indonesia), and in Eastern Java. His fields of interest include Southeast Asian Islam, kinship, ritual, cosmology, and the anthropology of the body. Current research is focussing on Islam in Eastern Java and the changing role of local pilgrimage sites. His publications are concerned with the history of anthropology, ritual and theatre, myth analysis, the body, and Southeast Asian Reformist Islam.

Almut Schneider is senior lecturer of social anthropology at the Institute of Ethnology, University of Münster. Previously, she had teaching engagements at the universities of Berlin, Heidelberg and Bolzano. She studied at the Free University of Berlin and did post-graduate studies at the Ecole des Hautes Etudes en Sciences Sociales (EHESS) in Paris, where she obtained her doctoral degree in 2011. Her fieldwork and ongoing research in the Mount Hagen area of the Western Highlands of Papua New Guinea focus on cosmology, kinship- and regional networks, exchange- and ritual relations. Among her publications is the monography *La vie qui vient d'ailleurs: Mouvement, échanges et rituel dans les Hautes-Terres de la Papouasie-Nouvelle-Guinée*

(2016). Her current work deals with social morphology, relationships between people, land and animals and with knowledge practices in the Alpine region.

Guido Sprenger is Professor of Social Anthropology at the Institute of Ethnology, University of Heidelberg since 2010. He has done research on ritual, cosmology and transculturality in the uplands of Laos since 2000. He was a postdoctoral fellow at the Academia Sinica, Taipei, from 2004 to 2007, and a Junior Professor in Münster, Germany, from 2007 to 2010. Among his publications is the monograph *Die Männer, die den Geldbaum fällten (The Men who cut the Money Tree: Concepts of Exchange and Society among Rmeet of Takheung, Laos)* (2006) and *Animism in Southeast Asia* (co-edited with Kaj Århem, 2016). His research interests include exchange, kinship and social morphology, human-environment relations, animism, cultural identity, gender and sexuality.

Christian Strümpell is research associate at the South Asia Institute, Heidelberg University and currently holds a fellowship at the research centre *Work and Human Lifecycle in a Global Perspective*, Humboldt University Berlin. His main interest is in the anthropology of caste and class, work and labour as well as gift and commodity exchange. He has undertaken long-term ethnographic research on the social transformations around heavy industries in Odisha and Bangladesh. Among his publications are articles in Contributions to Indian Sociology ('We work together, we eat together: Conviviality and Modernity in a Company Settlement in South Orissa' 42(3): 351-81), Economic and Political Weekly ('On the Desecration of Nehru's Temples: Bhilai and Rourkela Compared' 43(19): 47-57, co-authored with Jonathan P. Parry), and the special forum in Modern Asian Studies co-edited with Andrew Sanchez ('Class Matters: New Ethnographic Persp ectives on the Politics of Indian Labour' 48(3)). Christian Strümpell is currently writing a monograph on the Odishan steel town Rourkela.

Yulia Sugandi received academic training at the Sociology and Intercultural Counseling Programme at Gadjah Mada University, Indonesia and the University of Eastern Finland. After doing extensive ethnographic research among the Hubula in the highlands of Papua, in 2013 she obtained her PhD in Anthropology at the Institute of Ethnology of the University of Muenster, Germany. Her research interests include intercultural dialogue, ethno-development, sustainable peace building, collective ethno-ecology, and identity and dignity. She has worked for various organisations including the Asia Europe Foundation, Peace Brigades International, UNDP Papua, Misereor, etc. Currently, she works as Senior Fellow at the Department of Community Development and Communication Sciences at the Faculty of Human Ecology in Bogor Agricultural University, Indonesia.

Han F. Vermeulen is a research associate at the Max Planck Institute for Social Anthropology in Halle (Saale), Germany. Trained as a cultural anthropologist at Leiden University in the 1970s, he specialised in the history of anthropology, ethnology, and ethnography. He has conducted ethnographic fieldwork in North Africa (Tunisia) as well as library and archival research in the Netherlands, Germany, Austria, the United Kingdom, and Russia. He has published numerous articles and is co-editor of seven books, including *Fieldwork and Footnotes: Studies in the History of European Anthropology* (London: Routledge 1995), *Treasure Hunting? Collectors and Collections of Indonesian Artefacts* (Leiden: CNWS 2002), and the two-volume *Tales from Academia: History of Anthropology in the Netherlands* (Nijmegen/Saarbrücken 2002). His monograph *Before Boas: The Genesis of Ethnography and Ethnology in the German Enlightenment*_was published by the University of Nebraska Press in July 2015.

Torben A Vestergaard graduated as a mag.art. in social anthropology from Aarhus University, Denmark, in 1979. 1976-77 was spent at the Department of Cultural Anthropology in Leiden on a Dutch state scholarship. He has been a Curator of the Danish Museum of Fisheries in Grenaa, the Ethnographic Collections in Aarhus and the department of brickworks and industrial history at the Museum of Southern Jutland. He was through the 1990'ies a Senior Lecturer at the Department of Ethnography and Social Anthropology at Aarhus University. Vestergaard began his research among sheep farmers in the North Atlantic Faroe Islands, which in turn led his research interests into medieval studies of law, kinship, property and social structure, and contemporary studies of natural resource use, fisheries and industry. His publications include 'The System of Kinship in Early Norwegian Law', *Mediaeval Scandinavia.* Vol. 12, pp. 160-193, at one extreme, and 'The Fishermen and the Nation: The Identity of a Danish Occupational Group', *Maritime Anthropological Studies – MAST,* Vol 3(2) 1990, pp. 14-34, at the other.

Ingo Wallner graduated in engineering, second degree in economics and business administration; work in an eletricity company; at the same time manager of a group marketing company and a confederation of industries for the advancement of sustainable energies. Honorary consul of Madagascar. Degree in Social Anthropology/European Anthropology at the University of Münster, and subsequent PhD degree. Publication: Madagaskar: Nationale Identität in Sicht? Saarbrücken: Südwestdeutscher Verlag für Hochschulschriften, 2012.

Robert Wessing (ret.) received his PhD in anthropology in 1974 from the University of Illinois in Urbana, IL (USA). His research, conducted in West and East Java, Aceh, and Madura, has dealt primarily with symbolism, as well as with the social implications of belief systems. He has published numerous articles, as well as the books *Cosmology and Social Behavior in a West Javanese Settlement* (1978) and *The Soul of Ambiguity: the Tiger in Southeast Asia* (1986) and a lengthy article 'A Community of Spirits. People, Ancestors and Nature Spirits in Java', *Crossroads* (2006: 11-111).

Edwin P. Wieringa is Professor of Indonesian Philology with Special Reference to Islamic Cultures at the University of Cologne. From 1996-1997 he held a two-year Von Humboldt Fellowship at the *Institut für Ethnologie* in Münster with Prof. Platenkamp as his academic host. His most recent publication is a bilingual exhibition catalogue of Indonesian manuscripts at the State Library in Berlin (co-edited with T. Hanstein) *SchriftSprache; Aksara dan Bahasa. Ausstellungskatalog; Katalog Pameran* (Berlin; Jakarta, 2015).

Sjoerd Zanen studied Cultural Anthropology at Leiden University from 1969-1976. His major was Theoretical Anthropology and Structural Anthropology (Prof. P.E. de Josselin de Jong) and Methods and Techniques of Social Research (Prof Speckmann), his regional specialisation was South East Asia, and later the Middle East and West Africa. He conducted research in Lebanon, South India, Nepal, South Sudan, Burkina Faso and Mali. After graduation he worked for the Dutch Ministry of Foreign Affairs from 1978-1999. In 1996 he wrote a PhD thesis on Development Cooperation (Leiden) about the cultural dimension of development. Since 2000 he worked as a consultant for MDF T&C and specialised in development planning, monitoring and evaluation. He retired in 2015.